CARBOHYDRATE-BASED VACCINES AND IMMUNOTHERAPIES

Wiley Series in Drug Discovery and Development
Binghe Wang, Series Editor

Drug Delivery: Principles and Applications
Edited by Binghe Wang, Teruna Siahaan, and Richard A. Soltero

Computer Applications in Pharmaceutical Research and Development
Edited by Sean Ekins

Glycogen Synthase Kinase-3 (GSK-3) and Its Inhibitors: Drug Discovery and Development
Edited by Ana Martinez, Ana Castro, and Miguel Medina

Aminoglycoside Antibiotics: From Chemical Biology to Drug Discovery
Edited by Dev P. Arya

Drug Transporters: Molecular Characterization and Role in Drug Disposition
Edited by Guofeng You and Marilyn E. Morris

Drug–Drug Interactions in Pharmaceutical Development
Edited by Albert P. Li

Dopamine Transporters: Chemistry, Biology, and Pharmacology
Edited by Mark L. Trudell and Sari Izenswasser

Carbohydrate-Based Vaccines and Immunotherapies
Edited by Zhongwu Guo and Geert-Jan Boons

CARBOHYDRATE-BASED VACCINES AND IMMUNOTHERAPIES

Edited by

Zhongwu Guo

Wayne State University, Department of Chemistry

Geert-Jan Boons

University of Georgia, Complex Carbohydrate Research Center

A JOHN WILEY & SONS, INC., PUBLICATION

Copyright © 2009 by John Wiley & Sons, Inc. All rights reserved

Published by John Wiley & Sons, Inc., Hoboken, New Jersey
Published simultaneously in Canada

No part of this publication may be reproduced, stored in a retrieval system, or transmitted in any form or by any means, electronic, mechanical, photocopying, recording, scanning, or otherwise, except as permitted under Section 107 or 108 of the 1976 United States Copyright Act, without either the prior written permission of the Publisher, or authorization through payment of the appropriate per-copy fee to the Copyright Clearance Center, Inc., 222 Rosewood Drive, Danvers, MA 01923, (978) 750-8400, fax (978) 750-4470, or on the web at www.copyright.com. Requests to the Publisher for permission should be addressed to the Permissions Department, John Wiley & Sons, Inc., 111 River Street, Hoboken, NJ 07030, (201) 748-6011, fax (201) 748-6008, or online at http://www.wiley.com/go/permission.

Limit of Liability/Disclaimer of Warranty: While the publisher and author have used their best efforts in preparing this book, they make no representations or warranties with respect to the accuracy or completeness of the contents of this book and specifically disclaim any implied warranties of merchantability or fitness for a particular purpose. No warranty may be created or extended by sales representatives or written sales materials. The advice and strategies contained herein may not be suitable for your situation. You should consult with a professional where appropriate. Neither the publisher nor author shall be liable for any loss of profit or any other commercial damages, including but not limited to special, incidental, consequential, or other damages.

For general information on our other products and services or for technical support, please contact our Customer Care Department within the United States at (800) 762-2974, outside the United States at (317) 572-3993 or fax (317) 572-4002.

Wiley also publishes its books in a variety of electronic formats. Some content that appears in print may not be available in electronic formats. For more information about Wiley products, visit our web site at www.wiley.com.

Library of Congress Cataloging-in-Publication Data:

Carbohydrate-based vaccines and immunotherapies / [edited by] Zhongwu Guo, Geert-Jan Boons.
   p. ; cm.—(Wiley series in drug discovery and development)
  Includes bibliographical references and index.
  ISBN 978-0-470-19756-1 (cloth)
  1. Carbohydrates—Immunology. 2. Vaccines. 3. Glycoconjugates—Immunology. I. Guo, Zhongwu. II. Boons, Geert-Jan. III. Series: Wiley series in drug discovery and development.
  [DNLM: 1. Vaccines. 2. Carbohydrates—immunology. 3. Glycoconjugates. QW 805 C2645 2009]
  QR186.6.C37C373 2009
  616.07'9—dc22

2009015924

Printed in the United States of America

10 9 8 7 6 5 4 3 2 1

To my wife, Wen, our children, Yuchen and Lily, and my parents and family, for their loyal support and generous love.

Zhongwu Guo

CONTENTS

Preface xv

Contributors xvii

1 Glycobiology and Immunology 1
Udayanath Aich and Kevin J. Yarema

- 1.1 Introduction / 1
- 1.2 Glycobiology / 3
  - 1.2.1 Glycosylation—Is It Worth the Cost? / 3
  - 1.2.2 Glycan Biosynthesis—A Dauntingly Complex Process / 6
  - 1.2.3 Glycoproteins / 7
  - 1.2.4 Lipid-Based Glycans / 16
  - 1.2.5 Polysaccharides: Glycosaminoglycans and Bacterial Capsular Components / 18
- 1.3 The Immune System / 20
  - 1.3.1 Introductory Comments / 20
  - 1.3.2 Overview of the Immune System / 20
  - 1.3.3 Glycoimmunobiology / 23
  - 1.3.4 Interplay between Glycosylation and Sugars: a Two-Way Street / 27
- 1.4 Carbohydrate Antigens / 28
  - 1.4.1 Carbohydrate Antigens in Humans / 28
  - 1.4.2 Carbohydrates and Pathogens / 30
  - 1.4.3 Carbohydrate-Based Vaccines / 34
  - 1.4.4 Concluding Comments: Building on Success / 38
- Acknowledgment / 38
- References / 38

2 Preparation of Glycoconjugate Vaccines 55
Wei Zou and Harold J. Jennings

2.1 Introduction / 55
2.2 Capsular Polysaccharide–Protein Conjugates / 56
    2.2.1 *Haemophilus influenzae* Type b / 56
    2.2.2 *Streptococcus pneumoniae* / 59
    2.2.3 *Neisseria meningitidis* / 60
    2.2.4 *Salmonella typhi* Vi / 64
    2.2.5 Group B *Streptococcus* / 65
    2.2.6 *Staphylococcus aureus* Types 5 and 8 / 67
2.3 Lipopolysaccharide (LPS) and Lipooligosaccharide (LOS) Conjugates / 69
    2.3.1 *Escherichia coli* O157 / 69
    2.3.2 *Vibrio cholerae* O1 and O139 / 70
    2.3.3 *Shigella dysenteriae* Type 1, *sonnei* and *flexneri 2a* / 71
    2.3.4 *Neisseria meningitidis* and Nontypeable *Haemophilus influenzae* / 72
2.4 Total Synthetic Glycoconjugate Vaccines / 76
References / 79

3 Adjuvants for Protein- and Carbohydrate-Based Vaccines 89
Bruno Guy

3.1 Introduction / 89
3.2 Initiation and Stimulation of Adaptive Responses / 90
3.3 "Old" Adjuvants and Formulations / 92
    3.3.1 Aluminum / 92
    3.3.2 Emulsions / 93
    3.3.3 Saponins, QS21, and ISCOMS / 94
    3.3.4 Liposomes and Microparticles / 94
    3.3.5 Antigen/Formulation Targeting / 94
    3.3.6 Induction of CD8 CTLs with Soluble Antigens / 95
3.4 Renaissance of Innate Immunity / 95
    3.4.1 Toll-Like Receptors: Agonists and Roles / 95
    3.4.2 Non-TLRs Innate Receptors / 97
    3.4.3 Other Receptors Involved in Antigen Capture and Recognition / 97
3.5 From Basic Research to Practical Applications: Identification of New Adjuvants / 97
    3.5.1 TLR Synthetic Agonists / 97
    3.5.2 Combination of PRR Agonists / 98

3.6 Adjuvants for Carbohydrate-Based Vaccines / 98
    3.6.1 Td and Ti B-Cell Responses / 99
    3.6.2 Adjuvants for "Free" Polysaccharides (Ti Antigens) / 99
    3.6.3 Adjuvants for Glycoconjugate Vaccines (T-Dependent Antigens) / 100
3.7 Combinations of Adjuvants: Preclinical and Clinical Developments / 101
3.8 Immunomodulation of Existing Responses: Adjuvants for Therapeutic Vaccines / 101
3.9 Take Another Route / 102
    3.9.1 Adjuvants for Mucosal Immunization / 102
    3.9.2 Epidermal or Intradermal Routes / 102
3.10 Practical Aspects of Adjuvant Development / 103
    3.10.1 Regulatory Aspects / 103
    3.10.2 Safety versus Efficacy: Risk–Benefit Ratio / 103
3.11 Preclinical Models Used in Adjuvant Development / 104
    3.11.1 Animal Models / 104
    3.11.2 In vitro Models / 104
3.12 Conclusions and Perspectives / 106
Acknowledgment / 106
References / 106

4 Carbohydrate-Based Antibacterial Vaccines   117

Robert A. Pon and Harold J. Jennings

4.1 Introduction / 117
4.2 Polysaccharide and Glycoconjugate Immunobiology / 118
4.3 Deficiencies in the Human Immune Response to Polysaccharides / 120
4.4 Glycoconjugate Vaccines / 121
4.5 *Haemophilus influenzae* / 122
    4.5.1 Hib Polysaccharides / 122
    4.5.2 Hib Conjugate Vaccines / 123
4.6 *Neisseria meningitidis* / 125
    4.6.1 Meningococcal Polysaccharide Vaccines / 126
    4.6.2 Meningococcal Conjugate Vaccines / 126
4.7 *Streptococcus pneumoniae* / 133
    4.7.1 Impact on Invasive Pneumococcal Disease / 139
    4.7.2 Impact on Acute Otitis Media / 140
4.8 Group B *Streptococcus* / 140
4.9 *Salmonella typhi* / 145

4.10 Conjugate Vaccines: Future Concerns / 146
4.11 Summary / 147
References / 148

5 Carbohydrate-Based Antiviral Vaccines — 167
Benjamin M. Swarts and Zhongwu Guo

5.1 Introduction / 167
5.2 Viral Glycosylation / 168
    5.2.1 Viral N-glycosylation / 168
    5.2.2 Carbohydrates of HIV / 170
    5.2.3 Carbohydrates of Influenza A Virus / 172
    5.2.4 Carbohydrates in Hepatitis C Virus / 173
    5.2.5 Carbohydrates in Other Viruses / 174
5.3 Vaccine and Drug Development / 174
    5.3.1 Human Immune Deficiency Virus / 174
    5.3.2 Influenza A Virus / 181
    5.3.3 Hepatitis C Virus / 182
5.4 Conclusions / 182
Acknowledgment / 183
References / 183

6 Carbohydrate-Based Antiparasitic Vaccines — 195
Faustin Kamena, Xinyu Liu, and Peter H. Seeberger

6.1 Introduction / 195
6.2 GPI-Based Antimalarial Vaccine / 197
    6.2.1 GPI as a Malaria Toxin / 197
    6.2.2 Synthetic GPI as Antitoxic Malaria Vaccine Candidate / 198
    6.2.3 Synthetic GPI Microarray to Define Antimalarial Antibody Response / 200
6.3 LPG-Based Antileishmanial Vaccine / 201
    6.3.1 LPG in Leishmaniasis Pathogenesis / 201
    6.3.2 Synthetic Phosphoglycan Repeating Unit as Potential Antileishmanial Vaccine / 203
    6.3.3 Synthetic LPG Cap Oligosaccharide as Antileishmanial Vaccine Candidate / 204
6.4 Other Examples / 205
    6.4.1 Fucosylated N-glycan as Potential Vaccine Lead against Schistosomiasis / 205
    6.4.2 GPIs as Potential Vaccine Lead against Toxoplasmosis and Chagas' Disease / 207

6.5 Perspectives and Future Challenge / 208
Acknowledgment / 209
References / 209

7 Carbohydrate-Based Antifungal Vaccines 215

Magdia De Jesus, Liise-anne Pirofski, and Arturo Casadevall

7.1 Introduction / 215
7.2 Terminology / 216
    7.2.1 Vaccination versus Immunization / 216
    7.2.2 Toxoids / 216
    7.2.3 Glycoconjugates / 216
7.3 Antifungal Glycoconjugate Vaccines / 217
    7.3.1 *C. neoformans* Polysaccharide–Protein Conjugates / 217
    7.3.2 Development of Alternative Vaccines for *C. neoformans* / 220
    7.3.3 *C. albicans* Mannan–Protein Conjugates / 220
    7.3.4 β-Glucan–Protein Conjugates / 221
7.4 Antifungal Vaccines and the Immune System / 222
7.5 Summary / 223
Acknowledgment / 224
References / 224

8 Cancer-Associated and Related Glycosphingolipid Antigens 227

Steven B. Levery

8.1 Introduction / 227
8.2 Structural Classification of Antigens / 228
8.3 "Abnormal" Expression of Glycosphingolipid (GSL) Glycan Structures in Cancer Tissues / 228
8.4 Discussion of Selected Antigens / 234
    8.4.1 Globo-Series and Related Antigens / 234
    8.4.2 Ganglio-Series Antigens / 237
    8.4.3 Lacto-Series (Type 1 Chain; Lc_n) Antigens / 241
    8.4.4 Neolacto-Series (Type 2 Chain; nLc_x) Antigens / 242
8.5 Other Antigens / 248
    8.5.1 Le^a-Le^a and Le^b-Le^a / 248
    8.5.2 Le^a-Le^x / 249
Acknowledgment / 250
References / 250

9 Semisynthetic and Fully Synthetic Carbohydrate-Based Cancer Vaccines 263
Therese Buskas, Pamela Thompson, and Geert-Jan Boons

    9.1   Introduction to Cancer Vaccines / 263
    9.2   Tumor-Associated Carbohydrate Antigens (TACAs) / 264
    9.3   Carbohydrate-Based Cancer Vaccines / 267
    9.4   Humoral Immune Response to Carbohydrates / 267
    9.5   MHC-Mediated Immune Response to Glycopeptides / 269
    9.6   Toll-Like Receptors and the Link Between Innate and Adaptive Immunity / 270
    9.7   Chemical Synthesis of Tumor-Associated Carbohydrates and Glycopeptides / 271
    9.8   Semisynthetic Carbohydrate-Based Cancer Vaccines / 276
    9.9   Fully Synthetic Carbohydrate-Based Cancer Vaccines / 279
    9.10 B-Epitope and Receptor Ligand Di-Epitope Constructs / 279
    9.11 B- and T-Cell Di-Epitope Constructs / 284
    9.12 Tricomponent Vaccines / 288
    References / 292

10 Glycoengineering of Cell Surface Sialic Acid and Its Application to Cancer Immunotherapy 313
Zhongwu Guo

    10.1 Introduction / 313
    10.2 Engineering of Cell Surface Sialic Acids / 314
    10.3 Sialic Acid Engineering for Modulation of Cell Surface Reactivity / 318
    10.4 Sialic Acid Engineering for Cancer Immunotherapy / 321
    10.5 Summary / 325
    Acknowledgment / 326
    References / 326

11 Therapeutic Cancer Vaccines: Clinical Trials and Applications 333
Hans H. Wandall and Mads A. Tarp

    11.1 Introduction / 333
    11.2 Innate and Adaptive Immunity in Relation to Cancer Immunotherapy / 334
    11.3 Design Issues for Clinical Cancer Vaccine Trials / 337
    11.4 Clinical Development of Cancer Vaccines / 337
    11.5 Proof of Principle Trials / 338
          11.5.1 Toxicity and Pharmacokinetics / 339
          11.5.2 Dose and Administration Schedule / 339
          11.5.3 Endpoints: Biological Activity and Clinical Activity / 339

11.6 Efficacy Trials / 340
11.7 Clinical Endpoints in Efficacy Trials / 340
11.8 Challenges in Vaccine Development / 341
11.9 Defining the Target Tumor-Associated Antigens / 342
11.10 Production and Storage Issues / 344
11.11 Clinical Trials / 345
    11.11.1 Glycosphingolipid-Based Vaccines / 347
    11.11.2 *O*-glycan-Based Vaccines / 351
11.12 Conclusions / 354
Acknowledgment / 355
References / 355

12 Carbohydrates as Unique Structures for Disease Diagnosis   367
Kate Rittenhouse-Olson

12.1 Introduction / 367
12.2 Viruses / 369
    12.2.1 Infectious Mononucleosis / 369
    12.2.2 Influenza A and B / 370
12.3 Bacteria / 371
    12.3.1 *Streptococcus pyogenes* / 371
    12.3.2 Groups A, B, C, D, F, and G *Streptococcus* / 371
    12.3.3 *Streptococcus pneumoniae* / 373
    12.3.4 Meningitis / 373
    12.3.5 *Chlamydia trachomatis* / 374
    12.3.6 Future / 374
12.4 Fungi / 374
    12.4.1 *Aspergillus fumigatus* / 375
    12.4.2 Invasive Candidiasis / 375
    12.4.3 *Cryptococcus neoformans* / 375
    12.4.4 *Histoplasma capsulatum* / 376
12.5 Parasites / 377
    12.5.1 *Echinococcus multilocularis* / 377
    12.5.2 *Clonorchis sinensis* / 378
    12.5.3 *Trichinella* / 378
    12.5.4 *Schistosoma mansoni* / 378
12.6 Autoimmunity / 378
    12.6.1 Diabetes / 378
    12.6.2 Cold Agglutinin Disease / 379
    12.6.3 Inflammatory Bowel Disease / 380
12.7 Tumors / 380
    12.7.1 Bladder / 381
    12.7.2 Breast / 381

      12.7.3 Colon / 382
      12.7.4 Liver / 382
      12.7.5 Lung / 383
      12.7.6 Melanoma / 384
      12.7.7 Ovarian / 385
      12.7.8 Pancreatic / 386
      12.7.9 Prostate / 387
12.8 Inherited or Acquired Disorders of Glycosylation / 388
References / 388

Index   395

PREFACE

Carbohydrates, the most abundant and arguably the most structurally diverse organic molecules in nature, play a pivotal role in various biological processes. For example, the cell glycocalyx, which is the thick layer of carbohydrates covering cell surfaces, is involved in cell differentiation, recognition, adhesion, and many other important events, including pathological developments. Thus, glycosylation is fundamental for life, as briefly discussed in Chapter 1 of this volume. Meanwhile, the glycocalyx of each specific type of cell, containing both common and unique glycans, provides a splendid platform and a rich source of structural scaffolds for the design and development of novel therapeutic strategies.

The location of carbohydrates at the forefront of cell surface enables them to effectively interact with the immune system; thus, their role as cell antigenic determinants has been firmly established, and the marriage of glycobiology with immunology has actually founded the term *glycoimmunology*. Cell surface carbohydrates, including both large polysaccharides and short oligosaccharides, have been explored for vaccine development for decades. While some bacterial polysaccharides have been used as vaccines in clinic, they are ineffective in children; moreover, the majority of free carbohydrates are poorly immunogenic. To overcome these deficiencies, carbohydrate antigens are conjugated to proteins or other carrier molecules to form conjugate vaccines, which have significantly extended the scope and efficiency of carbohydrate-based vaccines. This book intends to introduce the recent progresses in this exciting field with each chapter contributed by established experts in the specified area.

Chapter 1 as an introduction to the whole volume provides a concise, yet quite complete, account of the major topics in glycoimmunology, with an emphasis on concepts related to carbohydrate-based vaccines. Chapter 2 provides an overview about the design strategies and synthetic methods for carbohydrate-based vaccines with an emphasis on carbohydrate–protein conjugation. Chapter 3 provides a complete review concerning the properties, functions, and applications of adjuvants, in use or in development, for improving the overall immunogenicity and/or immunological properties of vaccines, as well as the criteria for choosing and designing proper adjuvants for specific targets. Recent developments toward clinically useful vaccines against bacteria, viruses, parasites, and fungi are discussed in detail in Chapters 4, 5, 6, and 7, respectively. Among all, antibacterial vaccines have witnessed the greatest success, with several glycoconjugate vaccines already on the market and several others in clinical trials. Despite the difficulty, significant advancement has been made toward antiviral, antiparasitic, and antifungal glycoconjugate vaccines, thanks to the in-depth understanding of the glycosylation pathways of these pathogens, the characterization of novel carbohydrate antigens, and the recent progress in synthetic carbohydrate chemistry. Another exciting and expanding area covered by this volume is carbohydrate-based therapeutic cancer vaccine or cancer immunotherapy. Chapter 8 presents in detail the abnormal carbohydrates expressed by tumors, termed tumor-associated carbohydrate antigens (TACAs). The state-of-the-art strategies developed for TACA-based cancer vaccine design and synthesis are reviewed in Chapter 9, while the results of clinical trials on synthetic cancer vaccines are discussed in Chapter 11. Glycoengineered delivery of unnatural carbohydrate structures onto cell surfaces can further improve the immunogenicity of cancer and expand the scope of cancer immunotherapy, which is explored in Chapter 10. In addition to therapeutic values, cell surface carbohydrates are also successfully used for disease diagnosis, particularly immunodiagnosis using antibodies. New promising methods are emerging with the identification of novel carbohydrate antigens and with recent advances of biotechnology (see Chapter 12).

We hope that this volume will serve well as a brief introduction to various aspects of carbohydrate-based vaccines, a rich and ever growing field in biomedicine and glycobiology. Readers are advised to consult with the literature cited in individual chapters for further detailed information about a specific topic.

We are grateful for all contributors to this volume and for Dr. Binghe Wang for his help in planning this volume.

ZHONGWU GUO

Wayne State University

GEERT-JAN BOONS

University of Georgia

CONTRIBUTORS

Udayanath Aich, Department of Biomedical Engineering, The Johns Hopkins University, Baltimore, Maryland 21218

Geert-Jan Boons, Complex Carbohydrate Research Center, University of Georgia, Athens, Georgia 30602

Therese Buskas, Complex Carbohydrate Research Center, University of Georgia, Athens, Georgia 30602

Arturo Casadevall, Department of Microbiology and Immunology, Albert Einstein College of Medicine, Bronx, New York 10461

Magdia De Jesus, Department of Microbiology and Immunology, Albert Einstein College of Medicine, Bronx, New York 10461

Zhongwu Guo, Department of Chemistry, Wayne State University, Detroit, Michigan 48202

Bruno Guy, Research Department, Sanofi Pasteur, Campus Merieux, 69280 Marcy l'Etoile, France

Harold J. Jennings, Institute for Biological Sciences, National Research Council of Canada, Ottawa, Ontario, Canada K1A0R6

Faustin Kamena, Laboratory for Organic Chemistry, Swiss Federal Institute of Technology (ETH) Zurich, Zurich, Switzerland

Steven B. Levery, Department of Cellular and Molecular Medicine, Faculty of Health Sciences, University of Copenhagen, DK-2200 Copenhagen N, Denmark

Xinyu Liu, Laboratory for Organic Chemistry, Swiss Federal Institute of Technology (ETH) Zurich, Zurich, Switzerland

Liise-anne Pirofski, Department of Microbiology and Immunology, Albert Einstein College of Medicine, Bronx, New York 10461

Robert A. Pon, Institute for Biological Sciences, National Research Council of Canada, Ottawa, Ontario, Canada K1A0R6

Kate Rittenhouse-Olson, Department of Biotechnical and Clinical Laboratory Sciences and Department of Microbiology, University at Buffalo, Buffalo, New York 14214

Peter H. Seeberger, Laboratory for Organic Chemistry, Swiss Federal Institute of Technology (ETH) Zurich, Zurich, Switzerland

Benjamin M. Swarts Department of Chemistry, Wayne State University, Detroit, Michigan 48202

Mads A. Tarp, Department of Cellular and Molecular Medicine, Faculty of Health Sciences, University of Copenhagen, Copenhagen N, Denmark

Pamela Thompson, Complex Carbohydrate Research Center, University of Georgia, Athens, Georgia 30602

Hans H. Wandall, Department of Cellular and Molecular Medicine, Faculty of Health Sciences, University of Copenhagen, Copenhagen N, Denmark

Kevin J. Yarema, Department of Biomedical Engineering, The Johns Hopkins University, Baltimore, Maryland 21218

Wei Zou, Institute for Biological Sciences, National Research Council of Canada, Ottawa, Ontario, Canada K1A0R6

1

GLYCOBIOLOGY AND IMMUNOLOGY

Udayanath Aich and Kevin J. Yarema

Department of Biomedical Engineering, The Johns Hopkins University, Baltimore, Maryland 21218

1.1 INTRODUCTION

This chapter takes on the ambitious task of introducing two immensely complicated biological systems. By necessity, only the briefest outline of each topic will be provided while maintaining a focus on the surprisingly robust, albeit often overlooked, overlap between the two areas. The term *glycobiology* was first formally used two decades ago by Rademacher, Parekh, and Dwek to describe the merging of the traditional disciplines of carbohydrate chemistry, biochemistry, and cell biology [1]. In reality, the detailed study of biologically important sugars greatly predated the formal recognition of glycobiology as a distinct field of investigation as—almost a century earlier—Emil Fischer performed an elegant series of seminal experiments that described the isomeric nature of sugars and the stereochemical configuration of common monosaccharides. In the intervening years, carbohydrates have been established as the most abundant—and arguably the most structurally diverse—organic molecules found in nature. They play major structural roles in fungi, crustaceans, and plants and are often subject to postsynthetic modifications, especially in higher animals, that greatly increase their chemical diversity and biological activities.

In mammals, carbohydrates are quantitatively less abundant when measured by bulk mass compared to many lower organisms, and they are also skewed toward a lower size distribution, often occurring as oligosaccharide structures of 20 or fewer—sometimes

---

Carbohydrate-Based Vaccines and Immunotherapies. Edited by Zhongwu Guo and Geert-Jan Boons
Copyright © 2009 John Wiley & Sons, Inc.

only as 1 or 2—residues instead of as large polysaccharides. More than compensating for their modest size, however, mammalian oligosaccharides have critical biological functions derived from fundamental differences compared to other classes of biomacromolecules such as nucleic acids, proteins, and lipids. For example, the monomeric units of carbohydrates can connect to each other by several different linkages, resulting in branched structures that enable even relatively small oligosaccharides to exist in a profuse number of structural variations. Consequently, these molecules have immense information-carrying capacity so that, even when they are present in vanishingly small quantities, they have a profound impact on modulating the function of their host protein or lipid. With a growing arsenal of technical tools at their disposal to decipher the complexities of glycosylation with increasing precision, glycobiologists are now well positioned to tackle a formidable biological problem—unraveling the intersection of carbohydrates with immunity.

The realms of glycobiology and immunology have long overlapped in the form of the ABO(H) blood group antigens that were discovered in 1900 [2]. In the past two decades the molecular basis of these antigenic structures, which necessitate careful attention to blood-type compatibility during transfusions, have been found be an oligosaccharide structure where a small structural difference—the presence or absence of an *N*-acetyl group on galactose—separates the A and B epitopes. The ability of the immune system to recognize minor changes to the chemical structures of large carbohydrate structures has profound biomedical implications that go beyond the largely solved problem of blood typing in transfusions. For example, the trisaccharide α-gal antigen leads to hyperacute rejection of xenotransplanted organs and has spawned ongoing efforts to create α-gal knockout pigs [3]. Similarly, the antigenic Neu5Gc form of sialic acid has recently been found to "contaminate" human stem cell lines, raising concerns about the introduction of tissue-engineered organs into a recipient [4]. In contrast to these examples, in this book the impingement of the immune system into glycobiology is not regarded as a problem but rather as an enticing opportunity. In short, based on the century-old precedent that carbohydrates *are* antigenic, the logical—although not simple (hence, the need for an entire book on the topic!)—course of action is to exploit complex sugars as vaccines and for immunotherapy.

To provide a broader context for the subsequent chapters in this book, which delve into the nitty-gritty aspects of carbohydrate-based vaccine development, this chapter gives an overview of glycobiology and then presents a sampling of specific examples that exemplifies how this field connects with immunology. First, the major classes of mammalian carbohydrates, along with brief descriptions of their biosynthesis, are covered in Section 1.2. Next, in Section 1.3, an even briefer overview of immunology is provided along with a peek into several aspects of *glycoimmunobiology*, a field that is emerging as it becomes increasingly clear that links between glycobiology and immunology are a two-way street. Not only does the immune system reach into the realm of glycobiology by recognizing carbohydrate as antigens, but glycobiology also intrudes into immunology to the extent that glycans influence multiple levels of the immune response. Accordingly, selected specific examples of how carbohydrates tune the immune response will be given to provide a small window into the biological roles of complex sugar structures (i.e., into glycobiology). Finally, in Section 1.4 a brief overview of antigenic carbohydrate structures found in nature that hold potential value for vaccine development will be surveyed along with a brief discussion of unique challenges faced by—if we continue the trend of dubbing sugar-related areas of investigation with the *glyco* prefix—the "glycovaccinist."

1.2 GLYCOBIOLOGY

1.2.1 Glycosylation—Is It Worth the Cost?

1.2.1.1 Basic Considerations A newcomer to the field of glycobiology might ask the following question: Why do cells bother with glycosylation? This question arises from a quick accounting of the costs in energy and resources expended by a cell to produce glycans that include the hundreds of proteins that comprise the biosynthetic *glycosylation machinery*, the requirement for high-energy nucleotide sugars used as building blocks for multimeric carbohydrate structures, as well as the energy foregone by not using sugars for their "canonical" biological function—energy production! Even worse, after expending all of this effort, surface-displayed glycans can be co-opted by opportunistic molecular toxins, viral and microbial pathogens, and eukaryotic parasites (see Fig. 1.1 and Section 1.2.1.3); furthermore, the complexities of glycosylation provide ample opportunity for metabolic aberrations

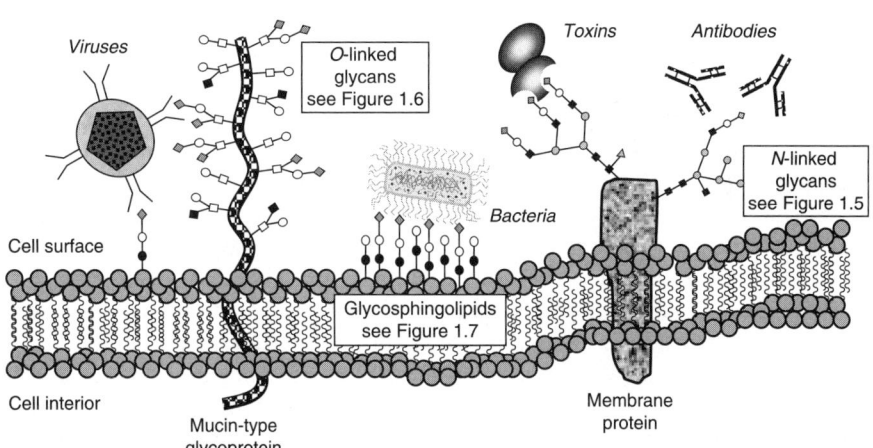

Figure 1.1 Landscape of the cell is dominated by carbohydrates. The surface of a mammalian cell is decorated by various classes of complex carbohydrates; these oligosaccharides are often referred to as *glycans* and collectively as the *glycocalyx*. An overview of glycan biosynthesis is provided in Figures 1.2–1.4 and more information on each class of these molecules is given in Figures 1.5–1.7 and the accompanying text. This graphic is not drawn to scale (as is evident from the relative sizes of toxins, antibodies, viruses, and bacteria, which are all pathogens that exploit surface sugars for entry into a cell). It is notable that even though the glycocalyx comprises only about 8–10% of the mass of the plasma membrane, in a typical mammalian cell it forms a continuous (albeit not uniform) layer ∼8 nm thick, occupying roughly the same volume as the lipid and protein constituents of the membrane. (See color insert.)

to arise that cause or exacerbate disease (Section 1.2.1.2). Clearly, against this rather dire backdrop, cells must have a compelling reason to produce glycans. As depicted in Figure 1.1, and discussed in detail for specific aspects of the immune system in Section 1.3.3, surface sugars comprise the interface between a cell and its outside environment and in a very real sense allow multicellular life to exist by enabling not only the detrimental pathogen binding events shown in the figure but also by facilitating cell–cell and cell–extracellular matrix (ECM) interactions, serving as receptors for hormones and lectins, and contributing to cell motility [5]. In the overall balance, after playing with sugars during hundreds of millions of years of evolution, nature has decided that they are well worth the pitfalls they create; we trust that this chapter will lead the reader toward a similar conclusion.

1.2.1.2 Glycans in Disease Even before getting into glycosylation per se where the complexities of the assembly of complex saccharide structures come into play, simply having sugars around is a dicey—albeit an absolutely necessary—proposition for a living organism. Indirectly, as evidenced by the large quantities of antioxidants consumed by the general public as nutritional supplements, oxygen used to liberate energy from sugars can be highly lethal to cells if reactive free radical by-products are not tightly controlled. Less well known, but just as insidious, the simple sugars that are the feedstock for energy-providing oxidation reactions can damage biological macromolecules through "advanced glycation end products" (AGE). From a chemical perspective, the nonenzymatic reaction between amino acids and reducing sugars was discovered by L. C. Maillard a century ago. The individual steps of the Maillard reaction have now been unraveled in detail starting with the formation of reversible Schiff bases, which are transformed into more stable aldoamines (Amadori products) or ketoamines (Heyns products) [6]. Following a chain of chemical rearrangements, the latter products are converted into terminal adducts that are irreversibly bound to the target molecules as AGEs. During the 1970s and 1980s, AGE products—present at particularly high levels in diabetes, renal failure, and amyloidosis because of increased levels of circulating sugar [7]—were implicated in disease complications primarily via adventitious crosslinking of proteins. Collagen, for example, is a prime target of AGE reactions, and the results are especially damaging to the vascular system. More recently, and appropriate to mention in a combined discussion of glycobiology and immunology, evidence is emerging that AGEs have antigenic properties leading to the hypothesis that AGE structures found in vivo may elicit autoimmune responses that contribute to the atherogenic processes associated with diabetes [8].

In addition to the hazards cells encounter by using sugars as an energy source, or simply by having them around, the assembly of monosaccharides into complex structures leads to a whole new set of pitfalls that are manifest at various levels of severity. At the most critical level, genes involved in the production of certain glycan structures are absolutely essential for life—for example, aberrations involved in sugar structures needed during fertilization would prevent formation of viable embryos from the very start of an organism's life span [9]. In other cases, exemplified by congenital disorders of glycosylation that arise from defects in N-glycan biosynthesis (see Section 1.2.3.2), disease—which is often fatal—is clinically manifest early in childhood [10]. Other times, illustrated by mutations to glucosamine (UDP-N-Acetyl)-2-epimerase/ N-acetylmannosamine kinase (GNE) (the key biosynthetic enzyme in sialic acid biosynthesis [11]) found in hereditary inclusion body myopathy (HIBM [12]), symptoms do not appear for decades and are not found in all individuals afflicted with the disease-causing mutation. Finally, continuing to move toward the benign end of the severity spectrum, certain congenital genetic abnormalities—demonstrated by a lack of phenotype in knockout mice—have no impact on development and any phenotypic effects that do occur are very subtle and are often only manifest in complex behavioral traits [13].

Glycosylation abnormalities acquired later in life upon somatic mutation also contribute to disease; the outstanding example is cancer, which is virtually always accompanied by aberrant glycan production [14–17]. Nascent efforts to exploit glycans in cancer immunotherapy are described briefly in Section 1.4 and in detail elsewhere in this book (see Chapters 8–11). Another interesting example of glycosylation in disease is provided by prions, a novel class of "protein-only" infectious pathogens that cause a group of invariably fatal neurodegenerative diseases. Although the protein-only description specifically refers to the surprising absence of nucleic acids found in conventional infectious agents, it is not strictly correct insofar as the prion protein has two highly conserved potential sites of N-glycosylation, allowing prion proteins to exist as three classes of molecules that differ in their degree of glycosylation. The unglycosylated (5%) and monoglycosylated (25%) isoforms are minor cell surface components dominated by the diglycosylated form (70%). Although the roles that N-linked glycans play in prion biology are not completely understood, a comparative analysis of the glycans on healthy and diseased proteins reveal differences that include the proportion of tri- and tetra-antennary structures as well as the amount of Lewis X (Le^X) and sialyl Lewis X (sLe^X) epitopes [18–21]. Ultimately, differences in glycan profiles may prove to be the critical determinant of prion disease progression and the prefix *glyco* will need to be added to the protein-only descriptor.

1.2.1.3 Toxins and Infectious Agents Bind to Surface Glycans Because they are easily accessible, sugars displayed on the surfaces of mammalian cells represent enticing binding opportunities for opportunistic pathogens ranging from molecular toxins to viruses and from primitive bacteria to sophisticated eukaryotic parasites. To give representative examples, at the molecular level ricin—a versatile and durable toxin (allegedly) used by KGB assassins during the cold war and by terrorists today—consists of two peptide chains crosslinked by a disulfide bond. One of the protein chains is a lectin (*lectin* is a generic name for a protein that binds to a carbohydrate) that recognizes terminal galactose on cell surface glycans and serves to ferry the other peptide chain into the lumen of secretory vesicles where a single molecule, upon translocation into the cytosol, can kill the cell via catalytic deactivation of ribosomes [22, 23]. Viruses also ubiquitously exploit surface glycans as binding epitopes; in a well-known example, influenza virus binds to sialic acid and does so with remarkable discrimination by distinguishing between $\alpha 2,3$-linked and $\alpha 2,6$-linked sialosides as well as eschewing the Neu5Gc form of sialic acid in preference to Neu5Ac [24–27]. Moving to microbes, many pathogenic bacteria bind to host cells through glycan recognition. A particularly interesting example is the adherence of *Escherichia coli* to epithelial cells of the gastrointestinal or urinary tract mediated by bacterial lectins present on fimbriae. These lectins preferentially bind to cell surface glycoproteins containing mannose and have counterintuitive—but very useful from the bug's point of view—catch bond behavior characterized by tighter binding as shear force increases [28, 29]. Among other benefits, this clever binding mode prevents elimination of the infecting bacterium during urination while allowing it to swim free in the bladder at other times.

In addition to comprising easily accessible binding epitopes, the information-rich "sugar code" provided by glycan diversity [30] contributes to host–pathogen specificity. For example, the gonorrhea organism *Neisseria gonorrhoeae* has long been known to adhere to human cells of the genital and oral epithelia but not to cells from other organs or other animal species, explaining why only humans are prone to gonorrhea. The molecular basis for host specificity of *N. gonorrhoeae* is in part explained by glycosylation patterns of CD66 [31] and reaches even greater importance in the exquisite sensitivity of influenza viral strains for different sialic acids, as mentioned above. For the nonspecialist to make sense of jargon such as $\alpha 2,3$-linked and $\alpha 2,6$-linked sialosides, to comprehend the subtle but important differences between similar sugars such as Neu5Ac and Neu5Gc that many pathogens can distinguish with precision, or to understand why a pathogenic bacteria would opt for mannose as a binding epitope, it is helpful to take a closer look at the biosynthetic processes for mammalian glycans and understand the chemical basis of the "building blocks" and the structural ramifications of how monosaccharides are assembled into complex structures. An investment in time to understand these issues (discussed in Sections 1.2.2–1.2.4) will benefit any researcher working with carbohydrate–based vaccines because the mammalian immune system is equally attuned to the subtleties of carbohydrate structure as the sampling of pathogens just mentioned.

1.2.2 Glycan Biosynthesis—A Dauntingly Complex Process

As a consequence of their structural complexity and ubiquitous nature, carbohydrates play important roles in many physiological and pathological cellular functions. Glycobiologists study these biological functions but are also keenly interested in the biosynthetic processes through which complex sugars are assembled. A molecular-level understanding of the production of glycans has greatly lagged nucleic acid and proteins because, unlike these other biopolymers, carbohydrate structures are not template based. An estimated 1–3% of the human genome as well as dozens of metabolic intermediates are involved in the biosynthetic process; furthermore, oligosaccharide synthesis and modification is spatially located across several subcellular compartments; these complexities have rendered the biosynthesis of glycans a substantial challenge to unravel.

Despite formidable challenges posed by the structural complexity and less than straightforward biosynthetic routes of glycans, substantial progress in several disciplines has converged over the past few years and has contributed to a greatly increased understanding of the assembly and identification of oligosaccharides [38, 39].

A critical advance was the sequencing of the human genome, which allowed all of the genes involved in glycosylation—that code for various enzymes and membrane transporters—to be at least tentatively identified. In tandem, increasingly sensitive analytic techniques are now readily available that allow intracellular metabolites and mature glycan structures to be characterized even in minute quantities [40, 41]. Efforts are also underway to develop computational models that use intracellular information (i.e., the genomic and small-molecule metabolite compositions) to predict surface glycan and to furthermore connect these parameters with development status, a disease state, or environmental insult [42–45].

The production of complex carbohydrates starts with basic building blocks—a set of monosaccharides—that are analogous to the nucleotides or amino acids used for nucleic acids or proteins biosynthesis, respectively. The sugary feedstock for glycan assembly is primarily glucose, which can be obtained from the diet, transported into a cell, converted into other monosaccharides through epimerization (and other) reactions, phosphorylated, and ultimately converted into a high-energy or "activated" nucleotide sugar donor. A summary of the reactions that convert monosaccharides obtained from the extracellular milieu into nucleotide sugar donors is shown in Figure 1.2; chemical structures of common mammalian monosaccharides and nucleotide sugars are provided in Figure 1.3.

The monosaccharide processing reactions shown in Figure 1.2 primarily occur in the cytosol, setting the stage for oligosaccharide assembly that can begin on the cytosolic face of the endoplasmic reticulum (ER), but more commonly the nucleotide sugars are transported into the lumen of the ER or a Golgi compartment where the majority of glycosyltransferases are localized. Glycosyltransferase reactions that assemble nucleotide sugars into multimeric macromolecules—exemplified by the suite of sialyltransferases shown in Figure 1.4—add another substantial level of complexity to the glycosylation machinery. In humans there are 20 different sialyltransferases that create 6 distinct types of glycosidic linkages (i.e., $\alpha2,3$, $\alpha2,6$, and $\alpha2,8$ on either protein or lipid-attached glycans), thus providing redundancy that depends on factors such as developmental stage or tissue type. Once the mature glycan is synthesized, a process that involves considerable nuance depending on whether the sugar structure is attached to a protein, lipid, glycosylphosphatidylinositol (GPI) anchor, or exists as a "free" polysaccharide (as described in the next sections), the glycosylated macromolecule is almost always secreted or moved to the cell surface; the major exception are *O*-GlcNAc-modified cytosolic and nuclear proteins. Finally, scavenging of glycans and recycling of their molecular components occurs through the action of glycosidases that degrade oligo- and polysaccharides and thereby liberate monosaccharides for reuse.

1.2.3 Glycoproteins

1.2.3.1 Glycosylation Is a Ubiquitous and Diverse Co- and Posttranslational Protein Modification In terms of diversity of structures as well as the sheer proportion of molecules affected, glycosylation is the most significant co-translational and posttranslational modification of proteins. Like all complex

8 GLYCOBIOLOGY AND IMMUNOLOGY

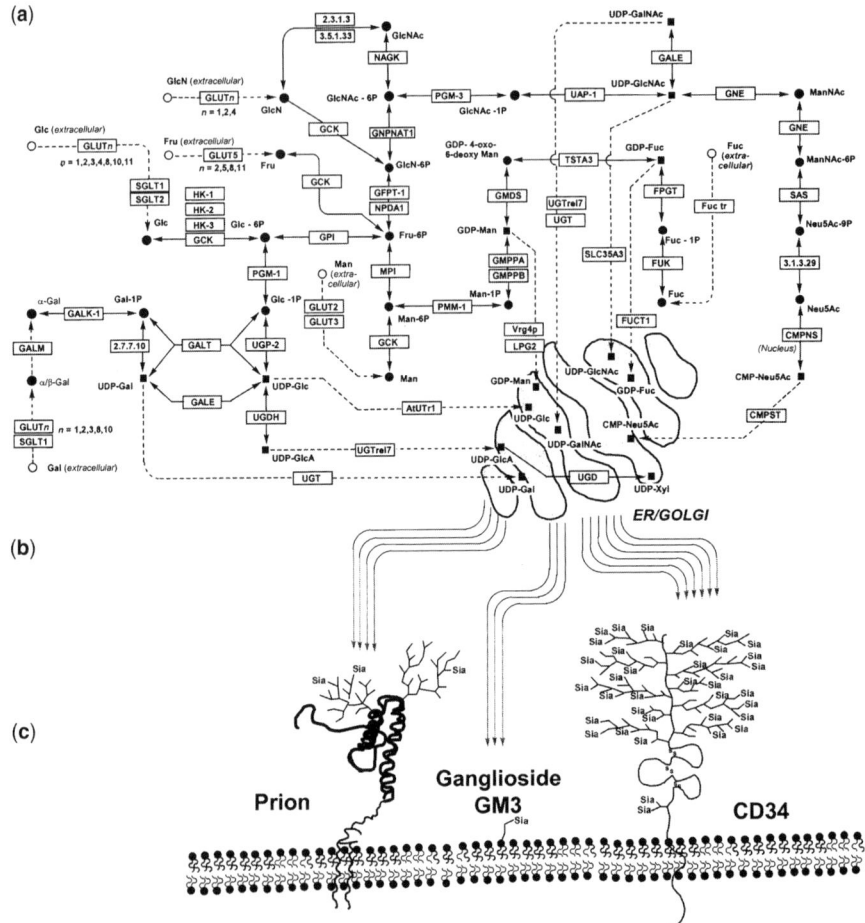

Figure 1.2 Overview of glycan biosynthesis. (a) Monosaccharide uptake and processing (see Fig. 1.3 for structures). Sugars obtained exogenously by cells such as Gal, Glc, and GlcN (full names and chemical structures of common mammalian monosaccharides are shown in Fig. 1.3a) are taken up by families of membrane transporters [32, 33] and converted to nucleotide sugar donors such as UDP-GlcNAc and CMP-Neu5Ac (Fig. 1.3c) by mostly cytosolic enzymatic reactions. Additional information on the enzymes is provided in other review articles [34] or from online search engines or databases such as Pubmed [35], KEGG [36], or HUGO [37]. (b) Glycan assembly (see Fig. 1.4 for details of sialylation). The nucleotide sugars are assembled by a suite of glycosyltransferases (illustrated in detail for sialyltransferases in Fig. 1.4) into structurally complex surface-displayed glycans. (c) Representative glycans include GPI-anchored prions (see Fig. 1.7b, which in turn bear N-linked glycans, Fig. 1.5), the glycosphingolipid ganglioside GM3 (see Figs. 1.7a and 1.10e), and CD34, a mucin-type glycoprotein that bears numerous O-linked glycans that often include TACAs (see Figs. 1.6a and 1.6b). (See color insert.)

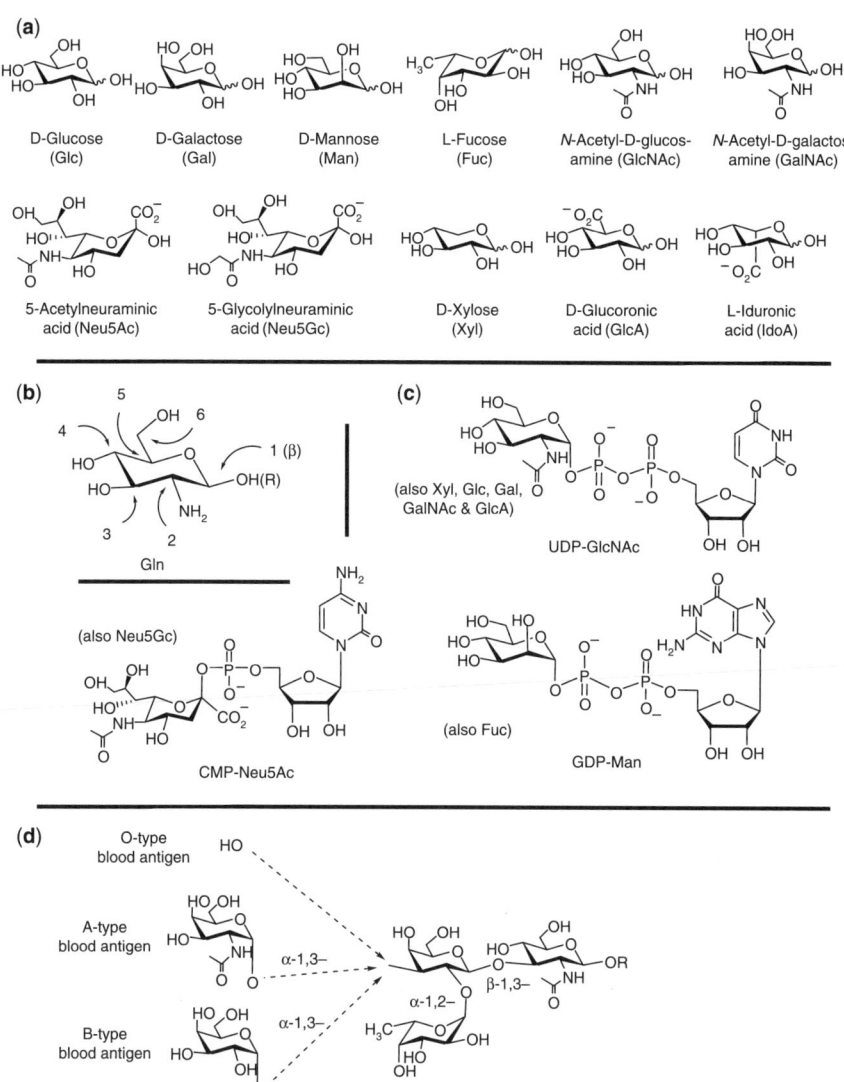

Figure 1.3 Chemical structures of monosaccharides, nucleotide sugars, and an explanation of α/β glycosidic linkages. (a) The 10 monosaccharides found in human glycans are shown; GlcA and IdoA are found exclusively in GAGs (see Fig. 1.6e), and the Neu5Gc form of sialic acid is found in animals other than humans and chickens. (b) Monosaccharides are converted to one of three types of nucleotide sugars, based either on UDP, GDP, or CMP. (c) Glucosamine, a sugar not found in human glycans without further modification (i.e., conversion to GlcNAc; see Figs. 1.2 and 1.4), is used as an example to show ring numbering and (α/β) linkage (in general β linkages are equatorial and α linkages are axial, except for sialic acid, which only has equatorial glycosidic linkages that are designated α; see Fig. 1.4). (d) Specific examples of glycosidic linkages are illustrated by the oligosaccharide structures comprising the ABO blood group antigens discussed in the Introduction.

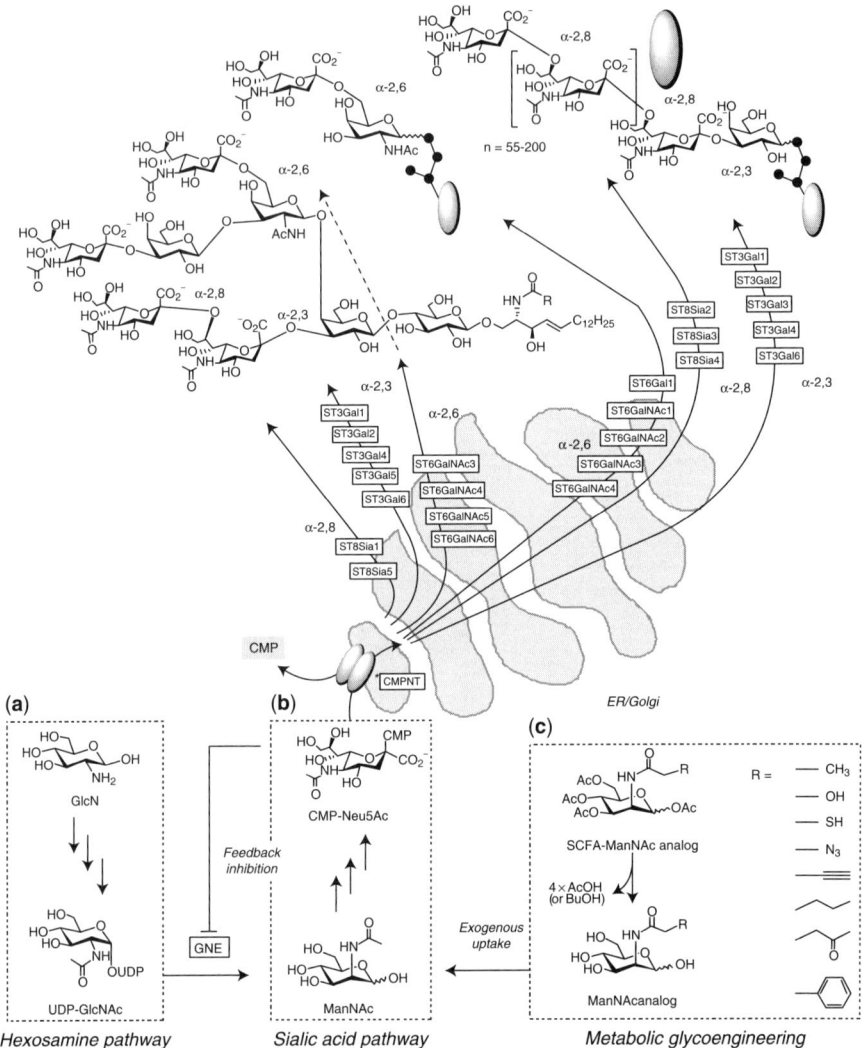

Figure 1.4 Details of sialylation. Sialic acid is one of the ~10 sugars found in mammalian glycans and can be produced from glucose or glucosamine (GlcN; see Figs. 1.2 and 1.3) obtained from exogenous sources that enter the hexosamine pathway [46]. (a) Passage of Gln through the hexosamine pathway results in the production of UDP-GlcNAc, which can be used directly for glycan assembly (or converted to UDP-GalNAc and used in a likewise manner) or for O-GlcNAc protein modification [47, 48]. (b) Another fate for UDP-GlcNAc is conversion to N-acetyl-D-mannosamine (ManNAc) by the bifunctional enzyme GNE [11]; this sugar does not appear in mammalian glycans but is instead converted to Neu5Ac [49] and installed into glycans by a set of 20 sialyltransferases (in humans) that have various overlapping linkage and substrate specificities (top). (c) In the predominant example of "metabolic glycoengineering," metabolic flux is supplied into the sialic acid pathway by nonnatural ManNAc analogs bearing abiotic "R" groups at the N-acyl position. These modifications include extended alkyl chains [50] and various chemical functional groups not usually found in sugars such as thiols, ketones, azides, or alkynes [51]. These analogs transit the biosynthetic pathway and appear in mature glycans in place of the Neu5Ac form of sialic acid most commonly found in humans.

carbohydrates, the glycan chains of glycoproteins are biosynthesized by the concerted action of a set of enzymes, rather than on the basis of a template akin to the way nucleotide sequence specifies primary amino acid structure. Instead, the final composition of a glycan is determined by multiple factors such as peptide sequence of the protein undergoing glycosylation, the availability of substrates in various subcellular locales, which in turn is determined by the expression and activity of membrane transporters, the localization of the particular glycosyltransferases to certain regions of the ER or Golgi, and competition between glycosyltransferases and glycosidases. This complex and indirect biosynthetic process results in heterogeneous glycosylation at two levels. First, any potential site of glycosylation—that is, the side chain of a candidate amino acid—may or may not be occupied. Different proteins vary considerably in the number of potential glycosylation sites—ranging from two sites where N-linked glycans can be attached to the prion protein to 26 possible sites for $\alpha3\beta1$ or $\alpha5\beta1$ integrin dimers [52]. Mucin-type proteins heavily invested with O-linked glycans—such as CD34 (see Fig. 1.2)—can have dozens of sites where sugars are attached to the peptide backbone [53]. It is clear that—even if the same oligosaccharide structure was attached, or not attached, to each site—a greatly diverse pool of glycoproteins would exist for each primary gene sequence.

The situation becomes exponentially more complex because each site of glycosylation can potentially be endowed with any one of dozens of different glycan structures. These range from a single monosaccharide to complex branching oligosaccharides of 20 or more residues in size to—more rarely because they are generally not covalently linked to surface elements—polysaccharides hundreds of residues in length. The diversity of glycans that can occur at each glycosylation site is referred to as microheterogeneity [54] and a "simple" case of microheterogeneity is provided by the prion protein that has two sites for N-linked glycan attachment. With 50 different glycan structures available for display at each site [21, 55], a prion can exist as $\sim 50^2$ or ~ 2500 distinct chemical entities. Clearly, this number is markedly greater than the three classes (un-, mono-, and di-glycosylated) of prions referred to earlier, making it transparent how sufficient diversity exists to allow significantly different profiles of glycans to be present on healthy and scrapie prion proteins [18].

Applying a similar analysis to CD34, each copy of this molecule in the human body could be chemically distinct; in fact, "ballpark" calculations indicate that each human CD34 that has ever existed could be unique. To elaborate briefly, if all of a person's approximately one trillion leukocytes bear an upper estimate of $\sim 100{,}000$ copies of CD34, a human body has $\sim 10^{17}$ of this molecule. Over a lifetime, assuming a turnover of once a day, a person would produce $\sim 2.5 \times 10^{21}$ copies of CD34. Next, extrapolating to the ~ 100 billion people estimated to have ever lived [56], nature has produced $\sim 2.5 \times 10^{32}$ human CD34 molecules. By comparison, based on an average CD34 molecule having ~ 50 sites where O-linked glycans can be attached, with a conservative estimate of 20 *different* glycans occurring at each site, the resulting number of chemically distinct forms of CD34 is 20^{50} or $\sim 10^{65}$. While this calculation is presented mainly for entertainment purposes in the vein of "every snowflake is different," it nonetheless vividly demonstrates how thoroughly nature has obliterated the "one gene–one protein" hypothesis and turned it into a "one gene–innumerable *glyco*proteins" reality through the clever use of sugars.

1.2.3.2 N-Linked Glycans Although many types of peptide–sugar linkages exist in nature [57], two classes of protein glycosylation—*N* and *O* linked—are dominant in mammals. We will discuss *N*-linked glycans, which are more abundant than *O*-linked glycans in most cells, first. This class of sugars is covalently attached to the peptide backbone of a protein through a 2-acetamido-2-deoxy-β-D-glucopyranosyl (GlcNAc) β linked to the amide nitrogen (hence, *N*-linked) of an asparagine (Asn) side chain (Fig. 1.5, inset). *N*-linked glycan structures occur at the consensus sequence Asn-Xaa-Ser/Thr, where Xaa is any amino acid other than Pro; despite the consensus sequence, any particular site may or may not be occupied by a glycan for reasons that remain obscure. The GlcNAc-β-Asn linkage is widely observed in glycoproteins isolated from eukaryotes and is also widely distributed through phyla ranging from archaea and eubacteria [58].

N-glycans can be subdivided into three distinct groups that include high-mannose-type, hybrid-type, and complex-type structures (Fig. 1.5). The groups share a common biosynthetic route where the first step is the co-translational transfer of the dolichol-linked oligosaccharide $Glc_3Man_9GlcNAc_2$ by oligosaccharyl transferase to an Asn residue of the growing polypeptide chain in the endoplasmic reticulum. As an aside, the production of this 14-mer is an interesting story in itself, beginning on

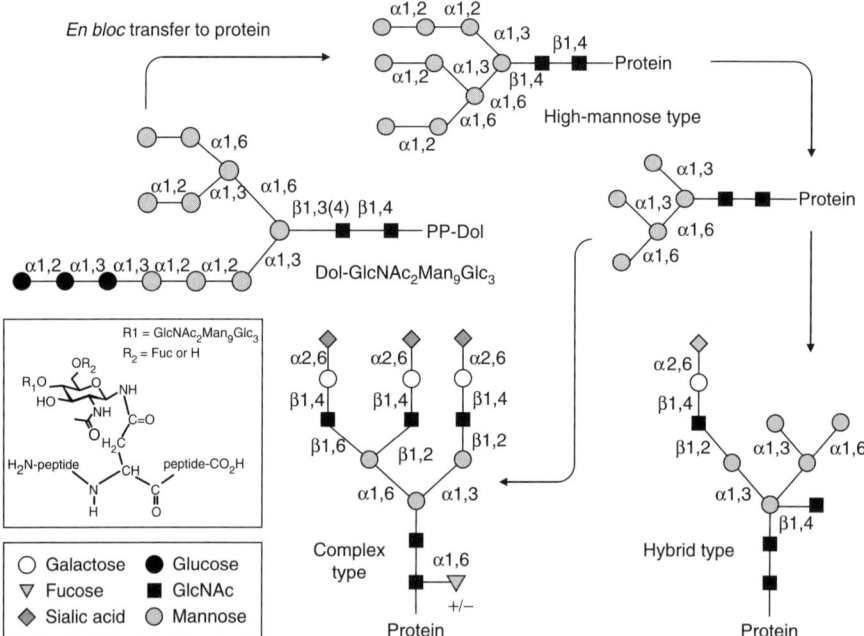

Figure 1.5 Biosynthesis and major classes of *N*-linked glycans. The $Dol-GlcNAc_2Man_9Glc_3$ 14-mer is assembled on the cytosolic face of the ER, flipped to the lumen, and transferred to a consensus sequence during translation of a nascent protein (the chemical linkage is shown in the inset along with the symbols used to depict the constituent monosaccharides). Subsequent trimming of glucose and mannose produces the high mannose-type glycan structure, which can be elaborated to form hundreds of hybrid or complex-type structures.

the cytosolic face of the ER where the sugars are linked to the lipid dolichol phosphate and ultimately involves the improbable translocation of the hydrophilic glycan structure across the membrane via a "flippase" [59, 60]. The importance of this preparatory process, reviewed in ample detail elsewhere by us [61] and many others [10, 62–65], is evidenced in a bevy of congenital disorders of glycosylation (CDG) that arise from biosynthetic missteps and that are usually fatal early in childhood (although a few can be overcome with fairly straightforward remedies such as dietary supplementation of rare sugars such as fucose [66]).

Once the $Glc_3Man_9GlcNAc_2$ structure is transferred to an Asn in nascent peptide chain, thereby creating the eponymous N-linked structure, a series of trimming events are set into motion. Initial trimming by the glucoside hydrolases I and II—in reasonably fast reactions that occur over a time span of a few minutes—remove the three glucose residues and yield the high-mannose-type glycoprotein. While it may seem odd that a cell is undoing a biosynthetic process in which it just invested a significant amount of energy, in reality the trimming process plays a vital role in protein folding and quality control along the secretory pathway [67–70]. The importance of this process is evident as a comparable processing mechanism for high mannose N-glycan chains is present in ancient pathways found in almost all eukaryotes including unicellular organisms such as yeast.

In mammals, further trimming and additions to N-linked glycans occur in the Golgi apparatus where sialic acid residues are added to yield hybrid-type or complex-type glycans (see Fig. 1.5) depending on the three-dimensional structure of the proteins, the cell type, and the organism [71]. From an evolutionary point of view, it is interesting to note that the primitive unicellular organisms do not have the mechanisms needed for the synthesis of hybrid and complex N-glycans. These organisms— which rely on early steps of N-glycan metabolism to ensure protein quality control—apparently do not require the complex intercellular recognition events needed to orchestrate multicellular life, which are in large part enabled by the sugar code [30] and thus could dispense with the effort needed to evolve the complete range of complexity found in the N-linked glycans of higher organisms.

1.2.3.3 O-Linked Glycans
A second major class of glycoproteins is known as O linked because the sugar moiety is attached to the oxygen of the hydroxyl group of either a serine (Ser) or a threonine (Thr) residue of the polypeptide chain [72]. The most prevalent type of O-glycan arises from mucin-type glycosylation, where N-acetyl-D-galactosamine (GalNAc) is attached to a Ser or Thr through an α linkage [73] (Fig. 1.6a). In general, O-linked glycans are an unruly bunch compared to their N-linked siblings; first, they are not confined to a consensus sequence but can seemingly occur on *any* Ser or Thr. Second, they are not limited to a single type of monosaccharide used to link the sugar to the peptide [57] but commonly use GalNAc, GlcNAc, and xylose-linked O-glycans (Fig. 1.6; additional O-linkages are provided in comprehensive reviews [57]). Nor are O-glycans based on an en bloc structure common to the entire class; rather, the mucin-type O-glycans are assembled into "core" structures by elaboration of an α-linked GalNAc (Fig. 1.6b). O-linked glycans have several more notable differences compared to N-linked structures, a

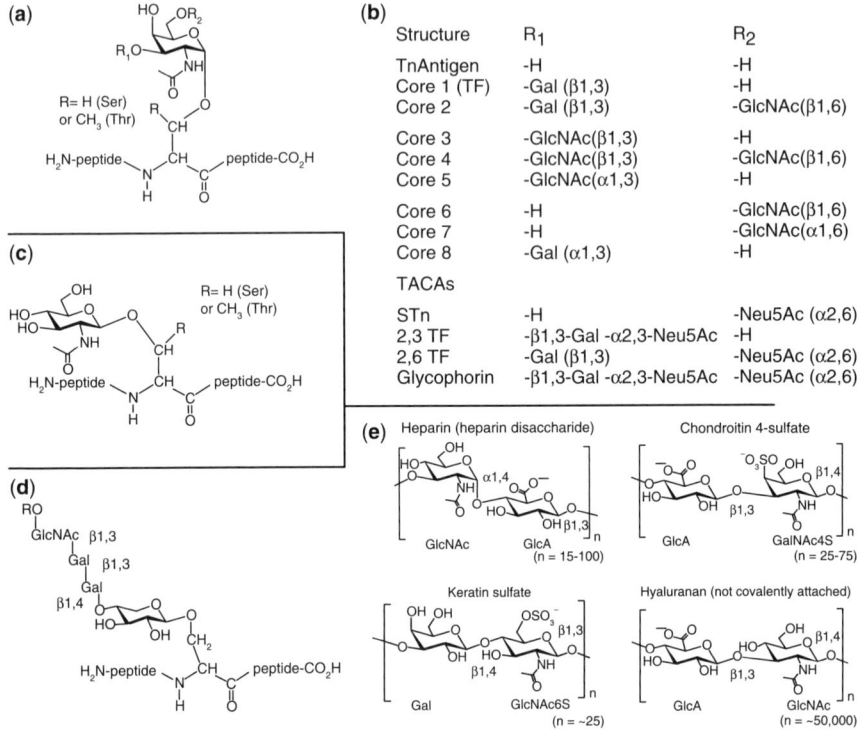

Figure 1.6 Structures of O-linked glycans. (a) The predominant surface displayed O-linked glycans in humans are the mucin-type characterized by an α-linked GalNAc attached to a serine or threonine that can be further elaborated at either the C4 or C6 hydroxyl group to give the "core" structures and tumor-associated carbohydrate antigens (TACAs) shown in (b). (c) The O-GlcNAc protein modification, remaining as a single monosaccharide residue, is a unique example of glycosylation found on nuclear and cytosolic proteins. (d) A xylose-originated tetrasaccharide is used to covalently link a subset of glycosaminoglycans (GAGs) to cell surface proteins such as syndecans or glypicans; the major classes of GAGs (the majority of which *are not* covalently linked to cell surface elements) are shown in (e).

major one being that they are added post- instead of co-translationally; consequently, they do not play a role in quality control during folding. Thus, instead of simply being an artifact of quality control during protein folding, which is postulated to be the case for a subset of *N*-linked glycoproteins, a cell presumably installs *O*-linked glycans on proteins only when they are critical modulators of biological activity.

Biosynthesis of the ubiquitous mucin-type *O*-linked glycoproteins (Fig. 1.6a) found in mammals and other eukaryotes is initiated by the family of GalNAc transferases (ppGalNAcTs) that utilize UDP-GalNAc as the nucleotide donor substrate to modify protein substrates [74]. There are ~24 unique ppGaNTase human genes that display tissue-specific expression in adult mammals as well as unique spatial and temporal patterns of expression during development. One explanation for why there are so many genes coding for similar biochemical activity is that this redundancy provides protection against defects in any one particular gene. An emerging picture is much more complex, however, as a subset of the ppGaNTases have overlapping substrate specificities and certain ppGaNTases require the prior addition of GalNAc to a peptide before they can catalyze sugar transfer to the substrate. Moreover, site-specific *O*-glycosylation by several ppGaNTases is influenced by the position and structure of previously added *O*-glycans [74]. The product of any ppGaNTase reaction, α-GalNAc on Ser/Thr residues, is termed the "Tn antigen" and is further elaborated by downstream glycosyltransferases to generate a series of core *O*-linked glycans [75] (Fig. 1.6b). These core structures are then further modified by other Golgi-resident glycosyltransferases to generate complex *O*-linked glycans that are involved in a variety of biological processes in health and disease [76]; for example, *O*-linked glycan abnormalities occur in cancer when relatively subtle changes to the normal core structures convert them into tumor-associated carbohydrate antigens (TACAs, Fig. 1.6b).

1.2.3.4 O-GlcNAc Modification of Nuclear and Cytosolic Proteins
For decades it was accepted wisdom that only cell-surface-displayed and secreted proteins were glycosylated in mammals. In 1984, however, Torres and Hart described a posttranslational modification where Ser and Thr residues found in nuclear and cytoplasmic proteins were *O* linked to GlcNAc through a β linkage (Fig. 1.6c) [47, 77]. *O*-GlcNAc had two novel aspects: First, as mentioned, it had a nucleocytoplasmic distribution, whereas "traditional" glycoproteins were localized to the cell surface and topologically equivalent intracellular compartments, such as the lumens of the endoplasmic reticulum and Golgi apparatus [78]. Second, with the possible exception of plant nuclear pore proteins [79], *O*-GlcNAc is not elongated into more complex structures but rather remains as a single monosaccharide on the peptide backbone. As such, the biosynthetic "machinery" for *O*-GlcNAc is extremely simple—with an important caveat mentioned below regarding the dynamic nature of this modification—compared to other types of glycosylation. For example, in marked contrast to dozens of enzymes involved in sialylation (Fig. 1.4), *O*-GlcNAc metabolism involves one enzyme that adds the sugar to the protein [uridine diphospho-*N*-acetyl-D-glucosamine: peptide β-*N*-acetylglucosaminyl transferase (OGT)] and another [*O*-β-*N*-acetylglucosaminidase hexosaminidase (*O*-GlcNAcase)] that removes it [80]. Another contrast is that these two enzymes work in concert in a highly dynamic fashion, remodeling the "*O*-GlcNAcome" in a matter of minutes rather than over the hours-to-days turnover rates of most glycans.

The dynamic nature of *O*-GlcNAc protein modification is one of several similarities that this form of glycosylation shares with phosphorylation in cellular regulation. In addition to the rapid cycling of *O*-GlcNAc in response to metabolic factors, extracellular signals, stress, or stages in cell cycle progress, sugar attachment often occurs at the very same amino acid side chains on the protein backbone that are modified by protein kinases [48]. Unlike protein phosphorylation, however, where over 650 genetically distinct enzymes regulate the addition and removal of phosphate, as just mentioned only two catalytic polypeptides catalyze the turnover of *O*-GlcNAc. Consequently, *O*-GlcNAc is much simpler than both of its comparable biochemical systems of glycosylation and phosphorylation. Having down played the complexity of *O*-GlcNAc, we emphasize that the apparent simplicity of *O*-GlcNAc protein modification is countered

16 GLYCOBIOLOGY AND IMMUNOLOGY

by links through the metabolic substrate UDP-GlcNAc—an indicator of flux through the hexosamine pathway [46]—to cell nutritional status and the greater complexity of sialylation and glycan biosynthesis (see Figs. 1.2 and 1.4).

1.2.4 Lipid-Based Glycans

1.2.4.1 Glycosphingolipids
Glycosphingolipids (GSLs) are components of the plasma membranes of all eukaryotic cells. Roughly speaking, approximately 1%—or about one billion—of the lipids found in the plasma membrane of a typical 20-μm mammalian cell is glycosylated [81]. Approximately 300 different GSL structures have been identified [82] where at least one monosaccharide residue is glycosidically linked to a hydrophobic ceramide or sphingoid long-chain aliphatic amino alcohol that is imbedded in the lipid bilayer. The presence of these molecules at the plasma membrane enriches the outer surface in a layer of carbohydrate that helps to protect the cell membrane from chemical and mechanical damage. Despite the relatively small overall contribution of glycosphingolipids ($\leq 8\%$) to the aggregate

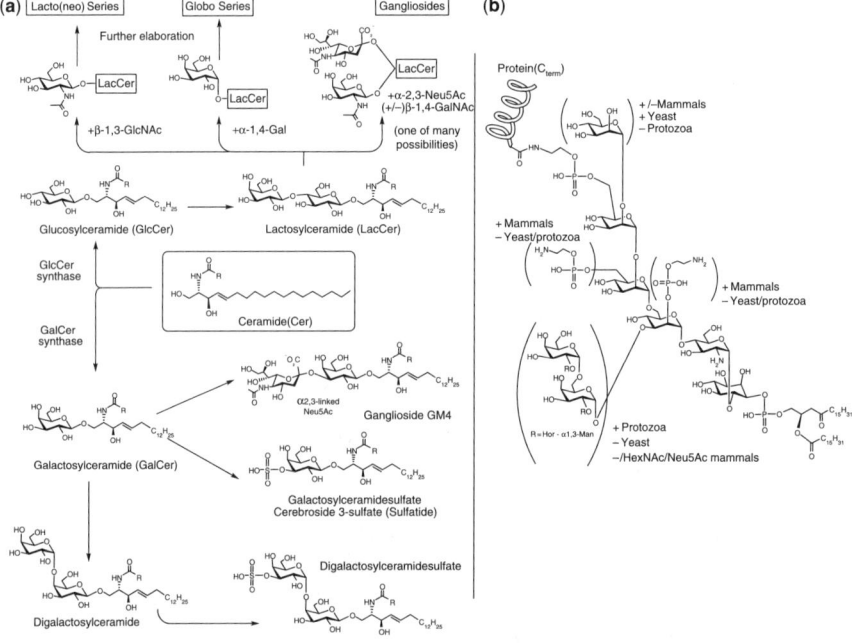

Figure 1.7 Lipid-based glycans. (a) Glycosphingolipid (GSL) biosynthesis begins with ceramide, which is most commonly elaborated with a Glc and a Gal to form LacCer, which in turn forms the "core" for three major classes of GSL [the lacto(neo) and globo series as well as gangliosides; top], each of which contain dozens (or hundreds) of structures. Alternately, and less commonly, a Gal instead of a Glc is added to Cer to form GalCer, a marker of autoimmune disease [89]. GalCer can be further elaborated to form a limited number of structures, including ganglioside GM4, sulfatide, and digalactosylceramidesulfate. (b) Glycosylphosphatidylinositol (GPI) membrane anchors, showing structural differences between various species.

mass of the plasma membrane, they play several critical functions including cell adhesion, cell growth regulation, and differentiation [83]. The critical importance of GSLs in development has been demonstrated by the embryonic lethality in the mouse resulting from disruption of the gene-encoding ceramide-specific glucosyltransferase, an enzyme that initiates the synthesis of all glycosphingolipids [84]. Continuing to mature organisms, GSL "lubricate" signaling pathways [85] through "the glycosynapse" [86] and play ongoing roles in the maintenance of health.

Glycosphingolipids are derived from a common biosynthetic pathway that starts with the condensation reaction between palmitoyl-CoA and serine that leads to the formation of ceramide, which is the basic lipid structure of GSL (Fig. 1.7a). One class of GSL results from the addition of a galactose residue via a galactosyltransferase-catalyzed reaction to form galactosylceramide (GalCer), while glucosylceramide (GlcCer) is a product of a glucosyltransferase-catalyzed reaction. The glycosyltransferases responsible for these two reactions do not reside in the same subcellular compartment or have similar structural features, which is surprising since both use the same acceptor (ceramide) and similar nucleotide sugar donors. The ceramide-specific galactosyltransferase is a type I transmembrane protein whose catalytic domain is localized to the lumen of the endoplasmic reticulum [87]. By contrast, the catalytic domain of the ceramide-specific glucosyltransferase faces the cytosol with the enzyme restricted to the Golgi membrane [88].

In order to accommodate the various luminal and cytosolic orientations of the processing enzymes, newly synthesized ceramide is able to translocate across the membrane during bulk flow to the plasma membrane due to the rapid and spontaneous interbilayer transfer (flip flop) [90]. In turn, glucosylceramide translocates across the Golgi membrane where it becomes the substrate for Golgi resident glycosyltransferases [91]. The addition of galactose to form a lactose unit on ceramide is mediated by a glucosylceramide-specific galactosyltransferase [92]. At this point in the biosynthetic pathway, there is considerable competition for common substrates because lactosylceramide (LacCer) is the acceptor for various transferases that generate distinct groups of complex GSLs that consist of hundreds of already known [82]—and likely many more yet-to-be discovered—structures.

Not surprisingly considering their important roles in cellular physiology, GSLs play many roles in pathological processes when their complicated biosynthetic processes go awry. For example, when catabolism of glycosphingolipids is impaired, several severe pathological conditions are manifest [93], often as glycolipid lysosomal storage disease typified by the well-known example of Tay Sachs disease [94]. Collectively, congenital glycolipid disease is estimated to occur at a frequency of $\sim$1 in 18,000 live births worldwide and is the most common cause of pediatric neurodegenerative disease. Aberrations in GSL also play significant roles in acquired disease—for example, in cancer, where changes in the relative expression of these molecules virtually always accompany oncogenic transformation [83, 95–98]. The close proximity of glycosphingolipids to the lipid bilayer—unlike farther outlying protein-associated glycans, in particular, polysaccharides associated with the extracellular matrix, that can more easily slough off and act as ineffectual binding decoys—is exploited by a number of viral and bacterial pathogens that have adapted to adhere selectively to these carbohydrate residues as a prelude to internalization and pathogenesis [99].

1.2.4.2 Glycosylphosphatidylinositol Membrane Anchors
Proteins were discovered to be anchored to membranes through a covalently attached glycosylphosphatidylinositol (GPI) moiety in the 1980s [100, 101]. Subsequent structural determination of these GPI "anchors" that have been found in protozoa, yeast, plants, and mammals has revealed the common core structure: ethanolamine-PO_4-6-Man-α-1,2-Man-α-1,6-Man-α-1,4-$GlcNH_2$-α-1,6-*myo*-inositol-1-PO_4-lipid (Fig. 1.7B) where the ethanolamine is amide bonded to the α-carboxyl group of the C-terminal amino acid of the mature protein. The conserved core structure may possess a variety of side-chain modifications (additional phosphoethanolamines and sugars such as GalNAc, galactose, mannose) that are protein, tissue, and species specific [102]. The lipid moieties range from ceramide in most yeast and slime mold GPI-anchored proteins to diacylglycerol in protozoa and (predominantly) 1-alkyl-2-acylglycerol in mammalian proteins [103]. In some cases, the inositol ring contains an additional lipid modification in the form of an ester-linked palmitic acid that renders the anchor resistant to the action of phospholipase C (PLC) [104].

The biosynthesis of GPI membrane anchors has been investigated in a broad range of eukaryotes ranging from protozoa and yeast to humans. Other than in mammals, GPI biosynthesis has been most extensively investigated in trypanosomes and many similarities—as well as a few differences (as described in detail elsewhere [105, 106])—have been found between mammals and bloodstream forms of the parasite *Trypnosoma brucei* [107]. In mammals, GPI anchors provide an alternative to hydrophobic transmembrane polypeptide anchors and also participate in intracellular sorting, in the endocytic process of potocytosis, in transmembrane signaling. They also facilitate the embedding of proteins into glycosphingolipid-rich lipid rafts and allow host proteins to be selectively released from the cell surface by the action of phospholipases. In trypanosomes, GPI anchors are present at up to 10–20 million copies ($\sim$100 times more than found on a mammalian cell) and dominate the cell surface molecular architecture of these organisms. The GPI anchors of trypanosomes ties in with the second major topic discussed in this chapter—immunity—because these molecules are well known for their ability to help the parasite avoid immune elimination by switching the immunodominant variant surface glycoprotein (VSG) coat during infection. The fact that trypanosomes have up to 1000 different VSG genes, combined with the ability of the parasite to rapidly release existing VSG by virtue of phospholipase C cleavage of proteins from the GPI moiety, affords the parasite extensive opportunity to escape host B- and T-cell responses by displaying new coat antigens [108].

1.2.5 Polysaccharides: Glycosaminoglycans and Bacterial Capsular Components

Although highly complex and collectively the dominant feature of the cell surface landscape, the majority of the *N*-, *O*- and lipid-linked glycans are relatively modest in size, consisting of 20 or fewer monosaccharide residues. By contrast, polysaccharides—primarily the glycosaminoglycans (GAGs; see Fig. 1.6e for examples) but also including specialized structures such as α2,8-linked homopolymeric polysialic acids (see Fig. 1.4b, top)—are much larger with sizes of 100 monosaccharide residues or greater being commonplace. In general, although some GAGs are linked covalently to the cell surface—for example, to glypican or syndecans [16, 109] via the xylose-linked O-glycan structure shown in Figure 1.6d—in the majority of cases GAGs are free of explicit surface entanglements and instead exist as part of the proteoglycan component of the extracellular matrix (ECM). The major GAGs of physiological significance are hyaluronic acid, dermatan sulfate, chondroitin sulfate, heparin, heparin sulfate, and keratan sulfate. While each has a distinctive molecular composition, all GAGs are based on disaccharide units that contain either GalNAc or GlcNAc combined with one of two uronic acids (glucuronate or iduronate). Structurally, GAGs are long unbranched polysaccharides and are the most abundant heteropolysaccharides in the human body. GAGs are highly negatively charged molecules, highly hydrated, and exist in extended conformations that give high viscosity to the ECM. The "hydrogel" properties of these molecules are evident by considering that while most tissues have between 2 and 50 μg/mg dry weight of GAG [110], when hydrated these polysaccharides can constitute 70% or more of the volume of the ECM.

Mammalian GAGs are being implicated as an increasingly broad repertoire of biological functions besides the strictly structural. To illustrate with hyaluronic acid (HA), this very large glycosaminoglycan (with molecular weights of 100,000–10,000,000) is expressed in virtually all tissues and has long been known to be a critical structural component in the tissue interstitium by providing the core backbone of proteoglycans. HA is unique among the GAGs in that it does not contain any sulfate and is not found covalently attached to proteins but rather solely forms noncovalent proteoglycan complexes in the ECM. The discovery of HA-binding proteins led to the hypothesis that HA also serves as an adhesive substrate for cellular trafficking [111] and, most recently, the finding that HA fragments can deliver maturational signals to dendritic cells (DCs) and high-molecular-weight HA polymers can deliver co-stimulatory signals to T cells has established HA as in important immunomodulatory molecule [112]. Immune complications witnessed from the use of low-molecular-weight heparin as an anticoagulant [113] established that the ability of GAGs to engage an immune response is by no means unique to HA.

Although mammalian GAGs are relatively conserved in their basic structures, considerable structural diversity is obtained by postsynthetic modifications such as sulfation and epimerization of GlcA to IdoA. The chemical diversity nonetheless lags bacterial polysaccharides used either in the cell wall or in capsules where numerous structures not found in mammals abound. The ability of bacteria to employ an expanded repertoire of immunogenic monosaccharides in their glycans has led to the glycovaccinist exploiting these molecules to combat pathogenic microbes. Indeed, some of the earliest and most successful examples of carbohydrate vaccines are targeted against these molecules (as discussed in Section 1.4 and in detail in Chapters 2 and 4).

1.3 THE IMMUNE SYSTEM

1.3.1 Introductory Comments

Now that glycosylation has been briefly outlined, we will next provide an even less thorough, but hopefully helpful for the nonspecialist, overview of the immune system. Clearly, immunity is a huge topic—with well over one million articles available through searches of computer databases such as PubMed [35]—therefore, at the outset we emphasize that we provide a "bare-bones" discussion just sufficient to place into context some of the intriguing connections between glycosylation and immunity, and we trust that the valued reader will not feel slighted if his or her favorite aspect of the immune system is omitted in this chapter.

1.3.2 Overview of the Immune System

1.3.2.1 Immune System Provides Protection The term *immune system* generically refers to a collection of mechanisms within an organism that protects against disease by identifying and killing invading pathogens and, in higher organisms, even providing protection against tumor cells. Virtually all multicellular organisms—including mollusks, worms, and insects—have primitive but effective immune systems; here we will primarily limit discussion to mammalian immunity. Of course, many features of immunity are broadly shared across phyla; on the other hand substantial differences separate humans from even mice, for example, complicating the biomedical researcher's efforts to investigate human disease in this widely used animal model. The immune system of a mammal has the daunting task of not only recognizing a wide variety of agents, from viruses to parasitic worms, but also needs to distinguish them from the organism's own healthy cells and tissues. In immunology, *self* molecules are those components of an organism's body that can be distinguished from foreign substances by the immune system. Conversely, *nonself* molecules are those recognized as foreign molecules. The class of nonself molecules that are the smallest unit that the immune system responds to by binding to specific immune receptors to elicit an immune response are called *antigens*, short for *anti*body *gen*erators.

The immune system protects an organism from infection through a multilayer blanket comprised of several different steps. First, physical or mechanical barriers—such as skin in humans—prevent pathogens from entering the body [114]. If a pathogen crosses a physical barrier, the innate immune system—which is found in all plants and animals—acts quickly responding to challenges in a few minutes but in a nonspecific manner [115]. If a pathogen thwarts this second layer, a third level of protection found in most vertebrates—the adaptive immune system—can be activated. In adaptive (also known as acquired or learned) immunity, whose existence in invertebrates has been postulated but remains controversial [116], the immune response improves its ability to deal with an infectious agent by retaining a "memory" of the pathogen. Immunological memory allows the adaptive immune system to work faster and stronger each time a particular pathogen is encountered; for primary infections an effective response can require up to 5–7 days, whereas responses to subsequent insults are mounted within 1–3 days [114]. Both innate and adaptive immunity are highly complex biological systems and each will now be discussed briefly (the innate immunity of carbohydrates is discussed throughout this book in greater detail, in particular in Chapters 3, 9, and 11).

1.3.2.2 Innate Immunity The mammalian innate immune system has two arms that are capable of mounting the complement and the inflammatory reponses. First, the complement system consists of over 20 different proteins capable of mounting a biochemical cascade that attacks the surfaces of foreign cells. It is named for its ability to "complement" the killing of pathogens by antibodies and functions most effectively as a front line of defense when preexisting, circulating antibodies are present. Less helpful, complement-mediated cell killing complicates medical intervention; for example, immune response directed at the α-gal epitope is the source of hyperacute rejection of xenotransplanted organs and tissues. Either way, the existence, or new production of these antibodies, in turn relies on the larger functioning of the immune system, as described below.

Inflammation, one of the first "active" responses of the immune system to infection, results when cytokines or eicosanoids are released by injured or infected cells. Common cytokines include interleukins that are responsible for communication between white blood cells, chemokines that promote chemotaxis, and interferons that have antiviral effects, such as shutting down protein synthesis in the host cell [117, 118]. Eicosanoids include prostaglandins that produce fever and the dilation of blood vessels associated with inflammation, and leukotrienes serve as chemoattractants for certain types of white blood cells. Despite overall coordination by the immune system, white blood cells—or the leukocytes—behave as independent, single-celled organisms with many different duties and are the functional workhorses of the innate immune system.

Leukocytes that actively participate in innate immunity include mast cells, eosinophils, basophils, natural killer cells, and the phagocytes (macrophages, neutrophils, and dendritic cells). Collectively, these cells identify and eliminate pathogens, either by engulfing and digesting smaller microorganisms or attacking larger pathogens through contact. To briefly describe the roles of specific cells, mast cells reside in connective tissues and mucous membranes, regulate the inflammatory response, and contribute to allergies and anaphylaxis [119]. Basophils and eosinophils secrete chemicals involved in defending against parasites and play a role in allergic reactions, such as asthma [120]. Natural killer (NK) cells are leukocytes that attack and destroy tumor cells or cells that have been infected by viruses [121, 122].

Phagocytosis performed by cells called phagocytes that engulf and eat pathogens or other threatening particles is an important part of innate immunity. Of the three main types of phagocytic cells, neutrophils and macrophages constantly patrol the body searching for pathogen but can also be summoned to specific locations by cytokines. During the acute phase of inflammation, during bacterial infection, for example, neutrophils migrate toward the site of insult in a process called chemotaxis and are usually the first cells to arrive at the scene. Once a pathogen has been engulfed by a phagocyte, it becomes trapped in an intracellular vesicle called a phagosome, which subsequently fuses with a lysosome to form a phagolysosome. The pathogen is killed by the activity of digestive enzymes or by a respiratory burst that releases free radicals into the phagolysosome [123]. Dendritic cells (DC) are phagocytes that reside in tissues that come into contact with the external environment such as the skin, nose, lungs, stomach, and intestines. These cells resemble—and derive their name from—neuronal dendrites, as both have many spinelike projections, but dendritic cells are not functionally connected to the nervous system. Instead, they function as a link between the innate and adaptive immune systems by presenting antigen to T cells, one of the key cell types of the adaptive immune system [124, 125].

1.3.2.3 Adaptive Immunity The adaptive immune system arose early in vertebrate evolution and provided a stronger immune response as well as immunological memory, where each pathogen is "remembered" by a signature antigen through the coordinated action of B and T lymphocytes. B cells identify pathogens when surface antibodies bind to foreign antigens [126] and in turn their chief function is to secrete antibodies into bodily fluids in what is known as the *humoral* response. This sequence of events occurs because after a B cell encounters its triggering antigen, it gives rise to many large cells known as plasma cells. Every plasma cell is essentially a factory for producing a specific antibody, for example, one produces antibody against this year's strain of influenza virus while another might produce antibody against the bacterium that causes pneumonia. In relatively short order, each plasma cell manufactures millions of identical antibody molecules and pours them into the bloodstream or lymph fluid.

An antibody matches an antigen much like a key matches a lock; some pairs form exact matches and bind with high affinity while others fit more loosely like a skeleton key. Whenever antigen and antibody successful form complexes, however, the antibody marks the antigen for destruction. Antibodies belong to a family of large molecules known as immunoglobulins and different family members play distinct—but sometimes overlapping roles—in host defense. Immunoglobulin G, or IgG, efficiently coats microbes speeding their uptake by phagocytic immune cells, whereas IgM is more effective at killing bacteria. Immunoglobulin A, or IgA, is concentrated in bodily fluids—tears, saliva, the secretions of the respiratory tract, and the digestive tract—and guards the entrances to the body. Immunoglobulin E, or IgE, has the beneficial natural job of protecting against parasitic infections but is also the villain responsible for the symptoms of allergy. Finally, immunoglobulin D, or IgD, remains attached to B cells and play a key role in initiating early B-cell response. Collectively, the five classes constitute about 25% of all serum proteins and, in any one individual, have a diversity of about 10^7 different binding specificities. Interestingly, a diversity of 10^8-10^9 binding specificities has been estimated to be found collectively in all humans but up to 10^{11} are theoretically possible as shown by combinatorial human antibody libraries [114]; these numbers may be of interest to the vaccine developer who will note that the human immune system has the (at least in theory) "extra" capacity to respond to novel antigens such as glycoconjugates.

The second component of adaptive immunity that complements the humoral response, known as the *cell-mediated response*, involves specialized white blood cells called thymus-derived cells, T lymphocytes, or T cells. Unlike B cells, T cells do not recognize free-floating antigens. Rather, their surfaces contain specialized antibody-like receptors—the well-known T-cell receptor (TCR)—designed to recognize fragments of antigens on the surfaces of damaged, invading, infected, or even cancerous cells. T cells contribute to immune defenses in two major ways: some direct and regulate immune responses; others directly attack diseased cells. *Helper T cells* regulate both the innate and adaptive immune responses and help determine which types of immune responses the body will make to a particular pathogen [127, 128]. These T cells are not cytotoxic and thus do not kill infected cells or clear pathogens directly. They instead control the immune response by directing other cells to perform these tasks. By contrast, *killer T cells*, which are also called *cytotoxic T lymphocytes*, or *CTLs*, are a subgroup of T cells that kill cells infected with viruses (and other pathogens) or are otherwise damaged or dysfunctional. CTLs directly attack other cells carrying foreign or abnormal molecules on their surfaces and are particularly valuable for their antiviral action because viruses often hide from other parts of the immune system while they reproduce inside infected cells. CTLs, however, can recognize small fragments of these viruses on the membranes of infected cells and can eradicate the diseased cell [129].

1.3.3 Glycoimmunobiology

Now that we have provided a rudimentary introduction to immunology, we will delve into three specific areas in more detail to provide a small window into just how complicated the immune system is and also to merge the two areas of discussion—glycosylation and immunology—into a brief discussion of glycoimmunobiology. These examples touch base on three integral functions of the immune system—antibody function, cell movement, and cell activation—and illustrate how carbohydrates play a critical role in various facets of immunity. The intent once again is not to cover these areas comprehensively but to provide a sampling of concrete examples (selected out of many) to hopefully whet the interest of the reader in the manifold roles that sugar enjoys in the functioning of virtually all complex biological systems.

1.3.3.1 Antibody Glycosylation Recombinant monoclonal antibodies (mAbs) are arguably the most important protein-based therapeutic agents. At the end of 2006, 18 mAb had been approved by the U.S. Food and Drug Administration (FDA) to treat a wide range of human diseases [130]; mAbs are also invaluable research tools for the biomedical community, which increases their biomedical and biotechnological importance yet further. For therapeutic purposes, mAbs are usually produced using mammalian cell lines, purified, concentrated, and subject to appropriate formulation for in vivo administration. In order to enhance biological activity and optimize pharmacologic properties, as well as to avoid deleterious off-target effects, increasing attention has been paid to the posttranslational modifications (PTM) of recombinant antibodies and, surely not surprising to a reader of this chapter, glycosylation is emerging as a critical PTM determinant of antibody structure and function [131–134].

The presence of various particular glycan structures on antibodies is crucial for antibody structure [135], interaction with Fc receptors [136, 137], binding to the complement component C1q, and the lack of particular sugar attachments has been implicated in autoimmune disease [136]. N-Glycosylation of immunoglobulin G (IgG) has been studied in detail and 32 different IgG glycoforms have been identified in human serum [138]. The N-linked oligosaccharides attached to antibodies produced by mammalian cells are mostly of the complex biantennary type, containing a mannosyl-chitobiose core and two N-acetylglucosamine (GlcNAc) residues, with variable additions of fucose, galactose, sialic acids, and bisecting GlcNAc (a bisecting GlcNAc is shown on the hybrid-type N-glycan in Fig. 1.5) [136]. At the submolecular scale, individual monosaccharides that comprise the glycans found on antibodies have now received intense scrutiny with galactose [139, 140], bisecting GlcNAc [141, 142], fucose [143], and sialic acid [144] all studied and found to have distinct contributions to antibody structure and function.

1.3.3.2 Leukocyte Extravasation

Inflammation is a process in which the body's white blood cells and chemical agents protect one from infection by foreign substances including pathogens such as bacteria, parasites, and viruses. But, consider the situation where a person—maybe you—steps on a rusty, tetanus-laden nail; hopefully, leukocytes will rush to the site of insult to your foot. However, if these protective cells—which actually spend only ∼2% of their time in the blood—were to wend their way through tissue or the ECM from where they are likely to be stationed in your body, which could be a meter (or more) away, at the top speed of 2.0 μm/h for this mode of locomotion [145], they would not reach the site of injury for 500,000 h—or about 57 years later. Clearly, such a situation is untenable because you would have long since died from the infection, or possibly old age. Leukocytes solve this problem by exploiting the blood as a rapid transit system capable of reaching any point in the body within a minute or so; but to do so in a way that successfully fights the infection, immune cells must have a way to exit the swiftly flowing blood at the site of injury. Carbohydrates play an absolutely critical role in the extravasation of leukocytes from the bloodstream into the underlying tissue at the site of insult. Later, when the crisis is over, cells with memory of the pathogen return to their home in lymph nodes—also far away—and use the same vascular transport and homing mechanisms.

Glycans play three distinct roles in leukocyte extravasation (Fig. 1.8). The first step—involving the endothelial glycocalyx layer (EGL)—is probably the least understood. What is known is that the EGL lines the lumens of blood vessels and, at up to half a micron in thickness, has an antiadhesive effect by preventing interaction between adhesion molecules on passing leukocytes (e.g., L-selectin, Fig. 1.8b) and their binding partners on the endothelium (e.g., CD34, Fig. 1.8c). The EGL has mechanical properties and structural rigidity such that the microvilli of leukocytes "tip-toe across it much like a Jesus Christ lizard can run across water" [146, 147]. Thus, based on the height that selectins and their binding partners extend above the lipid bilayer—a maximum of ∼50 μm—the EGL must collapse by close to 90% of its usual thickness of >400 μm to allow these molecules to interact and mediate the well-known "tethering and rolling" behavior characteristic of leukocyte extravastion. In this second step (Fig. 1.8a),

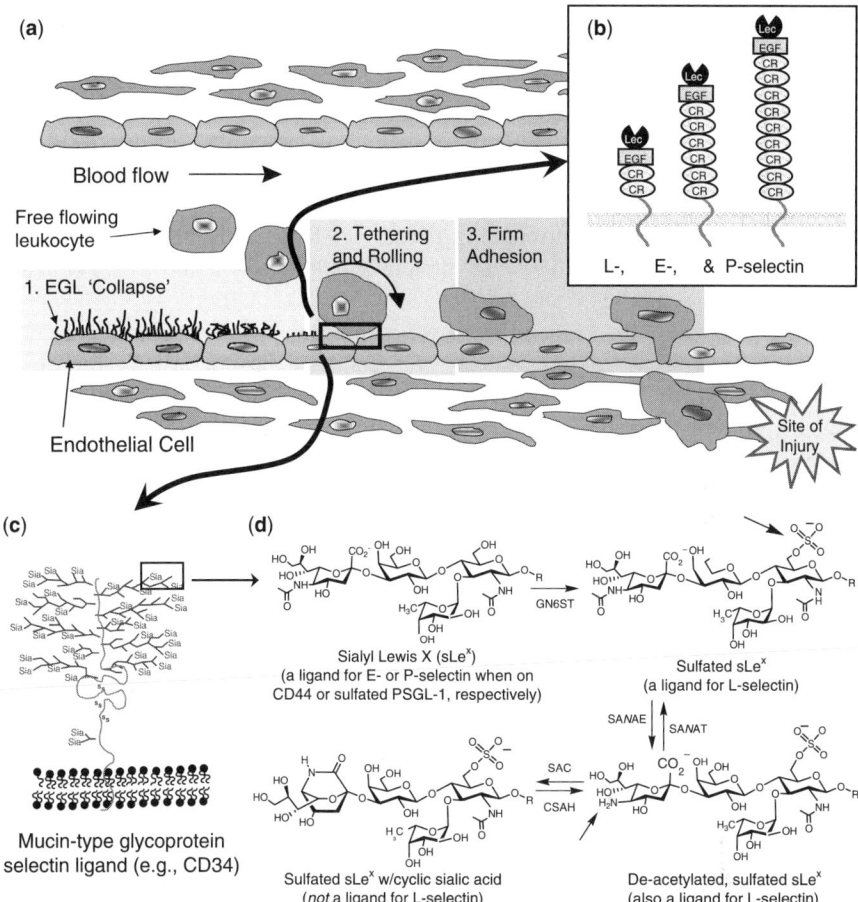

Figure 1.8 Role of glycans in leukocyte extravasation. (a) Free-flowing leukocyte is shown in a cross section of a venule (not to scale; the leukocytes are typically 6–12 μm in diameter, the vessels where extravasation takes place have a diameter of 30–80 μm, and the EGL is approximately 0.2–0.5 μm thick) where extravasation to sites of injury or homing in the lymph node occurs. (step 1) The EGL collapses to a height of less than 50 μm allowing (step 2) selectin-mediated tethering and rolling to take place followed (step 3) by integrin-mediated firm adhesion and extravasation. (b) Graphic of selectin structures where CR represents the cysteine-rich consensus repeat domains, EGF represents the EGF-like domain, and Lec represents the carbohydrate recognition (lectin) domain. (c) CD34 exemplifies mucin-type counter-receptor for selectins where the peptide forms a scaffold for multivalent glycan display; up to 80% of the mass of CD34 can be carbohydrate [53]. (d) Chemical modification of sLe^X determines physiological binding specificity between various selectins and counter-receptors. Of note, the reverse situation, shown where selectins are found on the epithelial substrate (e.g., P-selectin) and interact with ligands (e.g., PSGL-1) on the incoming cell, can also occur. (See color insert.)

selectins—a family of three structurally related lectins with affinity for the sialyl Lewis tetrasaccharide (sLe^X) [148] (Fig. 1.8b)—interact with mucin-type glycoprotein ligands (Fig. 1.8c). Selectin–mucin interactions lead to unique flow-dependent rolling behavior mediated by unusual catch bond-to-slip bond characteristics [149, 150] of these binding partners that allow rapidly moving cells to slow down and sample the local environment near the site of infection [151, 152]. If the threat is legitimate, evidenced by the presence of the appropriate chemokines, a transition to firm adhesion and extravasation into the surrounding tissue will take place, primarily mediated through integrin activation and protein–protein binding interactions that are nonetheless also tuned by carbohydrates.

Of the three steps involved in leukocyte extravasation—ESL collapse, selectin-mediated tethering and rolling, and integrin-facilitated firm adhesion—by far the most is known about the contributions of carbohydrates to the middle step. Selectins vividly demonstrate the exquisite control that nature has achieved through subtle variation in the structure of the sLe^X tetrasaccharide epitope. The binding partners for all selectins—L-selectin (e.g., GlyCAM-1 or CD34), P-selectin (e.g., PSGL-1), and E-selectin (e.g., CD44) all require the multivalent presentation of sLe^X but have little cross-reactivity under physiological conditions [153]. It is now clear that in vivo binding specificity for each selectin depends on chemical modifications to their counter receptor such as sulfation; for L-selectin binding partners sLe^X itself needs to be sulfated while for P-selectin ligands this modification occurs on the peptide backbone of PSGL-1 [154, 155]. Finally, nature has come up with a way to mask sLe^X—to essentially sequester it in an inactive form that can be rapidly mobilized—by cyclization and decyclization [156] (Fig. 1.8d). In summary, submolecular microsurgery of carbohydrate epitopes—exemplified by sLe^X—enables cells to move to where they are needed; of course, if they were not properly activated upon arrival, the entire process would be rather futile. Thus, in the next section we will discuss Siglecs, a family of molecules that participates in the activation of various leukocytes.

1.3.3.3 Cell- to Systems-Level Control Is Regulated through Siglecs In addition to providing cells with a precise homing mechanism as they move rapidly to specific locations throughout the body, sugars also assist immune cells with activitation through Siglecs (<u>s</u>ialic acid-binding <u>Ig</u>-like <u>lectins</u>). Siglecs belong to an immunoglobulin superfamily (IgSF) of about a dozen—in humans—of cell surface receptors that recognize sugar ligands and play a smattering of roles in coordinating the myriad activities of the immune system [157, 158]. The first Siglec discovered was sialoadhesin (Siglec-1/CD169), a lectinlike adhesion molecule found on macrophages [159]. Other members of the Siglec family subsequently described include CD22 (Siglec-2), which is restricted to B cells and has an important role in regulating their adhesion and activation [160], CD33 (Siglec-3) [161], and myelin-associated glycoprotein (MAG/Siglec-4) [162]. Several additional Siglecs (Siglecs 5–12) have been identified in humans that are highly similar in structure to CD33 and collectively referred to as CD33-related Siglecs [163, 164]. CD33-related Siglecs have two conserved immunoreceptor tyrosine-based inhibitory (ITIM)-like motifs in their cytoplasmic tails implicating their involvement in cellular activation [165].

The mammalian glycome contains numerous sialylated glycans that are potential ligands for Siglecs and are therefore candidates as modulators of these receptors as they regulate certain aspects of adhesion, cell signaling, and endocytosis in the immune system. A decade of painstaking work is now reaching fruition in deciphering the relative affinities of individual Siglecs for α2,3-, α2,6-, or α2,8-linked sialic acids [166–170], as well as their preference for the Neu5Ac or Neu5Gc forms of sialic acids [171, 172]. Moreover, the requirement of certain Siglec family members for sulfated carbohydrate ligands [173–175] is reminiscent of the binding habits of selectins involved in leukocyte extravasation.

The local concentration of sialic acids on surfaces of immune cells is very high; for example, on B cells it has been estimated to exceed 100 mM [176]. As a consequence, Siglec binding sites are typically "masked" by cis interactions with other glycan ligands expressed on the same cell [177–179]. In fact, Siglec-2 (CD22) has recently been shown to prefer itself as a binding partner [180]. In general, interactions with cis ligands dominate interactions with trans ligands in modulating the biological activities of Siglecs and these interactions tend to keep the cell quiescent. Important exceptions to the dominance of cis interactions in Siglec biology are provided by several examples of trans interactions that have the potential to regulate distant elements of the immune system. Sialoadhesin, for example, has an extended structure that projects its sialic-acid-binding site away from the plasma membrane, which reduces cis interactions. In other examples, trans interactions have been implicated in the activity of B cells [181] and in the suppression of Siglec-7-dependent natural killer cell activation [182, 183] in tissues such as those of the nervous system in which the inhibitory Neu5Ac-α-2,8-Neu5Ac-containing glycan ligands for this lectin are abundantly expressed [184, 185]. All in all, Siglecs comprise a versatile regulatory mechanism for the immune response.

1.3.4 Interplay between Glycosylation and Sugars: a Two-Way Street

Up to this point, we have covered briefly the biosynthesis of glycans, learning how cells invest substantial resources at considerable peril in the production of complex sugars. This effort does not go to waste, or impudently put an organism at risk of infection or metabolic disease because these glycans play innumerable roles in all aspects of the life a multicellular organism, with several specific examples—antibody structure and function, leukocyte homing, and Siglec regulation—of their critical roles in the functioning of the immune system described above. The reader who values fair play will find it heartening that glycans do not make all of these contributions without proper recognition from the immune system as it—quite literally but perhaps in an underappreciated manner—in turn recognizes carbohydrates as antigens. The fact that sugar structures are immunogenic runs counter to several generally held premises and leads to a number of questions. One of these is "how do carbohydrates fit into the conventional peptide processing system for protein-based antigens?" Another issue is protein–carbohydrate binding interactions typically conform to the *cluster glycoside effect* where multivalent carbohydrate presentation and multiple simultaneous binding interactions are needed to ensure high affinity and avidity as well as binding specificity [186, 187]. These requirements are unlike the highly specific and high-affinity interactions generally thought to occur between a single antigen and matched antibody. Despite these, and several other puzzles in various stages of being solved, it is now undisputable that carbohydrates comprise antigens important to human health, both from the perspective of human disease (discussed next in Section 1.4.1) and from interaction with pathogens lurking within the environment (Section 1.4.2).

1.4 CARBOHYDRATE ANTIGENS

1.4.1 Carbohydrate Antigens in Humans

1.4.1.1 Historically, the Antigenicity of Carbohydrate Structures Has Been a Biomedical Problem Carbohydrate-based antigens have long posed problems for the biomedical community. The outstanding historical example—dating back to 1900—is the ABO(H) blood group system [2]. Over the past two decades, the biochemical basis of these antigens has been unraveled with the discovery of a common structure—the "H antigen" found in individuals with the O blood type—that is elaborated with a GalNAc or galactose residue to produce the structures that specify the A and B blood types, respectively (see Fig. 1.3d) [188]. More recently, and firmly establishing that blood type antigens are not an outlier but rather that carbohydrate immunogenicity is a central issue in transplantation [189], a similar problem has arisen from efforts toward the xenotransplantation of nonprimate organs that bear the "α-gal trisaccharide" (190–192). When α-gal-specific natural antibodies bind to the endothelium of vascularized xenografts, the complement system is activated, which leads to the activation of the coagulation cascade and rapid (within minutes to hours) graft rejection. This hyperacute immune rejection has spawned efforts to create α-gal knockout pigs because their organs are similar to human organs in many respects and if not for immune incompatibility would be attractive [193, 194]. Interestingly—and somewhat distressingly—despite knocking out the α-gal transferase gene, the first go-round of engineered α-gal knockout pigs still express this trisaccharide epitope [195].

Recently, another example of sugar-based transplant antigenicity has arisen for stem cell research. Based on the use of murine feeder layers or animal products such as fetal bovine serum, human stem cells scavenge the nonhuman Neu5Gc form of sialic acid and present it on the cell surface [4]. Because humans have circulating antibodies to Neu5Gc [196], concerns arise that the implantation of Neu5Gc-bearing engineered tissues would be subject to immune rejection. While regenerative medicine is a little outside the scope of this chapter, this information does raise two points that are worth emphasizing. First, the sensitivity of the immune system to minute changes in the chemical structure of carbohydrates is highlighted where a single hydroxyl of Neu5Gc—a difference in mass of only $\sim 1\%$ when these sugars are in a decasaccharide—is sufficient to elicit an immune response (a similar response where discrimination between a hydroxyl and *N*-acetyl group of A and B blood type occurs). Second, the overexpression of Neu5Gc—presumably obtained from the diet—on human cancer cells [197] raises intriguing possibilities for new forms of cancer treatment. For example, the ability of cancer cells to scavenge nonhuman sugars and preferentially display them in their surface glycans provides impetus toward the development of metabolically glycoengineering strategies to develop cancer vaccines (see Section 1.4.3.4 and Chapter 10).

1.4.1.2 Abnormal Glycosylation and Cancer—An Opportunity to Exploit TACAs Therapeutically? Immunosurveillance mechanisms exist such that only a small fraction—perhaps as low as one in a million—of nascent tumor cells actually develop into full-fledged malignant disease [198–200]. The ability of the immune system to detect and eradicate cancer cells is in part due to the sometimes subtle and at other times quite dramatic changes to cell surface carbohydrates that take place during transformation [16, 201]. Since the 1980s numerous tumor-associated carbohydrate antigens (TACAs) expressed on both *N*- and *O*-linked glycoproteins as well as glycolipids have been cataloged [14, 201–203]. Common alterations— resulting from incomplete glycosylation or neoglycan production—occur in both *N*- and *O*-linked glycan structures and include increased size due to GlcNAc branching of core sequences close to the protein and variation in the terminal sequences [15]. Less commonly, but with significant repercussions particularly in breast cancers, mucins—such as MUC1—often have truncated glycans (such as the Tn antigen, Fig. 1.6b) that can be highly sialylated (e.g., sTn) and less sulfated than their normal counterparts [204–207]. TACAs expressed on glycolipids often involve changes to ganglio- or globo-series structures—such as glycolipid displayed fucosyl GM1—that are abundantly present in specific types of human cancers such as melanoma, Burkitt's lymphoma, neuroblastoma, and small-cell lung carcinoma [208–211]. TACAs are further discussed in Chapter 8.

Now, taking a step back to combine the two ideas presented above—that is, that the human immune system can eradicate cells based on surface carbohydrate antigens such as α-gal during xenotransplantation or TACAs in cancer—the logical route of deliberately targeting TACA for immunotherapy becomes attractive. This course of action is intended to assist nature in eradicating the occasional but potentially devastating cancer cell that was otherwise overlook or allowed to survive. Indeed, tumors that develop into full-fledged malignant disease appear to be immunosculpted specifically for immune tolerance and sometimes even suppression [199]. One approach to help the immune system identify these disease-causing outliers, interestingly enough, directly exploits circulating antibodies against carbohydrate antigens by seeking to use gene therapy to selectively express α-gal on cancer cells, thus restoring their immunicity and targeting them for eradication [3, 212]. Complementary to such biology-based approaches, elegant—but exhausting—efforts to use synthetic chemistry to produce complex TACAs are underway (see Section 1.4.3.3). Finally, as already mentioned, innovative approaches that combine the power of biology and chemistry into "chemical biology" approaches using metabolic glycoengineering (Fig. 1.4c) to install *neo*-TACA onto cancer cells are underway (Section 1.4.3.4).

Clearly, much effort—evidenced by several chapters in this book alone—is going into TACA vaccine development, which raises a somewhat philosophical but nonetheless relevant question of "if a cancer cell can 'trick' the immune system into leaving it alone once, couldn't its capacity for hypermutability come back into play and thwart a carbohydrate-based vaccine?" This concern is allayed by accumulating evidence that TACAs are not merely cancer "markers" (as an aside, carbohydrates "markers" are not meant to be diminished by this statement because of their value in immunodiagnosis; see Chapter 12) but play an active role in diseases progression. Consequently, if a cancer cell attempted to adapt to a vaccine by expressing less of a vaccine-targeted TACA, it may indeed be able to escape eradication by the therapeutic agent but in the process would become less virulent by virtue of no longer displaying the offending glycan. A specific example is provided by Lewis antigens, such as sLe^X that facilitates leukocyte extravasation (Fig. 1.8) but is also relevant to cancer because the homing mechanism used by leukocytes to travel through the body has been co-opted by invasive cancer cells during metastasis [213]. Hence, a vaccine targeted against sLe^X would be expected to reduce the invasive potential of low expressing cancer cell populations selected for survival by this therapeutic agent because any potentially metastatic cells would no longer be able to efficiently exit the bloodstream and establish secondary tumors.

1.4.2 Carbohydrates and Pathogens

1.4.2.1 Pathogens and Glycosylation—Inexorably Connected As discussed earlier, many pathogens exploit host glycans as binding epitopes during infection. Pathogens, immunology, and glycobiology, however, are interrelated in many additional ways as well. For example, on a fundamental level using the very broad definition of immunity to include protection achieved by physical barriers [114], bacteria that live in nutrient-rich but hostile environments—such as sewage or the human body—often utilize a carbohydrate capsule to protect themselves from bacteriophage and human immunity, respectively. As described in Section 1.4.2.2, bacterial glycans, which include these capsular polysaccharides as well as highly immunogenic lipopolysaccharides (LPSs), have already been used as vaccines to elicit protective immune responses. In a functionally analogous manner, eukaryotic parasites such as malaria and *Leishmania* (Section 1.4.2.3) also use surface carbohydrates to thwart host immunity and have likewise spawned rapidly maturing efforts to develop practical vaccines. Finally, viruses, which exploit of host glycans with exquisite binding specificity during infection, are ubiquitously glycosylated themselves, and viral glycans may someday be profitably exploited in vaccine development as well, as discussed briefly in Section 1.4.2.4.

1.4.2.2 Bacterial Carbohydrate-Based Antigens By contrast to TACAs whose glycoforms are co-opted from healthy cells, bacterial glycosylation presents a relatively easy target for vaccine development because of distinctive differences between microbial and human glycans. Even without outside intervention by a vaccine developer, humans and animals mount massive humoral responses against the LPSs of

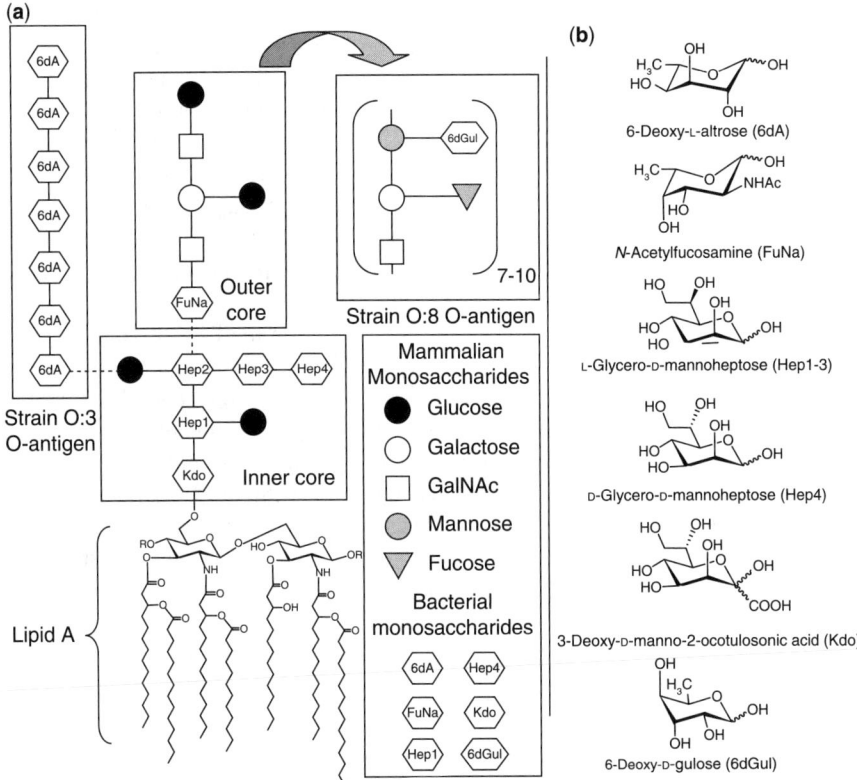

Figure 1.9 Bacterial O antigens and unique monosaccharide structures. (a) Lipopolysaccharide (LPS) structures are based on an inner core oligosaccharide that is attached to lipid A. The immunogenicity of the O antigen is determined by outer core glycan structures that vary both within and between strains and species (differences between *Yersinia enterocolitica* O:3 and O:8 strains [215] are shown here). The O antigen polysaccharides consist of monosaccharides that are common to the mammalian host (see Fig. 1.3 for structures) as well as those unique to the bacterium [representative bacterial monosaccharides that occur in *Y. enterocolitica* O:3 and O:8 are shown in (b)].

gram-negative bacteria. Features that contribute to the high inherent antigenicity of LPS include the bacterial O-antigen glycan structures recognized by the human immune system. Representative O antigens are shown in Figure 1.9a for two strains of *Yersinia* [214, 215], the bacteria responsible for bubonic plague; more generally, the O antigen can be comprised of numerous—often dozens from a single bacterial strain—of glycan epitopes attached to lipid A (in some cases, multiple antigens occur on a single LPS molecule). One reason that LPSs are highly immunogenic is that bacteria utilize many monosaccharide building blocks for their glycans that are not found in humans (a few are shown in Fig. 1.9b).

Despite the pronounced "non-self" features of bacterial LPS, producing an effective vaccine composed of nontoxic, immunogenic polysaccharides found in O antigens has been very challenging, as illustrated by pioneering efforts against *Pseudomonas aeruginosa* [216]. One difficulty arises from the chemical diversity found among the different O antigens representative of the 20 major serotypes of this pathogen (additional diversity comes from variant subtype O antigens) that translates into a large degree of serologic variability. Accordingly, a broad acting O-antigen targeted vaccine by necessity must consist of a highly complex mixture of often poorly characterized glycans. Further complications originate from the poor immunogenicity of the major protective epitope expressed by some O antigens, and a large degree of diversity in animal responses that preclude predicting the optimal vaccine formulation from in vitro experiments or studies with model organisms.

In a complementary approach that does not depend on the LPS of gram-negative bacteria, immunogenic capsular polysaccharides from various pathogenic species that include *Streptococcus pneumoniae*, *Neisseria meningitidis*, *Haemophilus influenzae*, *Salmonella typhi*, *Shigella dysenteriae*, and *Klebsiella pneumoniae* are sufficiently abundant to be isolated from large-scale fermentation cultures and used as vaccines [217]. Similar to the problem encountered with the O antigen of LPS, however, carbohydrate diversity within species and between strains hampers vaccine development because once again complex mixtures of glycan structures must be dealt with. The polysaccharide vaccine PPV23, for example, contains 23 antigenically distinct polysaccharides found on the surface capsules of *S. pneumoniae*. These 23 serotypes were selected for inclusion in the vaccine because at least one of the polysaccharides occurs in most clinical cases of pneumococcal infections [218]. Unfortunately, for the impoverished developing countries where the majority of the hundreds of thousands of annual deaths from bacterial infections occur, the complex process of bacterial fermentation and isolation from many strains make "conventional" carbohydrate-based vaccines prohibitively expensive for widespread public health programs.

1.4.2.3 Malaria and Leishmaniasis—Parasites
Malaria afflicts about 300 million people annually worldwide causing up to 2 million deaths, predominantly in children. Among the four different *Plasmodium* species that cause this disease, *P. falciparum* is the most common and the most virulent. In recent years, malaria has spread at an alarming rate owing to the increased resistance of the parasite to drugs and of carrier mosquitoes to insecticides, and new approaches to combat malaria are urgently needed. As evidence pointing to the importance of GPI anchor structures in this disease's morbidity and mortality mounts (the parasite's GPI anchors can activate PTK- and PKC-dependent signaling pathways to regulate REL-A, C-Rel, and NF-κB/*rel*-dependent expression of cytokines, cell adhesion molecules, and iNOS resulting in erythrocyte sequestration and immune dysregulation characteristic of malaria pathogenesis), carbohydrate-based vaccines offer an attractive approach toward the amelioration of this pathogen. Toward this end, the Seeberger group recently showed that a synthetic malaria vaccine candidate (Fig. 1.9a) dramatically increased survival in infected mice (from 0–9% to ~70% [219]). In a similar approach, this group explored using a portion of the GPI structure (Fig. 1.7b) as a vaccine candidate for leishmaniasis, which as discussed earlier, is caused by parasites

that have surfaces dominated by these anchoring structures [219]. Additional perspective and recent advances in antiparasitic vaccines are described in Chapter 6 of this book and a topic not otherwise covered in this introductory chapter—carbohydrate-based fungal vaccines—is covered in Chapter 7.

1.4.2.4 Viral Glycosylation Viruses co-opt host biosynthetic pathways to generate their genetic and structural material and use host glycosylation pathways to modify viral proteins with N-linked glycans. As occurs with the host's surface and secreted proteins, N-glycosylation of viral envelope proteins promotes proper folding through interactions with the host's cellular chaperones and facilitates proper trafficking through the secretory apparatus. In addition to these "quality control" functions, changes in glycosylation can reduce the ability of a virus to be recognized by the host's immune system; for example, HIV (human immunodeficiency virus) and influenza, two clear threats to human health, rely on expression of specific oligosaccharides to evade detection by the host immune system. In addition, N-glycosylation plays important roles in a diverse set of vital biological functions of viruses that are specific to various classes of these pathogens. A few examples include the heavy influence of glycosylation over infectivity and intracellular transport in the hepatitis C virus [220]. Similarly, the Ebola, Hantaan, Newcastle, Hendra, Nipah, metapneumovirus, and SARS-CoV viruses all have N-linked glycans that make vital contributions to infectivity, protein folding, tropism, proteolytic processing, and immune evasion [221–229]. Finally, the glycosylation status of the West Nile virus has recently been linked to neuroinvasiveness and replication efficiency in several strains [230–233] in a manner reminiscent to the role glycosylation plays in modulating the conformational changes to influenza HA protein during cellular uptake [234–237].

Because sugars borne on viruses are produced by host cells, they are similar to endogenous glycans and a clearly defined repertoire of viral glycan epitopes—akin to the TACA counterparts that accompany cancer—that can be targeted by vaccines is challenging to identify. Consequently, although vaccines that directly target viral glycans lag in development compared to bacterial or parasitic efforts, there are nevertheless compelling reasons to pursue this line of investigation (as is detailed in Chapter 5). To briefly summarize here, natural variability in the glycosylation status of many viruses exists that is exemplified by the human immunodeficiency virus-1 (HIV-1). This pathogen causes AIDS (acquired immunodeficiency syndrome) by recognizing host cells though the interaction of the viral glycoprotein, gp120 [238, 239] with CD4 present on the surface of human T-lymphocytes and a second "co-receptor" molecule on the host cell surface. A survey of global HIV gp120 showed that this glycoprotein had a range of N-linked glycosylation site occupancy of between 18 and 33 with a mean of 25 [240]. This variability is thought to be influenced by competing pressures on the virus, similar to those experienced by influenza [61], where the presence of glycans is driven by their indispensable contributions to viral infectivity and protection against neutralizing antibodies [241, 242]. On the other hand, excessive glycosylation masks the necessary receptor ligand binding contacts through steric hindrance and the nonspecifically antiadhesive nature of the glycocalyx, thereby supplying selective pressure that limits the upper range of glycosylation of the viral particle. Thus, in a manner similar to TACA vaccines used in cancer therapy as discussed above, although the production of a vaccine against a viral carbohydrate may be far from a panacea due to rapid mutation away from the targeted epitope, disruption of optimized glycosylation patterns through selective pressure imposed by the vaccine would render any surviving virus a less effective pathogen.

1.4.2.5 Translating Carbohydrate Antigens into Viable Vaccines

This chapter does not provide a comprehensive list of potential carbohydrate-based vaccines; instead its purpose is to make a compelling case for the development of these vaccines. A first necessary prerequisite—the fact that glycans are immunogenic—was met by discussing the downsides of longstanding challenges facing transplantation efforts and surveying carbohydrate aberrations characteristic of selected human disease and pathogens. The bottom line is that, now that it has been established that the immune system can detect detrimental glycan antigens ranging from the capsule polysaccharides of a pathogenic bacterium to the TACA of a cancer cell, the possibility exists that a vaccine can be developed to eradicate the offending entity. Then, because the development of carbohydrate-based vaccines is not a trivial undertaking, it is worth emphasizing one more time that these agents are sorely needed because they meet urgent health problems both in rich (cancer) and poor (infectious disease) nations and are well worth expending the effort needed to bring them to fruition.

As will be discussed in more detail below, unique challenges accompany carbohydrate-based vaccine development; for example, with the exception of recent discovery of zwitterionic bacterial polysaccharides that can elicit a T-cell response [243], glycans activate B cells in a thymus-independent type 2 (TI-2) manner [244]. Because carbohydrates typically engage antibodies B cells without the help of helper T cells, IgM is the predominant isotype produced, and there is negligible class switching, no affinity maturation, and little development of memory cells. Consequently, vaccines composed entirely of carbohydrate typically are only effective in children over the age of 18 months to 2 years, and their response in adults generally lasts for only 3–5 years [245]. These problems can be overcome by employing appropriate glycan conjugation and adjuvant strategies, which are a major topic of this book (covered in Chapters 2–7, 9, and 10).

1.4.3 Carbohydrate-Based Vaccines

1.4.3.1 Brief History of Vaccination

Although reports of people purposely inoculating themselves with other types of infections to protect themselves from disease date back to reports from 200 B.C. from China and India, the modern era of human vaccination is generally credited to Edward Jenner's efforts in 1796 to use the cowpox virus to prevent smallpox [246]. Almost 75 years later, Louis Pasteur first used the terms *immune* and *immunity* in the scientific sense but acknowledged Jenner's pioneering research by retaining the word *vaccination* (from the latin *vacca* for cow) to describe his own accomplishments in the prevention of rabies and anthrax. Since then, great success has been realized in the development of vaccines to manage and reverse infections caused by bacteria, viruses, and parasites, notably for diphtheria (von Behring), polio (Salk), and smallpox, which have been largely (or wholly) eliminated as threats to human health.

The field of carbohydrate vaccines, although blossoming tremendously of late, also has a venerable past dating back to the 1920s and 1930s when Landsteiner, Avery, and Goebel demonstrated that nonimmunogenic carbohydrates could become antigenic when covalently attached to proteins [247]. This early work reached clinical practice in ~1980 when Jennings and Roy derived polysialic acid from meningitis bacterial capsules, coupled this polysaccharide to a carrier protein to render it immunogenic, and ultimately produced a commercial vaccine [248]. Despite this (and now other examples of) initial success, immense challenges remain in bringing carbohydrates into the mainstream of vaccine development and immune therapies, which will be outlined in Section 1.4.3.3 after briefly describing general requirements that must be met during vaccine development.

1.4.3.2 General Requirements for Vaccines During the more than 200 years since successful vaccination was demonstrated by Jenner, a general set of conditions has become evident for the design and development of any vaccine. First and foremost, the identification and—to the extent possible—structural characterization of the antigen must be done. In the case of nonpeptide antigens, structural knowledge of the epitope is helpful in chemical synthesis of the antigen or a suitable mimetic. Once an appropriate antigen has been synthesized, or isolated from natural sources, a linker or spacer unit needs to be introduced for attachment to an immunogenic carrier protein or other immunostimulant while maintaining the immunological integrity of the antigen in order to produce a sufficiently potent vaccine. Once an appropriate conjugate has been obtained, immunological studies in animal models must be done to evaluate the vaccine's efficacy, and the antibodies elicited by the vaccine should be isolated for detailed study of their interaction with the target antigen. This latter step is particularly important for passive immunization strategies such as those being developed for cancer therapy against TACA. Finally, evaluation in human clinical trials, as discussed in Chapter 11, must be done before widespread use of the vaccine can begin.

1.4.3.3 Wrinkles Thrown at the Glycovaccinist Many aspects of carbohydrate-based vaccine construction, including the need to identify an appropriate antigen, conjugation to a suitable immunogenic carrier, and evaluation of various immunological adjuvants for co-administration are shared with general vaccine development efforts. By contrast, one uniquely difficult challenge that confronts the use of carbohydrate antigens in vaccine development is heterogeneity of naturally occurring glycans that renders the isolation and purification of these molecules to homogeneity a daunting task. Consequently, the development of fully synthetic antigens has become a large part of carbohydrate vaccine development efforts over the past decade. In addition to gaining homogeneous material, synthetic strategies allow linkers to be built into the carbohydrate structure appropriate for conjugation to a carrier. This is particularly important due to the unique processing of carbohydrate-only

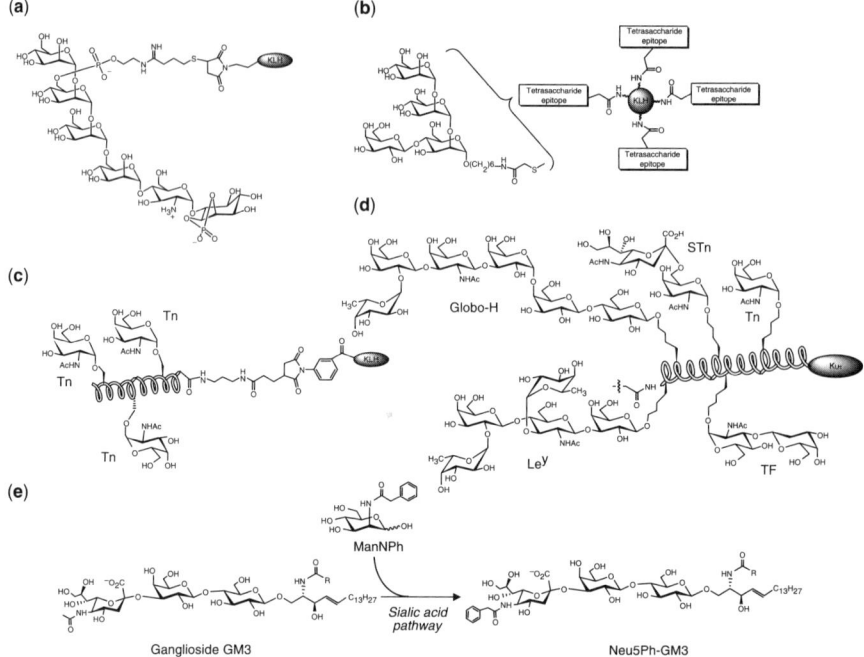

Figure 1.10 Synthetic carbohydrate vaccine candidates. (a) Malaria and (b) leishmaniasis vaccine candidates have been reported by the Seeberger group [219]. (c) Multivalent display of the Tn TACA and (d) pentavalent conjugates of several TACA have been reported by the Danishefsky laboratory [249]. (e) Installation of a phenyl-containing neo-TACA into ganglioside GM3 by metabolic glycoengineering (see Fig. 1.4c) [250].

antigens that, among other challenges, render sugars incapable of raising an immune response in infants (as mentioned in Section 1.4.2.5 and Chapter 2).

The complete synthesis of oligosaccharides in sufficiently large quantities for practical use in vaccine development has remained difficult despite the elegant and exhaustive efforts pioneered by the Danishefsky group who made Globo-H [251], KH-1 [252], and the Le^Y and Tn [253] TACAs (Figs. 1.10c and 1.10d). Fully synthetic carbohydrate vaccines have important advantages over those isolated from natural sources because synthetic glycans can, in theory, be produced as homogeneous compounds in a controlled manner with little or no batch-to-batch variability, thus making heroic synthetic efforts worthwhile. The safety of completely synthetic antigens is also higher than vaccines derived from live cultures where the danger of contaminating immunogens, or disease-causing microbes, is small but real. In addition, medicinal chemistry techniques can potentially be used to derivatize and modify synthetic carbohydrates to make vaccines that are more immunogenic than those based on natural carbohydrates. The present status of synthetic glycoconjugates used in vaccine development is provided in Chapters 2, 4–7, 9, and 10.

Steps toward solving a major limitation of conventional synthetic strategies—the insufficiently small amount of material obtained—are being taken by automated synthesizers being pioneered by the Seeberger group [254, 255] and "one pot" synthetic strategies reported by the Wong laboratory [256, 257]. Although not capable of producing any glycan structure on demand as automated DNA (deoxyribonucleic acid) synthesizers have long been able to do, this methodology provides a major boost toward several endpoints of major medical significance, including malaria (Fig. 1.10a) and leishmaniasis (Fig. 1.10b) [219]. Another important nuance of carbohydrate binding is to bring the cluster glycoside effect into play—demonstrated by the leishmaniasis vaccine candidate in Figure 1.10b and the Tn TACA in Figure 1.10c. A refinement of this technique is exemplified by multimeric antigenic constructs that target prostate and breast cancers with multiple TACA on the same molecular contruct (Fig. 1.10d) [249, 258].

1.4.3.4 Metabolic Glycoengineering—Enhancing Immunogenicity and the Therapeutic Window   Despite notable examples of potent immunogenicity, coupled with a growing number of successes at exploiting this for vaccine development, the immunogenicity of carbohydrate antigens is far from universally adequate and remains problematically weak in some of the most lucrative applications such as cancer. In cancer, two related problems arise. First, immune tolerance to TACAs—many of which are fetal-oncogenic markers—limits a robust immune response. Second, because many or most TACAs are expressed to some degree even on healthy cells, the "therapeutic window" remains a potential obstacle. An interesting approach to overcoming these problems is to use a metabolic glycoengineering strategy (see Fig. 1.4d) that was inspired by observations that human cancer cells displayed relatively high levels of the Neu5Gc form of sialic acid compared to normal cells [197]. The source of the high levels of "nonhuman" sialic acid on tumor cells was traced to their highly efficient ability to scavenge this sugar from a carnivorous diet and replace the commonly occurring Neu5Ac with the modified sialoside [259, 260].

Based on the ability of the sialic acid biosynthetic pathway to accept nonnatural metabolic substrates, primarily ManNAc analogs, and process them into the corresponding nonnatural cell surface displayed sialosides, the Jennings group demonstrated almost a decade ago that PSA-displaying cancer cells could be selectively killed by a passive immunity approach [261]. In this landmark study, ManNProp was used to incorporate Sia5Prop into polysialic acid and an antibody to the Prop form of PSA—which had been made by chemical methods [262]—was co-injected into animals providing a potent anticancer effect. Since then, the Guo group has expanded this novel approach by increasing the repertoire of immunogenic sugar analogs available to include those with highly immunogenic *N*-phenylacetyl groups and expanding from the relatively uncommon TACA polysialic acid to more broadly distributed markers such as GM3 found in melanomas [250, 263, 264]. Interestingly, pathogenic bacteria such as *Heamophylis ducyreii* can also display nonnatural sialic acids via metabolic glycoengineering [265, 266]. Thus, because these pathogens use natural sialic acids to fool the immune system into believing they have humanlike qualities, the analogs may function as Trojan horses and provide a vehicle for the microbe's demise by replacing their humanized sugars with abiotic, immunogenic counterparts.

1.4.4 Concluding Comments: Building on Success

The marriage of synthetic chemistry with established technologies is facilitating rapid progress in the development of a new generation of carbohydrate-based vaccines. Already, a synthetic vaccine that targets the bacterium *H. influenzae* type B was developed in 2004 in Cuba and is now part of that country's national vaccination program [267]; this project is a prime example of how a chemical approach is superior to fermenting pathogenic bacteria in giant vats, which is "messy, expensive, and inexact" [268]. A "chemical biology" approach is also paying off in the development of fully synthetic vaccines against TACAs capable of eliciting robust immune responses by combining a TLR2 agonist, a promiscuous petide T-helper epitope, and a tumor-associated glycopeptide into a single construct [269]. Increasingly sophisticated approaches of this kind, described in more depth in Chapter 9, portend a bright future for carbohydrate-based strategies to modulate immunity toward solving many of today's urgent health challenges.

ACKNOWLEDGMENT

The authors thank the National Institutes of Health for financial support (NCI CA112314-01A1).

REFERENCES

1. Rademacher TW, Parekh RB, Dwek RA (1988) Glycobiology. *Annu. Rev. Biochem.* 57:785–838.
2. Morgan WTJ, Watkins WM (2000) Unravelling the biochemical basis of blood group ABO and Lewis antigenic specficity. *Glycoconjug. J.* 17:501–530.
3. Galili U (2005) The α-gal epitope and the anti-Gal antibody in xenotransplantation and in cancer immunotherapy. *Immunol. Cell Biol.* 83:674–686.
4. Martin MJ, Muotri A, Gage F, Varki A (2005) Human embryonic stem cells express an immunogenic nonhuman sialic acid. *Nat. Med.* 11:228–232.
5. Lowe JB (2003) Glycan-dependent leukocyte adhesion and recruitment in inflammation. *Curr. Opin. Cell Biol.* 15:531–538.
6. Horvat S, Jakas A (2004) Peptide and amino acid glycation: New insights into the Maillard reaction. *J. Pept. Sci.* 10:119–137.
7. McKillop AM, Abdel-Wahab YH, Mooney MH, O'Harte FP, Flatt PR (2002) Secretion of glycated insulin from pancreatic β-cells in diabetes represents a novel aspect of beta-cell dysfunction and glucose toxicity. *Diabetes Metab.* 28:3S61–63S69.
8. Turk Z, Ljubic S, Turk N, Benkov B (2001) Detection of auto antibodies against advanced glycation endproducts and AGE-immune complexes in serum of patients with diabetes mellitus. *Clin. Chim. Acta* 303:105–115.
9. Dell A, Morris HR, Easton RL, Patankar M, Clark GF (1999) The glycobiology of gametes and fertilisation. *Biochim. Biophys. Acta* 1473:196–205.
10. Freeze HH (2001) Update and prespectives on congenital disorders of glycosylation. *Glycobiology* 11:129R–143R.

REFERENCES

11. Keppler OT, Hinderlich S, Langner J, Schwartz-Albiez R, Reutter W, Pawlita M (1999) UDP-GlcNAc 2-epimerase: A regulator of cell surface sialylation. *Science* 284:1372–1376.
12. Eisenberg I, Avidan N, Potikha T, Hochner H, Chen M, Olender T, Barash M, Shemesh M, Sadeh M, Grabov-Nardin G, Shmilevich I, Friedmann A, Karpati G, Bradley WG, Baumbach L, Lancet D, Ben Asher E, Beckmann JS, Argov Z, Mitrani-Rosenbaum S (2001) The UDP-N-acetylglucosamine 2-epimerase/N-acetylmannosamine kinase gene is mutated in recessive hereditary inclusion body myopathy. *Nat. Genet.* 29:83–87.
13. Dennis JW, Pawling J, Cheung P, Partridge E, Demetriou M (2002) UDP-*N*-acetylglucosamine: α6-mannoside β1,6 *N*-acetylglucosaminyltransferase V (*Mgat5*) deficient mice. *Biochim. Biophys. Acta* 1573:414–422.
14. Dennis JW (1992) Changes in glycosylation associated with malignant transformation and tumor progression. In Fukuda M, Ed. *Cell Surface Carbohydrates and Cell Development*. CRC Press, Boca Raton, FL, pp. 161–194.
15. Dennis JW, Granovsky M, Warren CE (1999) Glycoprotein glycosylation and cancer progression. *Biochim. Biophys. Acta* 1473:21–34.
16. Fuster MM, Esko JD (2005) The sweet and sour of cancer: Glycans as novel therapeutic targets. *Nat. Rev. Cancer* 5:526–542.
17. Dube DH, Bertozzi CR (2005) Glycans in cancer and inflammation—potential for therapeutics and diagnostics. *Nat. Rev. Drug Discov.* 4:477–488.
18. Rudd PM, Endo T, Colominas C, Groth D, Wheeler SF, Harvey DJ, Wormald MR, Serban H, Prusiner SB, Kobata A, Dwek RA (1999) Glycosylation differences between the normal and pathogenic prion protein isoforms. *Proc. Natl. Acad. Sci. U.S.A.* 96:13044–13049.
19. Endo T, Groth D, Prusiner SB, Kobata A (1989) Diversity of oligosaccharide structures linked to asparagines of the scrapie prion protein. *Biochemistry* 28:8380–8388.
20. Lawson VA, Collins SJ, Masters CL, Hill AF (2005) Prion protein glycosylation. *J. Neurochem.* 93:793–801.
21. Rudd PM, Merry AH, Wormald MR, Dwek RA (2002) Glycosylation and prion protein. *Curr. Opin. Struct. Biol.* 12:578–586.
22. Olsnes S (2004) The history of ricin, abrin and related toxins. *Toxicon* 44:361–370.
23. Sandvig K, Grimmer S, Iversen TG, Rodal K, Torgersen ML, Nicoziani P, van Deurs B (2000) Ricin transport into cells: Studies of endocytosis and intracellular transport. *Int. J. Med. Microbiol.* 290:415–420.
24. Suzuki Y, Ito T, Suzuki T, Holland J, Robert E, Chambers TM, Kiso M, Ishida H, Kawaoka Y (2000) Sialic acid species as a determinant of the host range of influenza A viruses. *J. Virol.* 74:11825–11831.
25. Shinya K, Ebina M, Yamada S, Ono M, Kasai N, Kawaoka Y (2006) Avian flu: Influenza virus receptors in the human airway. *Nature* 440:435–436.
26. Stevens J, Blixt O, Tumpey TM, Taubenberger JK, Paulson JC, Wilson IA (2006) Structure and receptor specificity of the hemagglutinin from an H5N1 influenza virus. *Science* 312:404–410.
27. Higa HH, Rogers GN, Paulson JC (1985) Influenza virus hemagglutinins differentiate between receptor determinants bearing *N*-acetyl-, *N*-glycolyl-, and *N,O*-diacetylneuraminic acids. *Virology* 144:279–282.

28. Thomas W, Forero M, Yakovenko O, Nilsson L, Vicini P, Sokurenko E, Vogel V (2006) Catch-bond model derived from allostery explains force-activated bacterial adhesion. *Biophys. J.* 90:753–764.

29. Nilsson LM, Thomas WE, Trintchina E, Vogel V, Sokurenko EV (2006) Catch bond-mediated adhesion without a shear threshold: Trimannose versus monomannose interactions with the FimH adhesin of *Escherichia coli*. *J. Biol. Chem.* 281:16656–16663.

30. Gabius H-J, Siebert H-C, André S, Jiménez-Barbero J, Rüdiger H (2004) Chemical biology of the sugar code. *ChemBioChem* 5:740–764.

31. Virji M, Evans D, Hadfield A, Grunert F, Teixeira AM, Watt SM (1999) Critical determinants of host receptor targeting by *Neisseria meningitidis* and *Neisseria gonorrhoeae*: Identification of Opa adhesiotopes on the N-domain of CD66 molecules. *Mol. Microbiol.* 34:538–551.

32. Kayano T, Burant CF, Fukumoto H, Gould GW, Fan YS, Eddy RL, Byers MG, Shows TB, Seino S, Bell GI (1990) Human facilitative glucose transporters. Isolation, functional characterization, and gene localization of cDNAs encoding an isoform (GLUT5) expressed in small instestine, kidney, muscle, and adipose tissue and an unusual glucose transporter pseudogene-like sequence (GLUT6). *J. Biol. Chem.* 265:13276–13282.

33. Wood IS, Trayhurn P (2003) Glucose transporters (GLUT and SGLT): Expanded families of sugar transport proteins. *Br J. Nutr.* 89:3–9.

34. Chen H, Wang Z, Sun Z, Kim EJ, Yarema KJ (2005) Mammalian glycosylation: An overview of carbohydrate biosynthesis. In Yarema KJ, Ed. *Handbook of Carbohydrate Engineering*. Francis & Taylor/CRC Press, Boca Raton, FL, pp. 1–48.

35. The Pubmed search engine and databases are available at: http://www.ncbi.nlm.nih.gov/sites/entrez?db = pubmed.

36. The KEGG (*Kyoto Encyclopedia of Genes and Genomes*) database is available at: http://www.genome.jp/kegg/.

37. The HUGO (Human Genome Organisation) database is available at: http://www.hugo-international.org/.

38. Campbell CT, Yarema KJ (2005) Large-scale approaches for glycobiology. *Gen. Biol.* 6:Article 236.

39. Bond MR, Kohler JJ (2007) Chemical methods for glycoprotein discovery. *Curr. Opin. Chem. Biol.* 11:52–58.

40. Tomiya N, Ailor E, Lawrence SM, Betenbaugh MJ, Lee YC (2001) Determination of nucleotides and sugar nucleotides involved in protein glycosylation by high-performance anion-exchange chromatography: Sugar nucleotide contents in cultured insect cells and mammalian cells. *Anal. Biochem.* 293:129–137.

41. Haslam SM, North SJ, Dell A (2006) Mass spectrometric analysis of *N*- and *O*-glycosylation of tissues and cells. *Curr. Opin. Struct. Biol.* 16:584–591.

42. Murrell MP, Yarema KJ, Levchenko A (2004) The systems biology of glycosylation. *ChemBioChem* 5:1334–1447.

43. Hossler P, Mulukutla BC, Hu W-S (2007) Systems analysis of *N*-glycan processing in mammalian cells. *PLoS One* 2:e713.

44. Krambeck FJ, Betenbaugh MJ (2005) A mathematical model of *N*-linked glycosylation. *Biotechnol. Bioeng.* 92:711–728.

45. Lau KS, Partridge EA, Grigorian A, Silvescu CI, Reinhold VN, Demetriou M, Dennis JW (2007) Complex N-glycan number and degree of branching cooperate to regulate cell proliferation and differentiation. *Cell* 129:123–134.
46. Hanover JA (2001) Glycan-dependent signaling: O-linked N-acetylglucosamine. *FASEB J.* 15:1865–1876.
47. Torres CR, Hart GW (1984) Topography and polypeptide distribution of terminal N-acetylglucosamine residues on the surfaces of intact lymphocytes. Evidence for O-linked GlcNAc. *J. Biol. Chem.* 259:3308–3317.
48. Zachara NE, Hart GW (2002) The emerging significance of O-GlcNAc in cellular regulation. *Chem. Rev.* 102:431–438.
49. Tanner ME (2005) The enzymes of sialic acid biosynthesis. *Bioorg. Chem.* 33:216–228.
50. Kayser H, Zeitler R, Kannicht C, Grunow D, Nuck R, Reutter W (1992) Biosynthesis of a nonphysiological sialic acid in different rat organs, using N-propanoyl-D-hexosamines as precursors. *J. Biol. Chem.* 267:16934–16938.
51. Campbell CT, Sampathkumar S-G, Weier C, Yarema KJ (2007) Metabolic oligosaccharide engineering: Perspectives, applications, and future directions. *Mol. Biosyst.* 3:187–194.
52. Gu J, Taniguchi N (2004) Regulation of integrin functions by N-glycans. *Glycoconjug. J.* 21:9–15.
53. Lanza F, Healy L, Sutherland DR (2001) Structural and functional features of the CD34 antigen: An update. *J. Biol. Regul. Homeost. Agents* 15:1–13.
54. Schachter H (1986) Biosynthetic controls that determine the branching and microheterogeneity of protein-bound oligosaccharides. *Biochem. Cell. Biol.* 64:163–181.
55. Rudd PM, Wormald MR, Wing DR, Prusiner SB, Dwek RA (2001) Prion glycoprotein: Structure, dynamics, and roles for the sugars. *Biochemistry* 40:3759–3766.
56. Ramsey T (1999) The number of people estimated to have ever lived was 96,100,000,000. Available at: http://www.math.hawaii.edu/~ramsey/People.html.
57. Spiro RG (2002) Protein glycosylation: Nature, distribution, enzymatic formation, and disease implications of glycopeptide bonds. *Glycobiology* 12:43R–56R.
58. Kornfeld R, Kornfeld S (1985) Assembly of asparagine-linked oligosaccharides. *Annu. Rev. Biochem.* 54:631–664.
59. Rush JS, van Leyen K, Ouerfelli O, Wolucka B, Waechter CJ (1998) Transbilayer movement of Glc-P-dolichol and its function as a glucosyl donor: Protein-mediated transport of a water-soluble analog into sealed ER vesicles from pig brain. *Glycobiology* 8:1195–1205.
60. Rush JS, Gao N, Lehrman MA, Waechter CJ (2008) Recycling of dolichyl monophosphate to the cytoplasmic leaflet of the endoplasmic reticulum after the cleavage of dolichyl pyrophosphate on the lumenal monolayer. *J. Biol. Chem.* 283:4087–4093.
61. Tong L, Baskaran G, Jones MB, Rhee JK, Yarema KJ (2003) Glycosylation changes as markers for the diagnosis and treatment of human disease. In Harding S, Ed. *Biochemical and Genetic Engineering Reviews*. Intercept, Andover, Hampshire, UK, pp. 199–244.
62. Freeze HH, Aebi M (2005) Altered glycan structures: The molecular basis of congenital disorders of glycosylation. *Curr. Opin. Struct. Biol.* 15:490–498.
63. Lehle L, Strahl S, Tanner W (2006) Protein glycosylation, conserved from yeast to man: A model organism helps elucidate congenital human diseases. *Angew. Chem. Int. Ed. Engl.* 45:6801–6818.

64. Helenius A, Aebi M (2004) Roles of N-linked glycans in the endoplasmic reticulum. *Annu. Rev. Biochem.* 73:1019–1049.
65. Aebi M, Hennet T (2001) Congenital disorders of glycosylation: Genetic model systems lead the way. *Trends Cell Biol.* 11:136–141.
66. Rush JS, Pannerselvam K, Waechter CJ, Freeze HH (2000) Mannose supplementation corrects GDP-mannose deficiency in cultured fibroblasts from some patients with congenital disorders of glycosylation (CDG). *Glycobiology* 10:829–835.
67. Helenius A, Aebi M (2001) Intracellular functions of N-linked glycans. *Science* 291:2364–2369.
68. Kato K, Kamiya Y (2007) Structural views of glycoprotein-fate determination in cells. *Glycobiology* 17:1031–1044.
69. Ruddock LW, Molinari M (2006) N-glycan processing in ER quality control. *J. Cell Sci.* 119:4373–4380.
70. Molinari M (2007) N-glycan structure dictates extension of protein folding or onset of disposal. *Nat. Chem. Biol.* 3:313–320.
71. Senger RS, Karim MN (2008) Prediction of N-linked glycan branching patterns using artificial neural networks. *Math. Biosci.* 211:89–104.
72. Peter-Katalinić J (2005) Methods in enzymology: O-glycosylation of proteins. *Meth. Enzymol.* 405:139–171.
73. Strous JG, Dekker J (1992) Mucin-type glycoproteins. *Crit. Rev. Biochem. Mol. Biol.* 27:57–92.
74. Ten Hagen KG, Fritz TA, Tabak LA (2003) All in the family: The UDP-GalNAc:polypeptide N-acetylgalactosaminyltransferases. *Glycobiology* 13:1R–16R.
75. Brockhausen I, Schachter H (1997) Glycosyltransferases involved in N- and O-glycan biosynthesis. In Gabius HJ, Gabius S, (Eds.) *Glycosciences: Status and Prespectives*. Chapman and Hall, Weinheim, pp. 78–113.
76. Hang HC, Bertozzi CR (2005) The chemistry and biology of mucin-type O-linked glycosylation. *Bioorg. Med. Chem.* 13:5021–5034.
77. Hanover JA, Cohen CK, Willingham MC, Park MK (1987) O-linked N-acetylglucosamine is attached to proteins of the nuclear pore. Evidence for cytoplasmic and nucleoplasmic glycoproteins. *J. Biol. Chem.* 262:9887–9894.
78. Hart GW (1992) Glycosylation. *Curr. Opin. Cell Biol.* 4:1017–1023.
79. Heese-Peck A, Cole RN, Borkhsenious ON, Hart GW, Raikhel NV (1995) Plant nuclear pore complex proteins are modified by novel oligosaccharides with terminal N-acetylglucosamine. *Plant Cell* 7:1459–1471.
80. Zachara NE, Hart GW (2006) Cell signaling, the essential role of O-GlcNAc!. *Biochim. Biophys. Acta* 1761:599–617.
81. Freitas RA Jr (1999) *Nanomedicine, Vol. I: Basic Capabilities*. Landes Bioscience, Georgetown, TX.
82. Chester MA (1999) IUPAC-IUB Joint Commission on Biochemical Nomenclature (JCBN) Nomenclature of glycolipids: Recommendations 1997. *Glycoconjug. J.* 16:1–6.
83. Hakomori S (1981) Glycosphingolipids in cellular interaction, differentiation, and oncogenesis. *Annu. Rev. Biochem.* 50:733–764.
84. Yamashita T, Wada R, Sasaki T, Deng C, Bierfreund U, Sandhoff K, Proia RL (1999) A vital role for glycosphingolipid synthesis during development and differentiation. *Proc. Natl. Acad. Sci. U.S.A.* 96:9142–9147.

85. Allende ML, Proia RL (2002) Lubricating cell signaling pathways with gangliosides. *Curr. Opin. Struct. Biol.* 12:587–592.
86. Hakomori S-I (2002) The glycosynapse. *Proc. Natl. Acad. Sci. U.S.A.* 99:225–232.
87. Sprong H, Kruithof B, Leijendekker R, Slot JW, van Meer G, van der Sluijs P (1998) UDP-galactose:ceramide galactosyltransferase is a class I integral membrane protein of the endoplasmic reticulum. *J. Biol. Chem.* 273:25880–25888.
88. Marks DL, Wu K, Paul P, Kamisaka Y, Watanabe R, Pagano RE (1999) Oligomerization and topology of the Golgi membrane protein glucosylceramide synthase. *J. Biol. Chem.* 274:451–456.
89. Van Kaer L (2005) α-Galactosylceramide therapy for autoimmune diseases: Prospects and obstacles. *Nat. Rev. Immunol.* 5:31–42.
90. Bai J, Pagano RE (1997) Measurement of spontaneous transfer and transbilayer movement of BODIPY-labeled lipids in lipid vesicles. *Biochemistry (Moscow)* 36:8840–8848.
91. Lannert H, Gorgas K, Meissner I, Wieland FT, Jeckel D (1998) Functional organization of the Golgi apparatus in glycosphingolipid biosynthesis. Lactosylceramide and subsequent glycosphingolipids are formed in the lumen of the late Golgi. *J. Biol. Chem.* 273:2939–2946.
92. Nomura T, Takizawa M, Aoki J, Arai H, Inoue K, Wakisaka E, Yoshizuka N, Imokawa G, Dohmae N, Takio K, Hattori M, Matsuo N (1998) Purification, cDNA cloning, and expression of UDP-Gal: Glucosylceramide β1,4-galactosyltransferase from rat brain. *J. Biol. Chem.* 273:13570–13577.
93. Kolter T, Doering T, Wilkening G, Werth N, Sandhoff K (1999) Recent advances in the biochemistry of glycosphingolipid metabolism. *Biochem. Soc. Trans.* 27:409–415.
94. Fernandes Filho JA, Shapiro BE (2004) Tay-Sachs disease. *Arch. Neurol.* 61:1466–1468.
95. Hakomori S, Kannagi R (1983) Glycosphingolipids as tumor-associated and differentiation markers. *J. Natl. Cancer Inst.* 71:231–251.
96. Hakomori S, Zhang Y (1997) Glycosphingolipid antigens and cancer therapy. *Chem. Biol.* 4:97–104.
97. Hakomori, S-I (1985) Aberrant glycosylation in cancer cell membranes as focused on glycolipids: Overview and perspectives. *Cancer Res.* 45:2405–2414.
98. Hakomori S-I (2002) Glycosylation defining cancer malignancy: New wine in an old bottle. *Proc. Natl. Acad. Sci. U.S.A.* 99:10231–10233.
99. Karlsson KA (1995) Microbial recognition of target-cell glycoconjugates. *Curr. Opin. Struct. Biol.* 5:622–635.
100. Low MG (1989) Glycosyl-phosphatidylinositol: A versatile anchor for cell surface proteins. *FASEB J.* 3:1600–1608.
101. Ferguson MA, Low MG, Cross GA (1985) Glycosyl-sn-1,2-dimyristylphosphatidylinositol is covalently linked to *Trypnosoma brucei* variant surface glycoprotein. *J. Biol. Chem.* 260:14547–14555.
102. Brewis IA, Ferguson MA, Mehlert A, Turner AJ, Hooper NM (1995) Structures of the glycosyl-phosphatidylinositol anchors of porcine and human renal membrane dipeptidase. Comprehensive structural studies on the porcine anchor and interspecies comparison of the glycan core structures. *J. Biol. Chem.* 270:22946–22956.
103. Hooper NM (1999) Detergent-insoluble glycosphingolipid/cholesterol-rich membrane domains, lipid rafts and caveolae. *Mol. Membr. Biol.* 16:145–156.

104. Ferguson MA (1992) Site of palmitoylation of a phospholipase C-resistant glycosylphosphatidylinositol membrane anchor. *Biochem. J.* 284:297–300.
105. Ferguson MA, Masterson WJ, Homans SW, McConville MJ (1992) *GPI Membrane Anchors.* Academic Press, San Diego.
106. Masterson WJ, Raper J, Doering TL, Hart GW, Englund PT (1990) Fatty acid remodeling: A novel reaction sequence in the biosynthesis of trypanosome glycosyl phosphatidylinositol membrane anchors. *Cell* 62:73–80.
107. Ferguson MA (1999) The structure, biosynthesis and functions of glycosylphosphatidylinositol anchors, and the contributions of trypanosome research. *J. Cell Sci.* 112:2799–2809.
108. Mansfield JM, Paulnock DM (2005) Regulation of innate and acquired immunity in African trypanosomiasis. *Parasite Immunol.* 27:361–371.
109. Häcker U, Nybakken K, Perrimon N (2005) Heparan sulphate proteoglycans: The sweet side of development. *Nat. Rev. Mol. Cell Biol.* 6:530–541.
110. Hodde JP, Badylak SF, Brightman AO, Voytik-Harbin SL (1996) Glycosaminoglycan content of small intestinal submucosa. *Tiss. Eng.* 2:209–217.
111. Pilarski LM, Pruski E, Wizniak J, Paine D, Seeberger K, Mant MJ, Brown CB, Belch AR (1999) Potential role for hyaluronan and the hyaluronan receptor RHAMM in mobilization and trafficking of hematopoietic progenitor cells. *Blood* 93:2918–2927.
112. Mummert ME (2005) Immunologic roles of hyaluronan. *Immunol. Res.* 31:189–206.
113. Fareed J, Leong WL, Hoppensteadt DA, Jeske WP, Walenga J, Wahi R, Bick RL (2004) Generic low-molecular-weight heparins: Some practical considerations. *Semin. Thromb. Hemost.* 30:703–713.
114. Freitas RA Jr (2003) *Nanomedicine, Vol. IIA: Biocompatibility.* Landes Bioscience, Georgetown, TX.
115. Litman GW, Cannon JP, Dishaw LJ (2005) Reconstructing immune phylogeny: New perspectives. *Nat. Rev. Immunol.* 5:866–879.
116. Hauton C, Smith VJ (2007) Adaptive immunity in invertebrates: A straw house without a mechanistic foundation. *Bioessays* 29:1138–1146.
117. Kawai T, Akira S (2006) Innate immune recognition of viral infection. *Nat. Immunol.* 7:131–137.
118. Le Y, Zhou Y, Iribarren P, Wang J (2004) Chemokines and chemokine receptors: Their manifold roles in homeostasis and disease. *Cell. Mol. Immunol.* 1:95–104.
119. Krishnaswamy G, Ajitawi O, Chi DS (2006) The human mast cell: An overview. *Methods Mol. Biol.* 315:13–34.
120. Weller PF, Lim K, Wan HC, Dvorak AM, Wong DT, Cruikshank WW, Kornfeld H, Center DM (1996) Role of the eosinophil in allergic reactions. *Eur. Respir. J. Suppl.* 22:109s–115s.
121. O'Connor GM, Hart OM, Gardiner CM (2006) Putting the natural killer cell in its place. *Immunology* 117:1–10.
122. Hsu KC, Dupont B (2005) Natural killer cell receptors: Regulating innate immune responses to hematologic malignancy. *Semin. Hematol.* 42:91–103.
123. Minakami R, Sumimotoa H (2006) Phagocytosis-coupled activation of the superoxide-producing phagocyte oxidase, a member of the NADPH oxidase (nox) family. *Int. J. Hematol.* 84:193–198.

REFERENCES

124. Steinman RM (2001) Dendritic cells and the control of immunity: Enhancing the efficiency of antigen presentation. *Mt. Sinai J. Med.* 68:160–166.
125. Rodriguez A, Regnault A, Kleijmeer M, Ricciardi-Castagnoli P, Amigorena S (1999) Selective transport of internalized antigens to the cytosol for MHC class I presentation in dendritic cells. *Nat. Cell Biol.* 1:362–368.
126. Siemasko K, Clark MR (2001) The control and facilitation of MHC class II antigen processing by the BCR. *Curr. Opin. Immunol.* 13:32–36.
127. Moser B, Ebert L (2003) Lymphocyte traffic control by chemokines: Follicular B helper T cells. *Immunol. Lett.* 85:105–112.
128. Abbas AK, Murphy KM, Sher A (1996) Functional diversity of helper T lymphocytes. *Nature* 383:787–793.
129. Mercer JC, Ragin MJ, August A (2005) Natural killer T cells: Rapid responders controlling immunity and disease. *Int. J. Biochem. Cell. Biol.* 37:1337–1343.
130. Schrama D, Reisfeld RA, Becker JC (2006) Antibody targeted drugs as cancer therapeutics. *Nat. Rev. Drug Discov.* 5:147–159.
131. Rehder DS, Dillon TM, Pipes GD, Bondarenko PV (2006) Reversed-phase liquid chromatography/mass spectrometry analysis of reduced monoclonal antibodies in pharmaceutics. *J. Chromatogr. A* 1102:164–175.
132. Morrison SL (2005) The role of glycosylation in engineered antibodies. In Yarema K, Ed. *Handbook of Carbohydrate Engineering.* CRC Press (Taylor & Francis Group), Boca Raton, FL.
133. Yoo EM, Morrison SL (2005) IgA: An immune glycoprotein. *Clin. Immunol.* 116:3–10.
134. Burton DR, Woof JM (1992) Human antibody effector function. *Adv. Immunol.* 51:1–84.
135. Krapp S, Mimura Y, Jefferis R, Huber R, Sondermann P (2003) Structural analysis of human IgG-Fc glycoforms reveals a correlation between glycosylation and structural integrity. *J. Mol. Biol.* 325:979–989.
136. Jefferis R, Lund J, Pound JD (1998) IgG-Fc-mediated effector functions: Molecular definition of interaction sites for effector ligands and the role of glycosylation. *Immunol. Rev.* 163:59–76.
137. Wright A, Morrison SL (1997) Effect of glycosylation on antibody function: Implications for genetic engineering. *Trends Biotechnol.* 15:26–32.
138. Arnold JN, Wormald MR, Sim RB, Rudd PM, Dwek RA (2007) The impact of glycosylation on the biological function and structure of human immunoglobulins. *Annu. Rev. Immunol.* 25:21–50.
139. Jefferis R, Lund J, Mizutani H, Nakagawa H, Kawazoe Y, Arata Y, Takahashi N (1990) A comparative study of the N-linked oligosaccharide structures of human IgG subclass proteins. *Biochem. J.* 268:529–537.
140. Wormald MR, Rudd PM, Harvey DJ, Chang SC, Scragg IG Dwek RA (1997) Variations in oligosaccharide-protein interactions in immunoglobulin G determine the site-specific glycosylation profiles and modulate the dynamic motion of the Fc oligosaccharides. *Biochemistry* 36:1370–1380.
141. Suzuki N, Khoo KH, Chen CM, Chen HC, Lee YC (2003) N-Glycan structures of pigeon IgG: A major serum glycoprotein containing Gala1-4 Gal termini. *J. Biol. Chem.* 278:46293–46306.

142. Hodoniczky J, Zheng YZ, James DC (2005) Control of recombinant monoclonal antibody effector functions by Fc N-glycan remodeling *in vitro*. *Biotechnol. Prog.* 21:1644–1652.
143. Shinkawa T, Nakamura K, Yamane N, Shoji-Hosaka E, Kanda Y, Sakurada M, Uchida K, Anazawa H, Satoh M, Yamasaki M, Hanai N, Shitara K (2003) The absence of fucose but not the presence of galactose or bisecting N-acetylglucosamine of human IgG1 complex-type oligosaccharides shows the critical role of enhancing antibody-dependent cellular cytotoxicity. *J. Biol. Chem.* 278:3466–3473.
144. Kaneko Y, Nimmerjahn F, Ravetch JV (2006) Anti-inflammatory activity of immunoglobulin G resulting from Fc sialylation. *Science* 313:670–673.
145. Miyahara S, Kiryu J, Miyamoto K, Katsuta H, Hirose F, Tamura H, Musashi K, Honda Y, Yoshimura N (2004) In vivo three-dimensional evaluation of leukocyte behavior in retinal microcirculation of mice. *Invest. Ophthalmol. Vis. Sci.* 45:4197–4201.
146. Weinbaum S, Zhang X, Han Y, Vink H, Cowin SC (2003) Mechanotransduction and flow across the endothelial glycocalyx. *Proc. Natl. Acad. Sci. U.S.A.* 100:7988–7995.
147. Weinbaum S, Tarbell JM, Damiano ER (2007) The structure and function of the endothelial glycocalyx layer. *Annu. Rev. Biomed. Eng.* 9:121–167.
148. Varki A (1994) Selectin ligands. *Proc. Natl. Acad. Sci. U.S.A.* 91:7390–7394.
149. Lou J, Zhu C (2007) A structure-based sliding-rebinding mechanism for catch bonds. *Biophys. J.* 92:1471–1485.
150. Zhu C, McEver RP (2005) Catch bonds: Physical models and biological functions. *Mol. Cell. Biomech.* 2:91–104.
151. Springer TA, Lasky LA (1991) Cell adhesion. Sticky sugars for selectins. *Nature* 349:196–197.
152. Puri KD, Chen S, Springer TA (1998) Modifying the mechanical property of and shear threshold of L-selectin adhesion independently of equilibrium properties. *Nature* 392:930–933.
153. Varki A (1997) Selectin ligands: Will the real ones please stand up? *J. Clin. Invest.* 99:158–162.
154. Bowman KG, Bertozzi CR (1999) Carbohydrate sulfotransferases: Mediators of extracellular communication. *Chem. Biol.* 6:R9–R22.
155. Bistrup A, Bhakta S, Lee JK, Belov YY, Gunn MD, Zuo F-R, Chiao-Chain H, Kannagi R, Rosen SD, Hemmerich S (1999) Sulfotransferases of two specificities function in the reconstitution of high endothelial cell ligands for L-selectin. *J. Cell Biol.* 145:899–910.
156. Kannagi R (2002) Regulatory roles of carbohydrate ligands for selectins in the homing of lymphocytes. *Curr. Opin. Struct. Biol.* 12:599–608.
157. Crocker PR (2002) Siglecs: Sialic acid-binding immunogloulin-like lectins in cell-cell interactions and signalling. *Curr. Opin. Struct. Biol.* 12:609–615.
158. Crocker PR, Varki A (2001) Siglecs in the immune system. *Immunology* 103:137–145.
159. Crocker PR, Gordon S (1986) Properties and distribution of a lectin-like hemagglutinin differentially expressed by murine stromal tissue macrophages. *J. Exp. Med.* 164:1862–1875.
160. Stamenkovic I, Seed B (1990) The B-cell antigen CD22 mediates monocyte and erythrocyte adhesion. *Nature* 345:74–77.
161. Freeman SD, Kelm S, Barber EK, Crocker PR (1995) Characterization of CD33 as a new member of the sialoadhesin family of cellular interaction molecules. *Blood* 85:2005–2012.

162. Sun J, Shaper NL, Itonori S, Heffer-Lauc M, Sheikh KA, Schnaar RL (2004) Myelin-associated glycoprotein (Siglec-4) expression is progressively and selectively decreased in the brains of mice lacking complex gangliosides. *Glycobiology* 14:851–857.
163. Angata T, Kerr SC, Greaves DR, Varki N, Crocker PR, Varki A (2002) Cloning and characterization of human Siglec-11. A recently evolved signaling that can interact with SHP-1 and SHP-2 and is expressed by tissue macrophages, including brain microglia. *J. Biol. Chem.* 277:24466–24474.
164. Foussias G, Taylor SM, Yousef GM, Tropak MB, Ordon MH, Diamandis EP (2001) Cloning and molecular characterization of two splice variants of a new putative member of the Siglec-3-like subgroup of Siglecs. *Biochem. Biophys. Res. Commun.* 284:887–899.
165. Crocker PR, Varki A (2001) Siglecs, sialic acids, and innate immunity. *Trends Immunol.* 22:337–342.
166. Bakker TR, Piperi C, Davies EA, Merwe PA (2002) Comparison of CD22 binding to native CD45 and synthetic oligosaccharide. *Eur. J. Immunol.* 32:1924–1932.
167. Blixt O, Collins BE, van den Nieuwenhof IM, Crocker PR, Paulson JC (2003) Sialoside specificity of the siglec family assessed using novel multivalent probes: Identification of potent inhibitors of myelin-associated glycoprotein. *J. Biol. Chem.* 278:31007–31019.
168. Powell LD, Sgroi D, Sjoberg ER, Stamenkovic I, Varki A (1993) Natural ligands of the B cell adhesion molecule CD22 β carry N-linked oligosaccharides with β2,6-linked sialic acids that are required for recognition. *J. Biol. Chem.* 268:7019–7027.
169. Kelm S, Schauer R, Manuguerra JC, Gross HJ, Crocker PR (1994) Modifications of cell surface sialic acids modulate cell adhesion mediated by sialoadhesin and CD22. *Glycoconjug. J.* 11:576–585.
170. Bochner BS, Alvarez RA, Mehta P, Bovin NV, Blixt O, White JR, Schnaar RL (2005) Glycan array screening reveals a candidate ligand for Siglec-8. *J. Biol. Chem.* 280:4307–4312.
171. Brinkman-Van der Linden EC, Sjoberg ER, Juneja LR, Crocker PR, Varki N, Varki A (2000) Loss of N-glycolylneuraminic acid in human evolution. Implications for sialic acid recognition by siglecs. *J. Biol. Chem.* 275:8633–8640.
172. Sonnenburg JL, Altheide TK, Varki A (2004) A uniquely human consequence of domain-specific functional adaptation in a sialic acid-binding receptor. *Glycobiology* 14:339–346.
173. Tateno H, Crocker PR, Paulson JC (2005) Mouse Siglec-F and human Siglec-8 are functionally convergent paralogs that are selectively expressed on eosinophils and recognize 6′-sulfo-sialyl Lewis X as a preferred glycan ligand. *Glycobiology* 15:1125–1135.
174. Campanero-Rhodes MA, Childs RA, Kiso M, Komba S, Le Narvor C, Warren J, Otto D, Crocker PR, Feizi T (2006) Carbohydrate microarrays reveal sulphation as a modulator of siglec binding. *Biochem. Biophys. Res. Commun.* 344:1141–1146.
175. Rapoport EM, Pazynina GV, Sablina MA, Crocker PR, Bovin NV (2006) Probing sialic acid binding Ig-like lectins (siglecs) with sulfated oligosaccharides. *Biochemistry (Moscow)* 71:496–504.
176. Collins BE, Blixt O, DeSieno AR, Bovin N, Marth JD, Paulson JC (2004) Masking of CD22 by cis ligands does not prevent redistribution of CD22 to sites of cell contact. *Proc. Natl. Acad. Sci. U.S.A.* 101:6104–6109.

177. Hanasaki K, Varki A, Powell LD (1995) CD22-mediated cell adhesion to cytokine-activated human endothelial cells. Positive and negative regulation by α2,6-sialylation of cellular glycoproteins. *J. Biol. Chem.* 270:7533–7542.

178. Razi N, Varki A (1998) Masking and unmasking of the sialic acid-binding lectin activity of CD22 (Siglec-2) on B-lymphocytes. *Proc. Natl. Acad. Sci. U.S.A.* 95:7469–7474.

179. Razi N, Varki A (1999) Cryptic sialic acid binding lectins on human blood leukocytes can be unmasked by sialidase treatment or cellular activation. *Glycobiology* 9:1225–1234.

180. Han S, Collins BE, Bengtson P, Paulson JC (2005) Homo-multimeric complexes of CD22 revealed by *in situ* photoaffinity protein-glycan crosslinking. *Nat. Chem. Biol.* 1:93–97.

181. Lanoue A, Batista FD, Stewart M, Neuberger MS (2002) Interaction of CD22 with α-2,6-linked sialoglycoconjugates: Innate recognition of self to dampen B cell autoreactivity?. *Eur. J. Immunol.* 32:348–355.

182. Falco M, Biassoni R, Bottino C, Vitale M, Sivori S, Augugliaro R, Moretta L, Moretta A (1999) Identification and molecular cloning of 75/AIRM1, a novel member of the sialoadhesin family that functions as an inhibitory receptor in human natural killer cells. *J. Exp. Med.* 190:793–802.

183. Nicoll G, Ni J, Liu D, Klenerman P, Munday J, Dubock S, Mattei MG, Crocker PR (1999) Identification and characterization of a novel siglec, siglec-7, expressed by human natural killer cells and monocytes. *J. Biol. Chem.* 274:34089–34095.

184. Avril T, North SJ, Haslam SM, Willison HJ, Crocker PR (2006) Probing the cis interactions of the inhibitory receptor Siglec-7 with α2,8-disialylated ligands on natural killer cells and other leukocytes using glycan-specific antibodies and by analysis of α-2,8-sialyltransferase gene expression. *J. Leukoc. Biol.* 80:787–796.

185. Nicoll G, Avril T, Lock K, Furukawa K, Bovin N, Crocker PR (2003) Ganglioside GD3 expression on target cells can modulate NK cell cytotoxicity via siglec-7-dependent and -independent mechanisms. *Eur. J. Immunol.* 33:1642–1648.

186. Lundquist JJ, Toone EJ (2002) The cluster glycoside effect. *Chem. Rev.* 102:555–578.

187. Kiessling LL, Pohl S (1996) Strength in numbers: Non-natural polyvalent carbohydrate derivatives. *Chem. Biol.* 3:71–77.

188. Yamamoto F, Hakomori S (1990) Sugar-nucleotide donor specificity of histo-blood group A and B transferases is based on amino acid substitutions. *J. Biol. Chem.* 265:19257–19262.

189. Chen X, Andreana PR, Wang PG (1999) Carbohydrates in transplantation. *Curr. Opin. Chem. Biol.* 3:650–658.

190. Yarema KJ, Bertozzi CR (1998) Chemical approaches to glycobiology and emerging carbohydrate-based therapeutic agents. *Curr. Opin. Chem. Biol.* 2:49–61.

191. Galili U (2001) The α-gal epitope (Galα1-3Galβ1-4GlcNAc-R) in xenotransplantation. *Biochimie* 83:557–563.

192. Galili U (2004) Immune response, accommodation, and tolerance to transplantation carbohydrate antigens. *Transplantation* 78:1093–1098.

193. Ramssondar JJ, Macháty Z, Costa C, Williams BL, Fodor WL, Bondioli KR (2003) Production of α1,3-galactosyltransferase-knockout cloned pigs expressing human α1,2-fucosylosyltransferase. *Biol. Reprod.* 69:437–445.

194. Dai Y, Vaught TD, Boone J, Chen S-H, Phelps CJ, Ball S, Monahan JA, Jobst PM, McCreath KJ, Lamborn AE, Cowell-Lucero JL, Wells KD, Colman A, Polejaeva IA, Ayares DL (2002) Targeted disruption of the α1,3-galactosyltransferase gene in cloned pigs. *Nat. Biotechnol.* 20:251–255.

195. Milland J, Christiansen D, Sandrin MS (2005) α1,3-Galactosyltransferase knockout pigs are available for xenotransplantation: Are glycosyltransferases still relevant? *Immunol. Cell Biol.* 83:687–693.

196. Tangvoranuntakul P, Gagneux P, Diaz S, Bardor M, Varki N, Varki A, Muchmore E (2003) Human uptake and incorporation of an immunogenic nonhuman dietary sialic acid. *Proc. Natl. Acad. Sci. U.S.A.* 100:12045–12050.

197. Malykh YN, Schauer R, Shaw L (2001) *N*-Glycolylneuraminic acid in human tumours. *Biochimie* 83:623–634.

198. Franco A (2008) Glycoconjugates as vaccines for cancer immunotherapy: Clinical trials and future directions. *Anticancer Agents Med. Chem.* 8:86–91.

199. Reiman JM, Kmieciak M, Manjili MH, Knutson KL (2007) Tumor immunoediting and immunosculpting pathways to cancer progression. *Semin. Cancer Biol.* 17:275–287.

200. Jakóbisiak M, Lasek W, Gołąb J (2003) Natural mechanisms protecting against cancer. *Immunol. Lett.* 90:103–122.

201. Sell S (1990) Cancer-associated carbohydrates identified by monoclonal antibodies. *Hum. Pathol.* 21:1003–1019.

202. Hakomori S (1991) Possible functions of tumor-associated carbohydrate antigens. *Curr. Opin. Immunol.* 3:646–653.

203. Singhal A, Hakomori S (1990) Molecular changes in carbohydrate antigens associated with cancer. *Bioessays* 12:223–230.

204. Burchell JM, Mungul A, Taylor-Papadimitiou J (2001) *O*-Linked glycosylation in the mammary gland: Changes that occur during malignancy. *J. Mammary Gland Biol. Neoplasia* 6:355–364.

205. Brockhausen I (1999) Pathways of *O*-glycan biosynthesis in cancer cells. *Biochim. Biophys. Acta* 1473:67–95.

206. Brockhausen I (2003) Glycodynamics of mucin biosynthesis in gastrointestinal tumor cells. *Adv. Exp. Med. Biol.* 535:163–188.

207. Brockhausen I (2003) Sulphotransferases acting on mucin-type oligosaccharides. *Biochem. Soc. Trans.* 31:318–325.

208. Nilsson O, Brezicka FT, Holmgren J, Sörenson S, Svennerholm L, Yngvason F, Lindholm L (1986) Detection of a ganglioside antigen associated with small cell lung carcinomas using monoclonal antibodies directed against fucosyl-GM1. *Cancer Res.* 46:1403–1407.

209. Brezicka FT, Olling S, Nilsson O, Bergh J, Holmgren J, Sörenson S, Yngvason F, Lindholm L (1989) Immunohistological detection of fucosyl-GM1 ganglioside in human lung cancer and normal tissues with monoclonal antibodies. *Cancer Res.* 49:1300–1305.

210. Vangsted AJ, Clausen H, Kjeldsen TB, White T, Sweeney B, Hakomori S, Drivsholm L, Zeuthen J (1991) Immunochemical detection of a small cell lung cancer-associated ganglioside (FucGM1) antigen in serum. *Cancer Res.* 51:2879–2884.

211. Brezicka T, Bergman B, Olling S, Fredman P (2000) Reactivity of monoclonal antibodies with ganglioside antigens in human small cell lung cancer tissues. *Lung Cancer* 28:29–36.

212. Galili U (2004) Autologous tumor vaccines processed to express α-gal epitopes: A practical approach to immunotherapy in cancer. *Cancer Immunol. Immunother.* 53:935–945.
213. McCarty OJT, Mousa SA, Bray PF, Konstantopoulos K (2000) Immobilized platelets support human colon carcinoma cell tethering, rolling, and firm adhesion under dynamic flow conditions. *Blood* 96:1789–1797.
214. Bengoechea JA, Najdenski H, Skurnik M (2004) Lipopolysaccharide O antigen status of *Yersinia enterocolitica* O:8 is essential for virulence and absence of O antigen affects the expression of other *Yersinia* virulence factors. *Mol. Microbiol.* 52:451–569.
215. Bruneteau M, Minka S (2003) Lipopolysaccharides of bacterial pathogens from the genus Yersinia: A mini-review. *Biochimie* 85:145–152.
216. Pier GB (2003) Promises and pitfalls of *Pseudomonas aeruginosa* lipopolysaccharide as a vaccine antigen. *Carbohydr. Res.* 338:2549–2556.
217. Vliegenthart JFG (2006) Carbohydrate based vaccines. *FEBS Lett.* 580:2945–2950.
218. AlonsoDeVelasco E, Verheul AF, Verhoef J, Snippe H (1995) *Streptococcus pneumoniae*: Virulence factors, pathogenesis, and vaccines. *Microbiol. Rev.* 59:591–603.
219. Seeberger PH, Werz DB (2005) Automated synthesis of oligosaccharides as a basis for drug discovery, *Nat. Rev. Drug Discov.* 4:751–763.
220. Beyene A, Basu A, Meyer K, Ray R (2004) Influence of *N*-linked glycans on intracellular transport of hepatitis C virus E1 chimeric glycoprotein and its role in pseudotype virus infectivity. *Virology* 324:273–285.
221. Eichler R, Lenz O, Garten W, Strecker T (2006) The role of single *N*-glycans in proteolytic processing and cell surface transport of the Lassa virus glycoprotein GP-C. *Virol. J.* 3:41.
222. Aguilar HC, Matreyek KA, Filone CM, Hashimi ST, Levroney EL, Negrete OA, Bertolotti-Ciarlet A, Choi DY, McHardy I, Fulcher JA, Su SV, Wolf MC, Kohatsu L, Baum LG, Lee B (2006) *N*-Glycans on Nipah virus fusion protein protect against neutralization but reduce membrane fusion and viral entry. *J. Virol.* 80:4878–4889.
223. Melanson VR, Iorio RM (2006) Addition of *N*-glycans in the stalk of the Newcastle disease virus HN protein blocks its interaction with the F protein and prevents fusion. *J. Virol.* 80:623–633.
224. Shi X, Brauburger K, Elliott RM (2005) Role of *N*-linked glycans on bunyamwera virus glycoproteins in intracellular trafficking, protein folding, and virus infectivity. *J. Virol.* 79:13725–13734.
225. Moll M, Kaufmann A, Maisner A (2004) Influence of *N*-glycans on processing and biological activity of the nipah virus fusion protein. *J. Virol.* 78:7274–7278.
226. Lin G, Simmons G, Pöhlmann S, Baribaud F, Ni H, Leslie GJ, Haggarty BS, Bates P, Weissman D, Hoxie JA, Doms RW (2003) Differential *N*-linked glycosylation of human immunodeficiency virus and Ebola virus envelope glycoproteins modulates interactions with DC-SIGN and DC-SIGNR. *J. Virol.* 77:1337–1346.
227. Bossart KN, Crameri G, Dimitrov AS, Mungall BA, Feng YR, Patch JR, Choudhary A, Wang LF, Eaton BT, Broder CC (2005) Receptor binding, fusion inhibition, and induction of cross-reactive neutralizing antibodies by a soluble G glycoprotein of Hendra virus. *J. Virol.* 79:6690–6702.
228. Oostra M, de Haan CA, de Groot RJ, Rottier PJ (2006) Glycosylation of the severe acute respiratory syndrome coronavirus triple-spanning membrane proteins 3a and M. *J. Virol.* 80:2326–2336.

229. Schowalter RM, Smith SE, Dutch RE (2006) Characterization of human metapneumovirus F protein-promoted membrane fusion: Critical roles for proteolytic processing and low pH. *J. Virol.* 80:10931–10941.
230. Shirato K, Miyoshi H, Goto A, Ako Y, Ueki T, Kariwa H, Takashima I (2004) Viral envelope protein glycosylation is a molecular determinant of the neuroinvasiveness of the New York strain of West Nile virus. *J. Gen. Virol.* 85:3637–3645.
231. Scherret JH, Mackenzie JS, Khromykh AA, Hall RA (2001) Biological significance of glycosylation of the envelope protein of Kunjin virus. *Ann. N.Y. Acad. Sci.* 951:361–362.
232. Beasley DW, Whiteman MC, Zhang S, Huang CY, Schneider BS, Smith DR, Gromowski GD, Higgs S, Kinney RM, Barrett AD (2005) Envelope protein glycosylation status influences mouse neuroinvasion phenotype of genetic lineage 1 West Nile virus strains. *J. Virol.* 79:8339–8347.
233. Lad VJ, Shende VR, Gupta AK, Koshy AA, Roy A (2000) Effect of tunicamycin on expression of epitopes on Japanese encephalitis virus glycoprotein E in porcine kidney cells. *Acta Virol.* 44:359–364.
234. Wagner R, Wolff T, Herwig A, Pleschka S, Klenk HD (2000) Interdependence of hemagglutinin glycosylation and neuraminidase as regulators of influenza virus growth: A study by reverse genetics. *J. Virol.* 74:6316–6323.
235. Klenk HD, Wagner R, Heuer D, Wolff T (2002) Importance of hemagglutinin glycosylation for the biological functions of influenza virus. *Virus Res.* 82:73–75.
236. Baigent SJ, McCauley JW (2001) Glycosylation of haemagglutinin and stalk-length of neuraminidase combine to regulate the growth of avian influenza viruses in tissue culture. *Virus Res.* 79:177–185.
237. Daniels R, Kurowski B, Johnson AE, Hebert DN (2003) *N*-Linked glycans direct the cotranslational folding pathway of influenza hemagglutinin. *Mol. Cell* 11:79–90.
238. Fenouillet E, Gluckman JC, Bahraoui E (1990) Role of *N*-linked glycans of envelope glycoproteins in infectivity of human immunodeficiency virus type 1. *J. Virol.* 64:2841–2848.
239. Montefiori DC, Robinson WEJ, Mitchell WM (1988) Role of protein *N*-glycosylation in pathogenesis of human immunodeficiency virus type 1. *Proc. Natl. Acad. Sci. U.S.A.* 85:9248–9252.
240. Korber B, Gaschen B, Yusim K, Thakallapally R, Kesmir C, Detours V (2001) Evolutionary and immunological implications of contemporary HIV-1 variation. *Br. Med. Bull.* 58:19–42.
241. Sagar M, Wu X, Lee S, Overbaugh J (2006) Human immunodeficiency virus type 1 V1-V2 envelope loop sequences expand and add glycosylation sites over the course of infection, and these modifications affect antibody neutralization sensitivity. *J. Virol.* 80:9586–9598.
242. Wolk T, Schreiber M (2006) *N*-Glycans in the gp120 V1/V2 domain of the HIV-1 strain NL4-3 are indispensable for viral infectivity and resistance against antibody neutralization. *Med. Microbiol. Immunol.* 195:165–172.
243. Tzianabos A, Wang JY, Kasper DL (2003) Biological chemistry of immunomodulation by zwitterionic polysaccharides. *Carbohydr. Res.* 338:2531–2538.
244. González-Fernández M, Carrasco-Marín E, Alvarez-Domínguez C, Outschoorn IM, Leyva-Cobián F (1997) Inhibitory effects of thymus-independent type 2 antigens on

MHC class II-restricted antigen presentation: Comparative analysis of carbohydrate structures and the antigen presenting cell. *Cell. Immunol.* 176:1–13.
245. Hirst RA, Kadioglu A, O'callaghan C, Andrew PW (2004) The role of pneumolysin in pneumococcal pneumonia and meningitis. *Clin. Exp. Immunol.* 138:
246. Lombard M, Pastoret PP, Moulin AM (2007) A brief history of vaccines and vaccination. *Rev. Sci. Tech.* 26:29–48.
247. Borman S (2004) Carbohydrate vaccines. *Chem. Eng. News* 82:31–35.
248. Jennings HJ (1983) Capsular polysaccharides as human vaccines. *Adv. Carbohydr. Chem. Biochem.* 41:155–208.
249. Warren JD, Geng X, Danishefsky SJ (2006) Synthetic glycopeptide-based vaccines. *Top. Curr. Chem.* 267:109–141.
250. Wang Q, Zhang J, Guo Z (2007) Efficient glycoengineering of GM3 on melanoma cell and monoclonal antibody-mediated selective killing of the glycoengineered cancer cell. *Bioorg. Med. Chem.* 15:7561–7567.
251. Allen JR, Allen JG, Zhang XF, Williams LJ, Zatorski A, Ragupathi G, Livingston PO, Danishefsky SJ (2000) A second generation synthesis of the MBr1 (Globo-H) breast tumor antigen: New application of the *N*-pentenyl glycoside method for achieving complex carbohydrate protein linkages. *Chemistry* 6:1366–1375.
252. Deshpande PP, Danishefsky SJ (1997) Total synthesis of the potential anticancer vaccine KH-1 adenocarcinoma antigen. *Nature* 387:164–166.
253. Allen JR, Harris CR, Danishefsky SJ (2001) Pursuit of optimal carbohydrate-based anticancer vaccines: Preparation of a multiantigenic unimolecular glycopeptide containing the Tn, MBr1, and LewisY antigens. *J. Am. Chem. Soc.* 123:1890–1897.
254. Seeberger PH (2003) Automated carbohydrate synthesis to drive chemical glycomics. *Chem. Commun.* 21:1115–1121.
255. Seeberger PH, Werz DB (2007) Synthesis and medical applications of oligosaccharides. *Nature* 446:1046–1051.
256. Koeller KM, Wong CH (2000) Complex carbohydrate synthesis tools for glycobiologists: Enzyme-based approach and programmable one-pot strategies. *Glycobiology* 10:1157–1169.
257. Koeller KM, Wong C-H (2000) Synthesis of complex carbohydrates and glycoconjugates: Enzyme-based and programmable one-pot strategies. *Chem. Rev.* 100:4465–4494.
258. Ragupathi G, Koide F, Livingston PO, Cho YS, Endo A, Wan Q, Spassova MK, Keding SJ, Allen J, Ouerfelli O, Wilson RM, Danishefsky SJ (2006) Preparation and evaluation of unimolecular pentavalent and hexavalent antigenic constructs targeting prostate and breast cancer: A synthetic route to anticancer vaccine candidates. *J. Am. Chem. Soc.* 128:2715–2725.
259. Oetke C, Brossmer R, Mantey LR, Hinderlich S, Isecke R, Reutter W, Keppler OT, Pawlita M (2002) Versatile biosynthetic engineering of sialic acid in living cells using synthetic sialic acid analogues. *J. Biol. Chem.* 277:6688–6695.
260. Bardor M, Nguyen DH, Diaz S, Varki A (2005) Mechanism of uptake and incorporation of the non-human sialic acid *N*-glycolylneuraminic acid into human cells. *J. Biol. Chem.* 280:4228–4237.
261. Liu T, Guo Z, Yang Q, Sad S, Jennings HJ (2000) Biochemical engineering of surface $\alpha 2,8$ polysialic acid for immunotargeting tumor cells. *J. Biol. Chem.* 275:32832–32836.

262. Hayrinen J, Jennings H, Raff HV, Rougon G, Hanai N, Gerardy-Schahn R, Finne J (1995) Antibodies to polysialic acid and its *N*-propyl derivative: Binding properties and interaction with human embryonal brain glycopeptides. *J. Infect. Dis.* 171:1481–1490.
263. Chefalo P, Pan Y, Nagy N, Guo Z, Harding CV (2006) Efficient metabolic engineering of GM3 on tumor cells by *N*-phenylacetyl-D-mannosamine. *Biochemistry* 45:3733–3739.
264. Chefalo P, Pan Y-B, Nagy N, Harding C, Guo Z-W (2004) Preparation and immunological studies of protein conjugates of *N*-acylneuraminic acids. *Glycoconjug. J.* 20:407–414.
265. Schilling B, Goon S, Samuels NM, Gaucher SP, Leary JA, Bertozzi CR, Gibson BW (2001) Biosynthesis of sialylated lipooligosaccharides in *Haemophilus ducreyi* is dependent on exogenous sialic acid and not mannosamine. Incorporation studies using *N*-acylmannosamine analogues, *N*-Glycolylneuraminic acid, and 13C-labeled *N*-acetylneuraminic acid. *Biochemistry* 40:12666–12677.
266. Goon S, Schilling B, Tullius MV, Gibson BW, Bertozzi CR (2003) Metabolic incorporation of unnatural sialic acids into *Haemophilus ducreyi* lipooligosaccharides. *Proc. Natl. Acad. Sci. U.S.A.* 18:3089–3094.
267. Verez-Bencomo V, Fernández-Santana V, Hardy E, Toledo ME, Rodríguez MC, Heynngnezz L, Rodriguez A, Baly A, Herrera L, Izquierdo M, Villar A, Valdés Y, Cosme K, Deler ML, Montane M, Garcia E, Ramos A, Aguilar A, Medina E, Toraño G, Sosa I, Hernandez I, Martínez R, Muzachio A, Carmenates A, Costa L, Cardoso F, Campa C, Diaz M, Roy R (2004) A synthetic conjugate polysaccharide vaccine against *Haemophilus influenzae* type b. *Science* 305:522–525.
268. Kaiser J (2004) Synthetic vaccine is a sweet victory for Cuban science. *Science* 305:460.
269. Ingale S, Wolfert MA, Gaekwad J, Buskas T, Boons G-J (2007) Robust immune responses elicited by a fully synthetic three-component vaccine. *Nat. Chem. Biol.* 3:663–667.

2

PREPARATION OF GLYCOCONJUGATE VACCINES

Wei Zou and Harold J. Jennings

Institute for Biological Sciences, National Research Council of Canada, Ottawa, Ontario, Canada K1A0R6

2.1 INTRODUCTION

Bacterial polysaccharides have been used successfully for some time as licensed vaccines against diseases caused by *Streptococcus pneumoniae, Neisseria meningitidis,* and *Haemophilus influenzae* type b. Their success is based on the fact that polysaccharides are the most conserved and accessible molecules on the bacterial surface. In this location they not only provide the epitopes that induce protective antibodies, but for some polysaccharides they also enable the bacteria to evade the human immune system. However, the continued use of polysaccharide vaccines has revealed a serious flaw in their performance, namely, that they are poorly immunogenic in infants [1–4], who unfortunately are the most susceptible cohort. These deficiencies have been largely overcome by conjugating the polysaccharides to protein carriers, and the resultant glycoconjugate vaccines have greatly extended the potential of polysaccharide vaccines as human vaccines. For a detailed review of the polysaccharide structures and the immunological and clinical aspects of both polysaccharide and glycoconjugate vaccines, the reader is referred to Chapter 4. As early as 1931, Goebel and Avery [5] made a conjugate of the *S. pneumoniae* type 3 polysaccharide by linking it to horse serum globulin, which was able to induce specific antibodies in rabbits,

---

Carbohydrate-Based Vaccines and Immunotherapies. Edited by Zhongwu Guo and Geert-Jan Boons
Copyright © 2009 John Wiley & Sons, Inc.

unresponsive to the pure polysaccharide, and found that even a cellobiuronic acid conjugate was able to confer immunity to challenge by live type 3 pneumococci in mice [6, 7]. But it was not until the late 1980s that glycoconjugate vaccines were first used in humans.

The potential of using glycoconjugates as human vaccines was first demonstrated in both adults and infants using *H. influenzae* type b polysaccharide–protein conjugates (see Chapter 4). Subsequently, these vaccines were licensed in the United States, and their use resulted in a more than 95% decline in the incidence of meningitis caused by *H. influenzae* type b. This phenomenal success led to the rapid development of conjugate vaccines against other prevalent diseases such as *N. menigitidis*, *S. pneumoniae*, group B *Streptococcus* and *S. typhi* [8, 9]. Although the immunogenicity of conjugates largely depends on the individual polysaccharides and carrier proteins selected, there is evidence that the method of conjugation employed can also be an important factor. In order to make a successful conjugate vaccine, the conjugation method should be simple, efficient, and result in minimal distortion of the individual components, and many different chemical techniques have been employed that satisfy these criteria.

This review focuses on the chemistry used in the synthesis of polysaccharide- and oligosaccharide-conjugate vaccines associated with bacterial diseases and is restricted to those currently used as human vaccines or those having potential as human vaccines. A previous relevant review by Jennings and Sood [10] describes the coupling chemistry associated with the different functional groups of capsular polysaccharides, and another by Pozsgay [11] describes the chemistry used in the conjugation of synthetic oligosaccharides associated with bacterial polysaccharides and lipopolysaccharides.

2.2 CAPSULAR POLYSACCHARIDE–PROTEIN CONJUGATES

2.2.1 *Haemophilus influenzae* Type b

Haemophilus influenzae type b (Hib) was a leading cause of bacterial meningitis in the United States prior to the introduction of Hib conjugate vaccines. The conjugate vaccines were developed in the 1980s and subsequently licensed in the 1990s by Wyeth Lederle, Merck, and Sanofi Pasteur. The Wyeth Lederle vaccine was produced by a method described by Anderson et al. [12, 13]. This method was based on an aldehyde-reductive amination reaction previously used in the synthesis of monovalent meningococcal conjugates [14], but because Hib capsular polysaccharide is a linear poly-ribosyl ribitol phosphate (PRP), extension of the period of periodate oxidation also led to depolymerization of PRP with the spontaneous generation of two terminal aldehydes per saccharide. Direct conjugation of the resultant PRP oligomer to protein carrier CRM197, a nontoxic variant of diphtheria toxin, was then achieved by reductive amination as shown in Scheme 2.1, and the degree of depolymerization was controlled by the amount of periodate used. Reductive amination involves the formation of Schiff's base between a saccharide aldehyde and a protein amine group, followed by its reduction using sodium cyanoborohydride as a reducing agent, and because in this case the activated saccharide was divalent, the resultant conjugate was moderately crosslinked.

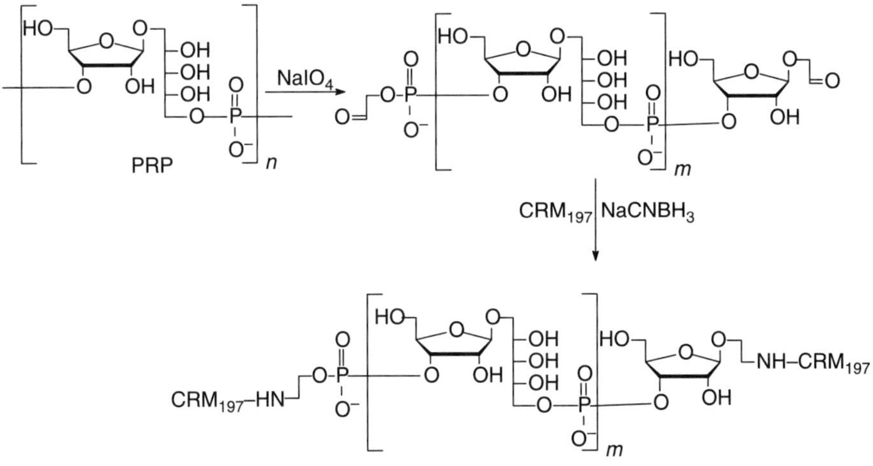

2.2 CAPSULAR POLYSACCHARIDE–PROTEIN CONJUGATES

Scheme 2.1

In contrast to terminal coupling, the other two Hib conjugate vaccines were produced by random activation of the depolymerized polysaccharide, whereby multiple couplings to protein carriers also occur in the inner region of polysaccharide chains. Thus this procedure has the potential to produce more extensively crosslinked conjugates. The Sanofi Pasteur procedure is based on a method developed by Robbins et al. [15] in which PRP, depolymerized by thermal hydrolysis, was activated using cyanogen bromide (CNBr) and coupled to adipic dihydrazide (ADH) activated tetanus toxoid (TT) through a covalent isourea linkage as depicted in Scheme 2.2 [16–18].

Under slightly alkaline conditions (pH 10–11), CNBr reacts rapidly with hydroxyl groups of polysaccharides to form cyanate esters [19–22], and the cyanate esters can further react with vicinal hydroxyl groups to form cyclic imidocarbonates. Although

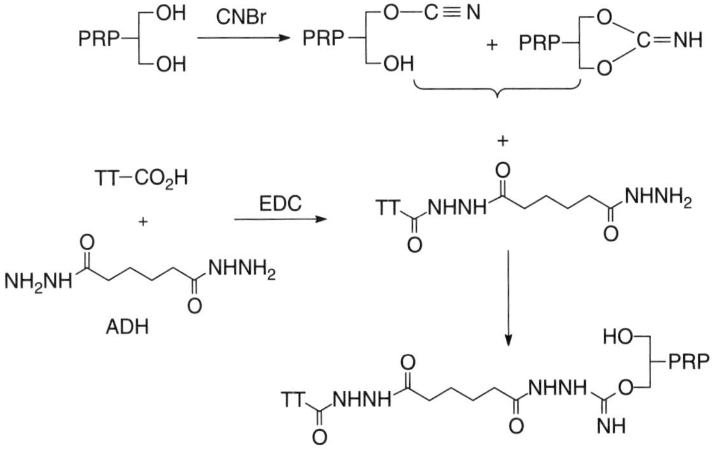

Scheme 2.2

imidocarbonate is less reactive than cyanate ester, both intermediates can effectively react with nucleophiles, for example, the lysine ε-amino groups of proteins at about pH 8.5 to form stable *O*-alkylisourea conjugates [23]. Alternatively, the conjugation sequence may be reversed by performing the nucleophilic addition of ADH to CNBr-activated polysaccharide followed by 1-ethyl-3-(3-dimethylaminopropyl)-carbodiimide (EDC)-mediated condensation with the carboxylic acid of proteins such as TT [24]. Another variation of this method involves using 6-aminocaproic acid as a bi-functional linker. Conjugation is accomplished by coupling of the linker's amino group to the CNBr-activated polysaccharide, followed by condensation of the linker's carboxylic acid to amino groups of the protein [25].

The Hib conjugate vaccine from Merck contains a disulfide linker. Multistep synthesis of the conjugate involves activation of both polysaccharide and the carrier protein, the latter being an outer membrane protein complex (OMPC) from *N. meningitidis* (Scheme 2.3) [26–28]. The polyanionic polysaccharide was masked using tetrabutylammonium as a counter ion in order to improve the solubility of polysaccharide in *N,N*-dimethylformamide (DMF). Mixing carbonyldiimidazole and polysaccharide led to an imidazolylurethane intermediate, which was then reacted with butyldiamine under slightly basic conditions (pH 10.35) to form a carbamate, which was then *N*-bromoacetylated. Protein activation was achieved by reacting the

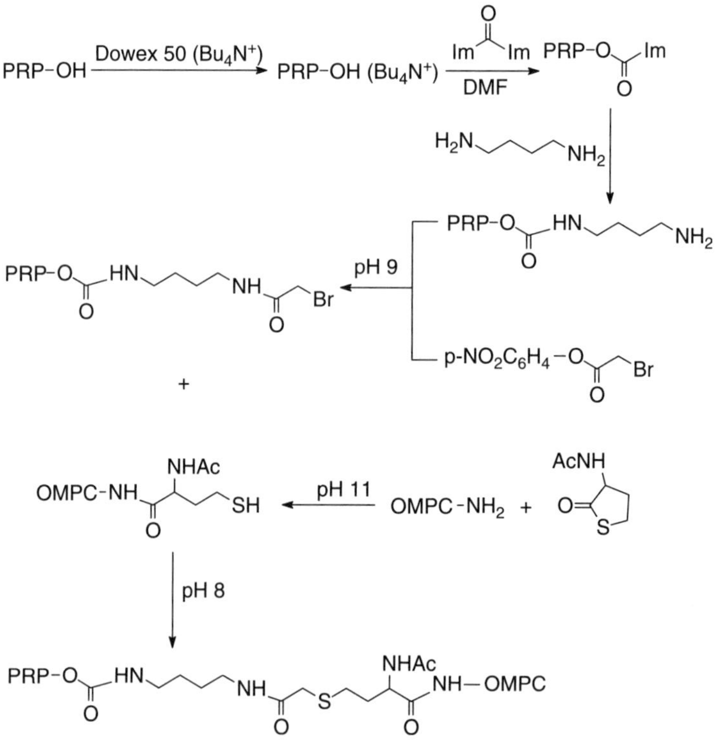

Scheme 2.3

protein with N-acetylhomocysteine thiolactone to introduce thiol groups. Finally, coupling between the bromoacetylated polysaccharide and thiolated protein led to the Hib–OMPC conjugate.

A comparative study on Hib conjugate vaccines by Schlesinger and Granoff [29] found that both Hib–CRM197 and Hib–TT conjugate vaccines elicited higher avidity antibodies, which correlate to protection against disease, than Hib–OMPC. In addition, production of Hib–CRM197 vaccine is the most practical and most cost effective because it combines depolymerization and activation of polysaccharide in a single step and requires neither linker nor protein activation. Moreover, Hib–CRM197 conjugate can be better defined structurally than those obtained by random activation.

In order to investigate the immunologic impact of polysaccharide chain length, terminal group, and polysaccharide/protein ratio in a conjugate, Anderson et al. [30] prepared conjugates consisting of *H. influenzae* type b oligosaccharides coupled to CRM197. The uniterminally coupled oligosaccharides in the conjugates varied in chain length, terminal residue (phosphate, ribose, or ribitol), and multiplicity of loading, as defined by the ribose/protein ratio. Because the glycosidic cleavage of PRP can be controlled by acid hydrolysis, oligomers of PRP with ribose at the reducing end and ribitol as the terminal group were obtained at pH 3 with 4, 6, and 12 repeating units, and their conjugates were prepared by direct reductive amination. Hydrolysis of PRP calcium salt at pH 1 cleaves mainly phosphate diester groups producing oligomers with monophosphate at the nonreducing end, and from these oligomers a conjugate with 7 repeating units was then made by selective periodate oxidation of terminal ribitol followed by reductive amination. Acid phosphatase treatment of the latter conjugate gave another conjugate having ribose at the nonreducing end. Immunological studies in both adults and infants indicated that the oligosaccharide chain length and terminal residue structure were less critical variables in the immune response than the extent of oligosaccharide loading. This result also suggests that an extended epitope is not required in the generation of a protective immune response by PRP.

2.2.2 *Streptococcus pneumoniae*

Streptococcus pneumoniae is the most common pathogen that causes bacterial pneumonia and is also one of the major causes of bacterial otitis media (middle ear infections), meningitis, and bacteremia. There are more than 100 types of pneumococci, each having a different capsular polysaccharide structure, and the capsular polysaccharides are their principal virulence factors. The polysaccharides are able to induce antibody responses in adults, but unfortunately they are poorly immunogenic in children of less than 2 years of age, a segment of the population that is the most prone to infection.

A study in 1995 [31] indicated that the most common pneumococcal serogroups isolated in developed countries were, in descending order, 14, 6, 19, 18, 9, 23, 7, 4, 1, and 15, while in developing countries they were 6, 14, 8, 5, 1, 19, 9, 23, 18, 15, and 17. Currently, two heptavalent pneumococcal conjugate vaccines are formulated to include serotypes 4, 6B, 9V, 14, 18C, 19F, and 23F, which provide over 80% coverage of pneumococcal disease in the developed world.

The heptavalent vaccines are formulated by combining seven individual polysaccharide conjugates. The pneumococcal conjugate vaccine from Wyeth Lederle, Prevnar, uses CRM197 as carrier protein and is produced by controlled periodate oxidation of the polysaccharide moiety to generate aldehyde groups, followed by reductive amination with protein. The conjugation procedure is essentially the same as that used for the Hib–CRM197 conjugate (see Scheme 2.1) except that for some serotype polysaccharides, periodate-sensitive vicinal hydroxyls, other than those associated with interchain polyols were targeted. The other heptavalent vaccine (PncOMPC) from Merck was manufactured by the same method used for the production of its Hib conjugate vaccine, and again the OMPC was used as a carrier (see Scheme 2.3) [26].

Sanofi Pasteur produced an 11-valent pneumococcal conjugate vaccine (PncD/T11) containing polysaccharides 1, 4, 5, 7F, 9V, 19F, and 23F (coupled to TT) and polysaccharides 3, 6B, 14, and 18C (coupled to diphtheria toxoid (DT)) by the same method used for its Hib conjugate vaccine (see Scheme 2.2) [32, 33]. GlaxoSmithKline is currently testing a 10-valent conjugate vaccine (Streptorix), which includes pneumococcal serotypes 1, 4, 5, 6B, 7F, 9V, 14, 18C, 19F, and 23F, conjugated to an *H. influenzae* protein D carrier [34]. In addition, Wyeth Lederle is developing a 13-valent vaccine (PCV13) including serotype 1, 3, 5, 6A, 7F, and 19A in addition to its heptavalent conjugate vaccine.

2.2.3 *Neisseria meningitidis*

Neisseria meningitidis became a leading cause of bacterial meningitis in North America following dramatic reductions in the incidence of *S. pneumoniae* and *H. influenzae* type b meningitis caused by the introduction of their respective conjugate vaccines. Endemic disease caused by different groups (A, B, C, Y, and W135) occurs with an incidence 1–5/100,000 and affects mainly infants 6 months to 2 years old. During epidemics, however, which are mainly due to group A, and which occur predominantly in developing countries, the incidences can be as high as 500/100,000. Group B, C, and more recently group Y in the United States are the major causes of meningococcal disease in developed countries. Since the immunization of all persons aged 12 months to 17 years with meningoccocal group C conjugate vaccine in the United Kingdom in 1999, the incidence of meningoccocal group C disease declined dramatically, even in the first year, with efficacy rates from 88 to 98% in the different age groups [35].

The three group C conjugate vaccines that are available are Meningtec from Wyeth Lederle, NeisVac-C from Baxter Health, and Menjugate from Novartis Chiron. The basic chemistry used in the production of the first two vaccines was pioneered by Jennings and Lugowski [14], whereby fragments of groups A, B, and C polysaccharides were selectively oxidized with periodate to yield activated fragments having terminal aldehyde groups, which subsequently linked directly to tetanus toxoid lysine groups by reductive amination. Variations of this simple coupling procedure have also been successfully employed in the synthesis of an inexpensive group A polysaccharide–TT conjugate for use in the developing world [36], and as described in this chapter and of many other polysaccharide and oligosaccharide conjugate vaccines, some of which are being marketed worldwide. As shown in Scheme 2.4, the only differences between the Meningtec and NeisVac-C vaccines are that in the former the O-acetylated C polysaccahride is coupled to CRM197 (Scheme 2.4), whereas in the latter the de-O-acetylated C polysaccharide is coupled to TT. It is interesting to note that removal of the O-acetyl groups from the C polysaccharide not only facilitated the conjugation procedure but also resulted in a better immune response in toddlers [35, 37].

Although reductive amination was also employed by Novartis in the synthesis of its group C conjugate vaccine, additional procedures were also used. Unlike the difunctional approach used by Wyeth Ledlerle and Baxter Health, where aldehydes were introduced at both termini of the group C polysaccharide, Novartis used a monofunctional approach involving a linker to couple the polysaccharide to CRM197 via its

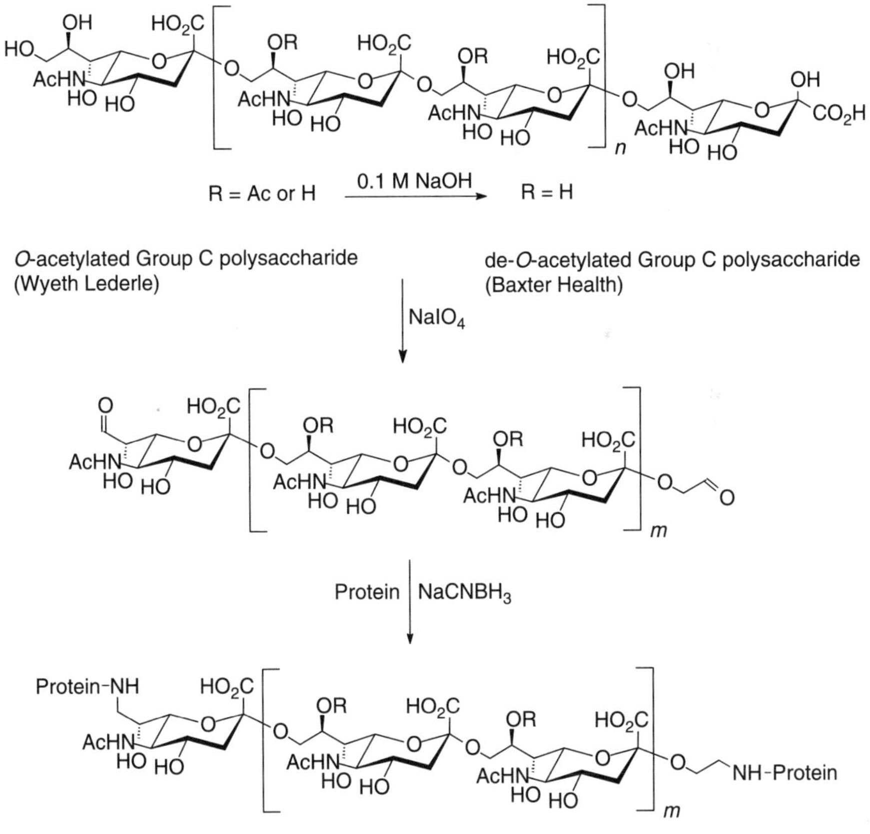

Wyeth Lederle vaccine: Protein = CRM_{197}; R = Ac or H
Baxter Health vaccine: Protein = TT; R = H

Scheme 2.4

reducing end. The polysaccharide was first depolymerized by acid hydrolysis to oligosaccharides, and the reducing end ketone was then converted to an amino group by reductive amination with aqueous ammonium chloride (see Scheme 2.5). The amino oligosaccharide was then subsequently dissolved in dimethyl sulfoxide (DMSO)-H$_2$O (9:1) and reacted with excessive N-hydroxysuccinimide diester of adipic acid. The resultant activated oligosaccharide was then precipitated with 1,4-dioxane and coupled to CRM197 in phosphate buffer via an amide linkage [38]. This procedure was also applied to the synthesis of a meningococcal A polysaccharide–CRM197 conjugate vaccine [38].

Recently, a tetravalent A, C, Y, W-135 conjugate vaccine (Menactra) from Sanofi Pasteur, containing capsular polysaccharides conjugated to DT, has been approved for use in persons aged 11–55 years [39]. The polysaccharide antigens are individually conjugated to DT, and the conjugation method used was the same as that previously

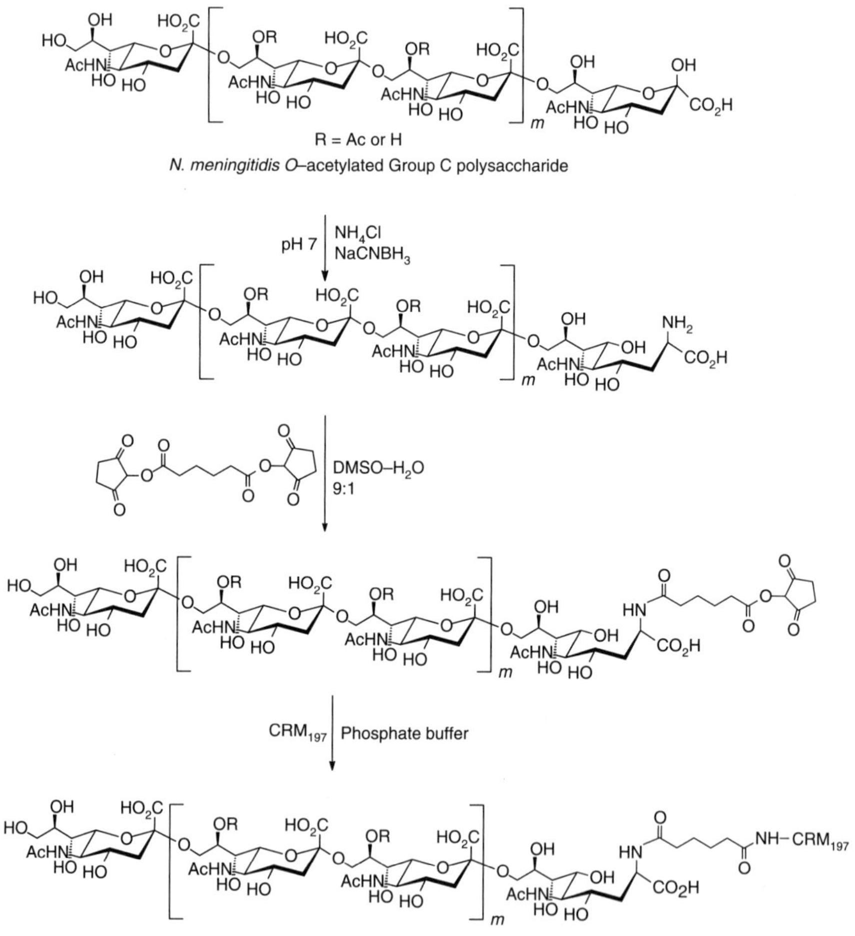

Scheme 2.5

used for the production of both Hib and *S. pneumoniae* polysaccharide conjugate vaccines by Sanofi Pasteur (see Scheme 2.2).

As the only conserved antigenic structure on the surface of group B *N. meningitidis*, its capsular polysaccharide is an ideal molecule on which to base a vaccine. However, even when conjugated to TT, it is poorly immunogenic [14], and, because it consists of α(2-8)-polysialic acid (PSA), a structure also found in human tissue, it carries the perceived risk of inducing autoimmunity. The immunogenicity of PSA can be greatly improved by constructing *N*-propionylated (NPr) PSA conjugates [40], which were synthesized by treating PSA with base to remove the *N*-linked acetyl groups and replacing them with propionyl groups. NPrPSA was then subjected to controlled periodate oxidation to introduce two terminal aldehydes into its structure (Scheme 2.6), which as shown for the group C conjugate (Scheme 2.4) were used to couple the saccharide to TT. This conjugate readily induced bactericidal IgG antibodies in mice, which were both NPrPSA-specific and PSA-cross reactive, but only the former were bactericidal for group B meningococci [40]. This evidence indicates that NPrPSA mimics a different protective epitope on the surface of group B meningococci than is presented by PSA alone [40]. The protective NPrPSA epitope is unique, consisting of at least

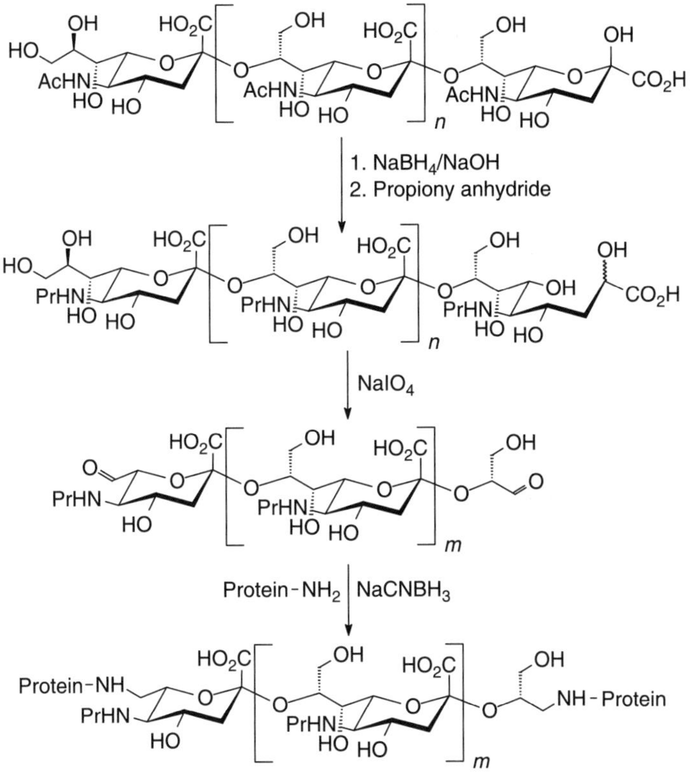

Scheme 2.6

eight contiguous NeuNPr residues located on extended helical domains of NPrPSA [41, 42]. A preclinical evaluation of an NPrPSA conjugate in baboons and rhesus monkeys, in which the carrier protein was a recombinant group B meningococcal porin (rPorB) protein, was shown to elicit high boostable NPrPSA antibody (IgG) titers having bactericidal activity both with monkey and human complement [43]. These positive results suggest that NPrPSA conjugates would be excellent vaccine candidates, but their development has been suspended because of the perceived risk of autoimmunity, and the poor performance of an NPrPSA–TT conjugate vaccine in a limited human trial [44].

2.2.4 Salmonella typhi Vi

The capsular polysaccharide of *S. typhi* Vi is a linear homopolymer of α-(1-4)-D-GalpA, *N*-acetylated at C-2 and *O*-acetylated at C-3, and although it has proven to be a protective antigen and has been used as a vaccine, it is poorly immunogenic in infants. Initially, Vi polysaccharide conjugates were prepared by linking the polysaccharide to cholera toxin (CT), or to its B subunit (CTB), through a disulfide linker (Scheme 2.7). Carboxylic acids of the polysaccharide were condensed with cystamine using EDC, and after blocking the thiol groups by iodoacetic acid, the amino groups of both cholera toxin and its B unit were activated using *N*-succinimidyl 3-(2-pyridyldithio)propionate (SPDP) at pH 7.55. Following treatment with dithiothreitol (DTT) the thiolated Vi polysaccharide was mixed with activated protein to form a conjugate having disulfide linkages [45, 46].

In some cases the amino groups of carrier protein available for conjugation are limited due to prior detoxification of the protein with formaldehyde. To avoid this problem, the carboxylic acid groups of the protein were used for conjugation of the Vi polysaccharide, and ADH was used as the linker of choice to couple the carboxylic

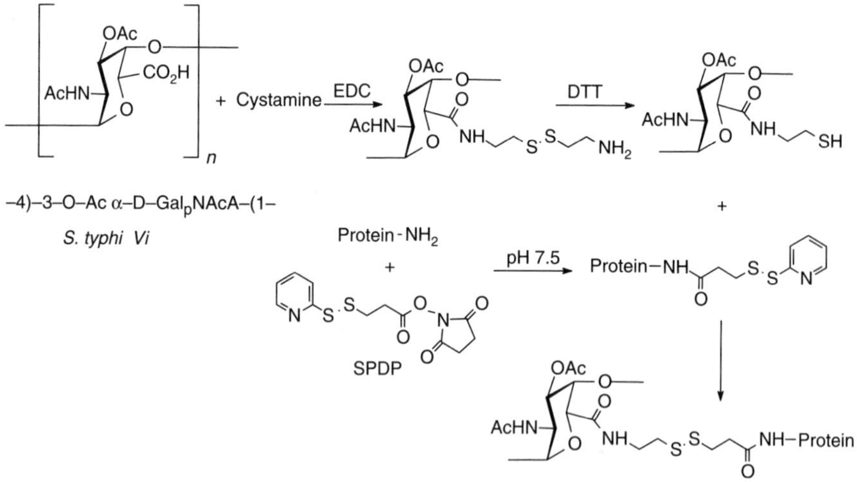

Scheme 2.7

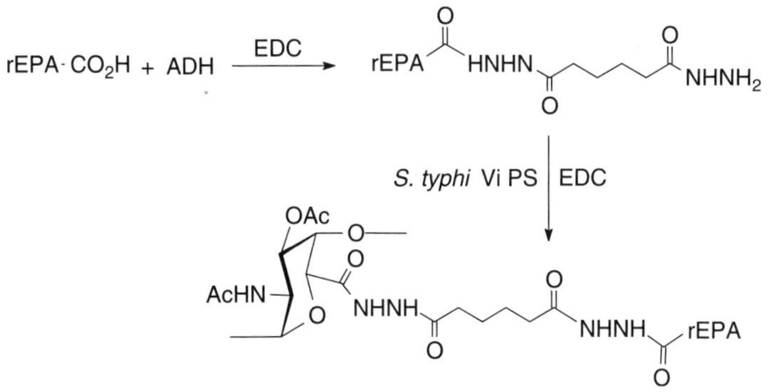

Scheme 2.8

acid groups from both protein and Vi polysaccharide using EDC. The safety and immunogenicity of Vi conjugates prepared by the above methods were compared, and Vi conjugates with ADH linkers were found to be more immunogenic in animals [47] and in humans [48] than those prepared with SPDP. Although ADH can be first coupled with either carboxylic acid from protein or Vi polysaccharide, the human Vi polysaccharide-rEPA (a recombinant exoprotein of *Pseudomonas aruginosa*) vaccine was prepared by ADH activation of rEPA prior to its coupling to carboxylic acid of Vi polysaccharide because the coupling of the ADH-activated Vi polysaccharide to rEPA was not successful (Scheme 2.8) [47].

2.2.5 Group B *Streptococcus*

Group B *Streptococcus* (GBS) is a gram-positive bacterium that causes meningitis and sepsis, and five serotypes have been identified based on their capsular polysaccharide structures (Ia, Ib, II, III, and V) that account for the majority of cases of neonatal meningitis. Because of their extensive structural mimicry with human tissue antigens, GBS polysaccharides are generally poor immunogens. However, when conjugated to a protein carrier, the GBS polysaccharides produce strong antibody responses in women that could provide protection against infection of neonates.

A unique structural feature of native GBS polysaccharides is the presence of terminal sialic acid in polysaccharides of all five serotypes (see Chapter 4), which depending on the strains, have been reported to have varying degree of *O*-acetylation [49]. However, in the purification of the GBS polysaccharides prior to conjugation, they were treated with a base to remove the contaminating GBS group antigen [50], a process that also results in their de-*O*-acetylation. Fortunately, there is strong accumulated evidence that the *O*-acetyl groups are not immunodominant, and that conjugates made from the de-*O*-acetylated GBS polysaccharides are still highly efficient (see Chapter 4). Because the side chains of sialic acid can be oxidized selectively by periodate to generate a terminal aldehyde groups at C7 or C8, the subsequent conjugation of the aldehyde groups of all five serotypes to the amino groups of protein can thus be achieved by reductive amination. The GBS III polysaccharide

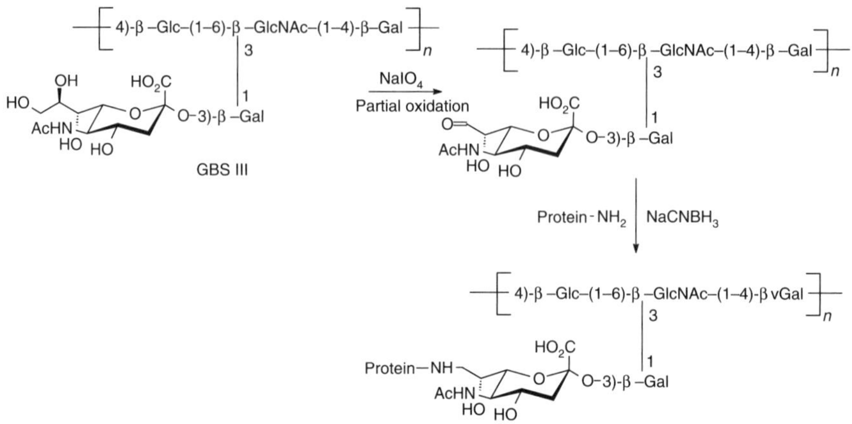

Scheme 2.9

conjugate is illustrated in Scheme 2.9 as a typical example. Since there are multiple aldehyde groups along the polysaccharide, the final conjugates can have varying degrees of crosslinking depending on the degree of oxidation, and too much crosslinking must be avoided as it may disrupt the conformational epitope required for the production of protective antibodies [51, 52]. The degree of oxidation can be measured by gas chromatography–mass spectroscopy (GC–MS) analysis on trimethylsilyl (TMS) derivatives of sialic acid (C-9) and oxidized sialic acids (C-7 and C-8) after they are released from polysaccharides by methanolysis.

Polysaccharide–TT conjugates of GBS Ia/Ib [53–55], II and III [56–60], and V [61, 62] were prepared as shown in Scheme 2.9, and their immunological properties were investigated in animals and humans. Promising preclinical results with GBS capsular polysaccharide conjugate vaccines led to phase 1 and phase 2 clinical trials (see Chapter 4). In all cases the GBS conjugate vaccines were well tolerated and induced a dose-dependent polysaccharide-specific IgG response in healthy nonpregnant adults. Other protein carriers such as α-C protein [63], modified diphtheria toxin [64], and a recombinant duck hepatitis B core antigen [64] were also used for making experimental conjugates.

Because of their large molecular size, the depolymerization of GBS polysaccharides prior to conjugation is preferred, but due to the lability of the terminal sialic acid residues, methods based on thermal, acid hydrolysis and ultrasonic irradiation are inapplicable. One of the first depolymerization methods applied to the type III GBS polysaccharides was the use of endo-β-galactosidase from *Citrobacter freundii* [65], which hydrolyzes the backbone galactopyranoside linkages. However, application of this method was very limited due to the fact that the enzyme preferentially hydrolyzes smaller saccharides, thus resulting in low yields of suitably sized fragment saccharides. Because type II and III GBS polysaccharides contain *N*-acetyl-glucosamine in their backbone, nitrous acid treatment of partially de-*N*-acetylated polysaccharides enables their fragmentation with concomitant formation of a 2,5-anhydro-D-mannose residue, which contains a terminal aldehyde

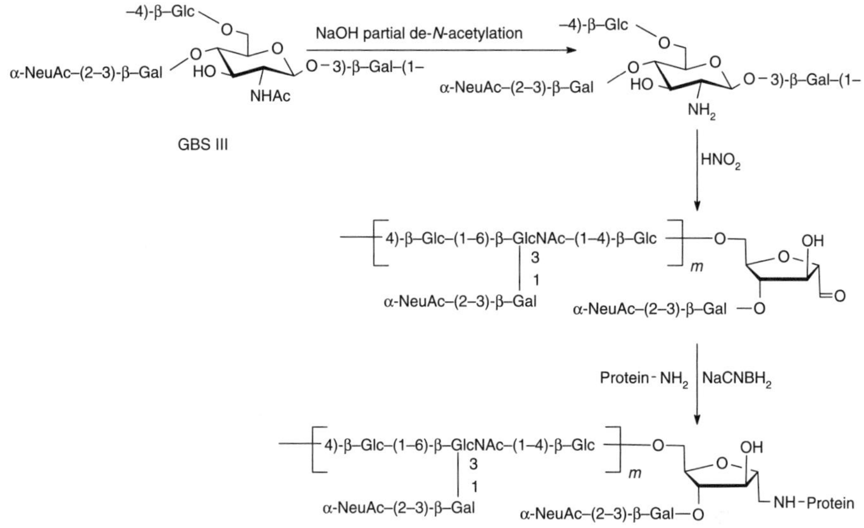

2.2 CAPSULAR POLYSACCHARIDE–PROTEIN CONJUGATES

Scheme 2.10

group ready for conjugation (Scheme 2.10) [66, 67]. The drawback with this method is that it is not applicable to all the GBS polysaccharides and that partial de-*N*-acetylation of the sialic acid residues also occurs, making it unsuitable for producing small oligosaccharide fragments, such as those having the structural integrity required for epitope mapping [52]. In the case of the type III GBS polysaccharide, this problem can be circumvented by using a sialyltransferase to introduce terminal sialic acid residues to the partially de-*N*-acetylated *S. pneumoniae* 14 polysaccharide prior to nitrous acid treatment, which results in small defined GBS III oligosaccharide repeating units [68].

Another potentially useful depolymerization method used for polysaccharides is controlled ozonolysis, and because backbone β-pyranoside linkages are the most susceptible to ozone treatment, this method is applicable to all the GBS polysaccharides. However, because all the GBS polysaccharides contain more than one backbone β-pyranoside, the method is not suitable for producing small defined saccharide fragments. Ozonolysis can be carried out in either aqueous solution [69] or in organic solvent following acetylation of the polysaccharides [70]. For example, in the depolymerization of the type III GBS polysaccharide shown in Scheme 2.11, the breakdown of the β-galactosidic linkage in organic solvent leads to the formation of an aldonic ester, which can then be spontaneously functionalized in the subsequent de-*O*-acetylation step using ethylene diamine. Thus as depicted in Scheme 2.11, a saccharide amide is formed having a free amine, which by means of a squarate linker was conjugated to a carrier protein [71].

2.2.6 *Staphylococcus aureus* Types 5 and 8

Staphylococcus aureus causes several diseases by different pathogenic mechanisms, of which the most frequent and serious is bacteremia, which causes complications

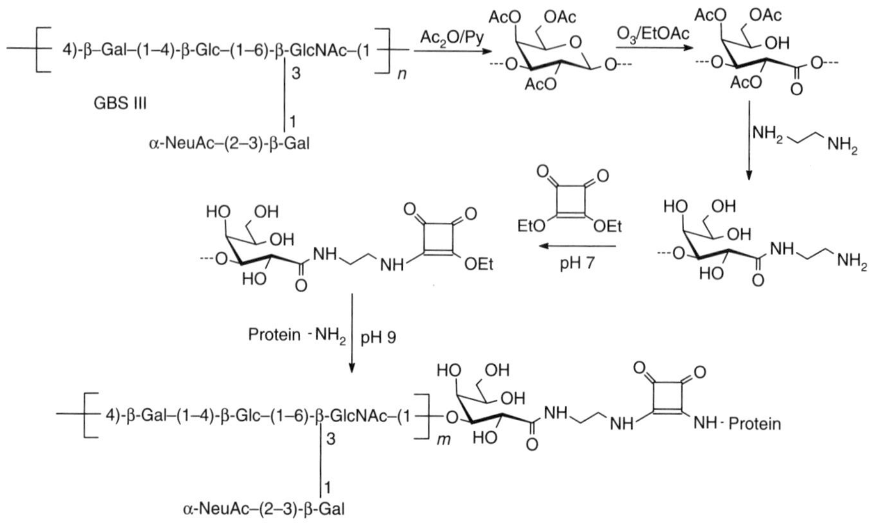

Scheme 2.11

in hospitalized patients. Of the more than 11 types, type 5 and 8 comprise the majority of clinical isolates. The capsular polysaccharide structures of type 5 and 8 are shown in Figure 2.1 [72, 73]. Because both polysaccharides have 2-acetamido-2-deoxy-D-mannuronic acid (ManNAcA) in their repeating units, the conjugation methods developed for *S. typhi* Vi polysaccharide (see Scheme 2.7) are also applicable to both *S. aureus* type 5 and 8 polysaccharides. Similar to Vi polysaccharide conjugates, the *S. aureus* type 8 capsular polysaccharide when coupled to carrier proteins such as rEPA and DT using a ADH linker, elicited higher levels of polysaccharide-specific antibodies, especially after the first immunization, than did those prepared with an SPDP linker. Similar levels of rEPA antibodies were elicited by both conjugates, and higher levels of DT antibodies were elicited by the conjugate prepared with SPDP than the one prepared with ADH [74]. In adult volunteers both *S. aureus* type 5 and 8 capsular polysaccharide rEPA conjugate vaccines induced polysaccharide-specific IgG antibodies with opsonophagocytic activities [75]. Depolymerization of the polysaccharides prior to conjugation was achieved by ultrasonic irradiation.

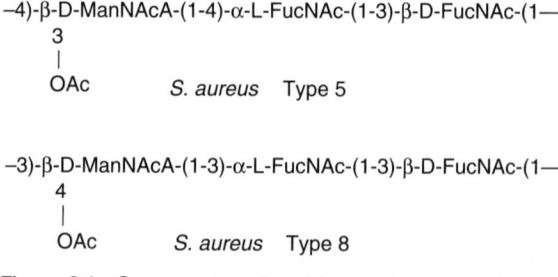

Figure 2.1 *S. aureus* type 5 and 8 capsular polysaccharide.

2.3 LIPOPOLYSACCHARIDE (LPS) AND LIPOOLIGOSACCHARIDE (LOS) CONJUGATES

Lipopolysaccharide is a major constituent of the outer membrane of the human pathogenic gram-negative bacteria that include *Haemophilus* and *N. meningitidis* and enteric pathogens such as *Salmonella*, *Shigella*, pathogenic *Escherichia coli*, and *Vibrio cholerae*. LPSs have four domains as illustrated in Figure 2.2. Lipid A contains two glucosamine residues with *O*- and *N*-acyl groups that are attached to the outer membrane of the bacteria, a conserved inner core region of 2-keto-3-deoxyoctulosonic acid (KDO) and heptose residues, an outer core oligosaccharide, and an *O*-specific polysaccharide having oligosaccharide repeating units. The *O*-specific polysaccharide acts as a capsule and is an essential virulence factor. Therefore, antibodies raised against *O*-specific polysaccharides have been shown to be protective against homologous strains of bacteria. Basically, *O*-polysaccharide conjugates can be made in the same way as those for capsular polysaccharides using methods such as cyanogen bromide activation, ADH activation, and reductive amination [76, 77]. However, some bacteria such as nontypable *Haemophilus* and *N. meningitidis* lack *O*-specific polysaccharides. In these cases, outer and inner core structures, termed LOS, are being investigated as possible vaccine candidates.

O-specific polysaccharides (O-SP) can be obtained from LPS by acid or hydrazine treatment. Partial acid hydrolysis cleaves the KDO glycosidic linkage producing O-SP having a partial core structure, whereas hydrazine treatment only removes *O*-acyl and *N*-acyl groups from the LPS (deLPS). Both procedures eliminate the toxicity caused by lipid A (fever, inflammation, and shock), but acid hydrolysis has the potential to hydrolyze other acid-labile glycosidic linkages in the structure.

2.3.1 *Escherichia coli* O157

Escherichia coli O157 is one of the most pathogenic enteric bacteria, which causes bloody diarrhea, leading occasionally to kidney failure, and is especially virulent in young children and elderly people. Currently, there is no vaccine available. In order to develop a conjugate vaccine *E. coli* O157 O-SP was conjugated to rEPA and other proteins [78] as well as to Shiga toxins [79] by the ADH method (see

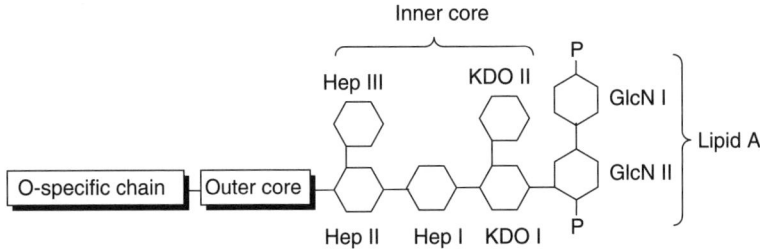

Figure 2.2 Schematic presentation of lipopolysaccharide (LPS).

Scheme 2.2). Alternatively, a conjugate was also made through 1-cyano-4-dimethylaminopyridinium tetrafluoroborate (CDAP) cyanogen activation of the O-SP followed by direct coupling with a protein amino group without a linker [80]. The conjugates elicited high titers of serum bactericidal activity in adults [81].

Although both CNBr and CDAP are able to activate polysaccharides to form reactive intermediates, CNBr is highly toxic and volatile when compared to CDAP. A recent study compared the two activation reagents in the conjugation of the group A *N. meningitidis* polysaccharide (GAMP) using ADH as a linker [82]. In the first experiments, ADH was bound to GAMP activated by either CNBr or CDAP, and the resultant activated polysaccharides were then coupled to bovine serum albumin (BSA) in the presence of EDC to form conjugates. In the second experiments, ADH was bound to BSA using EDC prior to coupling to CNBr- and CDAP-activated GAMP to form conjugates. When the properties of the conjugates were compared, it was found that, in both cases, CDAP activation resulted in a higher polysaccharide loading (35–40% vs. 5–20%) and elicited higher IgG anti-GAMP bactericidal antibodies in mice.

2.3.2 *Vibrio cholerae* O1 and O139

Protection against cholera has been correlated with the level of serum vibriocidal antibodies, the specificity of which are mostly to LPS. Because of the epidemic and pandemic potential of *V. cholerae* O1 and O139, there is an urgent need to vaccinate against these newly emerged *V. cholerae* serogroups.

Conjugates of hydrazine and acid detoxified *V. cholerae* O1 LPS with cholera toxin (CT) and other proteins were synthesized by two conjugation methods [83]. One method utilized SPDP to thiolate both the protein and the terminal amino groups on the deLPS, followed by the formation of a disulfide linker between the two activated molecules in a single-point conjugation (see Scheme 2.12). The other method utilized CNBr activation of deLPS followed by reaction with ADH and an EDC-mediated coupling of ADH to the carboxylic acid groups of proteins, which

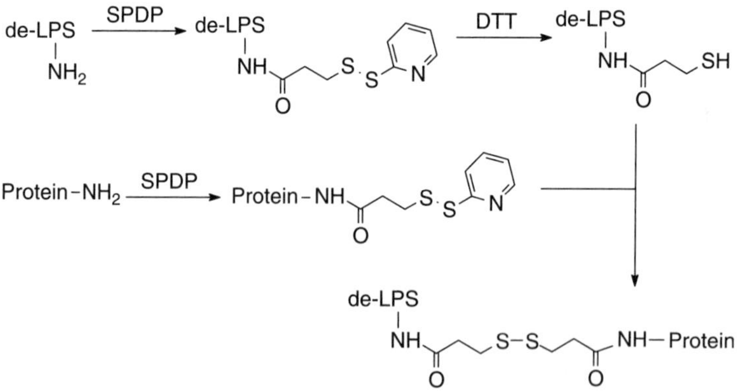

Scheme 2.12

resulted in multiple coupling of deLPS to protein (see Scheme 2.2). Conjugates obtained by multiple coupling of the deLPS to protein were more immunogenic than those with single-point attachment.

Conjugate vaccines prepared from LPS of *V. cholerae* O1, serotype Inaba, to CT variants CT-1 and CT-2 were tested in volunteers [84] and compared with the licensed whole-cell vaccine. The vibriocidal activities of sera from patients immunized with vaccines could be absorbed (>75%) by LPS but not by either CT-1 or CT-2. In addition, only the conjugates induced IgG vibriocidal antibodies, which persisted longer than those elicited by the whole-cell vaccine.

A conjugate made by linking the LPS core of *V. cholerae* O139 to TT was also prepared by ADH activation of the O-SP followed by EDC-mediated condensation (see Scheme 2.2) [85]. The immunologic properties of the conjugate were tested in BALB/c mice, and the conjugate elicited high levels of IgG antibodies. These antibodies, which recognized both the capsular polysaccharide and LPS from *V. cholerae* O139, were vibriocidal and were protective in a neonatal mouse model of cholera infection. Because *V. cholerae* O139 is also encapsulated, conjugates consisting of its capsular polysaccharide linked to the recombinant diphtheria toxin mutant (CRMH21G) were prepared using CDAP-activated polysaccharide, and ADH as a linker (see Scheme 2.2) [86]. The conjugates elicited IgG antibodies with vibriocidal activity and high levels of serum diphtheria toxin IgG in animals, but in contrast to *V. cholerae* O1 they induced vibriocidal antibodies that were mostly of the IgM class.

2.3.3 Shigella dysenteriae Type 1, sonnei and flexneri 2a

Shigellosis is a serious disease that is a major cause of growth retardation and death in children in developing countries, and the bacteria are now becoming resistant to most antibiotics. However, it has been demonstrated that serum IgG antibodies to its O-SP confer immunity to *Shigella*. Conjugates of the O-SP of *Shigella dysenteriae* type 1 [87], *Shigella sonnei*, and *Shigella flexneri 2a* [88] were prepared essentially by the method illustrated in Scheme 2.2. However, a modified method was used for the conjugation of *Shigella flexneri 2a* O-SP to CRM9 and to rEPA. The CNBr-activated O-SPs were first reacted with ADH. Instead of direct coupling to the carboxylate groups of protein, the protein was first treated with succinic anhydride at neutral pH to convert its amino groups to amides, and to introduce into it additional carboxylate groups, which improved its coupling to the ADH-activated polysaccharides (Scheme 2.13) [89]. The conjugate of *S. flexneri 2a* O-SP to CRM9 elicited better immune responses in mice, and the safety and immunogenicity of the conjugates have been evaluated in 1 to 4-year-old children [90].

In general, the conjugates were stable and elicited high levels of boostable IgG O-SP antibodies in young mice when injected subcutaneously in saline at 1/10 of the proposed human dose. Adsorption onto alum or concurrent administration with monophosphoryl lipid A (MPL) enhanced both the IgG and IgM antibody responses to the O-SP of the conjugate, and both the nonadsorbed and adsorbed conjugates elicited higher levels of IgG than of IgM antibodies.

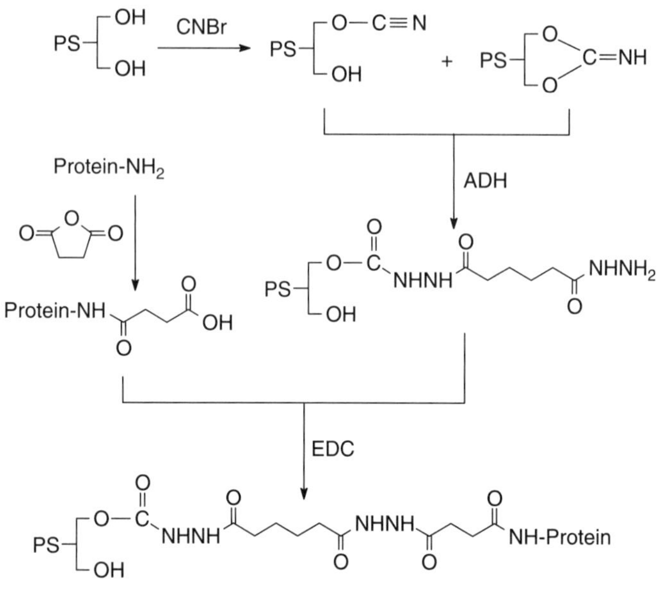

Scheme 2.13

2.3.4 *Neisseria meningitidis* and Nontypeable *Haemophilus influenzae*

Some pathogens such as *N. meningitidis* and nontypable *Haemophilus* lack O-SP, and, because their core structures are relatively conserved, they have been explored as vaccine candidates. Mild acid hydrolysis of LOS removes the lipid A moiety by cleavage of the susceptible KDO glycosidic linkages. Although direct coupling between the ketone group of KDO and amino group of protein by reductive amination is possible [91, 92], the carboxylic acid group of KDO in the LOS has also been more frequently used. Because there is evidence that phosphoethanolamine (PEA) substituents of the meningococcal LOS are required for the production of protective antibodies, a coupling method that keeps PEA substituents intact was developed by Verheul et al. [93]. The carboxylic acid group of the terminal KDO was utilized to introduce a thiol group into meningococcal immunotypes L2 and L3, 7, and 9 lipopolysaccharide-derived oligosaccharides. The thiol-group-containing oligosaccharides were subsequently coupled to bromoacetylated TT (Scheme 2.14). Both the immunotype L2 and immunotype L3, 7, 9 conjugates evoked high IgG antibody titers in rabbits following the first booster injection. These conjugates also displayed an ability to induce long-lasting IgG antibody levels, which could be detected 9 months after a third booster injection. A conjugate prepared from the dephosphorylated L3, 7, 9 oligosaccharides evoked a significantly lower IgG response than a similar PEA-containing conjugate. Furthermore, antisera elicited with the complete bacteria contained antibodies directed against PEA-containing epitopes, which stressed the importance of the inclusion of unmodified PEA groups in meningococcal

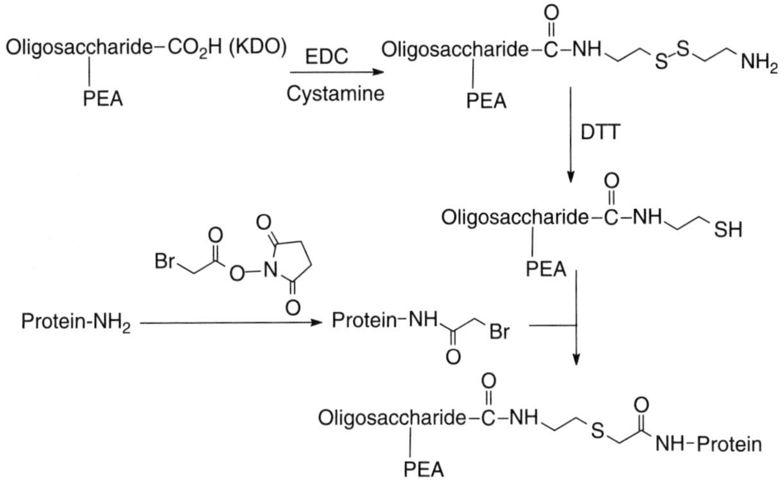

2.3 LIPOPOLYSACCHARIDE (LPS) AND LIPOOLIGOSACCHARIDE (LOS) CONJUGATES

Scheme 2.14

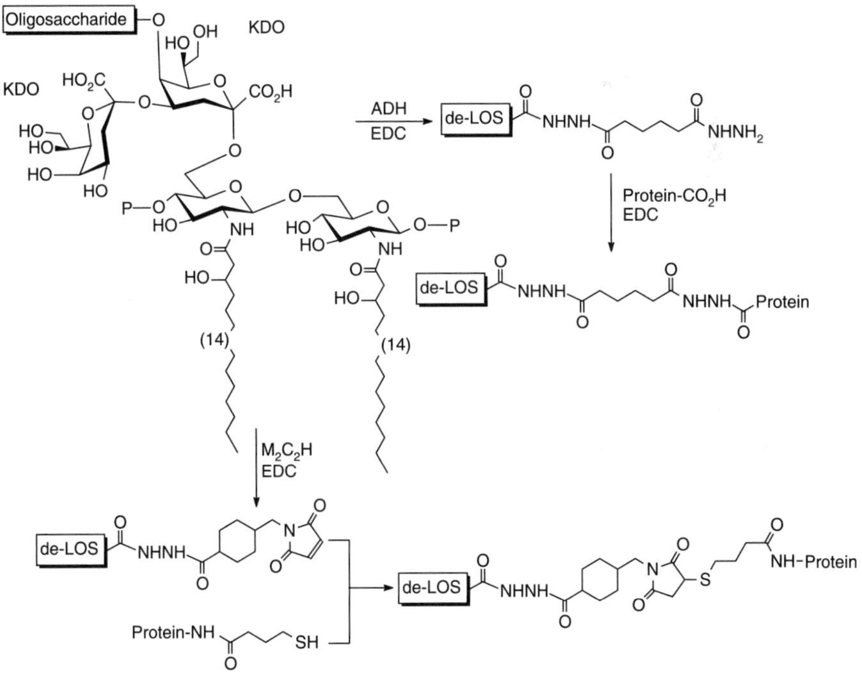

Scheme 2.15

LOS–protein conjugates. The procedure therefore could prove to be useful in the development of a vaccine against group B *N. meningococci.*

The LOS containing lipid A may also been conjugated to protein via the carboxylic acid group of KDO, but prior to conjugation, detoxification of the LOS using anhydrous hydrazine is required to remove *O*-acyl groups of lipid A. Gu et al. coupled the *N. meningitidis,* nontypeable *H. influenzae,* and *M. catarrhalis* LOS to TT and other high-molecular-weight proteins [94–102] by coupling ADH to the carboxylate group of KDO, followed by EDC-mediated conjugation to the carboxylic acid groups of proteins (Scheme 2.15). These conjugates were tested in animals and humans for protection against otitis media. The nontypeable *H. influenzae* LOS–TT conjugate

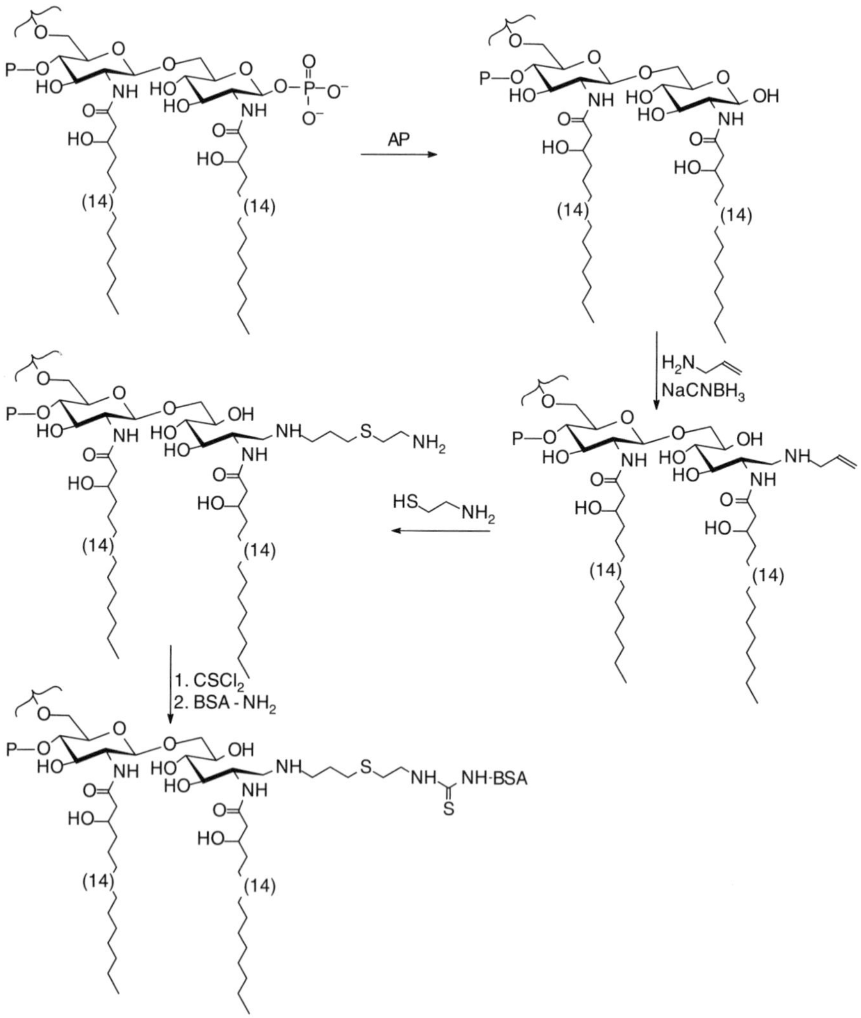

Scheme 2.16

induced LOS-specific antibodies in healthy adults but did not yield significant bactericidal activity in vitro [98]. Other linkers were also successfully used for preparing *N. meningitis* LOS conjugates (see Scheme 2.15) [103] where thiolated protein was efficiently coupled with maleimide-derived LOS.

Because KDO could be an important part of the LOS epitope that induces protective antibodies, it is advantageous to use the reducing terminal lipid A for conjugation. Brade et al. [104] were able to remove the 1-*O*-phosphate group from the lipid A of a recombinant *E. coli* LPS that expressed *Chlamydia tracchomatis* KDO transferase by using alkaline phosphatase (AP). The terminal hemiacetal released was then coupled to allylamine by reductive amination, which was followed by spacer elongation with cystamine. This molecule was then treated with thiophosgene prior to its being coupled with the amino groups of protein (BSA) (Scheme 2.16). In order to prepare nontypeable *H. influenzae* LOS conjugates via the lipid A reducing end, the AP-treated deLOS was coupled with cystamine at the reducing end by reductive amination by Cox et al. (unpublished results). DTT treatment reduced the disulfide to a thiol group, which was conjugated to bromoacetylated protein to give desired conjugates (see Scheme 2.17).

Direct coupling of the reducing end of glucosamine with protein amino groups by reductive amination was also carried out (Scheme 2.18) [103, 105]. Using this method *N. meningitidis* LOS of the L3 and L7 serotype were conjugated to TT and CRM197, respectively. Studies in animal models demonstrated that conjugates with terminal

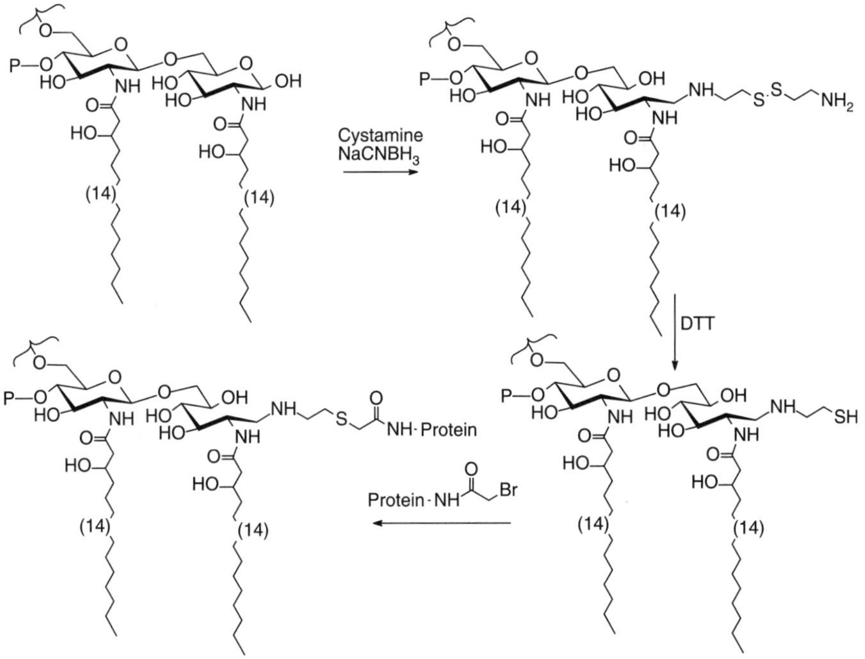

Scheme 2.17

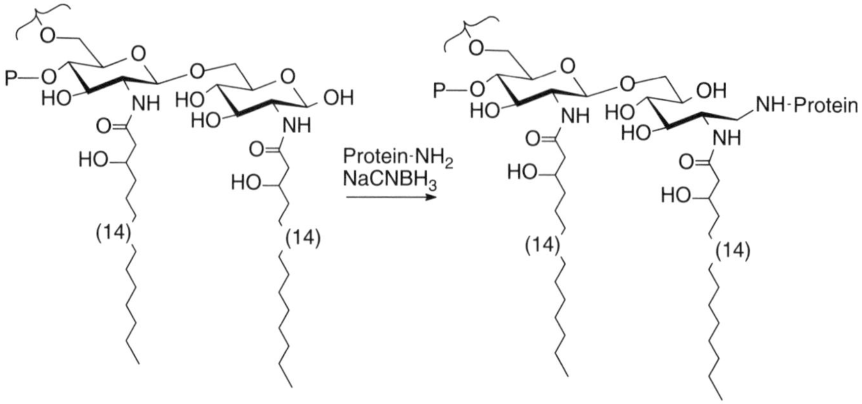

Scheme 2.18

coupling via glucosamine produced superior immune responses to those coupled through KDO, confirming the importance of the KDO residues [103, 105].

2.4 TOTAL SYNTHETIC GLYCOCONJUGATE VACCINES

Due to recent advances in synthetic methodology, pure oligosaccharides can now be obtained in satisfactory yields, and the possibility of using totally synthetic glycoconjugate vaccines has become feasible. The first and only licensed synthetic oligosaccharide vaccine was against Hib and was produced in Cuba. Bencomo et al. [106, 107] synthesized an eight-repeating unit oligosaccharide having an amine spacer attached to phosphate at its nonreducing end. The amino group was coupled to a maleimide-activated ester, which then allowed the saccharide to be conjugated to thiolated proteins (Scheme 2.19). When compared to depolymerized Hib polysaccharide conjugate vaccines, it was shown that the synthetic oligosaccharide versions induced similar antibody response patterns in terms of titer, specificity, and functional capacity.

Prior to the Cuban Hib vaccine, oligosaccharides consisting of trimeric and tetrameric repeating units were coupled to proteins (TT and DT) via a thioether linkage (Scheme 2.20) [108]. Unlike the Cuban vaccine, these included a linker at the reducing end. De-S-acetylation and coupling to bromoacetylated protein were achieved in a one-pot reaction, in the presence of hydroxylamine, without isolating the thiol intermediate. Both conjugates elicited anti-PRP antibody responses with increasing IgG/IgM ratios in adult mice and monkeys, but the antibody responses induced by the trimer conjugates were lower than those of the corresponding tetramer conjugates. In adult monkeys it was also observed that these antibody responses were equivalent to those induced by a commercial Hib–CRM197 conjugate vaccine.

Synthesis of oligosaccharide conjugate vaccines against *S. pneumoniae* has been another focus, including types 3, 6B, 14, 17F, and 23F. Type 3 di-, tri-, and tetra-oligosaccharides with 3-aminopropyl spacers were synthesized and conjugated to

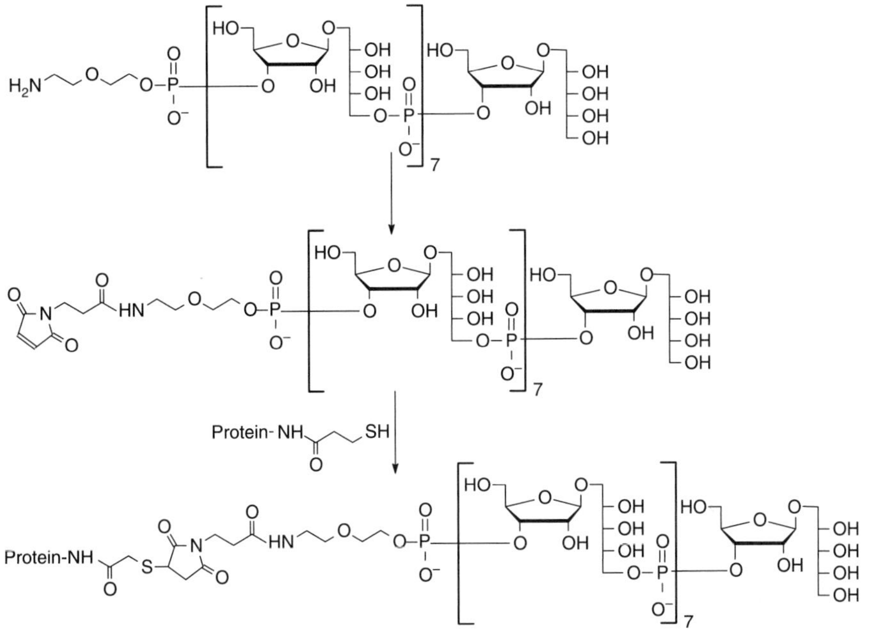

2.4 TOTAL SYNTHETIC GLYCOCONJUGATE VACCINES 77

Scheme 2.19

CRM197, KLH, and TT using squarate as a crosslinker (Scheme 2.21) [109–111]. The conjugates elicited polysaccharide-specific IgG antibodies in mice, and those having polysaccharide-specific antibodies survived intraperitoneal challenge with type 3 pneumococcoci.

Conjugates of synthetic *S. pneumoniae* 6B capsular polysaccharide repeating units, for example, phosphate-containing disaccharide (Rha-ribitol-*P*-), trisaccharide

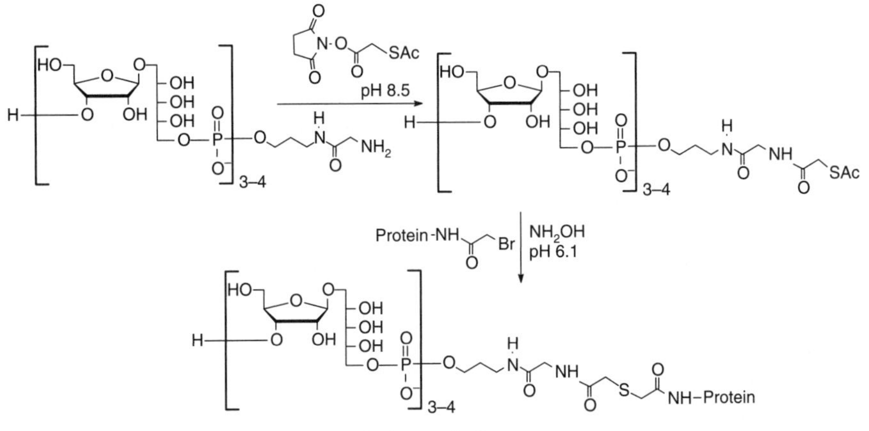

Scheme 2.20

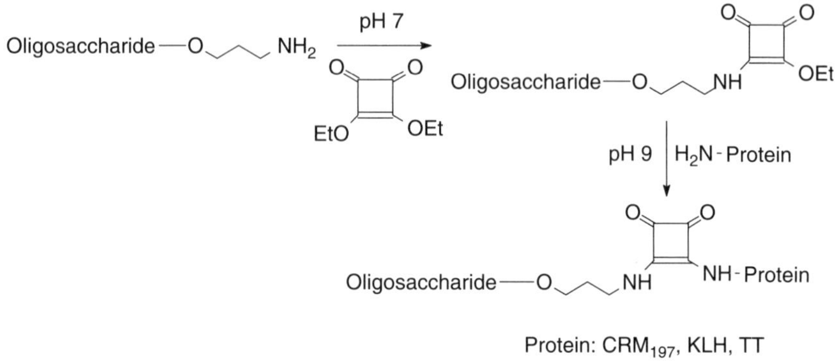

Scheme 2.21

(ribitol-*P*-Gal-Glc-), and tetrasaccharide (Rha-ribitol-*P*-Gal-Glc-) linked to keyhole limpet hemocyanin (KLH) were prepared for immunogenicity and protection studies [112]. The conjugation included *S*-acetylation of the saccharide amino group and bromoacetylation of amino groups of protein (see Scheme 2.20). Similar methods were also used in the synthesis of type 17F [113] and 23F conjugates [114]. The method provides an alternate way of coupling between two amino groups. It was found that the disaccharide and tetrasaccharide *S. pneumoniae* 6B conjugates contain epitopes capable of inducing homologous antibodies in rabbits and mice, which were fully protective in mice.

In addition to coupling through a squarate linker (see Scheme 2.21) synthetic pneumococcal type 14 oligosaccharides, consiting of a tetrasaccharide corresponding to one repeating unit of the Pn14PS, a hexasaccharide, and an octasaccharide fragment obtained by Pn14PS depolymerization, were conjugated to CRM197 using an adipic acid diester linker by a procedure similar to the one illustrated in Scheme 2.6. The anomeric allyl groups of the oligosaccharides were converted to β-aminoethylthio-extended glycosides by the light-induced free-radical addition of cysteamine suitable for coupling to a protein carrier (see Scheme 2.16) [115]. The conjugate with the tetrasaccharide chains induced anti-Pn14PS antibodies when injected subcutaneously into mice, the antibody titers of which increased with oligosaccharide loading, and which were higher than those prepared using a squarate spacer (see Scheme 2.21). Thus, the CRM197 conjugate of a single repeat unit of the Pn14PS could be a potential vaccine candidate [116].

The use of squarate as a crosslinker was also employed in the synthesis of BSA conjugates of *V. cholerae* O1 hexasaccharides related to serotype Ogawa and Inaba (see Scheme 2.21) [117]. Protective immunity was observed for *V. cholerae* O1 Ogawa [118] but not for *V. cholerae* O1 Inaba [119].

Two modified methods to couple oligosaccharides to proteins through their reducing end were introduced by Pozsgay et al. Instead of EDC-mediated coupling between hydrazide and protein carboxylic acid groups, oligosaccharides with an increasing number of repeating units of *Shigella dysenteriae* type 1 O-SP were conjugated to protein human serum albumin (HSA) by reductive amination through a latent aldehyde linker (Scheme 2.22) [11, 120]. The synthetic vaccine elicited higher levels

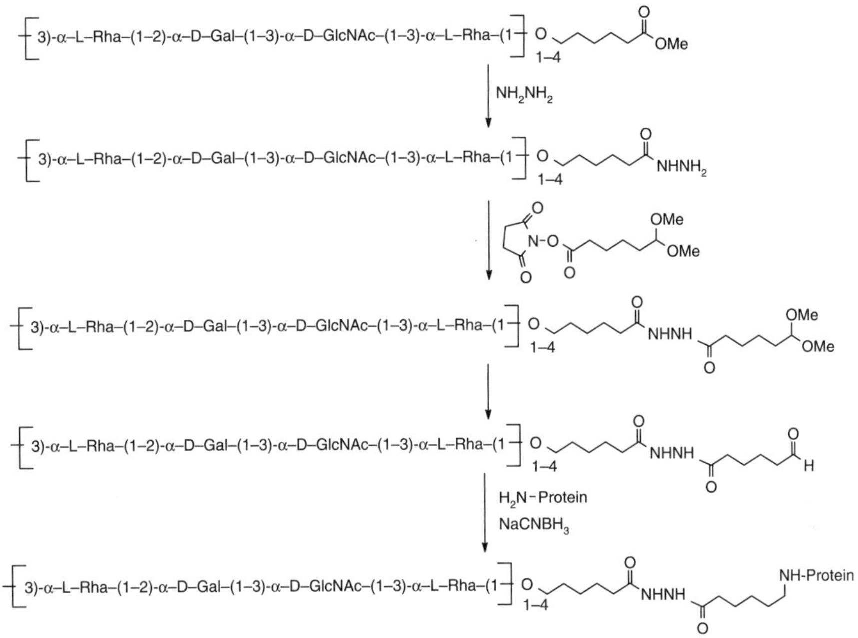

Scheme 2.22

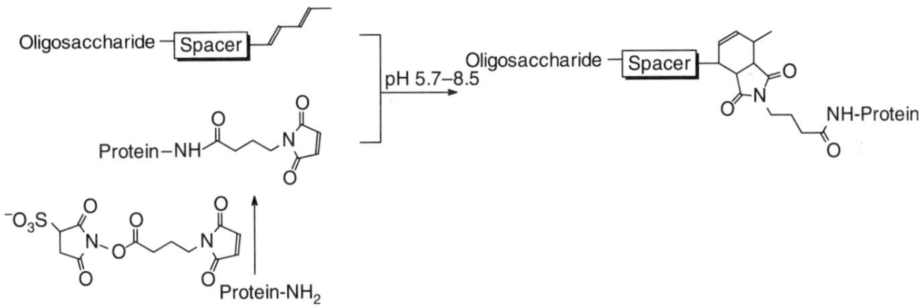

Scheme 2.23

of serum IgG LPS-specific antibodies in mice than the O-SP [121]. Another conjugation method employed a Diels–Alder cycloaddition under neutral conditions in water, where the synthetic oligosaccharide was functionized with a conjugated diene and the protein with an activated double bond (Scheme 2.23) [122, 123].

REFERENCES

1. Gotschlich EC, Liu TY, Artenstein MS (1969) Human immunity to the *meningococcus*. 3. Preparation and immunochemical properties of the group A, group B, and group C meningococcal polysaccharides. *J. Exp. Med.* 129:1349–1365.

2. Goldschneider I, Gotschlich EC, Artenstein MS (1969) Human immunity to the *meningococcus*. I. The role of humoral antibodies. *J. Exp. Med.* 129:1307–1326.
3. Gotschlich EC, Goldschneider I, Lepow ML, Gold R (1977) The immune responses to bacterial polysaccharides in man. In Haber E, Krause RM, Eds. *Antibodies in Human Diagnosis and Therapy*. Raven, New York, pp. 391–402.
4. Peltola H, Häyhty H, Sivonen A, Mäkelä PH (1977) *Haemophilus influenzae* type b capsular polysaccharide vaccine in children: A double blind field study of 100,000 vaccine in children 3 months to 5 years of age in Finland. *Pediatrics* 60:730–737.
5. Goebel WF, Avery OT (1931) Chemo-immunological studies on conjugated carbohydrate-proteins. IV. The synthesis of p-aminobenzyl ether of the soluble specific substance of type III *pneumococcus* and its coupling with protein. *J. Exp. Med.* 54:431–436.
6. Avery OT, Goebel WF (1931) Chemo-immunological studies on conjugated carbohydrate-proteins: V. The immunological specifity of an antigen prepared by combining the capsular polysaccharide of type III *pneumococcus* with foreign protein. *J. Exp. Med.* 54:437–447.
7. Goebel WF (1940) Study on antibacterial immunity induced by artificial antigens. II. Immunity to experimental pneumococcal infection with antigens containing saccharides of synthetic origin. *J. Exp. Med.* 72:33–48.
8. Jennings HJ (1983) Capsular polysaccharides as human vaccines. *Adv. Carbohy. Chem. Biochem.* 41:155–208.
9. Robbins JB, Schneerson R (1990) Polysaccharide protein conjugates: A new generation of vaccines. *J. Infect. Dis.* 161:821–832.
10. Jennings HJ, Sood RK (1994) Synthetic glycoconjugates as human vaccines. In Lee YC, Lee RT, (Eds) *Neoglycoconjugates. Preparation and Applications*. Academic, London, pp. 325–371.
11. Pozsgay V (2001) Oligosaccharide-protein conjugates as vaccine candidates against bacteria. *Adv. Carbohy. Chem. Biochem.* 56:153–199.
12. Anderson PW, Pichichero ME, Insel RA, Betts R, Eby R, Smith DH (1986) Vaccines consisting of periodate-cleaved oligosaccharides from the capsule of *Haemophilus influenzae* type b coupled to a protein carrier: Structural and temporal requirements for priming in the human infant. *J. Immunol.* 137:1181–1186.
13. Anderson PW (1983) Immunogenic conjugates. U.S. Patent 4,673574.
14. Jennings HJ, Lugowski C (1981) Immunochemistry of group A, B, and C meningococcal polysaccharide-tetanus toxoid conjugates. *J. Immunol.* 127:1011–1018.
15. Schneerson R, Barrera O, Sutton A, Robbins JB (1980) Preparation, characterization, and immunogenicity of *Haemophilus influenzae* type b polysaccharide-protein conjugates. *J. Exp. Med.* 152:361–376.
16. Gordon LK (1985) *Haemophilus influenzae* b polysaccharide-diphtheria toxoid conjugate vaccine. U.S. Patent 4,496,538.
17. Gordon LK (1987) *Haemophilus influenzae* B polysaccharide-diptheria toxoid conjugate vaccine. U.S. Patent 4,644,059.
18. Claesson BA, Schneerson R, Trollfors B, Lagergard T, Taranger J, Robbins JB (1990) Duration of serum antibodies elicited by *Haemophilus influenzae* type b capsular polysaccharide alone or conjugated to tetanus toxoid in 18- to 23-month-old children. *J. Pediat.* 116:929–931.

19. Axen R, Porath J, Ernback S (1967) Chemical coupling of peptides and proteins to polysaccharides by means of cyanogen halides. *Nature* 214:1302–1304.
20. Porath J (1974) General methods and coupling procedures. In Jokoby WB, Wilchek M, (Eds) *Methods in Enzymology*. Academic, New York, pp. 13–30.
21. Parikh I, Martin S, Cuatrecasas P (1974) Topics in the methodology of substitution reaction. In Jokoby WB, Wilchek M, Eds. *Methods in Enzymology*, Academic, New York, pp. 77–102.
22. Wilchek M, Miron T, Kohn J (1984) Affinity chromatography. In Jokoby WB, Ed. *Methods in Enzymology*, Academic, New York, pp. 3–55.
23. Kagedal L, Akisanya A (1971) Binding of covalent proteins to polysaccharides bycyanogen bromide and organic cyanates. Preparation of soluble glycine-, insulin- and ampicillin-dextran. *Acta Paediatr Jpn.* 25:1855–1859.
24. Chu CY, Schneerson R, Robbins JB, Rastogi SC (1983) Further studies on the immunogenicity of *Haemophilus influenzae* type b and pneumococcal type 6A polysaccharide-protein conjugates. *Infect. Immun.* 40:245–256.
25. Hilleman MR, Tai JY, Tolman RL, Vella PP (1984) Coupled *H. influenzae* type b vaccine. U.S. Patent 4,459,286.
26. Marburg S, Jorn D, Tolman RL, Arison B, McCauley J, Kniskern PJ, Hagopian A, Vella PP (1986) Bimolecular chemistry of macromolecules: Synthesis of bacterial polysaccharide conjugates with *Neisseria meningitidis* membrane protein. *J. Am. Chem. Soc.* 108:5282–5287.
27. Marburg S, Tolman RL, Kniskern PJ (1987) Covalently-modified polyanionic bacterial polysaccharides, stable covalent conjugates of such polysaccharides and immunogenic proteins with bigeneric spacers, and methods of preparing such polysaccharides and conjugates and of confirming covalency. U.S. Patent 4,695,624.
28. Marburg S, Tolman RL, Kniskern PJ, Miller WJ, Hagopian A, Ip CC, Hennessey Jr JP, Kubek DJ, Burke PD (1997) Pneumococcal polysaccharide conjugate vaccine. U.S. Patent 5,623,057.
29. Schlesinger Y, Granoff DM (1992) Avidity and bactericidal activity of antibody elicited by different *Haemophilus influenzae* type b conjugate vaccines. *J. Am. Med. Assoc.* 267:1489–1494.
30. Anderson PW, Pichichero ME, Stein EC, Porcelli S, Betts RFCDM, Korones D, Insel RA, Zahradnik JM, Eby R (1989) Effect of oligosaccharide chain length, exposed terminal group, and hapten loading on the antibody response of human adults and infants to vaccines consisting of *Haemophilus influenzae* type b capsular antigen unterminally coupled to the diphtheria protein CRM197. *J. Immunol.* 142:2464–2468.
31. Sniadack DH, Schwartz B, Lipman H, Bogaerts J, Butler JC, Dagan R, Echaniz-Avilas G, Lloyd-Evans N, Fenoll A, Girgis NI, Henrichsen J, Klugman K, Lehmann D, Takala AK, Vandepitte J, Gove S, Breiman RF (1995) Potential interventions for the prevention of childhood pneumonia: Geographic and temporal differences in serotype and serogroup distribution of invasive pneumococcal isolates from children—Implications for vaccine strategies. *Pediat. Infect. Dis. J.* 14:503–510.
32. Dagan R, Kayhty H, Wuorimaa T, Yaich M, Bailleux F, Zamir O, Eskola J (2004) Tolerability and immunogenicity of an eleven valent mixed carrier *Streptococcus pneumoniae* capsular polysaccharide-diphtheria toxoid or tetanus protein conjugate vaccine in Finnish and Israeli infants. *Pediat. Infect. Dis. J.* 23:91–98.

33. Dagan R, Goldblatt D, Maleckar JR, Yaich M, Eskola J (2004) Reduction of antibody response to an 11-valent pneumococcal vaccine coadministered with a vaccine containing acellular *Pertussis* components. *Infect. Immun.* 72:5383–5391.
34. Prymula R, Peeters P, Chrobok V, Kriz P, Novakova E, Kaliskova E, Kohl I, Lommel P, Prieels J-P, Schuerman L (2006) Pneumococcal capsular polysaccharides conjugated to protein D for prevention of acute otitis media caused by both *Streptococcus pneumoniae* and non-typable *Haemophilus influenzae*: A randomised double-blind efficacy study. *Lancet* 367:740–748.
35. Maiden MC, Stuard JM (2002) Carriage of serogroup C *meningococci* 1 year after meningococcal C conjugate polysaccharide vaccination. *Lancet* 359:1829–1830.
36. Jessouroun E, Da Silveira IAF, Bastos RC, Frasch CE, Lee JC (2004) Process for preparing polysaccharide-protein vaccines. Patent PCT/US2004/026431 (WO/2005/037320).
37. Richmond P, Borrow R, Goldblatt D, Findlow J, Martin S, Morris R, Cartwright K, Miller E (2001) Ability of 3 different meningococcal C conjugate vaccines to induce immunologic memory after a single dose in UK toddlers. *J. Infect. Dis.* 183:160–163.
38. Costantino P, Viti S, Podda A, Velmonte MA, Nencioni L, Rappuoli R (1992) Development and phase 1 clinical testing of a conjugate vaccine against *meningococcus* A and C. *Vaccine* 10:691–698.
39. Bilukha OO, Rosenstein N (2005) Prevention and control of meningococcal disease: Recommendations of the Advisory Committee on Immunization Practice (ACIP). *MMWR* 54(RR-7):1–21.
40. Jennings HJ, Roy R, Gamian A (1986) Induction of meningococcal group B polysaccharide-specific IgG antibodies in mice by using an *N*-propionylated B polysaccharide-tetanus toxoid conjugate vaccine. *J. Immunol.* 137:1708–1713.
41. Pon RA, Lussier M, Yang QL, Jennings HJ (1997) *N*-Propionylated group B meningococcal polysaccharide mimics a unique bactericidal capsular epitope in group B *Neisseria meningitidis*. *J. Exp. Med.* 185:1929–1938.
42. Jennings HJ (2003) Polysialic acid vaccines. In Wong CH, Ed. *Carbohydrate-Based Drug Discovery*, Vol. 1. Wiley-VCH, Weinheim, pp. 357–380.
43. Fusco PC, Michon F, Tai JY, Blake MS (1997) Preclinical evaluation of a novel group B meningococcal conjugate vaccine that elicits bactericidal activity in both mice and nonhuman primates. *J. Infect. Dis.* 175:364–372.
44. Bruge J, Bouveret LC, Danve B, Rougon G, Schulz D (2004) Clinical evaluation of a group B meningococcal *N*-propionylated polysaccharide conjugate vaccine in adult, male volunteers. *Vaccine* 22:1087–1096.
45. Szu SC, Stone AL, Robbins JD, Schneerson R, Robbins JB (1987) Vi capsular polysaccharide-protein conjugates for prevention of typhoid fever. Preparation, characterization, and immunogenicity in laboratory animals. *J. Exp. Med.* 166:1510–1524.
46. Szu SC, Li XR, Schneerson R, Vickers JH, Bryla D, Robbins JB (1989) Comparative immunogenicities of Vi polysaccharide-protein conjugates composed of cholera toxin or its B subunit as a carrier bound to high- or lower-molecular-weight Vi. *Infect. Immun.* 57:3823–3827.
47. Kossaczka Z, Bystricky S, Bryla DA, Shiloach J, Robbins JB, Szu SC (1997) Synthesis and immunological properties of Vi and di-O-acetyl pectin protein conjugates with adipic acid dihydrazide as the linker. *Infect. Immun.* 65:2088–2093.

48. Kossaczka Z, Lin F-YC, Ho VA, Thuy NTT, Bay PV, Thanh TC, Khiem HB, Trach DD, Karpas A, Hunt S, Bryla DA, Schneerson R, Robbins JB, Szu SC (1999) Safety and immunogenicity of Vi conjugate vaccines for typhoid fever in adults, teenagers, and 2- to 4-year-old children in Vietnam. *Infect. Immun.* 67:5806–5810.
49. Lewis AL, Nizet V, Varki A (2004) Discovery and characterization of sialic acid O-acetylation in group B *Streptococcus*. *Proc. Natl. Acad. Sci. U.S.A.* 101:11123–11128.
50. Michon F, Brisson JR, Dell A, Kasper DL, Jennings HJ (1988) Multiantennary group-specific polysaccharide of group B streptococcus. *Biochemistry* 27:5341–5351.
51. Jennings HJ, Lugowski C, Kasper DL (1981) Conformational aspects critical to the immunospecificity of the type III group B streptococcal polysaccharide. *Biochemistry* 20:4511–4518.
52. Zou W, Mackenzie R, Therien L, Hirama T, Yang QL, Gidney MAJ, Jennings HJ (1999) Conformational epitope of the type III group B *Streptococcus* capsular polysaccharide. *J. Immunol.* 163:820–825.
53. Wessels MR, Paoletti LC, Rodewald AK, Michon F, DiFabio J, Jennings HJ, Kasper DL (1993) Stimulation of protective antibodies against type Ia and Ib group B *Streptococci* by a type Ia polysaccharide-tetanus toxoid conjugate vaccine. *Infect. Immun.* 61:4760–4766.
54. Wessels MR, Paoletti LC, Guttormsen HK, Michon F, D'Ambra AJ, Kasper DL (1998) Structural properties of group B streptococcal type III polysaccharide conjugate vaccines that influence immunogenicity and efficacy. *Infect. Immun.* 66:2186–2192.
55. Baker CJ, Paoletti LC, Wessels MR, Guttormsen HK, Rench MA, Hickman ME, Kasper DL (1999) Safety and immunogenicity of capsular polysaccharide-tetanus toxoid conjugate vaccines for group B streptococcal types Ia and Ib. *J. Infect. Dis.* 179:142–150.
56. Wessels MR, Paoletti LC, Kasper DL, DiFabio J, Michon F, Holme K, Jennings HJ (1990) Immunogenicity in animals of a polysaccharide-protein conjugate vaccine against type III group B *Streptococcus*. *J. Clin. Invest.* 86:1428–1433.
57. Paoletti LC, Wessels MR, Michon F, DiFabio J, Jennings HJ, Kasper DL (1992) Group B *Streptococcus* type II polysaccharide-tetanus toxoid conjugate vaccine. *Infect. Immun.* 60:4009–4014.
58. Kasper DL, Paoletti LC, Wessels MR, Guttormsen HK, Carey VJ, Jennings HJ, Baker CJ (1996) Immune response to type III group B streptococcal polysaccharide-tetanus toxoid conjugate vaccine. *J. Clin. Invest.* 98:2308–2314.
59. Baker CJ, Paoletti LC, Rench MA, Guttormsen HK, Carey VJ, Hickman ME, Kasper DL (2000) Use of capsular polysaccharide-tetanus toxoid conjugate vaccine for type II group B *Streptococcus* in healthy women. *J. Infect. Dis.* 182:1129–1138.
60. Baker CJ, Rench MA, Fernandez M, Paoletti LC, Kasper DL, Edwards MS (2003) Safety and immunogenicity of a bivalent group B streptococcal conjugate vaccine for serotypes II and III. *J. Infect. Dis.* 188:66–73.
61. Wessels MR, Paoletti LC, Pinel J, Kasper DL (1995) Immunogenicity and protective activity in animals of a type V group B streptococcal polysaccharide-tetanus toxoid conjugate vaccine. *J. Infect. Dis.* 171:879–884.
62. Baker CJ, Paoletti LC, Rench MA, Guttormsen H-K, Edwards MS, Kasper DL (2004) Immune response of healthy women to 2 different group B streptococcal type V capsular polysaccharide-protein conjugate vaccines. *J. Infect. Dis.* 189:1103–1112.

63. Gravekamp C, Kasper DL, Paoletti LC, Madoff LC (1999) Alpha C protein as a carrier for type III capsular polysaccharide and as a protective protein in group B streptococcal vaccines. *Infect. Immun.* 67:2491–2496.
64. Paoletti LC, Peterson DL, Legmann R, Collier RJ (2002) Preclinical evaluation of group B streptococcal polysaccharide conjugate vaccines prepared with a modified diphtheria toxin and a recombinant duck hepatitis B core antigen. *Vaccine* 20:370–376.
65. Wessels MR, Munoz A, Kasper DL (1987) A model of high-affinity antibody binding to type III group B *Streptococcus* capsular polysaccharide. *Proc. Natl. Acad. Sci. U.S.A.* 84:9170–9174.
66. Michon F, Uitz C, Tai JY (2002) Antigenic group B *Streptococcus* type II and type III polysaccharide fragments having a 2,5-anhydro-D-mannose terminal structure and conjugate vaccine thereof. U.S. Patent 6,372,222.
67. Michon F, Uitz C, Tai JY (2003) Method of immunization using a group B *Streptococcus* type II and type III polysaccharide conjugate vaccine. U.S. Patent 6,602,508.
68. Zou W, Laferriere CA, Jennings HJ (1998) Oligosaccharide fragments of the type III group B streptococcal polysaccharide derived from *S. pneumoniae* type 14 capsular polysaccharide by a chemoenzymatic method. *Carbohydr. Res.* 309:297–301.
69. Wang Y, Hollingsworth RI, Kasper DL (2001) Method for generating antibodies to saccharide fragments. U.S. Patent 6,274,144.
70. Wang Y, Hollingsworth RI, Kasper DL (1998) Ozonolysis for selectively depolymerizing polysaccharides containing ß-aldosidic linkages. *Proc. Natl. Acad. Sci. U.S.A.* 95:6584–6589.
71. Wang JY, Chang AHC, Guttormsen H-K, Rosas AL, Kasper DL (2003) Construction of designer glycoconjugate vaccines with size-specific oligosaccharide antigens and site-controlled coupling. *Vaccine* 21:1112–1117.
72. Moreau M, Richards JC, Fournier JM, Bryla DA, Karakawa WW, Vann WF (1990) Structure of the type 5 capsular polysaccharide of *Staphylococcus aureus*. *Carbohydr. Res.* 201:285–297.
73. Fournier JM, Vann WF, Karakawa WW (1984) Purification and characterization of *Staphylococcus aureus* type 8 capsular polysaccharide. *Infect. Immun.* 45:87–93.
74. Fattom A, Shiloach J, Bryla D, Fitzgerald D, Pastan I, Karakawa WW, Robbins JB, Schneerson R (1992) Comparative immunogenicity of conjugates composed of the *Staphylococcus aureus* type 8 capsular polysaccharide bound to carrier proteins by adipic acid dihydrazide or *N*-succinimidyl-3-(2-pyridyldithio)propionate. *Infect. Immun.* 60:584–589.
75. Fattom A, Schneerson R, Watson DC, Karakawa WW, Fitzgerald D, Pastan I, Li X, Shiloach J, Bryla DA, Robbins JB (1993) Laboratory and clinical evaluation of conjugate vaccines composed of *Staphylococcus aureus* type 5 and type 8 capsular polysaccharides bound to *Pseudomonas aeruginosa* recombinant exoprotein A. *Infect. Immun.* 61:1023–1032.
76. Robbins JB, Chu CY, Schneerson R (1992) Hypothesis for vaccine development: Protective immunity to entric diseases caused by nontyphoidal salmonellae and shigellae may be conferred by serum IgG antibodies to the O-specific polysaccharide of their lipopolysaccharides. *Clin. Infect. Dis.* 15:346–361.
77. Cryz SJ, Furer EP (1994) *Escherichia coli* O-polysaccharide-protein conjugate vaccine. U.S. Patent 5,370,872.

78. Konadu E, Robbins JB, Shiloach J, Bryla DA, Szu SC (1994) Preparation, characterization, and immunological properties in mice of *Escherichia coli* O157 O-specific polysaccharide-protein conjugate vaccines. *Infect. Immun.* 62:5048–5054.
79. Konadu E, Donohue-Rolfe A, Calderwood SB, Pozsgay V, Shiloach J, Robbins JB, Szu SC (1999) Syntheses and immunologic properties of *Escherichia coli* O157 O-specific polysaccharide and shiga toxin 1 B subunit conjugates in mice. *Infect. Immun.* 67:6191–6193.
80. Lees A, Nelson BL, Mond JJ (1995) Activation of soluble polysaccharides with 1-cyano-4-dimethylaminopyridinium tetrafluoroborate for use in protein-polysaccharide conjugate vaccines and immunological reagents. *Vaccine* 14:190–198.
81. Konadu EY, Parke JC, Tran HT, Bryla DA, Robbins JB, Szu SC (1998) Investigational vaccine for *Escherichia coli* O157: Phase I study of O157 O-specific polysaccharide *Pseudomonas aeruginosa* recombinant exoprotein A conjugates in adults. *J. Infect. Dis.* 177:383–387.
82. Jin Z, Chu C, Robbins JB, Schneerson R (2003) Preparation and characterization of group A meningococcal capsular polysaccharide conjugates and evaluation of their immunogenicity in mice. *Infect. Immun.* 71:5115–5120.
83. Gupta RK, Szu SC, Finkelstein RA, Robbins JB (1992) Synthesis, characterization, and some immunological properties of conjugates composed of the detoxified lipopolysaccharide of *Vibrio cholerae* O1 serotype Inaba bound to cholera toxin. *Infect. Immun.* 60:3201–3208.
84. Gupta RK, Taylor DN, Bryla DA, Robbins JB, Szu SSC (1998) Phase 1 evaluation of *Vibrio cholerae* O1, serotype Inaba, polysaccharide-cholera toxin conjugates in adult volunteers. *Infect. Immun.* 66:3095–3099.
85. Boutonnier A, Villeneuve S, Nato F, Dassy B, Fournier JM (2001) Preparation, immunogenicity, and protective efficacy, in a murine model, of a conjugate vaccine composed of the polysaccharide moiety of the lipopolysaccharide of *Vibrio cholerae* O139 bound to tetanus toxoid. *Infect. Immun.* 69:3488–3493.
86. Kossaczka Z, Shiloach J, Johnson V, Taylor DN, Finkelstein RA, Robbins JB, Szu SC (2000) *Vibrio cholerae* O139 conjugate vaccines: Synthesis and immunogenicity of *V. cholerae* O139 capsular polysaccharide conjugates with recombinant diphtheria toxin mutant in mice. *Infect. Immun.* 68:5037–5043.
87. Chu CY, Liu BK, Watson D, Szu SC, Bryla D, Shiloach J, Schneerson R, Robbins JB (1991) Preparation, characterization, and immunogenicity of conjugates composed of the O-specific polysaccharide of *Shigella dysenteriae* type 1 (Shiga's bacillus) bound to tetanus toxoid. *Infect. Immun.* 59:4450–4458.
88. Ashkenazi S, Passwell JH, Harlev E, Miron D, Dagan D, Farzan N, Ramon R, Majadly F, Bryla DA, Karpas AB, Robbins JB, Schneerson R, the Israel Pediatric Shigella Study Group (1999) Safety and immunogenicity of *Shigella sonnei* and *Shigella flexneri* 2a O-specific polysaccharide conjugates in children. *J. Infect. Dis.* 179:1565–1568.
89. Pavliakova D, Chu C, Bystricky S, Tolson NW, Shiloach J, Kaufman JB, Bryla DA, Robbins JB, Schneerson R (1999) Treatment with succinic anhydride improves the immunogenicity of *Shigella flexneri* type 2a O-specific polysaccharide-protein conjugates in mice. *Infect. Immun.* 67:5526–5529.
90. Passwell JH, Ashkenazi S, Harlev E, Miron D, Ramon R, Farzam N, Lerner GL, Levi Y, Chu CY, Shiloach J, Robbins JB, Schneerson R (2003) Safety and immunogenicity of

Shigella sonnei-CRM9 and *Shigella flexneri* type 2a-rEPA(succ) conjugate vaccines in one- to four-year-old children. *Pediat. Infect. Dis. J.* 22:701–706.

91. Jennings HJ, Lugowski C, Ashton FE (1984) Conjugation of meningococcal lipopolysaccharide R-type oligosaccharides to tetanus toxoid as route to a potential vaccine against group B *Neisseria meningitidis*. *Infect. Immun.* 43:407–412.

92. Grimmecke HD, Brade H (1998) Studies on the reductive amination of 3-deoxy-D-mannooctulosonic acid (Kdo). *Glycoconjugate J.* 15:555–562.

93. Verheul AF, Braat AK, Leenhouts JM, Hoogerhout P, Poolman JT, Snippe H, Verhoef J (1991) Preparation, characterization, and immunogenicity of meningococcal immunotype L2 and L3,7,9-phosphoethanolamine group-containing oligosaccharide-protein conjugates. *Infect. Immun.* 59:843–851.

94. Gu XX, Tai CM (1993) Preparation, characterization, and immunogenicity of meningococcal lipooligosaccharide-derived oligosaccharide-protein conjugates. *Infect. Immun.* 61:1873–1880.

95. Gu XX, Chen J, Barenkamp SJ, Robbins JB, Tsai CM, Lim DJ, Battey J (1998) Synthesis and characterization of lipooligosaccharide-based conjugates as vaccine candidates for *Moraxella (Branhamella) catarrhalis*. *Infect. Immun.* 66:1891–1897.

96. Gu XX, Tsai CM, Ueyama T, Barenkamp SJ, Robbins JB, Lim DJ (1996) Synthesis, characterization, and immunologic properties of detoxified lipooligosaccharide from nontypeable *Haemophilus influenzae* conjugated to proteins. *Infect. Immun.* 64:4047–4053.

97. Gu XX, Tai CM, Lim DJ, Robbins JB (2001) Conjugate vaccine for nontypeable *Haemophilus influenzae*. U.S. Patent 6,207,157.

98. Gu XX, Rudy SE, Chu CY, McCullagh L, Kim HN, Chen J, Li JP, Robbins JB, Van Waes C, Battey JE (2003) Phase I of a lipooligosaccharide-based conjugate vaccine against nontypeable *Haemophilus influenzae*. *Vaccine* 21:2107–2114.

99. Gu XX, Tsai CM, Lim DJ, Robbins JB (2003) Conjugate vaccine for nontypeable *Haemophilus influenzae*. U.S. Patent 6,607,725.

100. Gu XX, Tai CM (2003) Conjugate vaccine for *Neisseria meningitidis*. U.S. Patent 6,531,131.

101. Yu S, Gu XX (2005) Synthesis and characterization of lipooligosaccharide-based conjugate vaccines for serotype B *Moraxella catarrhalis*. *Infect. Immun.* 73:2790–2796.

102. Yu S, Gu XX (2007) Biological and immunological characteristics of lipooligosaccharide-based conjugate vaccines for serotype C *Moraxella catarrhalis*. *Infect. Immun.* 75:2974–2980.

103. Cox AD, Zou W, Gidney MAJ, Lacelle S, Plested JS, Makepeace K, Wright JC, Coull PA, Moxon ER, Richards JC (2005) Candidacy of LPS-based glycoconjugates to prevent invasive meningococcal disease: Developmental chemistry and investigation of immunological responses following immunization of mice and rabbits. *Vaccine* 23:5045–5054.

104. Müller-Loennies S, Grimmecke D, Brade L, Lindner B, Kosma P, Brade H (2002) A novel strategy for the synthesis of neoglycoconjugates from deacylated deep rough lipopolysaccharides. *J. Endotoxin Res.* 8:295–305.

105. Mieszala M, Kogan G, Jennings HJ (2003) Conjugation of meningococcal lipooligosaccharides through their lipid A terminus conserves their inner epitopes and results in conjugate vaccines having improved immunological properties. *Carbohydr. Res.* 338:167–175.

106. Verez-Bencomo V, Fernandez-Santana V, Hardy E, Toledo ME, Rodriguez MC, Heynngnezz L, Rodriguez A, Baly A, Herrera L, Izquierdo M, Villar A, Valdes Y, Cosme K, Deler ML, Montane M, Garcia E, Ramos A, Aguilar A, Medina E, Torano G, Sosa I, Hernandez I, Martinez R, Muzachio A, Carmenates A, Costa L, Cardoso F, Campa C, Diaz M, Roy R (2004) A synthetic conjugate polysaccharide vaccine against *Haemophilus influenzae* type b. *Science* 305:522–525.
107. Fernandez-Santana V, Cardoso F, Rodriguez A, Carmenate T, Pena L, Valdes Y, Hardy E, Mawas F, Heynngnezz L, Rodriguez MC, Figueroa I, Chang J, Toledo ME, Musacchio A, Hernandez I, Izquierdo M, Cosme K, Roy R, Verez-Bencomo V (2004) Antigenicity and immunogenicity of a synthetic oligosaccharide-protein conjugate vaccine against *Haemophilus influenzae* type b. *Infect. Immun.* 72:7115–7123.
108. Peeters CC, Evenberg D, Hoogerhout P, Käyhty H, Saarinen L, Van Boeckel CA, van der Marel GA, Van Boom JH, Poolman JT (1992) Synthetic trimer and tetramer of 3-beta-D-ribose-(1-1)-D-ribitol-5-phosphate conjugated to protein induce antibody responses to *Haemophilus influenzae* type b capsular polysaccharide in mice and monkeys. *Infect. Immun.* 60:1826–1833.
109. Lefeber DJ, Kamerling JP, Vliegenthart JFG (2001) Synthesis of *Streptococcus pneumoniae* type 3 neoglycoproteins varying in oligosaccharide chain length, loading and carrier protein. *Chem. Eur. J.* 7:4411–4421.
110. Benaissa-Trouw B, Lefeber DJ, Kamerling JP, Vliegenthart JFG, Kraaijeveld K, Snippe H (2001) Synthetic polysaccharide type 3-related di-, tri-, and tetrasaccharide-CRM197 conjugates induce protection against *Streptococcus pneumoniae* type 3 in mice. *Infect. Immun.* 69:4698–4701.
111. Lefeber DJ, Benaissa TB, Vliegenthart JFG, Kamerling JP, Jansen WTM, Kraaijeveld K, Snippe H (2003) Th1-directing adjuvants increase the immunogenicity of oligosaccharide-protein conjugate vaccines related to *Streptococcus pneumoniae* type 3. *Infect. Immun.* 71:6915–6920.
112. Jansen WTM, Hogenboom S, Thijssen MJ, Kamerling JP, Vliegenthart JFG, Verhoef J, Snippe H, Verheul AF (2001) Synthetic 6B di-, tri-, and tetrasaccharide-protein conjugates contain pneumococcal type 6a and 6b common and 6B- specific epitopes that elicit protective antibodies in mice. *Infect. Immun.* 69:787–793.
113. Alonso De Velasco E, Verheul AF, Veeneman GH, Gomes LJF, Van Boom JH, Verhoef J, Snippe H (1993) Protein-conjugated synthetic di- and trisaccharides of pneumococcal type 17F exhibit a different immunogenicity and antigenicity than tetrasaccharide. *Vaccine* 11:1429–1436.
114. Alonso De Velasco E, Verheul AF, Van Steijn AMP, Dekker HAT, Feldman RG, Fernandez IM, Kamerling JP, Vliegenthart JFG, Verhoef J, Snippe H (1994) Epitope specificity of rabbit immunoglobulin G (IgG) elicited by pneumococcal type 23F synthetic oligosaccharide- and native polysaccharide-protein conjugate vaccines: Comparison with human anti-polysaccharide 23F IgG. *Infect. Immun.* 62:799–808.
115. Lee RT, Lee YC (1974) Synthesis of 3-(2-aminoethylthio)propyl glycosides. *Carbohydr. Res.* 37:193–201.
116. Mawas F, Niggemann J, Jones C, Corbel MJ, Kamerling JP, Vliegenthart JFG (2002) Immunogenicity in a mouse model of a conjugate vaccine made with a synthetic single repeating unit of type 14 pneumococcal polysaccharide coupled to CRM197. *Infect. Immun.* 70:5107–5114.

117. Chernyak A, Karavanov A, Ogawa Y, Kovac P (2007) Conjugating oligosaccharides to proteins by squaric acid diester chemistry: Rapid monitoring of the progress of conjugation, and recovery of the unused ligand. *Carbohydr. Res.* 330:479–486.
118. Chernyak A, Kondo S, Wade TK, Meeks MD, Alzari PM, Fournier JM, Taylor RK, Kovac P, Wade WF (2002) Induction of protective immunity by synthetic *Vibrio cholerae* hexasaccharide derived from *V. cholerae* O1 Ogawa lipopolysaccharide bound to a protein carrier. *J. Infect. Dis.* 185:950–962.
119. Meeks MD, Saksena R, Ma XQ, Wade TK, Taylor RK, Kovac P, Wade WF (2004) Synthetic fragments of *Vibrio cholerae* O1 Inaba O-specific polysaccharide bound to a protein carrier are immunogenic in mice but do not induce protective antibodies. *Infect. Immun.* 72:4090–4101.
120. Pozsgay V (1998) Synthetic Shigella vaccines: A carbohydrate-protein conjugate with totally synthetic hexadecasaccharide haptens. *Angew. Chem. Inter. Ed. Eng.* 37:138–142.
121. Pozsgay V, Chu C, Pannell L, Wolfe J, Robbins JB, Schneerson R (1999) Protein conjugates of synthetic saccharides elicit higher levels of serum IgG lipopolysaccharide antibodies in mice than do those of the O-specific polysaccharide from *Shigella dysenteriae* type 1. *Proc. Natl. Acad. Sci. USA* 96:5194–5197.
122. Pozsgay V, Vieira NE, Yergey A (2002) A method for bioconjugation of carbohydrates using Diels-Alder cycloaddition. *Org. Lett.* 4:3191–3194.
123. Pozsgay V (2007) Conjugation of biomolecules using Diels-Alder cycloaddition. U.S. Patent 7211445.

3

ADJUVANTS FOR PROTEIN- AND CARBOHYDRATE-BASED VACCINES*

Bruno Guy

Research Department, Sanofi Pasteur, Campus Merieux, 69280 Marcy l'Etoile, France

3.1 INTRODUCTION

Adjuvants are usually defined as compounds able to increase and/or modulate the intrinsic immunogenicity of an antigen. Adjuvants are literally vaccine helpers, as this term is from Latin *adjuvare*, meaning to help. In order to reduce reactogenicity, new vaccines have a better defined composition (highly purified antigens, recombinant proteins, synthetic peptides, etc.), often linked to a lower immunogenicity when compared to previous whole-cell/virus-based vaccines. Adjuvants should thus help new vaccines to induce potent and persistent immunity, with the additional benefit of needing less antigen and injections. Moreover, new vaccine targets often require the induction of strong cellular responses in addition to antibodies, including T helper (Th) and sometimes cytotoxic T cells (CTLs). As aluminum salts, so far the most used adjuvants in humans, induce almost only antibody responses, discovering new adjuvants capable of delivering additional signals with co-administered antigens represents a critical need (for reviews on vaccines and adjuvants, see Refs. [1–3]).

---

*Some sections of this chapter have been previously published in 2007 in *Nature Reviews in Microbiology* [1].

Carbohydrate-Based Vaccines and Immunotherapies. Edited by Zhongwu Guo and Geert-Jan Boons
Copyright © 2009 John Wiley & Sons, Inc.

Although past developments generated potent products, there is nevertheless a need to develop a new generation of adjuvants, rationally designed on the basis of recent findings in basic immunology, particularly in innate immunity. It is important to stress at this stage that adjuvants form a very heterogeneous group of compounds. Adjuvant is often a synonym of immunostimulant, but if the former term can indeed encompass single compounds having intrinsic immunostimulant and/or immunomodulatory properties, adjuvants can also be composed of different constituents bearing different functions and activities, such as carrier/depot, targeting molecule, immunostimulant, and/or immunomodulator. One challenge of adjuvant research is, in fact, to define optimal and safe formulations, whose different components will be not only additive but synergistic and that will eventually drive the desired type of immune response. The present review will attempt to present the last developments obtained in this rapidly moving area, addressing both protein- and carbohydrate-based vaccines.

3.2 INITIATION AND STIMULATION OF ADAPTIVE RESPONSES

Innate immunity was until recently mostly seen as a nonspecific first line of defense, giving more time for adaptive immunity to develop. It is now clear that adaptive responses largely depend on the level and specificity of the initial "danger" signals [4], perceived by innate cells upon infection (and vaccination). Some innate or adaptive cells have long been known as mandatory for initiating immune responses by presenting antigen to T cells, such as the dendritic cells (DCs), which are the key antigen-presenting cells (APCs) for priming naïve T cells [5]. In fact, APCs, together with T (CD4 T helper and CD8 CTL) and B cells constitute the critical *ménage à trois* of immunology, as presented in Figure 3.1.

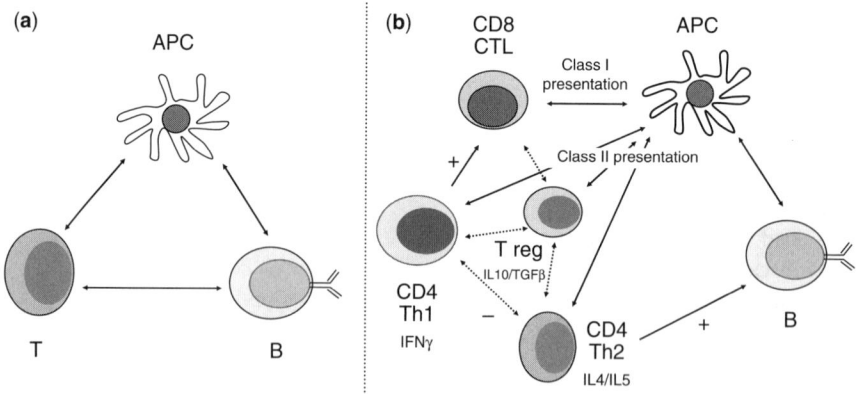

Figure 3.1 Immunological *ménage à trois*. (a) Simplified view of T, B, and APCs interactions. (b) Extended view, showing the major orientations of Th responses triggered by initial APC activation. APC class II presentation to CD4 cells will stimulate Th1, Th2, or Treg responses according to the cytokine environment. Th2 responses will further help B-cell responses predominantly while Th1 responses will support CTL C8 responses, should Ag also be presented by class I antigens to these latter cells. Tregs can regulate all different types of responses. (See color insert.)

The past decades characterized in many details adaptive responses and the respective roles of B cells, CD4, and CD8 T cells (Fig. 3.1). In particular CD4 T helper subsets (Th1, 2, 3, 17, etc.) were defined through their cytokine profile and their ability to modulate B and cytotoxic CD8 responses [6–8]. T suppressor cells were also rediscovered as regulatory T cells (Tregs), capable of inhibiting or balancing the activation of both Th1 and/or Th2 responses induced by infections [9]. Finally, memory responses were also dissected, allowing the identification of different means of inducing different subsets of memory cells bearing distinct abilities [10, 11].

More recently, the role of innate cells in shaping adaptive responses has been deeper investigated. Initiation of T-helper responses requires different signals from APCs. Signal 1 is triggered by specific peptide presentation by class II molecules to

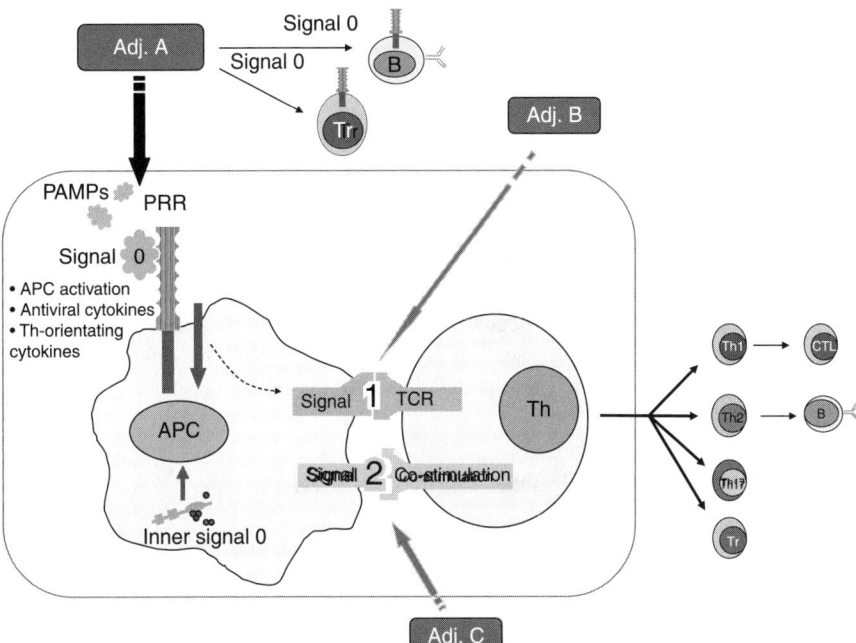

Figure 3.2 Where can adjuvants act? Initiation of T-helper responses needs three signals, as signals 0, 1, and 2. Adjuvants/formulations can in theory act, alone or in combination, on each of these three signals and can be referred as adjuvants A, B, and C, respectively. Most specific adjuvants developed recently, such as TLR agonists, can be considered as type A adjuvants: They act on signal 0, and indirectly on signal 2, by activating antigen-presenting cells (APCs) and triggering the secretion of cytokines such as IL12 to orientate subsequent Th1 responses, for instance. Some TLR agonist can also directly act on B cells and Tregs. In addition, TLR agonists can act on signal 1 by favoring an efficient presentation of the co-administered antigen (Ag). Adjuvants/formulations targeting APCs or favoring Ag capture can be viewed as type B adjuvants, acting on signal 1, as their effect is eventually mediated by enhanced Ag presentation to T cells. Liposomes, microspheres, and some emulsions can be found in this category. As stressed in this review, targeting signal 1 is not sufficient and an immunostimulant signal should be co-delivered for an optimal response. Finally, some specific ligands of co-stimulatory molecules can directly enhance signal 2, acting as type C adjuvants; such compounds must, however, be used with some caution, most likely in therapeutic applications. (Modified from Ref. [1].) (See color insert.)

T cell receptor (TCR). However, unless another signal is given, anergy and abortive responses are induced: co-stimulatory signal 2 is needed, through receptor–ligand interactions between APC/T antigens.

It appeared afterward that an additional—and prior—signal, called signal 0, was mandatory to activate APCs and orientate subsequent Th responses, for instance, through IL12 for Th1 responses. Signal 0 is mostly induced through recognition of pathogen-associated molecular patterns (PAMPs) by pathogen recognition receptors (PRR), including Toll-like receptors (TLRs, see next paragraphs). These signals may also apply to class I presentation by APCs for CD8 stimulation. It is in fact important to notice that while Th1 adjuvants promote $CD4^+$ interferon gamma (IFNγ)-secreting cells, this is not sufficient to trigger $CD8^+$ cytotoxic T cells, which require in addition class I antigen presentation.

These parallel discoveries linked more tightly initial innate and subsequent adaptive responses. Critical steps and signals required to induce T- and B-cell responses were defined, demonstrating that such innate signals modulate not only the magnitude of adaptive responses but also their quality such as Th1 polarization through IL12 secretion: Innate cells deliver to adaptive T (and B) cells a license to grow and differentiate.

These findings are summarized in Figure 3.2, which presents the different signals required to initiate potent responses, often referred to as signals 0, 1, and 2. Adjuvants can act on each of these three signals. Although targeting directly co-stimulatory molecules (signal 2) through antibodies, cytokines or chemokines represents an interesting possibility [12]; one needs, however, to be cautious, as this might result in nonfocused overreactions (see Section 3.10.2). This approach may rather be applied in therapeutic settings and will not be developed in the present review.

3.3 "OLD" ADJUVANTS AND FORMULATIONS

As stated in the introduction, carrier activities can be different from immunostimulant/ immunomodulatory ones, and each formulation is particular in this respect. Numerous formulations were developed before specific PRR ligands, including mineral salts (aluminum), emulsions, and immunostimulatory complexes (ISCOMs), which have both immunostimulant and carrier activities. On the other hand, liposomes and microparticles rather represent inert carriers, unless they present a specific composition or carry immunostimulants. The following discussion will address some main properties of these TLR-independent "classical" adjuvants, whose combinations with TLR agonists could thus be synergistic.

3.3.1 Aluminum

Paradoxically, while aluminum salts are by far the most widely used in humans, their mechanism of action is less characterized than for TLR agonists. They induce antibodies and Th2 responses, which was initially thought to require antigen surface adsorption (depot effect), transforming a soluble antigen into a particulate one to favour APC uptake. Most antigens are, however, rapidly desorbed from aluminum salts following exposure to interstitial fluid [13], and adsorption is not required even prior to administration for adjuvanticity [14]. However, adsorption or entrapment in aggregates may favor a high local concentration and improved uptake by APCs. Macrophages are activated by aluminum to present antigen [15], and a previously unknown population of IL-4-producing, $Gr1^+$, cells was shown to be required for in vivo priming and expansion of antigen-specific B cells [16].

Recently, the involvement of aluminum in the induction of IL1 and IL18 release through Caspase-1 activation was demonstrated [17]. These findings, as well as some similarities existing between aluminum and monosodium urate (MSU) crystals incited some authors to investigate whether aluminum salts could activate cells through the Nacht Domain-, Leucine-Rich Repeat-, and PYD-Containing Protein 3 (NALP3) inflammasome as MSU crystals do, and they demonstrated that this was indeed the case [18]. This activation was shown to depend neither on myeloid differentiation factor 88 (MYD88) (a TLR adaptor; see Section 3.4.1) nor on uric acid. These findings present, however, some discrepancies, possibly linked to different experimental models, with another recent study showing that aluminum adjuvant stimulates cells through uric acid in a MYD88-dependent way [19]. NALP3 is a member of specific cytosolic sensors called nucleotide binding oligomerization domain (NOD)-like receptors (NLRs) (see Section 3.4.2), and it is interesting to notice that aluminum salts, which were for a long time considered as prototype of "nonspecific" adjuvants, act in fact on well-defined and specific cellular activation pathways, and thus are closer to the recently developed TLR agonists (see Section 3.4.1). In any case, the exact mechanism by which extracellular/intravacuolar aluminum precipitates act on a cytosolic receptor is still unknown, and these recent findings open a new and very interesting area of research.

3.3.2 Emulsions

Emulsions are among the most used adjuvants after aluminum, both in humans and animals. An emulsion is a mixture of two immiscible substances. One substance (the dispersed phase) is dispersed in the other (the continuous phase) and stabilized with one or several surfactant(s) or emulsifier(s), present at the interface between the two phases.

Emulsions used in vaccines can be water-in-oil emulsions (W/O) or oil-in-water emulsions (O/W), and this depends on the volume fraction of both phases and on the type of surfactant. In the former case, aqueous droplets will be dispersed in the continuous oily phase, while the opposite will be true in the latter case. Antigens will usually be contained in water phases while some amphiphilic immunostimulants can be at the W/O interface, like surfactants.

As examples of W/O emulsions, one can mention Freund's adjuvant, named after J.T. Freund, which is a solution emulsified in mineral oil with manide monooleate surfactant. The complete form (CFA) contains inactivated and dried mycobacteria, usually *Mycobacterium tuberculosis*, while the incomplete form (IFA) does not contain mycobacteria. CFA cannot be used in humans because of its very high reactogenicity and should even be replaced in animals by other adjuvants when possible. Another type of W/O emulsion, Montanide ISA51 is a mix of a mineral oil with a mannide monooleate family surfactant. It gives W/O emulsions (50 : 50 O : W ratio). Montanide ISA720, which is a mix of vegetable oil with the same type of surfactant, also gives W/O emulsions (70 : 30 O : W ratio [20]).

On the other hand, a prototype O/W emulsion is MF59, which is a microfluidized oil-in-water emulsion, consisting of squalene oil and Tween 80 and Span 85 as surfactants [21]. It gives small (<250 nm) uniform and stable droplets. MF59 is a component of an adjuvanted flu vaccine marketed in Europe (Fluad). As for aluminum, adjuvanticity was proposed to be linked in part to the depot effect, but this does not seem to apply to O/W emulsions [21].

3.3.3 Saponins, QS21, and ISCOMS

Saponins from the bark of the South American tree *Quillaja saponaria Molina* have been tested and used in both veterinary and human fields for decades. Q saponins, including QS21, efficiently drive Th1 and antibody responses, but their large application has sometimes been hampered by their reactogenicity. Progress has been made in this respect [22], and one can also mention unfractionnated code name for a modified Quillaja saponin-based adjuvant (GPI-0100) containing semisynthetic derivatives saponins (DS saponins), two fractions of which stimulate Th1 and Th2 responses, respectively [23]. Saponins also constitute active components of particulate formulations such as immunostimulating complex or ISCOMs, which are cagelike structures containing antigen, cholesterol, phospholipid, and saponin. ISCOM-based vaccines promote both antibody and cellular immune responses including CTLs [24, 25], which may depend in part on the tight antigen binding to ISCOMs.

3.3.4 Liposomes and Microparticles

Such particulate compounds can either encapsulate antigen or carry it on their surface through adsorption or covalent linkage [26, 27]. In any case, should an antibody response be required against conformational epitopes, conformation and accessibility of antigen should be maintained. This also holds true for specific immunostimulants/ligands, which need to interact with their target properly. Actually, liposomes and microparticles need to co-deliver activatory signals. This can be linked to their composition (e.g., through cationic or anionic charges) [28, 29] or by addition of an immunostimulant. The particulate nature of these compounds may favor some targeting to APCs [30], although a real targeting requires specific ligands.

3.3.5 Antigen/Formulation Targeting

Targeting APCs, in particular through C-type lectin receptors (CLRs), may indeed enhance subsequent responses, should this occur in the presence of a co-stimulatory signal to avoid tolerance induction [31]. For instance, delivering proteins by antibodies to Dendritic and thymic Epithelial Cell-205 (DEC-205) in the presence of poly(I : C) and αCD40 combines targeting, stimulation of signals 0 and 2, overcomes suppression, and induces strong immunity [32]. Using very tightly defined sizes of nanobeads (40 nM) may also favor DC targeting [33].

3.3.6 Induction of CD8 CTLs with Soluble Antigens

Some means may target specifically class I antigen presentation pathway to stimulate CD8 cells (signal 1). Actually, Th1 responses are necessary but not sufficient for inducing CD8$^+$ CTLs. Achieving this with "inert" antigen remains a challenge, and different options have been proposed, including coupling to *Bordetella pertussis* adenylate cyclase (CyaA, [34]) or to negative charges [35], incorporation into ISCOMs, co-administration with TLR9 ligands [36], and with a cationic peptide such as in IC31 [37], or use of liposomes of particular composition [29, 38].

3.4 RENAISSANCE OF INNATE IMMUNITY

As stated before, the past years have seen a renaissance of innate immunity, which thus regained a first role, not only chronologically, but also through its critical importance for shaping adaptive responses, and applied research directly benefited from these basic findings, in particular thanks to the discovery of TLRs and other PRRs (for a review see Ref. [39]). These discoveries helped the scientists develop new and more focused adjuvants and/or to understand better the way "old" adjuvants, as those presented in the previous section, were acting.

3.4.1 Toll-Like Receptors: Agonists and Roles

Upon host invasion, infectious agents expressing PAMPs are recognized by innate cells through specific and nonpolymorphic PRR. Among them, TLRs were shown to play a critical role in the early steps of immune responses. These receptors were discovered in human cells [40] and named after the Drosophila fruit fly *Toll* gene, whose product is involved in innate immunity and development [41]. The importance of *Toll* signaling in mammalians was then confirmed by the observation that TLR4 was involved in lipopolysaccharide (LPS) recognition [42]. These seminal findings opened the way to a still expanding number of investigations regarding the different roles of TLRs in APC activation and T/B cell responses (for reviews see Refs. [43, 44]). TLRs are mostly sensed by foreign antigens, although it has been shown that endogenous proteins such as heat-shock proteins (HSPs) could also sense some of them. Evidence of direct binding of agonists to their corresponding TLRs is lacking in most cases; for instance, TLR4 has been shown not to bind directly to LPS [45], which interacts with myeloid differentiation 2 (MD2) [46].

Toll-like receptors 1, 2, 4, 5, 6, and 11 are surface exposed, while TLR 3, 7, 8, and 9 are located within endosomes. In addition to differences in cellular subcompartmentalization, differences exist regarding their cellular distribution: in humans, myeloid DCs express all TLRs except TLR7 and 9, which are selectively expressed by plasmacytoid DCs [47]. Epidermal Langerhans cells (LC) express mRNA (messenger ribonucleic acid) encoding TLR1, 2, 3, 5, 6, and 10, but do not directly sense TLR7/8 or TLR4 agonists [48].

Sensing the different TLRs with their corresponding natural or synthetic agonists can orientate responses toward Th1 and/or Th2 responses [49, 50] and also promote

Treg activity [51–53]. In particular, LPS agonists, imidazoquinolines, and unmethylated CpG oligonucleotides induce Th1 responses after sensing TLR4, 7/8, and 9, respectively.

In this regard, all TLRs were initially considered to transduce similar transactivation and phosphorylation pathways through their common adapter protein MyD88. This pathway is critical for inducing antigen-specific Th1 responses, as observed in MyD88-deficient mice. However, although most TLRs trigger MyD88-dependent pathways, TLR3 stimulates TRIF-dependent pathways, while TLR4 can signal through both pathways; interestingly, some TLR4 agonists have been shown to trigger preferentially one or the other pathway [54]. In addition, TLR negative regulators have progressively been identified, helping to keep inflammation under control.

Importantly, TLRs not only trigger APC activation (signals 0 and 2) but also play a role in antigen presentation (signal 1): The presence in the same cargo of both antigen and TLR is required for an optimal presentation upon activation, as seen with TLR4 and TLR11, leading in the latter case to an immunodominant response against the toxoplasma TLR11 ligand profilin [55, 56]. This highlights the need for having antigen and TLR agonists delivered simultaneously in the same vehicle for inducing an optimal response, which may also apply for CTLs.

Toll-like receptor stimulation also regulates Treg activity, directly through TLR recognition on Tregs or indirectly through APC–Treg interactions [51]. The consequence of the former direct stimulation is to activate Tregs as shown, for instance, for TLR2, TLR4 [52, 53], or to reverse their inhibitory effect through TLR8 [57], while TLR-linked APC stimulation rather inhibits Treg differentiation [58]. This shows that while TLR recognition induces inflammatory responses, it can also regulate such responses by triggering the expansion of Tregs, which will limit via a feedback mechanism potential harmful responses, once the infection is resolved. Finally, it has been shown that TLR-induced signals were important for the generation of memory CD4 T cells but not for their activation [59].

Moreover, direct TLR stimulation was also shown to be needed to fully activate B cells [60, 61]. However, the absolute requirement for TLR signaling in the generation of potent antibody responses has been questioned by using TLR signaling-deficient mice ($MyD88^{-/-}$; $Trif^{Lps2/Lps2}$) and chemically modified antigens, TNP-Hy and TNP-KLH [62]. It was shown under these conditions that TLR recognition might not account for all of the adjuvanticity of some compounds, including the TLR4 agonist monophosphoryl lipid A (MPL). While TLR-independent mechanisms were somehow expected for "classical" adjuvants such as alum and emulsions, these studies showed that even for some TLR agonists, alternative and possibly complementary pathways exist to stimulate antibody responses. Nevertheless, Gavin and collaborators [62] confirmed that TLRs control T-helper orientation and antibody isotypes, while neither the nature/duration of T and B memory nor the antibody functionality was investigated in their study. In any case, these findings support the proposal that associating TLR-dependent and independent adjuvants could be beneficial, by triggering different and potentially synergistic pathways. TLR-independent "classical" adjuvants would increase the global level of the immune response, while TLR agonists would in particular modulate its quality (Th1/Th2 bias). These aspects will be developed in the following sections.

3.4.2 Non-TLRs Innate Receptors

Toll-like receptors sense signals coming from outside the cell, and it was thus logical that intracellular cytosolic counterparts should sense inner signals. Such sensors include NOD-like receptors (NLRs) and retinoic acid inducible gene (RIG)-like helicases (RLHs), sensing signals from intracellular bacteria and viruses, respectively [63]. As for TLRs, agonists of such receptors could in theory be used as adjuvants, although their effect would be less focused as NLRs, and RLHs are more ubiquitous than TLRs.

3.4.3 Other Receptors Involved in Antigen Capture and Recognition

APCs express other receptors such as scavenger receptors, CLRs, and triggering receptors expressed on myeloid cells (TREMs). Scavenger receptors bind polyanionic ligands and can internalize pathogens but also host components such as apoptotic cells and modified low-density lipoproteins [64]. CLRs including DC-SIGN and mannose receptors can bind a wide range of viruses and bacteria through recognition of sugar moieties such as N-acetyl-glucosamine, mannose, N-acetyl-mannosamine, fucose, and glucose [65]. However, it seems that neither scavengers nor CLRs' first role is pathogen recognition. Pathogens rather seem to use to their own benefit in these host–host recognition systems. In contrast to TLRs whose primary function seems to discriminate self and non-self, targeting these latter receptors thus need a costimulatory signal to induce immunity rather than anergy, as shown with the DEC-205 receptor [31]; a yin/yang regulation has been proposed to describe the crosstalk between TLRs and CLRs [66]. These receptors have nevertheless a practical interest in vaccination, as their targeting may focus antigen and adjuvant to the proper cells, increasing both immunogenicity and safety. Finally, TREMs are characterized as amplifiers of immune responses, but so far their specific ligands have not been identified [67].

Beside receptors, one can mention two additional players acting at the border between innate and adaptive immunity: Natural killer (NK) T cells and $\gamma\delta$T cells. The former cells can be activated in particular by a synthetic glycolipid, α-galactosylceramide (α-GalCer). This compound and derivatives have demonstrated immunomodulatory properties in various settings, including cancer models [68]. The $\gamma\delta$T cells are specifically activated by small antigens, such as organic phosphoesters, which can also represent potential adjuvants. Moreover, some $\gamma\delta$T cells (Vδ2) can act themselves as APCs (for a review see Ref. [69]).

3.5 FROM BASIC RESEARCH TO PRACTICAL APPLICATIONS: IDENTIFICATION OF NEW ADJUVANTS

3.5.1 TLR Synthetic Agonists

The identification of natural TLR agonists promoted the design of synthetic ligands targeting TLRs more precisely and safely than pathogen-derived agonists do: Chemists and biochemists turned "dirty little secrets" of previous poorly purified (but immunogenic) vaccines into clean well-characterized adjuvants [70].

Interestingly, some adjuvants/immunostimulants previously identified where shown afterward to be the actual agonists of some TLRs: unmethylated viral or bacterial CpG DNA (deoxyribonucleic acid) and oligonucleotides for TLR9, poly(I:C) for TLR 3, lipopeptides and Pam3Cys for TLR2, LPS and their derivatives for TLR4, and imidazoquinolines for TLR 7/8 [71].

Additional TLR agonists have since been synthesized and tested in different models, such as second-generation oligonucleotide TLR9 agonists, whose sequences have been optimized with respect to species (e.g., mouse or human), type of immune response [36, 72], and stability. Similar approaches were followed to build second-generation TLR4 agonists, such as aminoalkyl glucosaminide 4-phosphates (lipid A mimetics), including RC529 adjuvant [73] and other new TLR4 synthetic agonists [74]. New TLR8 and/or TLR7 synthetic ligands such as 3M-019 were also synthesized [75].

3.5.2 Combination of PRR Agonists

Different PRR agonists can synergize and/or balance each other's immunomodulatory activity. This synergy has first been shown for different combinations of TLR agonists: TLR2 and TLR4 [76], and TL3 and 4 with TLR7, 8 [77], and 9 [78]. A recent report showed that synergy was induced by combining agonists acting on MyD88-dependent *and* independent pathways, while combining agonists acting on a single pathway induced tolerance [79]. A vaccine could then stimulate with some benefit different complementary TLR pathways to broaden Th responses, as shown with live YF17D vaccine [80], although viral replication itself plays some role in this latter case.

Crosstalks exist not only between TLRs but also between TLRs and NODs [81]. Synergy was observed between NOD1/2 and subactive doses of TLR4 agonists [82], as well as between NOD2 and TLR9 agonists. A full immune response may thus require engagement of member(s) of not only one PRR family, constituting an additional security against undesired responses, for instance, toward host components. Combining TLR agonists with NLR or RLH agonists in the same formulation, while attractive, thus requires documenting all consequences of such associations. Recent developments in clinical adjuvant research rather combined single TLR agonists within carrier/depot formulations, and this aspect will be detailed in a following section.

3.6 ADJUVANTS FOR CARBOHYDRATE-BASED VACCINES

Most of what has been described so far, both at the immunological and adjuvant levels, has been investigated in the case of "classical" T-dependent (Td) responses against protein antigens. However, carbohydrates, including polysaccharides and oligosaccharides, are T-cell-independent (Ti) antigens, unless they have been coupled to carrier proteins, as the present book describes in many details. The amount of work devoted to the testing of adjuvants with free polysaccharides or with glycoconjugates vaccines is much lower than for protein-based vaccines. Nevertheless, some authors have investigated the potential of some compounds, in particular TLR agonists, to enhance the immunogenicity of free or conjugated carbohydrate antigens.

3.6.1 Td and Ti B-Cell Responses

The initial humoral response to many pathogens is in fact based on the rapid Ti antibody production (mainly IgM) elicited by polysaccharide-like bacterial cell wall components or by repetitive structures of the viral capsid. B cells involved in Ti and Td responses present different phenotypes and locations, Ti-associated B cells being in particular located in the marginal zone of the spleen (MZ B cells; for a review see Ref. [83]) and in the pleural and peritoneal cavities in mice (B1 cells). In humans, these CD27+ IgD+ IgM+ MZ B cells appear to recirculate [84]. Ti antigens can also be distinguished from Td antigens by their inability to elicit memory—although some recent findings suggest it could be the case [85]—and to elicit a humoral immune response in an immature host (young infants). Neonatal B cells indeed exhibit various intrinsic defects in this respect. There are two types of Ti Ags, Ti-1 and Ti-2, the former being polyclonal B-cell activators while the second involve, for instance, capsular polysaccharides, which do not directly stimulate B cells. Ab responses [mostly IgM and secondarily IgG2 (in humans) to a lower extent] against Ti Ags usually are more cross-reactive and of weaker affinity than against Td ones (mostly IgG). Nevertheless, antibodies to Ti Ags are capable of fixing complement and of mediating, for instance, serum bactericidal activity.

At the cellular level, crosslinking of B-cell surface Ig (B-cell receptor, BCR) by repetitive/multimeric Ti Ags provides a first potent signal, which is insufficient for TCR stimulation, thus requiring additional signals from the microenvironment to trigger the multistep differentiation of Ti antigen-specific B cells into Ig-secreting plasma cells [60]. In this respect, the second co-stimulation can be provided by pathogen-derived B-cell activating factors (CpG, lipoproteins, many of them being TLR agonists) and/or by cytokines (IL-12, IFNγ, GM-CSF) produced by innate immune cells (macrophages and NK cells), activated in the early stages of infection [86–88]. Finally, although Ti Ags do not induce a cognate MHC class II/TCR-mediated interaction between T and B cells as for Td antigens, T lymphocytes play an indirect role in helping Ti B cells to mount immune responses, involving CD40-CD40L interactions with APCs [89, 90] as well as T-cell-derived factors (cytokines).

3.6.2 Adjuvants for "Free" Polysaccharides (Ti Antigens)

As stated above, Ti responses directed toward free polysaccharides (PS) involve B cells different from those mobilized for Td responses. Although a larger number of studies have investigated the role of co-activatory signals in Td B-cell responses than in Ti B-cell ones, activatory signals from both BCR and some other receptors—such as TLRs—are required on the same (MZ) B cell for optimal activation and plasmocyte differentiation, as stated above [86–88]. In this respect, some TLR agonists have been tested for their ability to enhance responses against Ti antigens, such as capsular PS or artificial antigens such as TNP-ficoll. It is to be stressed that in this regard, in particular, one has to be cautious when extrapolating animal (mouse) data to humans, as TLR expression by naïve or memory B cells can be different between animal and human B cells. For instance, human B cells do not express TLR4, while mouse B cells do [88, 91–93].

However, B cells of both species express TLR9, whose level can moreover be upregulated on naïve B cells by BCR triggering [86–88]. Agonists of TLR9 such as CpG oligonucleotides have indeed been evaluated with Ti antigens, sometimes leading to conflicting results. While different studies showed an increased IgG responses (including Th1-dependant IgG isotypes in mice), by combining Ti antigens with CpGs [94–97], it seems that the nature of the Ti antigen may be an important factor. For instance, anti-TNP-ficoll responses were enhanced by CpG in mice, while responses induced against free capsular PS were not [94]. The situation may also be different for type 1 Ti, which may have themselves an activatory function [98, 99]. The presence of TLR2 and 4 agonists in some commercial PS pneumococcal vaccines have also been shown to play a critical role in their immunogenicity, at least in a mouse model [100].

Other adjuvants/agonists have also been demonstrated to enhance Ti responses, including NOD agonist MDP [101], and α-toxin [102]. Coupling capsular polysaccharide to complement factor C3d was also shown to be effective in increasing immunogenicity and in inducing isotype switching [103]. In any case, even if antibody responses can be enhanced, by direct or indirect B cell co-stimulation thanks to these adjuvants, it seems unlikely that this would transform a Ti antigen into a Td one; and so far only conjugate vaccines succeeded in achieving such a goal, as detailed and exemplified in the different chapters of this book.

3.6.3 Adjuvants for Glycoconjugate Vaccines (T-Dependent Antigens)

While it is usually considered that conjugate vaccines do not require additional adjuvants, some of these compounds have nevertheless been used, including aluminum salts. The adjuvant role of these latter compounds seems more or less critical depending on the conjugates [104], and, moreover, it seems that their presence may have a detrimental effect of conjugate stability in some settings, such as with *Haemophilus influenzae* type B conjugates [105]. TLR9 agonists have also been used as adjuvants for conjugate vaccines, allowing to compensate some immune defects such as those encountered in early or late life [106–108]. Adjuvants could also mitigate immune interferences potentially present in multivalent conjugate vaccines, this being caused by B- or T-cell-dependent carrier suppression or by another mechanism [109–111].

As stated previously, conjugation turns a Ti antigen into a Td one, requiring cognate T-cell help [112], and in any case, it seems logical to consider that the numerous adjuvant candidates tested with protein antigens would also work, at least to some extent, with conjugate vaccines. The very nature of the B cells involved in the responses triggered by conjugates—and their identity with "classical" B cells involved in Td responses—remains nevertheless to be formally established.

3.7 COMBINATIONS OF ADJUVANTS: PRECLINICAL AND CLINICAL DEVELOPMENTS

Adjuvant activity and/or safety can be enhanced by combining formulations and immunostimulants. Associating classical TLR-independent adjuvants [62] with TLR agonists would trigger different and potentially synergistic pathways, as stated above. Some of the former ones would, for instance, increase the global level of immune response, while the latter ones would also modulate its quality (Th1/Th2 bias). However, finding the optimal association is not obvious as some antagonism/anergy can occur rather than synergy [79]. Co-formulating antigen and immunostimulant may also be required, as shown by the need for TLRs and antigen to be in the same compartment for optimal subsequent peptide presentation [55, 56]. Moreover, stimulating TLRs without antigen activates APCs and decreases their cross-presentation ability [113].

Combining antigens in the same formulation also allows associating in multivalent vaccines otherwise noncompatible antigens [114] and/or ensures that all antigens will be presented simultaneously. This may reduce risks of immunodominance should an antigen be presented before the others and take the lead [115]. As stated before, both antigen(s) and immunostimulants should keep a native conformation and their ability to interact with their ligands (e.g., TLRs or membrane-bound antibodies on B cells). In this respect, the antigen/adjuvant couple can be critical, as the antigen itself may increase or on the contrary counteract adjuvant efficacy or Th bias. For instance, some antigens trigger Th1 responses even with alum [116], and different antigens induce Th1 or mixed Th1/Th2 responses, although present in identical formulations [117].

Combined formulations have been developed sometimes before TLRs identification, reaching clinical trials and market approval. One can mention in particular a combination of aluminum salts and TLR4 agonist MPL, present in Cervarix [118], a vaccine recently developed against human papilloma virus (HPV) infection (for more details see reviews [3, 119]). Many of them aim at introducing a cellular Th1 arm in addition to antibodies to increase vaccine efficacy against targets such as certain viruses, for instance, through TLR4 or TLR9 agonists. One can mention in this respect the partial efficacy in HSV1 and HSV2 seronegative women of a subunit herpes vaccine adjuvanted with AS04 combining Alum and TLR4 agonist MPL, while a similar vaccine adjuvanted with MF59 emulsion was not protective [120].

3.8 IMMUNOMODULATION OF EXISTING RESPONSES: ADJUVANTS FOR THERAPEUTIC VACCINES

Adjuvants able to trigger selectively Th1 or Th2 responses offer the possibility to redirect Th responses once they have been established, although this is more difficult than in naïve nonbiased individuals. This has been applied successfully in human allergy to revert Th2 responses, using "Th1" agonists such as MPL [121] or by coupling allergen to CpG oligonucleotides [122]. This gives hope that immunotherapy could contribute to treat already established immune disorders. In the case of chronic infectious diseases, reorientating or awaking Th responses may also favor more efficient elimination mechanisms. For instance, TLR agonists can stimulate directly or indirectly T cells [123], targeting in particular Tregs to enhance or counteract their action [124]. Targeting the inhibitory PD1/PDL1 pathway can also reactivate exhausted effector cells [125], such as human immunodeficiency virus (HIV)-specific CD8 T-cells [126, 127]. Induction of stronger T-cell responses is also the goal of therapeutic vaccination against hepatitis B [128] and the addition of new adjuvants to the existing vaccines could be helpful in this respect. Adjuvants whose reactogenicity would be problematic in prophylaxis may also be more easily applied in therapy.

3.9 TAKE ANOTHER ROUTE

3.9.1 Adjuvants for Mucosal Immunization

Although most vaccines are systemic, numerous efforts concern mucosal vaccines for practical and target-driven reasons (for a review see [129]). While mucosal routes may require specific formulations, some "systemic" immunostimulants and carriers can be used as well. Properties required for a mucosal formulation are in fact very similar to those required for systemic immunization: carry antigen, bind and stimulate APCs, bearing in mind that a Th2/Th3 bias is usually associated to mucosal inductive sites. Some cells are specialized in particulate antigen uptake in such sites, called M or M-like cells [130]. Although liposomes and microparticles can encapsulate and protect antigens from harsh conditions encountered in some mucosal environments, they do not really target mucosal inductive sites. This may be achieved by coating formulations with specific M-cell ligands, and different candidates have been identified [131–133], including lectins and some viral proteins (e.g., reovirus σ1 protein).

Regarding cell activation, cholera toxin (CT, from *Vibrio cholerae*) is the prototype gold-standard mucosal adjuvant, along with the related heat labile toxin from *Escherichia coli* (LT; for a review see [134]). These two toxins almost perfectly fulfill two criteria defined previously: B subunit (CTB) first binds to the widely expressed GM1 ganglioside, and in a second step the A subunit (CTA) activates the cell by catalysing adenosine diphosphate (ADP) ribosylation of the regulatory protein Gs α of adenylate cyclase, increasing in turn the synthesis of $3'$, $5'$-cyclic adenosine monophosphate (cAMP) second messenger. The major problem linked to CT and LT is their high toxicity in humans, while it is not the case in mice. Different approaches have been followed to solve it, including in particular the construction of nontoxic mutants retaining adjuvanticity [134].

3.9.2 Epidermal or Intradermal Routes

Such routes are being explored as they may by themselves "adjuvate" vaccines, thanks to the high density of DCs present in the epidermis (Langerhans cells) and dermis. One can mention the topical application of antigen onto the skin through a nonabrasive patch, as this usually requires the use of adjuvants such as mucosal CT/LT toxins (for a review see [135]). Langerhans cells express a selective array of TLRs [48], which should guide in part the choice of adjuvants in this respect. CT/LT derivatives can also be used as adjuvants by classical subcutaneous route, inducing protection in an animal model of *Helicobacter pylori* infection [136]. Topical application of CpG oligonucleotides has also been shown to enhance cross-presentation and CTL responses against injected proteins [137].

3.10 PRACTICAL ASPECTS OF ADJUVANT DEVELOPMENT

3.10.1 Regulatory Aspects

Beside efficacy and immunogenicity, other factors that should be emphasized from the beginning critically impact the ability of any adjuvant/formulation to go beyond preclinical studies: (a) Constituents should be produced according to good manufacturing practices (GMP), defined chemically and biologically, stable, assuring safe and consistent responses in vaccinees; (b) the adjuvant combined with antigen(s) should induce a stronger protective immunity than the same antigen(s) alone; (c) adjuvant safety must be documented, including immediate and long-term side effects; and (d) the adjuvant should be intrinsically nonimmunogenic, biodegradable, and biocompatible. Apart some exceptions, adjuvants are not developed as drugs on their own, and toxicology studies need to be designed on a case-by-case basis, as the adjuvant/antigen couple also conditions safety and reactogenicity. Some specific guidelines already exist and others are under evaluation [138].

3.10.2 Safety versus Efficacy: Risk–Benefit Ratio

The benefit of using an adjuvant must indeed outweigh its risk. Most adjuvants' developments are stopped or hampered because of acute reactogenicity/toxicity issues. However, induction of long-term immunopathology such as autoimmunity is more difficult to document, and this has been widely discussed for TLR agonists, in particular TLR9 and TLR7 agonists [139] (for a review see [140]). However, TLR agonists may by themselves limit the risk of uncontrolled reactions by triggering both activatory and inhibitory signals: as stated previously, TLR4 and TLR2 agonists can initiate the immune reaction but at the same time stimulate Tregs, and thus limit via a feedback mechanism potential autoimmunity resulting from overactivated cells, once infection has been controlled. In addition, one has to consider the theoretical risks linked to a few immunizations and short-term stimulation with TLR agonists, in comparison to life-long stimulation through the same receptors by environmental pathogens or commensals.

Situations may be different when directly triggering specific T-cell co-stimulatory pathways. The Food and Drug Administration (FDA) recently approved CTLA4-Ig, an antagonist of CD28 co-stimulation, for the treatment of rheumatoid arthritis [141], which opened the way for similar compounds. However, it is certainly safer to inhibit than to directly activate, and the recent negative outcome of a trial using a super-CD28 agonist (TGN1412 antibody [142]) requires some prudence when using such strong signal 2 agonists, as stated previously. This antibody, which triggered T-cell activation without needing antigen-TCR stimulation (signal 1), satisfied preclinical evaluations in different species. Its goal was in fact to stimulate Tregs, which was achieved in reconstituted immunodeficient mice with a related antibody (5.11A1 [143]). Moreover, TG1412 had a positive safety record in *Cynomolgus* monkeys, whose sequences of CD28 external and internal domains are identical to human CD28 [140]. Predictability of animal models is thus questionable in some circumstances, and this has been in fact further addressed in the TGN1412 case [144].

3.11 PRECLINICAL MODELS USED IN ADJUVANT DEVELOPMENT

3.11.1 Animal Models

Mice are the most widely used preclinical model, but although they often bring very helpful information, some significant differences exist between immune systems of mice and humans (for a review see [145]). For instance, TLR9 is expressed in different DC subsets in mice and humans and optimal CpG sequences are also species specific. TLR8 responsiveness is also different in mice, which seem to respond poorly to TLR8 agonists, although a recent report showed that the murine receptor could be functional [146]. As stated above, TLR4 is expressed at significant levels in mouse B cells but not in human B cells under normal conditions, and human and mouse TLR4 respond differently to some LPSs [147]. TLR expression may vary even between different strains of mice [148].Testing in mice TLR4, 7/8 or TLR9 agonists, thus requires careful interpretation of results.

The preimmune status is another important point to consider. Most adjuvant evaluation is conducted in naïve animals, in which a dramatic benefit is often observed for many adjuvants. However, differences between adjuvanted and unadjuvanted vaccines can be decreased or absent in primed animals [149]. This could be consistent, for instance, with the fact that TLR stimulation is required for generating memory CD4 T cells, but not for their activation [59]. One thus needs to consider if the target human or animal population is preimmune for the vaccine antigen(s), and primed preclinical models could be used in this respect.

3.11.2 In vitro Models

Investigating adjuvant activity in human cells in vitro can therefore add valuable information to animal models. Transformed cells expressing specific arrays of TLRs can also be used to evaluate new agonists. Numerous publications have used primary human cells to evaluate adjuvants/vaccines, in particular monocyte-derived DCs [150]. Flow cytometry may allow, in particular, differentiating activation of cells having captured a formulation versus surrounding cells. Analyzing how focused stimulation is can be an indirect mean for evaluating safety, in addition to immunogenicity. This has been used with live vaccines [151] and can be applied as well for inert vaccines, should formulation constituents be directly labeled or tracked by fluorescent antibodies. Confocal microscopy also allows tracking the subcellular

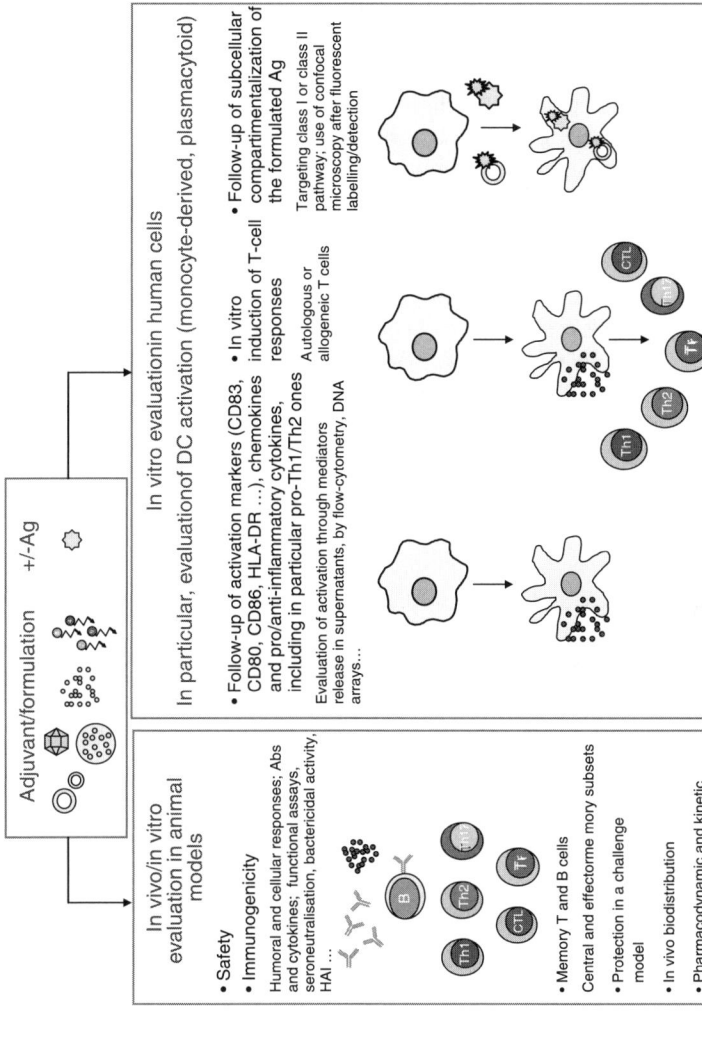

Figure 3.3 In vivo and in vitro preclinical evaluation of adjuvants. In addition to preclinical in vivo evaluation in animal models, it can be of interest to investigate in parallel adjuvants/formulations in vitro in human cells, in particular in APCs such as dendritic cells (DCs). The combination of information obtained in vivo and in vitro may increase our knowledge on both immunogenicity and safety of these formulations before going into clinics. (Modified from Ref. [1].) (See color insert.)

localization of antigen and/or formulations [45], and identifying which pathways are followed upon antigen/formulation capture, for instance, class I or class II presentation pathways [38]. DNA microarrays may also identify a "stimulation" signature uncoupled from a "toxicity" one. Figure 3.3 summarizes which in vitro assays can be performed for adjuvant evaluation.

3.12 CONCLUSIONS AND PERSPECTIVES

Observation of the different ways pathogens interact with our innate and adaptive immune system has led adjuvant researchers to create more focused adjuvants and formulations, defined optimally on a case-by-case basis.

Across the expanding universe of immunostimulants and formulations, one can now shift from yesterday's empiricism and rationally identify, characterize, and combine together those that will give vaccines the perfect help. Our choice is guided by the nature of the antigen(s) and the known or expected correlates of protection, keeping in mind that induction of immune responses is not a black or white process. The development of these sometimes complex products follows a long and often winding road, but, thanks to more and more basic and applied tools, we can now work out potential issues and propose new solutions against still unmet critical targets. Sustained research is, however, still needed to identify the safest and most efficient combination with respect to each considered pathogen.

ACKNOWLEDGMENT

I acknowledge Jeffrey Almond, Nicolas Burdin, and Jean Haensler for their support, helpful discussions, and critical reading of the manuscript. Adjuvant research is a very large field and I apologize for not having mentioned numerous and important works because of space constraint.

REFERENCES

1. Guy B (2007) The perfect mix: Recent progress in adjuvant research. *Nat. Rev. Microbiol.* 5:505.
2. Ulmer JB, Valley U, Rappuoli R (2006) Vaccine manufacturing: Challenges and solutions. *Nat. Biotechnol.* 24:1377.
3. Hunter RL (2002) Overview of vaccine adjuvants: Present and future. *Vaccine* 20(Suppl 3):S7.
4. Bianchi ME (2007) DAMPs, PAMPs and alarmins: All we need to know about danger. *J. Leukoc. Biol.* 81:1.
5. Steinman RM, Hemmi H (2006) Dendritic cells: Translating innate to adaptive immunity. *Curr. Top. Microbiol. Immunol.* 311:17.
6. Mosmann TR, Cherwinski H, Bond MW, Giedlin MA, Coffman RL (1986) Two types of murine helper T cell clone. I. Definition according to profiles of lymphokine activities and secreted proteins. *J. Immunol.* 136:2348.

7. Mosmann TR, Sad S (1996) The expanding universe of T-cell subsets: Th1, Th2 and more. *Immunol. Today* 17:138.
8. Harrington LE, Mangan PR, Weaver CT (2006) Expanding the effector CD4 T-cell repertoire: The Th17 lineage. *Curr. Opin. Immunol.* 18:349.
9. Belkaid Y, Rouse BT (2005) Natural regulatory T cells in infectious disease. *Nat. Immunol.* 6:353.
10. Lanzavecchia A, Sallusto F (2005) Understanding the generation and function of memory T cell subsets. *Curr. Opin. Immunol.* 17:326.
11. Pulendran B, Ahmed R (2006) Translating innate immunity into immunological memory: Implications for vaccine development. *Cell* 124:849.
12. Barr TA, Carlring J, Heath AW (2006) Co-stimulatory agonists as immunological adjuvants. *Vaccine* 24:3399.
13. HogenEsch H (2002) Mechanisms of stimulation of the immune response by aluminum adjuvants. *Vaccine* 20(Suppl 3):S34.
14. Romero Méndez IZ, Shi Y, HogenEsch H, Hem SL (2007) Potentiation of the immune response to non-adsorbed antigens by aluminum-containing adjuvants. *Vaccine* 25:825.
15. Rimaniol AC, Gras G, Verdier F, Capel F, Grigoriev VB, Porcheray F, Sauzeat E, Fournier JG, Clayette P, Siegrist CA, Dormont D (2004) Aluminium hydroxide adjuvant induces macrophage differentiation towards a specialized antigen-presenting type. *Vaccine* 22:3127.
16. Jordan MB, Mills DM, Kappler J, Marrack P, Cambier JC (2004) Promotion of B cell immune responses via an alum-induced myeloid cell population. *Science* 304:1808.
17. Li H, Nookala S, Re F (2007) Aluminum hydroxide adjuvants activate caspase-1 and induce IL-1beta and IL-18 release. *J. Immunol.* 178:5271.
18. Eisenbarth SC, Colegio OR, O'Connor W, Sutterwala FS, Flavell RA (2008) Crucial role for the Nalp3 inflammasome in the immunostimulatory properties of aluminium adjuvants. *Nature* 453:1122.
19. Kool M, Soullié T, van Nimwegen M, Willart MA, Muskens F, Jung S, Hoogsteden HC, Hammad H, Lambrecht BN (2008) Alum adjuvant boosts adaptive immunity by inducing uric acid and activating inflammatory dendritic cells. *J. Exp. Med.* 205:869.
20. Aucouturier J, Dupuis L, Deville S, Ascarateil S, Ganne V (2002) Montanide ISA 720 and 51: A new generation of water in oil emulsions as adjuvants for human vaccines. *Expert Rev. Vaccines* 1:111.
21. Podda A, Del Giudice G (2003) MF59-adjuvanted vaccines: Increased immunogenicity with an optimal safety profile. *Expert Rev. Vaccines* 2:197.
22. Kim YJ, Wang P, Navarro-Villalobos M, Rohde BD, Derryberry J, Gin DY (2006) Synthetic studies of complex immunostimulants from Quillaja saponaria: Synthesis of the potent clinical immunoadjuvant QS-21Aapi. *J. Am. Chem. Soc.* 128:11906.
23. Marciani DJ, Reynolds RC, Pathak AK, Finley-Woodman K, May RD (2003) Fractionation, structural studies, and immunological characterization of the semi-synthetic Quillaja saponins derivative GPI-0100. *Vaccine* 21:3961.
24. Sanders MT, Brown LE, Deliyannis G, Pearse MJ (2005) ISCOM-based vaccines: The second decade. *Immunol. Cell Biol.* 83:119.

25. Ennis FA, Cruz J, Jameson J, Klein M, Burt D, Thipphawong J (1999) Augmentation of human influenza A virus-specific cytotoxic T lymphocyte memory by influenza vaccine and adjuvanted carriers (ISCOMS). *Virology* 259:256.
26. Leserman L (2004) Liposomes as protein carriers in immunology. *J. Liposome Res.* 14:175.
27. O'Hagan DT, Singh M (2003) Microparticles as vaccine adjuvants and delivery systems. *Expert Rev. Vaccines* 2:269.
28. Singh M, Kazzaz J, Ugozzoli M, Malyala P, Chesko J, O'Hagan DT (2006) Polylactide-co-glycolide microparticles with surface adsorbed antigens as vaccine delivery systems. *Curr. Drug Deliv.* 3:115.
29. Guy B, Pascal N, Françon A, Bonnin A, Gimenez S, Lafay-Vialon E, Trannoy E, Haensler J (2001) Design, characterization and preclinical efficacy of a cationic lipid adjuvant for influenza split vaccine. *Vaccine* 19:1794.
30. Reddy ST, Swartz MA, Hubbell JA (2006) Targeting dendritic cells with biomaterials: Developing the next generation of vaccines. *Trends Immunol.* 27:573.
31. Bonifaz L, Bonnyay D, Mahnke K, Rivera M, Nussenzweig MC, Steinman RM (2002) Efficient targeting of protein antigen to the dendritic cell receptor DEC-205 in the steady state leads to antigen presentation on major histocompatibility complex class I products and peripheral CD8+ T cell tolerance. *J. Exp. Med.* 196:1627.
32. Boscardin SB, Hafalla JC, Masilamani RF, Kamphorst AO, Zebroski HA, Rai U, Morrot A, Zavala F, Steinman RM, Nussenzweig RS, Nussenzweig MC (2006) Antigen targeting to dendritic cells elicits long-lived T cell help for antibody responses. *J. Exp. Med.* 203:599.
33. Fifis T, Gamvrellis A, Crimeen-Irwin B, Pietersz GA, Li J, Mottram PL, McKenzie IF, Plebanski M (2004) Size-dependent immunogenicity: Therapeutic and protective properties of nanovaccines against tumors. *J. Immunol.* 173:3148.
34. Moron G, Dadaglio G, Leclerc C (2004) New tools for antigen delivery to the MHC class I pathway. *Trends Immunol.* 25:92.
35. Yamasaki Y, Ikenaga T, Otsuki T, Nishikawa M, Takakura Y (2007) Induction of antigen-specific cytotoxic T lymphocytes by immunization with negatively charged soluble antigen through scavenger receptor-mediated delivery. *Vaccine* 25:85.
36. McCluskie MJ, Krieg AM (2006) Enhancement of infectious disease vaccines through TLR9-dependent recognition of CpG DNA. *Curr. Top. Microbiol. Immunol.* 311:155.
37. Schellack C, Prinz K, Egyed A, Fritz JH, Wittmann B, Ginzler M, Swatosch G, Zauner W, Kast C, Akira S, von Gabain A, Buschle M, Lingnau K (2006) IC31, a novel adjuvant signaling via TLR9, induces potent cellular and humoral immune responses. *Vaccine* 24:5461.
38. Taneichi M, Ishida H, Kajino K, Ogasawara K, Tanaka Y, Kasai M, Mori M, Nishida M, Yamamura H, Mizuguchi J, Uchida T (2006) Antigen chemically coupled to the surface of liposomes are cross-presented to CD8+ T cells and induce potent antitumor immunity. *J. Immunol.* 177:2324.
39. Medzhitov R (2007) Recognition of microorganisms and activation of the immune response. *Nature* 449:819.
40. Medzhitov R, Preston-Hurlburt P, Janeway CA Jr (1997) A human homologue of the Drosophila Toll protein signals activation of adaptive immunity. *Nature* 388:394.

41. Lemaitre B, Nicolas E, Michaut L, Reichhart JM, Hoffmann JA (1996) The dorsoventral regulatory gene cassette spatzle/Toll/cactus controls the potent antifungal response in Drosophila adults. *Cell* 86:973.
42. Poltorak A, He X, Smirnova I, Liu MY, Van Huffel C, Du X, Birdwell D, Alejos E, Silva M, Galanos C, Freudenberg M, Ricciardi-Castagnoli P, Layton B, Beutler B (1998) Defective LPS signaling in C3H/HeJ and C57BL/10ScCr mice: Mutations in Tlr4 gene. *Science* 282:2085.
43. Iwasaki A, Medzhitov R (2004) Toll-like receptor control of the adaptive immune responses. *Nat. Immunol.* 5:987.
44. Akira S (2006) TLR signaling. *Curr. Top. Microbiol. Immunol.* 311:1.
45. Dunzendorfer S, Lee HK, Soldau K, Tobias PS (2004) TLR4 is the signalling but not the lipopolysaccharide uptake receptor. *J. Immunol.* 173:1166.
46. Ohto U, Fukase K, Miyake K, Satow Y (2007) Crystal structures of human MD-2 and its complex with antiendotoxic lipid IVa. *Science* 316:1632.
47. Jarrossay D, Napolitani G, Colonna M, Sallusto F, Lanzavecchia A (2001) Specialization and complementarity in microbial molecule recognition by human myeloid and plasmacytoid dendritic cells. *Eur. J. Immunol.* 31:3388.
48. Flacher V, Bouschbacher M, Verronèse E, Massacrier C, Sisirak V, Berthier-Vergnes O, de Saint-Vis B, Caux C, Dezutter-Dambuyant C, Lebecque S, Valladeau J (2006) Human langerhans cells express a specific TLR profile and differentially respond to viruses and gram-positive bacteria. *J. Immunol.* 177:7959.
49. Agrawal S, Agrawal A, Doughty B, Gerwitz A, Blenis J, Van Dyke T, Pulendran B (2003) Cutting edge: Different Toll-like receptor agonists instruct dendritic cells to induce distinct Th responses via differential modulation of extracellular signal-regulated kinase-mitogen-activated protein kinase and c-Fos. *J. Immunol.* 171:4984.
50. Dillon S, Agrawal A, Van Dyke T, Landreth G, McCauley L, Koh A, Maliszewski C, Akira S, Pulendran B (2004) A Toll-like receptor 2 ligand stimulates Th2 responses in vivo, via induction of extracellular signal-regulated kinase mitogen-activated protein kinase and c-Fos in dendritic cells. *J. Immunol.* 172:4733.
51. Sakaguchi S (2003) Control of immune responses by naturally arising CD4+ regulatory T cells that express toll-like receptors. *J. Exp. Med.* 197:397.
52. Sutmuller RP, den Brok MH, Kramer M, Bennink EJ, Toonen LW, Kullberg BJ, Joosten LA, Akira S, Netea MG, Adema GJ (2006) Toll-like receptor 2 controls expansion and function of regulatory T cells. *J. Clin. Invest.* 116:485.
53. Caramalho I, Lopes-Carvalho T, Ostler D, Zelenay S, Haury M, Demengeot J (2003) Regulatory T cells selectively express toll-like receptors and are activated by lipopolysaccharide. *J. Exp. Med.* 197:403.
54. Mata-Haro V, Cekic C, Martin M, Chilton PM, Casella CR, Mitchell TC (2007) The vaccine adjuvant monophosphoryl lipid A as a TRIF-biased agonist of TLR4. *Science* 316:1628.
55. Blander JM, Medzhitov R (2006) Toll-dependent selection of microbial antigens for presentation by dendritic cells. *Nature* 440:808.
56. Yarovinsky F, Kanzler H, Hieny S, Coffman RL, Sher A (2006) Toll-like receptor recognition regulates immunodominance in an antimicrobial CD4+ T cell response. *Immunity* 25:655.

57. Peng G, Guo Z, Kiniwa Y, Voo KS, Peng W, Fu T, Wang DY, Li Y, Wang HY, Wang RF (2005) Toll-like receptor 8-mediated reversal of CD4+ regulatory T cell function. *Science* 309:1380.
58. Pasare C, Medzhitov R (2003) Toll pathway-dependent blockade of CD4 + CD25+ T cell-mediated suppression by dendritic cells. *Science* 299:1033.
59. Pasare C, Medzhitov R (2004) Toll-dependent control mechanisms of CD4 T cell activation. *Immunity* 21:733.
60. Ruprecht CR, Lanzavecchia A (2006) Toll-like receptor stimulation as a third signal required for activation of human naive B cells. *Eur. J. Immunol.* 36:810.
61. Pasare C, Medzhitov R (2005) Control of B-cell responses by Toll-like receptors. *Nature* 438:364.
62. Gavin AL, Hoebe K, Duong B, Ota T, Martin C, Beutler B, Nemazee D (2006) Adjuvant-enhanced antibody responses in the absence of Toll-like receptor signaling. *Science* 314:1936.
63. Meylan E, Tschopp J, Karin M (2006) Intracellular pattern recognition receptors in the host response. *Nature* 442:39.
64. Peiser L, Mukhopadhyay S, Gordon S (2002) Scavenger receptors in innate immunity. *Curr. Opin. Immunol.* 14:123.
65. Robinson MJ, Sancho D, Slack EC, LeibundGut-Landmann S, Reis e Sousa C (2006) Myeloid C-type lectins in innate immunity. *Nat. Immunol.* 7:1258.
66. 't Hart BA, van Kooyk Y (2004) Yin-Yang regulation of autoimmunity by DCs. *Trends Immunol.* 25:353.
67. Klesney-Tait J, Turnbull IR, Colonna M (2006) The TREM receptor family and signal integration. *Nat. Immunol.* 7:1266.
68. Fujii S, Shimizu K, Hemmi H, Fukui M, Bonito AJ, Chen G, Franck RW, Tsuji M, Steinman RM (2006) Glycolipid alpha-C-galactosylceramide is a distinct inducer of dendritic cell function during innate and adaptive immune responses of mice. *Proc. Natl. Acad. Sci. U.S.A.* 103:11252.
69. Moser B, Brandes M (2006) Gammadelta T cells: An alternative type of professional APC. *Trends Immunol.* 27:112.
70. Dougan G, Hormaeche C (2006) How bacteria and their products provide clues to vaccine and adjuvant development. *Vaccine* 24:S2–13.
71. Kaisho T, Akira S (2002) Toll-like receptors as adjuvant receptors. *Biochim. Biophys. Acta* 1589:1.
72. Kandimalla ER, Yu D, Agrawal S (2002) Towards optimal design of second-generation immunomodulatory oligonucleotides. *Curr. Opin. Mol. Ther.* 4:122.
73. Persing DH, Coler RN, Lacy MJ, Johnson DA, Baldridge JR, Hershberg RM, Reed SG (2002) Taking toll: Lipid A mimetics as adjuvants and immunomodulators. *Trends Microbiol.* 10:S32
74. Przetak M, Chow J, Cheng H, Rose J, Hawkins LD, Ishizaka ST (2003) Novel synthetic LPS receptor agonists boost systemic and mucosal antibody responses in mice. *Vaccine* 21:961.
75. Johnston D, Zaidi B, Bystryn JC (2007) TLR7 imidazoquinoline ligand 3M-019 is a potent adjuvant for pure protein prototype vaccines. *Cancer Immunol. Immunother.* 56:1133.

76. Sato S, Nomura F, Kawai T, Takeuchi O, Mühlradt PF, Takeda K, Akira S (2000) Synergy and cross-tolerance between toll-like receptor (TLR) 2- and TLR4-mediated signaling pathways. *J. Immunol.* 165:7096.
77. Roelofs MF, Joosten LA, Abdollahi-Roodsaz S, van Lieshout AW, Sprong T, van den Hoogen FH, van den Berg WB, Radstake TR (2005) The expression of toll-like receptors 3 and 7 in rheumatoid arthritis synovium is increased and costimulation of toll-like receptors 3, 4, and 7/8 results in synergistic cytokine production by dendritic cells. *Arthritis Rheum.* 52:2313.
78. Napolitani G, Rinaldi A, Bertoni F, Sallusto F, Lanzavecchia A (2005) Selected Toll-like receptor agonist combinations synergistically trigger a T helper type 1-polarizing program in dendritic cells. *Nat. Immunol.* 6:769.
79. Bagchi A, Herrup EA, Warren HS, Trigilio J, Shin HS, Valentine C, Hellman J (2007) MyD88-dependent and MyD88-independent pathways in synergy, priming, and tolerance between TLR agonists. *J. Immunol.* 178:1164.
80. Querec T, Bennouna S, Alkan S, Laouar Y, Gorden K, Flavell R, Akira S, Ahmed R, Pulendran B (2006) Yellow fever vaccine YF-17D activates multiple dendritic cell subsets via TLR2, 7, 8, and 9 to stimulate polyvalent immunity. *J. Exp. Med.* 203:413.
81. Takada H, Uehara A (2006) Enhancement of TLR-mediated innate immune responses by peptidoglycans through NOD signaling. *Curr. Pharm. Des.* 12:4163.
82. Fritz JH, Girardin SE, Fitting C, Werts C, Mengin-Lecreulx D, Caroff M, Cavaillon JM, Philpott DJ, Adib-Conquy M (2005) Synergistic stimulation of human monocytes and dendritic cells by Toll-like receptor 4 and NOD1- and NOD2-activating agonists. *Eur. J. Immunol.* 35:2459.
83. Martin F, Kearney JF (2002) Marginal-zone B cells. *Nat. Rev. Immunol.* 2:323.
84. Weller S, Braun MC, Tan BK, Rosenwald A, Cordier C, Conley ME, Plebani A, Kumararatne DS, Bonnet D, Tournilhac O, Tchernia G, Steiniger B, Staudt LM, Casanova JL, Reynaud CA, Weill JC (2004) Human blood IgM "memory" B cells are circulating splenic marginal zone B cells harboring a prediversified immunoglobulin repertoire. *Blood* 104:3647.
85. Obukhanych TV, Nussenzweig MC (2006) T-independent type II immune responses generate memory B cells. *J. Exp. Med.* 203:305.
86. Bernasconi NL, Traggiai E, Lanzavecchia A (2002) Maintenance of serological memory by polyclonal activation of human memory B cells. *Science* 298:2199.
87. Bernasconi NL, Onai N, Lanzavecchia A (2003) A role for Toll-like receptors in acquired immunity: Up-regulation of TLR9 by BCR triggering in naive B cells and constitutive expression in memory B cells. *Blood* 101:4500.
88. Bourke E, Bosisio D, Golay J, Polentarutti N, Mantovani A (2003) The toll-like receptor repertoire of human B lymphocytes: Inducible and selective expression of TLR9 and TLR10 in normal and transformed cells. *Blood* 102:956.
89. Khan AQ, Lees A, Snapper CM (2004) Differential regulation of IgG anti-capsular polysaccharide and antiprotein responses to intact *Streptococcus pneumoniae* in the presence of cognate CD4+ T cell help. *J. Immunol.* 172:532.
90. Jeurissen A, et al. (2006) CD4+ T lymphocytes expressing CD40 ligand help the IgM antibody response to soluble pneumococcal polysaccharides via an intermediate cell type. *J. Immunol.* 176:529.

91. Genestier L, Taillardet M, Mondiere P, Gheit H, Bella C, Defrance T (2007) TLR agonists selectively promote terminal plasma cell differentiation of B cell subsets specialized in thymus-independent responses. *J. Immunol.* 178:7779.
92. Hornung V, Rothenfusser S, Britsch S, Krug A, Jahrsdörfer B, Giese T, Endres S, Hartmann G (2002) Quantitative expression of toll-like receptor 1-10 mRNA in cellular subsets of human peripheral blood mononuclear cells and sensitivity to CpG oligodeoxynucleotides. *J. Immunol.* 168:4531.
93. Hayashi EA, Akira S, Nobrega A (2005) Role of TLR in B cell development: Signaling through TLR4 promotes B cell maturation and is inhibited by TLR2. *J. Immunol.* 174:6639.
94. Kovarik J, Bozzotti P, Tougne C, Davis HL, Lambert PH, Krieg AM, Siegrist CA (2001) Adjuvant effects of CpG oligodeoxynucleotides on responses against T-independent type 2 antigens. *Immunology* 102:67.
95. Li WM, Bally MB, Schutze-Redelmeier MP (2001) Enhanced immune response to T-independent antigen by using CpG oligodeoxynucleotides encapsulated in liposomes. *Vaccine* 20:148.
96. Chelvarajan RL, Raithatha R, Venkataraman C, Kaul R, Han SS, Robertson DA, Bondada S (1999) CpG oligodeoxynucleotides overcome the unresponsiveness of neonatal B cells to stimulation with the thymus-independent stimuli anti-IgM and TNP-Ficoll. *Eur. J. Immunol.* 29:2808.
97. Landers CD, Bondada S (2005) CpG oligodeoxynucleotides stimulate cord blood mononuclear cells to produce immunoglobulins. *Clin. Immunol.* 116:236.
98. Zanin-Zhorov A, Tal-Lapidot G, Cahalon X, Cohen-Sfady M, Pevsner-Fischer M, Lider O, Cohen IR (2007) Cutting edge: T cells respond to lipopolysaccharide innately via TLR4 signaling. *J. Immunol.* 179:41.
99. Zughaier S, Steeghs L, van der Ley P, Stephens DS (2007) TLR4-dependent adjuvant activity of *Neisseria meningitidis* lipid A. *Vaccine* 25:4401.
100. Sen G, Khan AQ, Chen Q, Snapper CM (2005) In vivo humoral immune responses to isolated pneumococcal polysaccharides are dependent on the presence of associated TLR ligands. *J. Immunol.* 175:3084.
101. Brogden KA, DeBey B, Audibert F, Lehmkuhl H, Chedid L (1995) Protection of ruminants by *Pasteurella haemolytica* A1 capsular polysaccharide vaccines containing muramyl dipeptide analogs. *Vaccine* 13:1677.
102. Han HR, Park HM (2000) Effects of adjuvants on the immune response of staphylococcal alpha toxin and capsular polysaccharide (CPS) in rabbit. *J. Vet. Med. Sci.* 62:237.
103. Test ST, Mitsuyoshi J, Connolly CC, Lucas AH (2001) Increased immunogenicity and induction of class switching by conjugation of complement C3d to pneumococcal serotype 14 capsular polysaccharide. *Infect. Immun.* 69:3031.
104. Paoletti LC, Rench MA, Kasper DL, Molrine D, Ambrosino D, Baker CJ (2001) Effects of alum adjuvant or a booster dose on immunogenicity during clinical trials of group B streptococcal type III conjugate vaccines. *Infect. Immun.* 69:6696.
105. Sturgess AW, Rush K, Charbonneau RJ, Lee JI, West DJ, Sitrin RD, Hennessy JP Jr (1999) *Haemophilus influenzae* type b conjugate vaccine stability: Catalytic depolymerization of PRP in the presence of aluminum hydroxide. *Vaccine* 17:1169.
106. Chu RS, McCool T, Greenspan NS, Schreiber JR, Harding CV (2000) CpG oligodeoxynucleotides act as adjuvants for pneumococcal polysaccharide-protein conjugate vaccines

and enhance antipolysaccharide immunoglobulin G2a (IgG2a) and IgG3 antibodies. *Infect. Immun.* 68:1450.

107. Sen G, Chen Q, Snapper CM (2006) Immunization of aged mice with a pneumococcal conjugate vaccine combined with an unmethylated CpG-containing oligodeoxynucleotide restores defective immunoglobulin G antipolysaccharide responses and specific CD4 +- T-cell priming to young adult levels. *Infect. Immun.* 74:2177.

108. Olafsdottir TA, Hannesdottir SG, Giudice GD, Trannoy E, Jonsdottir I (2007) Effects of LT-K63 and CpG2006 on phenotype and function of murine neonatal lymphoid cells. *Scand. J. Immunol.* 66:426.

109. Schutze MP, Deriaud E, Przewlocki G, LeClerc C (1989) Carrier-induced epitopic suppression is initiated through clonal dominance. *J. Immunol.* 142:2635.

110. Barington T, Skettrup M, Juul L, Heilmann C (1993) Non-epitope-specific suppression of the antibody response to *Haemophilus influenzae* type b conjugate vaccines by preimmunization with vaccine components. *Infect. Immun.* 61:432.

111. Fattom A, Cho YH, Chu C, Fuller S, Fries L, Naso R (1999) Epitopic overload at the site of injection may result in suppression of the immune response to combined capsular polysaccharide conjugate vaccines. *Vaccine* 17:126.

112. Guttormsen HK, Sharpe AH, Chandraker AK, Brigtsen AK, Sayegh MH, Kasper DL (1999) Cognate stimulatory B-cell-T-cell interactions are critical for T-cell help recruited by glycoconjugate vaccines. *Infect. Immun.* 67:6375.

113. Wilson NS, Behrens GM, Lundie RJ, Smith CM, Waithman J, Young L, Forehan SP, Mount A, Steptoe RJ, Shortman KD, de Koning-Ward TF, Belz GT, Carbone FR, Crabb BS, Heath WR, Villadangos JA (2006) Systemic activation of dendritic cells by Toll-like receptor ligands or malaria infection impairs cross-presentation and antiviral immunity. *Nat. Immunol.* 7:165.

114. Sanchez V, Gimenez S, Haensler J, Geoffroy C, Rokbi B, Seguin D, Lissolo L, Harris B, Rizvi F, Kleanthous H, Monath T, Cadoz M, Guy B (2001) Formulation of single or multiple *H. pylori* antigens with DC Chol adjuvant induce protection by systemic route in mice. Optimal prophylactic combinations are different from therapeutic ones. *FEMS Immunol. Med. Microbiol.* 30:157.

115. Yewdell JW (2006) Confronting complexity: Real-world immunodominance in antiviral CD8+ T cell responses. *Immunity* 25:533.

116. Pattnaik P, Shakri AR, Singh S, Goel S, Mukherjee P, Chitnis CE (2007) Immunogenicity of a recombinant malaria vaccine based on receptor binding domain of Plasmodium falciparum EBA-175. *Vaccine* 25:806.

117. Joseph A, Louria-Hayon I, Piis-Finarov A, Zeira E, Zakay-Rones Z, Raz E, Hayashi T, Takabayashi K, Barenholz Y, Kedar E (2002) Liposomal immunostimulatory DNA sequence (ISS-ODN): An efficient parenteral and mucosal adjuvant for influenza and hepatitis B vaccines. *Vaccine* 20:3342.

118. Harper DM, Franco EL, Wheeler CM, Moscicki AB, Romanowski B, Roteli-Martins CM, Jenkins D, Schuind A, Costa Clemens SA, Dubin G; HPV Vaccine Study Group (2006) Sustained efficacy up to 4.5 years of a bivalent L1 virus-like particle vaccine against human papillomavirus types 16 and 18: Follow-up from a randomised control trial. *Lancet* 367:1247.

119. Burdin N, Guy B, Moingeon P (2004) Immunological foundations to the quest for new vaccine adjuvants. *BioDrugs* 18:79.

120. Aurelian L (2004) Herpes simplex virus type 2 vaccines: New ground for optimism? *Clin. Diagn. Lab. Immunol.* 11:437.
121. Drachenberg KJ, Wheeler AW, Stuebner P, Horak F (2001) A well-tolerated grass pollen-specific allergy vaccine containing a novel adjuvant, monophosphoryl lipid A, reduces allergic symptoms after only four preseasonal injections. *Allergy* 56:498.
122. Creticos PS, Schroeder JT, Hamilton RG, Balcer-Whaley SL, Khattignavong AP, Lindblad R, Li H, Coffman R, Seyfert V, Eiden JJ, Broide D; Immune Tolerance Network Group (2006) Immunotherapy with a ragweed-toll-like receptor 9 agonist vaccine for allergic rhinitis. *N. Engl. J. Med.* 355:1445.
123. Krieg AM (2006) Therapeutic potential of Toll-like receptor 9 activation. *Nat. Rev. Drug Discov.* 5:471.
124. Kretschmer K, Apostolou I, Hawiger D, Khazaie K, Nussenzweig MC, von Boehmer H (2005) Inducing and expanding regulatory T cell populations by foreign antigen. *Nat. Immunol.* 6:1219.
125. Barber DL, Wherry EJ, Masopust D, Zhu B, Allison JP, Sharpe AH, Freeman GJ, Ahmed R (2006) Restoring function in exhausted CD8 T cells during chronic viral infection. *Nature* 439:682.
126. Trautmann L, Janbazian L, Chomont N, Said EA, Gimmig S, Bessette B, Boulassel MR, Delwart E, Sepulveda H, Balderas RS, Routy JP, Haddad EK, Sekaly RP (2006) Upregulation of PD-1 expression on HIV-specific CD8+ T cells leads to reversible immune dysfunction. *Nat. Med.* 12:1198.
127. Freeman GJ, Wherry EJ, Ahmed R, Sharpe AH (2006) Reinvigorating exhausted HIV-specific T cells via PD-1-PD-1 ligand blockade. *J. Exp. Med.* 203:2223.
128. Pol S, Michel ML (2006) Therapeutic vaccination in chronic hepatitis B virus carriers. *Expert Rev. Vaccines* 5:707.
129. Freytag LC, Clements JD (2005) Mucosal adjuvants. *Vaccine* 23:1804.
130. Jones B, Pascopella L, Falkow S (1995) Entry of microbes into the host: Using M cells to break the mucosal barrier. *Curr. Opin. Immunol.* 7:474.
131. Clark MA, Blair H, Liang L, Brey RN, Brayden D, Hirst BH (2001) Targeting polymerised liposome vaccine carriers to intestinal M cells. *Vaccine* 20:208.
132. Lavelle EC (2001) Targeted delivery of drugs to the gastrointestinal tract. *Crit. Rev. Ther. Drug Carrier Syst.* 18:341.
133. Nochi T, Yuki Y, Matsumura A, Mejima M, Terahara K, Kim DY, Fukuyama S, Iwatsuki-Horimoto K, Kawaoka Y, Kohda T, Kozaki S, Igarashi O, Kiyono H (2007) A novel M cell-specific carbohydrate-targeted mucosal vaccine effectively induces antigen-specific immune responses. *J. Exp. Med.* 204:2789.
134. Pizza M, Giuliani MM, Fontana MR, Monaci E, Douce G, Dougan G, Mills KH, Rappuoli R, Del Giudice G (2001) Mucosal vaccines: Non toxic derivatives of LT and CT as mucosal adjuvants. *Vaccine* 19:2534.
135. Glenn GM, Kenney RT, Ellingsworth LR, Frech SA, Hammond SA, Zoeteweij JP (2003) Transcutaneous immunization and immunostimulant strategies: Capitalizing on the immunocompetence of the skin. *Expert Rev. Vaccines* 2:253.
136. Weltzin R, Guy B, Thomas WD Jr, Giannasca PJ, Monath TP. (2000) Parenteral adjuvant activities of Escherichia coli heat-labile toxin and its B subunit for immunization of mice against gastric *Helicobacter pylori* infection. *Infect. Immun.* 68:2775.

137. Najar HM, Dutz JP (2007) Topical TLR9 agonists induce more efficient cross-presentation of injected protein antigen than parenteral TLR9 agonists do. *Eur. J. Immunol.* 37:2242.
138. Sesardic D (2006) Regulatory considerations on new adjuvants and delivery systems. *Vaccine* 24:S2–86.
139. Deane JA, Pisitkun P, Barrett RS, Feigenbaum L, Town T, Ward JM, Flavell RA, Bolland S (2007) Control of Toll-like receptor 7 expression is essential to restrict autoimmunity and dendritic cell proliferation. *Immunity* 27:801.
140. Marshak-Rothstein A (2006) Toll-like receptors in systemic autoimmune disease. *Nat. Rev. Immunol.* 6:823.
141. Bluestone JA, St Clair EW, Turka LA (2006) CTLA4Ig: Bridging the basic immunology with clinical application. *Immunity* 24:233.
142. Suntharalingam G, Perry MR, Ward S, Brett SJ, Castello-Cortes A, Brunner MD, Panoskaltsis N (2006) Cytokine storm in a phase 1 trial of the anti-CD28 monoclonal antibody TGN1412. *N. Engl. J. Med.* 355:1018.
143. Hanke T (2006) Lessons from TGN1412. *Lancet* 368:1569.
144. Stebbings R, Findlay L, Edwards C, Eastwood D, Bird C, North D, Mistry Y, Dilger P, Liefooghe E, Cludts I, Fox B, Tarrant G, Robinson J, Meager T, Dolman C, Thorpe SJ, Bristow A, Wadhwa M, Thorpe R, Poole S (2007) "Cytokine storm" in the phase I trial of monoclonal antibody TGN1412: Better understanding the causes to improve preclinical testing of immunotherapeutics. *J. Immunol.* 179:3325.
145. Mestas J, Hughes CC (2004) Of mice and not men: Differences between mouse and human immunology. *J. Immunol.* 172:2731.
146. Gorden KK, Qiu XX, Binsfeld CC, Vasilakos JP, Alkan SS (2006) Cutting edge: Activation of murine TLR8 by a combination of imidazoquinoline immune response modifiers and polyT oligodeoxynucleotides. *J. Immunol.* 177:6584.
147. Nahori MA, Fournié-Amazouz E, Que-Gewirth NS, Balloy V, Chignard M, Raetz CR, Saint Girons I, Werts C (2005) Differential TLR recognition of leptospiral lipid A and lipopolysaccharide in murine and human cells. *J. Immunol.* 175:6022.
148. Liu T, Matsuguchi T, Tsuboi N, Yajima T, Yoshikai Y (2002) Differences in expression of toll-like receptors and their reactivities in dendritic cells in BALB/c and C57BL/6 mice. *Infect. Immun.* 70:6638.
149. Potter CW, Jennings R (2003) Effect of priming on subsequent response to inactivated influenza vaccine. *Vaccine* 21:940.
150. Spisek R, Brazova J, Rozkova D, Zapletalova K, Sediva A, Bartunkova J (2004) Maturation of dendritic cells by bacterial immunomodulators. *Vaccine* 22:2761.
151. Sanchez V, Hessler C, DeMonfort A, Lang J, Guy B (2006) Comparison by flow cytometry of immune changes induced in human monocyte-derived dendritic cells upon infection with dengue 2 live-attenuated vaccine or 16681 parental strain. *FEMS Immunol. Med. Microbiol.* 46:113.

4

CARBOHYDRATE-BASED ANTIBACTERIAL VACCINES

Robert A. Pon and Harold J. Jennings
Institute for Biological Sciences, National Research Council of Canada, Ottawa, Ontario, Canada K1A0R6

4.1 INTRODUCTION

Throughout time humans have engaged in a struggle for survival against the most insidious of enemies, namely microorganisms, and in particular the bacterium. Significant strides have been achieved over the past century; none more important than the development of the field of modern vaccinology, together with the discovery of penicillin that introduced the perceived magic bullet era of antibiotics. Despite these advances, bacterial infections remain a leading cause of death and morbidity particularly in the developing world for reasons that include the unavailability of effective vaccines against many bacterial pathogens. Perhaps more disconcerting is that the most vulnerable segment of our population suffering from high bacterial morbidity and mortality is children aged less than 5 years old. Furthermore, technological and medical advances resulting in median life span increases, together with the increased incidence of disease associated with immune deficiencies such as AIDS (acquired immune deficiency syndrome) and cancers, have placed a higher degree of need for more effective antibacterial prophylaxis within these burgeoning sectors.

Carbohydrate antigens were recognized early in the twentieth century as potential vaccine determinants when capsular polysaccharides isolated from *Streptococcus pneumoniae* were shown to be immunogenic in both rodents [1] and humans [2]. Following on these observations came the seminal human trial by MacLeod et al. in

Carbohydrate-Based Vaccines and Immunotherapies. Edited by Zhongwu Guo and Geert-Jan Boons
Copyright © 2009 John Wiley & Sons, Inc.

1945, which illustrated for the first time, that a nonprotein vaccine containing capsular polysaccharides derived from four different pneumococcal serotypes was found to be 90% effective in preventing adult pneumonia [3] and that resulted in its licensure in 1946. It was not until several decades later that there was resurgence in interest in capsular polysaccharide vaccines when enthusiasm for antibiotics waned with the recognition that they were not the panacea for all bacterial infections as was originally hoped.

The capsular polysaccharides of pathogenic bacteria are attractive vaccine candidates because they constitute the most highly conserved and most exposed bacterial surface antigens. These polymers, composed of oligosaccharide repeating units, are defined and characterizable structures that are easily purified from bacterial cultures and are amenable to chemical modifications. Due to their relative simplicity, safety, and affordability, licensed capsular polysaccharide vaccines have proven effective against human disease caused by encapsulated *S. pneumoniae* (23 serotypes), *Haemophilus influenzae* type b, *Neisseria meningitidis* serogroups A, C, W-135, and Y, and *Salmonella typhi*.

Stand-alone capsular polysaccharide vaccines, although generally effective in healthy adult populations, remain deficient in children less than five years old and virtually ineffective below the age of 2, the segment of the population most vulnerable to bacterial infection due to their relative immature immune systems [4]. Application of the principles first demonstrated by Avery and Goebel [5], whereby coupling pneumococcus serotype 3 oligosaccharides to protein rendered them immunogenic, has led to a rapid development over the last 20 years of the glycoconjugate vaccine field. Glycoconjugate vaccines have quickly been accepted and incorporated into routine childhood immunization programs within the developed world, and their impact has been dramatic; oftentimes virtually eliminating disease caused by the vaccine-related microorganism.

The intent of this review is to explore the current state of polysaccharide and glycoconjugate vaccines with a focus on bacterial diseases of the highest priority and whose vaccines have been licensed or are at an advanced stage of clinical evaluation. This review will cover the current state of polysaccharide-based vaccines but will focus on the newer generation glycoconjugate vaccines developed against *H. influenzae*, *N. meningitidis*, *S. pneumoniae*, *S. typhi*, and group B *Streptococcus* glycoconjugate vaccines, together with the lessons learned from over two decades of human use and their future prospects. The reader is referred to Chapter 2 of this volume and Jennings and Sood [6] for a detailed review of the strategies and technologies used in the rational design and synthesis of antibacterial glycoconjugate vaccines as well as to other comprehensive reviews of this subject [7–9].

4.2 POLYSACCHARIDE AND GLYCOCONJUGATE IMMUNOBIOLOGY

Humoral immunity, or the ability to mount a timely, specific, and effective antibody response, is the single most important adaptive measure to protect against invasive diseases caused by encapsulated bacteria. The level of circulating antibody specific to the capsular polysaccharides is the immune correlate most often cited for protective efficacy whereby invasive bacteria decorated with specific antibody and complement are cleared through complement-mediated bacteriolysis or through ingestion by phagocytic cells. Historically, polysaccharides are considered as T-cell independent type 2 (TI-2) antigens [10] due to their ability to specifically stimulate B cells, without the active participation of helper T cells, leading to their differentiation into antibody secreting, polysaccharide-specific plasma cells. It is the high valency of the oligosaccharide repeat units found within high-molecular-weight capsular structures that bestows the ability to stimulate B cells directly. This occurs by virtue of focused multivalent signaling through polysaccharide-specific B-cell receptor clustering (membrane-bound Ig clustering) that is sufficient to surpass the B-cell activation threshold. Physiochemically, polysaccharides require molecular masses in excess of 100 kDa with a minimum of 20 epitopes per polymer spatially separated over 100 Å of cell surface to elicit B-cell responses [11]. Typically, polysaccharide antigens stimulate the production of predominantly low-affinity, non-class-switched IgM antibody stemming from a limited clonal pool of B cells [12]. Furthermore, B cells responding to TI-2 antigens do not characteristically form germinal centers, undergo affinity maturation through somatic mutation, or generate a memory B-cell response. The lack of these beneficial features is attributed to the lack of polysaccharide-induced cognate T-cell help; however, TI-2 responses are influenced by T and other immune cells that can affect antibody production and quality. Secretion of cytokines by bystander T cells, antigen-independent cognate interactions between T and polysaccharide-activated B cells (e.g., CD40L–CD40 interaction) [13], as well as direct polysaccharide stimulation of innate immune system cells [14, 15], all play a role in augmenting the resultant antibody response and promoting an inefficient class switch to other Ig isotypes [16]. TI-2 responses as described are, however, quite effective in certain segments of the population (i.e., healthy adolescents and adults), as they generally produce long-lived circulating and functional antibody [17] that is well suited to protect against invasive bacterial infection.

In contrast, covalent coupling of polysaccharide to protein adds a T-dependent (TD) dimension to the polysaccharide component of the glycoconjugate by virtue of the TD properties of the carrier protein resulting in cognate T-cell help. The key mechanistic step involves the B-cell repertoire that expresses surface Ig specific for polysaccharide elements within the glycoconjugate. Internalization of the polysaccharide component via receptor-mediated endocytosis also internalizes the carrier protein, which is subsequently digested and expressed on the B-cell surface complexed with major histocompatibility complex (MHC) class II molecules. Carrier-specific T cells activated in a separate event by professional dendritic cells (DCs) are now capable of recognizing the specific peptide/MHC complexes on the B cells that also possess polysaccharide-specific Ig receptors. The T cell, through upregulated B-cell adhesion, focal delivery of cytokines, and CD40L–CD40/CD28–B7.1/.2 mediated signaling (cognate T-cell help), induces affinity-matured, isotype-switched antibody from germinal center-derived, high-frequency antibody-secreting plasma cells and the generation of long-lived memory B cells [18, 19]. There is, however, an important qualitative distinction between a classical TD response to protein vaccines and glycoconjugates.

Glycoconjugate antibody responses rely on a restricted oligoclonal repertoire of immunoglobulin heavy and light chains [20] that possess limited genetic differences with germ line genes [21, 22] more reminiscent of TI-2 polysaccharide responses. This suggests that somatic hypermutations associated with affinity maturation occur relatively inefficiently with polysaccharide antigens regardless of their presentation in a TD form. Furthermore, the diversity of polysaccharide-specific antibody isotype is more restricted, at least in younger age groups [23] compared to protein antigens. The TD properties of a glycoconjugate have the advantage of generating memory B cells, a topic of great current interest, which can respond to polysaccharide antigen with greater kinetics and antibody levels during secondary immune responses. Indeed, it is well established that humans primed with a polysaccharide conjugate vaccine and later challenged with the corresponding plain polysaccharide induces a memory B-cell booster response characterized by high levels of polysaccharide-specific antibody produced within 5–7 days of immunization [24]. The breadth of glycoconjugate memory B-cell responses, however, requires careful consideration since the protein component of conjugate vaccines are generally different from the pathogen they were designed to protect us from. Commonly used carrier proteins are tetanus, diphtheria, and diphtheria mutant toxoids that do not share T-cell epitopes with capsular pathogens of note; hence the generation of carrier-specific memory T-cell populations will be nonresponsive upon exposure to the given pathogen due to a lack of recognition. This phenomenon is observed when anamnestic responses to plain polysaccharide challenges are compared to those in which memory T-cell help is operational (e.g., boosting with conjugate vaccine), where the latter antibody response can be orders of magnitude greater thereby directly influencing the long-term effectiveness of conjugate vaccines [25, 26]. Although we now understand how T-cell help can initiate and strengthen the subsequent B-cell response to the polysaccharide component of glycoconjugates, it is its subsequent role in B-cell maturation and persistence of plasma cells and the events leading up to the formation of polysaccharide-specific memory cells that remains to be elucidated and is key to more effective conjugate vaccines as will become evident in later discussions.

4.3 DEFICIENCIES IN THE HUMAN IMMUNE RESPONSE TO POLYSACCHARIDES

Over three decades of data have now been collected on the efficacy of capsular polysaccharide vaccines in humans beginning with the first licensed 14-valent *S. pneumoniae* vaccine in 1977 [27]. In addition, capsular polysaccharide vaccines against *H. influenzae*, *N. meningitidis*, and *S. typhi* have been in routine use together with several clinical studies using capsular polysaccharides of group B *Streptococcus* and *K. pneumoniae* [28]. Significant limitations of the polysaccharide vaccines have now been clearly identified, with the most prominent being a general lack of antibody responsiveness in infants and young children, the population most at risk for invasive bacterial disease. This effect is not restricted to the use of polysaccharide vaccines as children <2 years of age frequently fail to mount a specific antibody response following invasive bacterial infection with encapsulated bacteria [29]. This hyporesponsiveness has been attributed to the TI-2 nature of the polysaccharide antigens coupled to an immature immune system that has not yet had the benefit of multiple exposures with cross-reactive natural flora [30], and it is estimated that maximal responsiveness to polysaccharides occurs only after the age of 6 [8]. This age dependency has been correlated with the maturity status of splenic marginal zone B cells [31, 32], thought to be a key cell type for TI-2 polysaccharide antigens [24, 33] but not for TD antigens that have the benefit of T-cell help, which is more resilient at younger ages. For reasons that are still not clear, some capsular polysaccharides such as group A meningococcal [34] or pneumococcal type 3 polysaccharides [35] are immunogenic in young infants despite their immature immune architecture; albeit their immune responses are not as robust as in older children. In fact, the group A polysaccharide behaves with TD-like properties eliciting booster responses upon reexposure to the polysaccharide [36], suggesting that they may not be as reliant on marginal zone B cells as proposed above.

There are also polysaccharide response deficiencies in those populations suffering from some form of immunodeficiency, the largest being those persons with ages >65. There are fundamental differences in the immune systems of the latter group compared with infants such that mechanistic correlations are difficult despite similar outcomes (e.g., poor immunogenicity). Notably, the elderly do not suffer from immature immune system architectures and have had the benefit of a lifetime of exposure to capsular structures on commensal organisms. A possible reason for deficient polysaccharide responses in this group are changes in the quality of the antibody produced [37] such as reduced opsonizing activity due to the production of lower avidity antibody [38]. In addition, multiple exposure to cross-reactive capsular structures on prevalent microorganisms can often lead to hyporesponsiveness such as the constitutive poor response to the pneumococcal polysaccharide vaccine in patients prone to pneumococcal pneumoniae [39].

Other than the elderly, polysaccharide vaccination is also less than adequate for individuals with underlying immune disorders resulting in immunodeficiency such as the human immunodeficiency virus (HIV), splenectomized individuals, sickle cell disease, chronic diabetes, and blood-borne cancers [40]. Polysaccharide vaccination is also poorly effective among various ethnic groups such as Alaskan Natives and American Indians [41] where genetic and environmental factors preclude susceptibility.

4.4 GLYCOCONJUGATE VACCINES

Prophylactic vaccines against invasive bacteria are in the midst of a renaissance due to the need to protect our most vulnerable segments of the population, beginning with the first licensed conjugate vaccine for *H. influenzae* type b (Hib) in 1987. Over the course of the past 20 years, conjugate vaccines have been developed and brought to late-stage clinical trial or market for many of the invasive pathogens for which we have had the greatest need, and significant efforts are being directed to those that are still outstanding. Currently, there are efficacious conjugate vaccines that have dramatically reduced or potentially could reduce the incidence of invasive disease against Hib, selected serotypes of *S. pneumoniae*, *N. meningitidis* serogroups A, C, W-135, and Y, *S. typhi*, and group B *Streptococcus*. Aside from the challenge of developing novel conjugate vaccines to those diseases for which vaccines would be most beneficial, such as group B meningococcal meningitis or enteric diseases, the current focus and one that places tremendous reliance and demands on the clinical setting, is in the formulation of combination and/or multivalent vaccines [42].

Extensive preclinical work has been performed investigating the slew of chemical and physical parameters necessary for the optimal design of effective conjugate vaccines, and an overview of this large body of work can be found in Chapter 2 as well as other comprehensive reviews on the matter [6, 7, 28]. The difficulty lies in the assessment of how these parameters may influence immunogenicity and ultimately efficacy in humans since there is often poor translation from animal studies to human use. Compounding this evaluation is the difficulty, expense, and ethics of performing comparative human studies to tease out the optimal strategy. Largely based on the performance of newly developed conjugate vaccines in humans, coupled with extensive postlicensure follow-up studies, immune correlates for vaccine efficacy have been estimated that have allowed for further refinements of conjugate vaccine technology. These important parameters are discussed in the context of the individual vaccines studies in the subsequent sections.

4.5 *Haemophilus influenzae*

Haemophilus influenzae is a gram-negative bacillus that regularly inhabits the nasopharynx and is serologically defined into six types (a–f) based on the elaborated capsular polysaccharide. Invasive disease, however, is primarily caused by *H. influenzae* type b (Hib) (>95% of cases) where it remains a major cause of meningitis, septicemia, pneumonia, epiglottis, and septic arthritis, particularly in the developing world. Hib disease occurs predominantly in children younger than 5 years of age (80% of cases) where ~40% of Hib-related disease occurs in children <4 months of age [4]. Prior to the introduction of effective Hib vaccines, 1 in 200 children developed invasive Hib disease, of which 60% developed meningitis, and where it and pneumonia were responsible for the majority of Hib-related deaths [43]. Among survivors, 20–30% had significant and permanent sequelae ranging from hearing loss to neurological deficits.

4.5.1 Hib Polysaccharides

A distinct advantage in the battle against Hib-related invasive disease is the restriction to only one serotype, which in turn renders the capsular polysaccharide a prime candidate for vaccine studies. The first generation Hib vaccine was based on the purified type b capsular polysaccharide, composed of ribose ribitol phosphate repeating units (PRP; Table 4.1), which became available in 1985.

Table 4.1 Structures of Polysaccharides from *Haemophilus influenzae*, *Neisseria meningitidis*, and *Salmonella typhi*

Group/Type	Structure	Reference
H. influenzae		
b	→3) βD-Rib$f$ (1 → 1) D-Ribitol-(5-OPO$_3$—	249
N. meningitidis		
A	→6) αD-Man$p$NAc (1-OPO$_3$—	250
B	→8) αD-Neu$p$NAc (2→	251
C	→9) αD-Neu$p$NAc(2→ 7/8 ↑ OAc	251
W-135	→6) αD-Gal$p$ (1 → 4) αD-Neu$p$NAc (2→	252
Y	→6) αD-Glc$p$ (1 → 4) αD-Neu$p$NAc (2 → (+OAc)	252
S. typhi		
Vi	→4) αD-Gal$p$NAcA(1→ 3 ↑ OAc$_{(0.6-0.9)}$	253
Di-*O*-acetylated pectin	→4) αD-Gal$p$A(1→ 2, 3 ↑ OAc	227

After preliminary trials in adults that demonstrated safety and the production of long-lived bactericidal antibodies [44], trials in children [45, 46], including a large-scale efficacy trial for children age 3 months to 5 years, were performed. The study results were disappointing in that the vaccine proved ineffective in children <18 months of age, the target population most in need of protection. Furthermore, specific antibody levels, which are the accepted immune correlate of protection for invasive Hib disease, was found to wane quickly as a function of younger ages and produced no evidence of memory upon polysaccharide boosting [47]. We now rationalize the poor immune response of the polysaccharide vaccine among children due to the TI-2 nature of the antigen as discussed in the previous section. However, despite the limitations of the polysaccharide vaccine, due to the severity of Hib-related invasive disease, the vaccine was licensed for use in children >24 months of age.

4.5.2 Hib Conjugate Vaccines

The development of Hib polysaccharide conjugate vaccines, which are the prototypical examples of this new generation of vaccine, was an example of a "perfect storm" for vaccines for the following reasons: (1) The high incidence and severity of Hib disease in industrial countries was the driving force behind its development. (2) Since invasive disease was caused by only 1 serotype, a monovalent vaccine was sufficient as opposed to the challenges presented by organisms such as pneumococcus with >90 different serotypes. (3) The simplicity of PRP allowed for easy chemical

Table 4.2 Hib Conjugate Vaccines in the United States

Vaccine	Trade Name	Manufacturer	Protein	Linkage	Saccharide Size	Efficacy (%)[a]
PRP-D	ProHibit	Connaught	Dt	Random	Moderate	(a) 94 (b) 32
HbOC	HibTiter	Wyeth	CRM_{197}	Defined	Small	86
PRP-T	ActHib	Sanofi-Pasteur	TT	Random	Large	82
PRP-OMP	PedvaxHib	Merck	OMP	Random	Moderate	86

[a]See Ref. [56].

degradation, activation, and final vaccine characterization (see Chapter 2). (4) Finally, efficacy trials with the plain PRP vaccine had previously established benchmark immunogenicity correlates (i.e., specific antibody levels >1.0 μg/ml) for protection allowing for quicker evaluation of a candidate vaccine's potential at earlier time points.

Connaught Laboratories produced the first licensed Hib conjugate vaccine in 1987 after the pioneering work by Schneerson et al. [48] through random conjugation of PRP to diphtheria toxoid (DT). Following on the heels of this vaccine, three other Hib conjugate vaccines, all differing in their carrier proteins and method of conjugation, have received Food and Drug Administration (FDA) approval, and these three conjugates currently form the backbone for routine Hib vaccination in North America (Table 4.2).

Recently, a novel Hib-TT vaccine was produced and licensed in Cuba where the Hib component was produced synthetically (average of seven repeating units) and performed adequately in comparative clinical trials with a commercially available Hib-cross-reactive mutant diphtheria toxin (CRM_{197}) conjugate [49]. It is unknown at this time whether synthetic vaccines such as this will receive more widespread use due to their associated production costs. The current recommended Hib conjugate vaccine immunization schedule for children in the United States is a two- or three-dose primary regime at 2, 4, and 6 months (HbOC, PRP-T) or 2 and 6 months (PRP-OMP), followed by a booster dose in the second year [50]. Using this schedule, all conjugate vaccines, with the exception of the original PRP-D vaccine (see below), resulted in the production of specific PRP antibodies at orders of magnitude higher levels than compared to baseline. The circulating level of anti-PRP antibody is generally recognized as the immune correlate for protective efficacy based largely on the previous studies with pure Hib polysaccharide vaccines. Levels of >0.15 μg/ml are considered protective in the short term, whereas levels >1.0 μg/ml provide long-term protection primarily through bactericidal activity [44]. That these levels could be maintained over time was demonstrated by tracking anti-PRP antibody levels in a Swedish cohort that had received the PRP-T conjugate vaccine during infancy and subsequently showed >4 μg/ml levels after 10 years [51].

All four Hib conjugates have run the gauntlet of extensive safety and efficacy trials in children, including trials in Finland, the United States, Gambia, and the United Kingdom [52–55]. The efficacies of Hib conjugate vaccines were recently the subject of a meta-analysis with data spanning 8 randomized trials and 365,368 children under the age of 5 [56]. Overall, the protective efficacy of Hib vaccination was found to be

84%, with individual trials reaching as high as 100% point efficacies (1 year) [54, 57]. The original PRP-D Hib conjugate showed disparate results being efficacious in a Finnish trial (94%) [52], but at the same time, found to be only 32% effective in a population with high endemic Hib disease rates at young ages [58]. Based largely on the poor efficacy of the PRP-D vaccine in this population and the generally low levels of antibody in infants under 6 months (0.2–0.42 μg/ml) [43], this vaccine originally licensed for older children is no longer in use. This subpar performance, relative to the other Hib conjugate vaccines, most likely lies in the conjugation procedure that resulted in a poorly defined vaccine [7]. Notably, there were only minor differences in vaccine efficacy among the three remaining and current Hib conjugate vaccines (83, 86, and 86% for PRP-T, HbOC, and PRP-OMP, respectively) [59, 60]. The introduction of routine Hib childhood vaccines in the United States has had a dramatic effect on the incidence of invasive Hib disease. Prevaccine rates were ~37/100,000 (1990), which precipitously fell by >99% to 0.3 cases/100,000 in 2000 [61]. Similar spectacular declines have been reported in all countries, including industrialized countries that have instituted routine childhood Hib immunization programs [59]. The fall off of Hib-related disease is in large part due to the surprising and dramatic reduction of Hib carriage in children within vaccinated populations, directly facilitating herd immunity [62, 63].

Pre- and postlicensure studies, particularly with the prototypical Hib conjugates, have revealed some important effects inherent to conjugate design (e.g., carrier protein influences) and implementation (scheduling, combination vaccines) that influence protective immunity and are often only revealed through long-term follow-up (carriage, herd immunity). Some of these issues have been discussed (see Chapter 2) or will be covered in subsequent sections.

4.6 Neisseria meningitidis

Neisseria meningitidis (Nm) is a leading cause of bacterial meningitis and sepsis and a frequent cause of endemic and epidemic disease worldwide. Epidemiology for meningococcal disease worldwide is extremely complex with regional differences related to the various serogroups/types, the propensity for epidemic or endemic disease, and the expression and migration of hypervirulent clones. In the United States, Nm has now become the leading cause of bacterial meningitis, reflecting the success of both the Hib and the 7-valent pneumococcal conjugate vaccines that have through routine use dramatically decreased meningitis due to these organisms. Nm disease rate is estimated at 1–6 cases per 100,000 for the United States and Europe [64], representing about 3000 individuals annually in the United States. As with most other invasive bacterial diseases, the rate of meningococcal meningitis is highest within the first 2 years of life, but unlike other invasive bacteria, the incidence of Nm disease is biphasic with resurgence in the adolescent years.

Neisseria meningitidis is a natural commensal of humans and can often be found asymptomatically colonizing the nasopharyngeal cavity of individuals (5–10% or higher depending on the environment), which upon penetration and invasion can lead to rapid progression of disease with high rates of morbidity and mortality. Overall case fatality rates are approximately 10–12% but can reach as high as 20% in children aged 1–4 years, 25% aged 5–14 years [65], and 40% with associated sepsis [66], whereas up to 20% of survivors suffer from some form of serious sequelae [64]. The gram-negative encapsulated diplococcus exists with at least 13 different serogroups, as defined by the composition of their capsular polysaccharide; however, disease is mainly restricted to serogroups A, B, C, W-135, and Y (Table 4.1).

4.6.1 Meningococcal Polysaccharide Vaccines

The pioneering work by Gotschlich et al. [67] demonstrated that the capsular polysaccharides of groups A and C Nm were immunogenic in humans and induced specific serum bactericidal antibody against the respective organisms, the immune correlate most often used to assess efficacy of both the polysaccharide and the newer generation conjugate vaccines. This vaccine was developed in response to epidemics among military recruits in the United States and has proven to be highly effective in this population (89%), controlling outbreaks in Egypt [68] and in Burkina Faso [69]. As a consequence, a bivalent A/C vaccine was licensed in 1976 in the United States that was subsequently replaced by the advent of the current tetravalent A, C, W-135, and Y vaccine (Menomune-Sanofi Pasteur) [70]. Other polysaccharide combinations covering the range from mono-, bi-, tri, and, tetravalent polysaccharide vaccines are available outside of the United States. The efficacy of the W-135 and Y components are not known at this time [71] due to the low disease incidence during trials, although a trivalent ACW-135 vaccine was used successfully to control an epidemic in Burkina Faso where NmC and NmA and for the first time W-135 were the causative agents [69].

The uses of meningococcal polysaccharide vaccines have identified some key limitations: (1) The group C polysaccharide is poorly immunogenic in young children consistent with other capsular polysaccharides, but surprisingly group A polysaccharide is modestly immunogenic in infants (see Section 4.3). (2) Vaccine effectiveness drops quickly in children <4 years old as compared to ages >4 [72]. (3) Booster doses of polysaccharide induce hyporesponsiveness [73]. (4) O-acetylation of the group A is a critical component for the generation of effective bactericidal responses [74]. (5) The current tetravalent formulation excludes serogroup B, which remains a major contributor to Nm disease worldwide, and in some countries like the United Kingdom it is now responsible for up to 90% of Nm disease [75].

4.6.2 Meningococcal Conjugate Vaccines

4.6.2.1 Serogroup A Group A meningococci (NmA) are responsible for large-scale epidemics in developing countries—in particular the sub-Saharan region of Africa known as the meningitis belt—but are rarely causes of disease in North America or Europe. Attack rates during epidemics can be as high as 1000/100,000 with recorded reports of upward of 200,000 cases in 1996 with 20,000 associated deaths [76]. In nonepidemic periods, the rate of endemic disease is still approximately 20-fold higher relative to the West at 30/100,000.

The capsular polysaccharide vaccine (A/C or A/C/W-135/Y) is made available to limit NmA outbreaks [77] and has had an impact in curbing major outbreaks [78]; however, its use has been more reactive than proactive. The drawback is the relatively poor response and longevity of the bactericidal antibody response in young children as opposed to older children and adults [79]; however, this negative is offset by the TD-like nature of the polysaccharide, rendering it somewhat immunogenic and boostable at young ages.

Two early conjugate vaccines encompassing NmA have been synthesized and evaluated in adults and children. The first vaccine produced by Novartis (Chiron) was a bivalent A/C conjugate containing short oligosaccharides of each serogroup (~6 residues) coupled to CRM_{197}. The second vaccine produced by Sanofi Pasteur was originally a bivalent A + C conjugate but has since been expanded into a tetravalent ACW-135 and Y conjugate vaccine with diphtheria toxoid (DT) as the carrier (Menactra). The salient points drawn from the early clinical trials in young children [80–85], which compared the bivalent conjugate vaccines with a bivalent polysaccharide (PS) vaccine, were: (1) The bivalent conjugate vaccine induced similar levels of antibody as the bivalent polysaccharide vaccine. (2) Antibodies generated using an accelerated immunization schedule (2, 3, 4 months) did not persist beyond 14 months. (3) Regardless of the absolute level of specific antibody produced, which was almost equivalent for the bivalent conjugate and polysaccharide vaccines alike, the bivalent conjugate vaccines induced qualitatively more functional activity (increased bactericidal activity [80] and higher avidities [86]). This latter result is one of the first indications that circulating antibody levels may not be sufficient as a predictor of effectiveness. (4) By age 5, there were no apparent advantages of the conjugates over the bivalent polysaccharide vaccine [83]. Thus, it would appear as though the bivalent conjugate vaccines may be somewhat more efficacious in the early years compared to the bivalent polysaccharide vaccine, but this benefit is lost by late childhood when natural immunological priming to NmA has occurred. From our gained collective knowledge with conjugate vaccine design, it is now apparent that the use of short oligosaccharides was probably a major factor for the performance characteristics of these initial bivalent A/C conjugate vaccines [87].

A third NmA conjugate vaccine, currently in phase II trials, is being produced in a unique manner solely for developing countries. With the goal of producing a monovalent A conjugate vaccine for less than $0.50/dose, the Meningitis Vaccine Project (MVP) with funding from the Gates Foundation has partnered a source of clinical-grade NmA polysaccharide (SynCo Bio Partners) with a developing world vaccine manufacturer (Serum Institute of India) and with a U.S. public research center (CBER/FDA) to develop and transfer conjugation technology. NmA polysaccharide has been conjugated to tetanus toxoid (TT) by reductive amination [88] using a modified hydrazide procedure [89]. Phase 1 trials in adults [90] have demonstrated safety, with a moderately high percentage of individuals after 1 year with greater than fourfold rises in serum bactericidal activities (SBA) that were significantly superior to levels induced by a stand-alone bivalent A/C polysaccharide vaccine. Phase II trials are currently underway in toddlers, and it is expected the vaccine will be licensed in 2008–2009 (India). The vaccine will be introduced into mass vaccination programs as a single dose for ages 1–2 years and a two-dose regime (14 and 36 weeks) for infants.

4.6.2.2 Serogroup C *Neisseria meningitidis* serogroup C (NmC) along with serogroup B (NmB) and more recently serogroup Y (NmY) in the United States are the primary etiological agents for invasive meningococcal disease in the developed world. NmC is most often present at endemic levels in the United States (0.3 cases/ 100,000) and is responsible for about one-third of all cases of invasive Nm disease primarily in older children and adolescents. The United Kingdom, Canada, and other regions of Europe have been exhibiting a steady rise in the incidence of NmC-related disease from 26 to 34% of cases in 1994–1998 [91] that has been associated with a hypervirulent and hyperinvasive clone belonging to the ST11 complex [92]. Although two bivalent A/C conjugate vaccines (described in the previous section) were developed but not licensed in the 1980s, it was this disconcerting rise in virulent NmC disease that prompted the development of three new monovalent NmC conjugate vaccines (Table 4.3).

Incentive for the vaccine manufactureers was that licensing was to be based on safety, noninterference with current pediatric regimes, and most importantly, immunogenicity correlates for vaccine effectiveness without the requirement for long-term and costly efficacy studies. Based on insights gained with Nm polysaccharide vaccines (A/ C; ACW-135Y), experts in the field estimated that bactericidal activity using rabbit source complement (rSBA) of >8 would be a predictor of susceptibility and >32 a predictor of protection [93]. Additionally, the new conjugates needed to demonstrate more than fourfold rises in rSBA titers and evidence for immunological memory either through polysaccharide boost studies or demonstration of avidity increases. Prelicensure studies in adults [94–96], children [97], and infants [98–100] demonstrated that all three NmC conjugates were highly immunogenic after 10 days, resulting in the production of rSBA levels >1 : 8 in more than 90% of individuals within all age

Table 4.3 NmC Conjugate Vaccines

Characteristic	Time Point	Novartis (NmC-N)	Wyeth (NmC-W)	Baxter (NmC-B)
Protein		CRM_{197}	CRM_{197}	TT
Polysaccharide		NmC-OAc	NmC-OAc	NmC-de-OAc
Bactericidal titer	Day 0	<4	<4	<4
(Antibody level μg/ml)		(0.81)	(0.84)	(0.88)
	Postprimary	123	141	564
		(8.03)	(11.1)	(13.3)
	Preboost	19	51	166
		(0.94)	(1.70)	(1.84)
	Postboost	1318	979	5272
		(10.8)	(9.0)	(14.2)

Source: Data taken from [97].

groups, and primed for an NmC polysaccharide memory boost in the younger age groups. Only 1 early trial compared the 3 NmC conjugate vaccines head to head in UK toddlers (Table 4.3) [97]. On average after 1 dose of vaccine, bactericidal geometric mean titers (GMT) increased from <4 to 215 and 91–100% of children achieved rSBA titers >1:8. Significant differences in immunogenicity were observed between the NmC-TT vaccine compared to the NmC-CRM$_{197}$ series. The NmC-TT vaccine induced significantly higher rSBA titers and proportionally had more individuals with rSBA titers >1:8 than the 2 other NmC-CRM$_{197}$ conjugates. This was observed both in postprimary immunization and in post-NmC polysaccharide boosting. Antibody titers were not significantly different for each conjugate nor were their avidity indices regardless of time point, which suggests the NmC-TT conjugate induced a greater proportion of functional antibody.

Three major differences exist in the formulation of the three NmC conjugate vaccines, which may have influenced their immunogenicity profiles: (1) The protein carrier used in both the Novartis (Chiron) Menjugate product (NmC-N) and Wyeth Meningitec (NmC-W) was the nontoxic mutant diphtheria toxin CRM$_{197}$, whereas the Baxter NeisVac C product (NmC-B) employed TT. However, based on previous experience with Hib conjugate vaccine comparisons with both of these proteins [60], the expectation was that they would perform similarly. (2) Both the NmC-CRM$_{197}$ vaccines (Wyeth and Novartis) use short-chain oligosaccharide NmC components, whereas the NmC-TT employed longer polysaccharide units [97]. The poorer immunogenicity in toddlers may therefore be a reflection of the chain length, which had been previously observed with Hib protein conjugates [87]. (3) NmC-B was formulated using fully de-*O*-acetylated NmC polysaccharide, whereas the NmC-N and NmC-W conjugates used polysaccharides with similar *O*-acetylation levels. The degree of *O*-acetylation may prove to be significant since only 12% of NmC isolates in the United Kingdom possess de-*O*-Ac NmC polysaccharide [101], and this analysis has been added into the ongoing postmarketing surveillance taking place in the United Kingdom. Studies comparing the immunogenicity of de-O-Ac and OAc + NmC polysaccharide vaccines have shown that the de-O-Ac polysaccharide is superior in adults [102] and children [103]; consistent with the comparative results of the conjugate vaccines. In terms of functional antibody NmC-TT (de-*O*-Ac) induced progressively higher SBAs against both NmC-OAc$^+$ and NmC-OAc$^-$ strains alike following a course of three injections in infants [104]. The SBA titers were significantly higher against OAc$^-$ NmC strains relative to OAc$^+$ strains, which paralleled the higher recognition of de-OAc versus the OAc$^+$ NmC polysaccharides. There is some evidence that less cross-reaction occurs with antibodies generated to OAc$^+$ polysaccharide components since the OAc epitope is dominant [105, 106]. Thus for broader spectrum protection, the NmC-TT vaccine generating improved cross-reactive SBA responses may indicate superiority.

The NmC vaccination program was introduced in late 1999 in the United Kingdom based on the encouraging prelicensure data using a three-dose accelerated infant schedule at 2, 3, and 4 months of age and a single dose catch up for ages 1–20. The UK experience has become a remarkable case study on the effects of vaccine scheduling, correlations between immune memory versus circulating bactericidal antibodies, herd immunity effects, and vaccine-induced pressure on microbial genetics, which were all made possible by the extensive ongoing postlicensing surveillance. Program coverage was extensive at >90% for infant immunizations and 85% for the catch-up program [107]. One year into the program, vaccine effectiveness was high at >90% across all ages and associated with a large reduction in NmC-related deaths. Short-term efficacy was estimated at 92% in infants and 89% in toddlers, and there was evidence of herd immunity among nonvaccinated individuals with reductions in disease incidence ranging from 34 to 61% in this nonvaccinated cohort [108]. Similar reductions (~97%) were obtained in Spain [109] and in Canada 97% [110] after introduction of the NmC conjugate. Initial optimism was tempered, however, when follow-up studies revealed a rapid decline in bactericidal antibody levels within the second year of immunized infants and toddlers, with an associated decrease in vaccine effectiveness from 93% to noneffective for infants and 88 to 61% in toddlers [111]. The rapid drop in SBA levels was previously encountered with infants [112], but the thesis at the time (now disproved) was that the strong memory response associated with immunological priming would compensate [113]. Consequently, the United Kingdom has now implemented a booster program for infants in the second year and a catch-up booster for previous three-dose immunized children. The interplay between waning SBA levels, memory responses, and herd immunity are discussed in a subsequent section.

4.6.2.3 Serogroup B
Meningitis and septicaemia due to *N. meningitidis* serogroup B (NmB) remains a major health burden in both the developing and the developed world. NmB can be the source of prolonged epidemics or hyperendemic disease and is the most common cause of endemic meningitis in industrial countries. It accounts for 30–40% of cases in North America but can be much higher in Europe accounting for up to 80–90% of cases [75]. Unlike NmC, which primarily affects adolescents, fully one-half of all NmB disease occurs in infants <1 year of age. Prolonged outbreaks such as the >10-year New Zealand epidemic [114] tend to be clonal, whereas sporadic disease is much more heterogeneous [115]. Due to this heterogeneity of NmB, the most attractive vaccine candidate is the common capsular polysaccharide composed of $\alpha(2-8)$ *N*-acetyl neuraminic acid repeating units (Table 4.1). However, it is the only serogroup in which a capsular polysaccharide-based vaccine is ineffective. This is due to the poor immunogenicity of the NmB polysaccharide most often attributed to the presence of similar structures in developing human brain tissues [116] and in which a state of immune tolerance is invoked. Attempts to improve upon its immunogenicity using protein conjugate strategies have also not proven fruitful [88, 117], dispelling hopes that this would be a viable option. Complicating the capsular polysaccharide vaccine strategy is the possibility that autoimmune disease may arise by breaking tolerance to the NmB structure, however, a comprehensive review of naturally occurring levels of human anti-NmB antibody has found no correlation with any immune pathology and thus appears to downplay this scenario [118]. Chemical modification of the NmB polysaccharide by replacing the *N*-acetyl functionalities with *N*-propionyl groups followed by conjugation to protein renders the modified NmB polysaccharide more immunogenic in mice [119–121]. More importantly and surprisingly, the antibody produced is primarily specific to NPrNmB with little cross-reaction

to the native NmB polysaccharide but still retains the ability to bind to and kill NmB organisms, suggesting the NPrNmB polysaccharide is mimicking a bacterial surface epitope [120, 122, 123]. It has also been suggested that this surface epitope may contain de-*N*-acetylated residues in common with incomplete chemical modification of NPrNmB [124]; however, more rigorous studies with completely derivatized NPrNmB polysaccharide suggests otherwise [125]. Evaluation of a similar vaccine coupled to a recombinant NmB porin (rPorB) in preclinical rodent and nonhuman primate studies demonstrated 20- to 40-fold increases in SBA titers over preimmune values [126]; however, more recent studies suggest some of this bactericidal activity lies in cross-reactive NmB antibody populations [127].

The results of these preclinical studies have prompted the evaluation of an NPrNmB-TT conjugate vaccine in adult male volunteers [128]. The vaccine was well tolerated without any undue signs of immunopathology or evidence of antibody binding to the polysialylated form of the human neural cell epitope (PSA-NCAM). The vaccine generated anamnestic levels of NPrNmB-specific IgGs but against a background of a substantial preimmune population NmB polysaccharide-specific IgM antibody. No evidence of further bactericidal activity as a result of the NPrNmB-TT vaccine was observed, and only modest protection occurred upon passive transfer into an infant rat model [128]. Although these results were discouraging, several factors have been proposed that may have compromised the trial: (1) It is not known what immune correlate is associated with NmB protection. (2) The assays employed are relatively insensitive. (3) The already elevated SBA levels in adults prior to vaccination, which contrasts what is normally found with infants. (4) Preclinical work employed strong adjuvants or adjuvanting proteins (rPorB) [126] to generate bactericidal responses, and these adjuvants were lacking in the human trial. Due to the ongoing difficulties in developing a comprehensive and effective NmB vaccine (see below), the capsular polysaccharide still remains an attractive target and warrants further investigations to build on this pilot trial in humans.

Alternative strategies for an NmB vaccine include the use of lipooligosaccharide conjugate vaccines [129–131] and protein-based meningococcal vaccines, particularly the outer membrane proteins. Of note, this latter approach, which is beyond the scope of this review (see [132, 133] for more specific information), has met with some success against clonal NmB disease and "designer" formulations of multivalent proteins to control NmB epidemics in Cuba [134], New Zealand [135], and Norway [136] may have some merit. The problem associated with this approach is the high genetic variability possessed by NmB requiring an ever-changing composition that would only be effective regionally.

4.6.2.4 Serogroups A, C, W-135, and Y Despite the success of the monovalent NmC conjugate vaccines in Europe and Canada, this vaccine was not made available in the United States even though serogroups C disease is abundant. One reason behind this omission is that the United States also possesses a significant amount of Nm serogroup Y disease (almost one-third of all Nm cases) [137], and, consequently, they were waiting on an anticipated conjugate vaccine containing this serogroup. Following from the success of Hib, NmC, and the pneumococcal conjugate vaccines, Sanofi-Pasteur developed a tetravalent A, C, W-135, and Y diphtheria toxoid (DT) vaccine (MCV-4; Menactra) that was licensed by the FDA in 1995 for individuals >11 years of age. The vaccine consists of 4 μg of each capsular polysaccharide conjugated to ~48 μg DT and was approved based on unpublished studies that demonstrated noninferiority to the existing tetravalent polysaccharide vaccine (MPSV4; Menomune) in over 5000 persons [137], but to date no efficacy trials have been performed. The results of these unpublished studies demonstrated that approximately 90% of persons age 11–18 and >83% of persons age 18–55 achieved >4-fold increases in rSBA titers 28 days postvaccination with >97% achieving rSBA titers >128 for all serogroups when using either the MCV-4 conjugate or the MPSV-4 polysaccharide [137]. Addressing the persistence of antibody in 3-year follow-up studies with immunized adolescents, individuals immunized with MCV-4 had substantially higher persisting rSBA titers for all serogroups relative to those who had received the plain polysaccharide vaccine (MPSV4) [138]. Immunological memory responses were demonstrated by a large anamnestic response to an MCV-4 booster after 3 years where all individuals possessed rSBA titers well above >128 [138].

A series of studies were conducted to assess the safety and immunogenicity of MCV-4 in 2- to 10-year-old children and in infants. The latter small study with 90 infants [139] demonstrated that MCV-4 given at 2, 4, and 6 months of age generated only small fold increases and low SBA titers of antibody (~50% with >1:8), indicating no significant benefit of the vaccine in infants. However, a pivotal study for children ages 2–10 [140] yielded data with similar trends as was seen for adolescents and adults and thus formed the basis for the 2007 licensing of the MCV4 vaccine for individuals >2 years of age [141].

A second meningococcal tetravalent vaccine from Novartis (MenACWY) has recently completed phase II trials in infants and appears to be promising [142]. This vaccine composed of 10 μg NmA and 5 μg each of NmC, W-135, and Y coupled to CRM_{197} demonstrated that a 2-, 4-, and 6-month schedule was optimal leading to >92% of infants generating hSBA titers >1:4 (estimate of protection with human complement) for all serogroups. Serum bactericidal titers (hSBA), similar to those observed in the United Kingdom experience with NmC vaccines in infants, declined rapidly after the first year. Following a booster dose at 12 months, at least 95% of infants developed assumed protective levels for C, W-135, and Y and ~85% protection for A, which indicated the infants were sufficiently primed for memory. It is projected that this vaccine should become available for adolescents in 2008 and the first such vaccine for infants in 2009 [143].

Some important issues remain before implementation of widespread use of these tetravalent meningococcal conjugate vaccines (a third version A, C, W-135, Y-TT from GSK is in phase II trials) [143]. (1) There has been an association between Guillain-Barré syndrome (GBS) and Menactra use (17 cases over 15 months), and consequently this requires close surveillance. (2) Questions remain about how the increased use of the CRM_{197} carrier protein will influence current CRM_{197}-based vaccines (e.g., pneumococcal 7-valent-CRM_{197}, Hib- CRM_{197}). (3) Postmarketing surveillance is required to assess efficacy, disease protection duration, the generation of herd immunity, and influence of Nm microbial population genetics. (4) All immunization programs must carry out careful cost–benefit analyses to decide whether use of a tetravalent meningococcal conjugate vaccine is warranted especially in regions that are lacking in one or more Nm serotype disease [144].

4.7 Streptococcus pneumoniae

Streptococcus pneumoniae (Pn) is a leading cause of serious illness in young children, older adults, and the immunocompromised, being responsible for both invasive disease with a high associated morbidity and mortality, as well as a leading cause of noninvasive diseases such as otitis media. The World Health Organization (WHO) estimated (2005) that 1.6 million people, including up to 1 million children under the age of 5, will die from pneumococcal pneumonia, meningitis, and sepsis, the bulk of which are in developing countries [145]. The incidence rate of invasive Pn disease is biphasic with children <2 years old carrying the heaviest burden and then peaking once again for adults >65 years of age. In this latter group, community-acquired pneumonia is at its highest level (10–100 cases per 100,000 in the United States) with a 5–7% case fatality rate, which increases up to 60% with corresponding bacteremic disease [145]. Prior to the introduction of Pn interventions, 70% of Pn-invasive disease in children <2 years was the result of bacteremia (nonfocal) with another 12–16% due to bacteremic pneumonia and ~10 cases/100,000 population would develop acute meningitis with fatality rates of ~30%. In addition to Pn-induced invasive disease, Pn is a major contributor to noninvasive mucosal infections such as otitis media, which is the leading cause of childhood hospital treatments and, as a result, a major contributing factor to the increase in antibiotic resistance found among the pneumococci [146]. An estimated 8 million cases of Pn-related otitis media occur each year among children <5 years old [147]. Similar to meningococcal and Hib disease, incidence rates for invasive Pn disease can be up to 50 times higher (5–9000/100,000) than average for those individuals with some form of immunodeficiency and is much more frequent in certain ethnic groups such as Alaskan Natives and American Indians [148, 149].

Pneumococci are gram-positive encapsulated organisms separated into >90 different serotypes based on the expression of their capsular polysaccharide. Similar to Hib and NmA, C, W-135, and Y, the capsular polysaccharides are immunogenic in adults and form the basis of a long-standing polysaccharide vaccine. Based on structural homologies and serological cross-reactivities, 23 different Pn serotypes (Table 4.4) have been formulated into a polysaccharide vaccine to theoretically cover ~88% of the Pn-related disease found in the United States [150] and has been used successfully in certain age categories since 1981. Similar to all capsular polysaccharides (except NmA), Pn polysaccharides are poorly immunogenic in the two age groups most in need with estimates of between 56 and 81% efficacy in the elderly [151, 152] and virtually ineffective in infants.

The synthesis of a protective Pn conjugate vaccine was immensely more challenging than the strategy used to tackle monovalent Hib disease due to the sheer number of Pn serotypes. It was already determined that the 23 serotypes

Table 4.4 Structures of Capsular Polysaccharides of *Streptococcus pneumoniae*

Type	Structure	Reference
1	→3) α-AATp (1 → 4) αD-Gal$p$A (1 → 3) αD-Gal$p$A (1 → AAT = 2-acetamido-4-amino-2,4,6-trideoxy-D-galactose	254
2	→4) βD-Glc$p$ (1 → 3) αL-Rha$p$ (1 → 3) αL-Rha$p$ (1 → 3) βL-Rha$p$ (1 → $\quad\quad\quad\quad\quad\quad\quad\quad\quad\quad\quad\quad$ 2 $\quad\quad\quad\quad\quad\quad\quad\quad\quad\quad\quad\quad$ ↑ $\quad\quad\quad\quad\quad\quad\quad\quad\quad\quad\quad\quad$ 1 $\quad\quad\quad\quad\quad\quad\quad\quad\quad\quad\quad$ αD-Glc$p$A (1 → 6) αD-Glc$p$	255
3	→4) βD-Glc$p$ (1 → 3) βD-Glc$p$A (1 →	256
4	→4) βD-Man$p$NAc (1 → 3) αL-Fuc$p$NAc (1 → 3) αD-Gal$p$NAc (1 → 4) αD-Gal$p$ (1 → $\quad\quad\quad\quad\quad\quad\quad\quad\quad\quad\quad\quad\quad\quad\quad\quad\quad\quad\quad\quad\quad\quad\quad\quad\quad$ 3 $\;$ 2 $\quad\quad\quad\quad\quad\quad\quad\quad\quad\quad\quad\quad\quad\quad\quad\quad\quad\quad\quad\quad\quad\quad\quad\quad\quad\quad$ X $\quad\quad\quad\quad\quad\quad\quad\quad\quad\quad\quad\quad\quad\quad\quad\quad\quad\quad\quad\quad\quad\quad\quad\quad$ H$_3$C $\;$ CO$_2$H	257
5	→4) βD-Glc$p$ (1 → 4) αL-Fuc$p$NAc (1 → 3) βD-Sug$p$ (1 → $\quad\quad\quad\quad$ 4 $\quad\quad\quad\quad$ ↑ $\quad\quad\quad\quad$ 1 $\quad$ αL-P$n$ep$p$NAc (1 → 4) βD-Glc$p$A Sug = 2-acetamido-2,6–dideoxy – D–$xylo$ – hexos – 4 – ulose	258
6B	→2) αD-Gal$p$ (1 → 3) αD-Glc$p$ (1 → 3) αL-Rha$p$ (1 → 4) Ribitol (5-OPO$_3$ →	259
7F	→6) αD-Gal$p$ (1 → 3) βL-Rha$p$ (1 → 4) βD-Glc$p$ (1 → 3) βD-Gal$p$NAc (1 → $\quad\quad\quad\quad\quad\quad\quad\quad$ 2 $\quad\quad\quad\quad\quad\quad\quad\quad\quad\quad\quad\quad\quad\quad\quad$ 4 $\quad\quad\quad\quad\quad\quad\quad\quad$ ↑ $\quad\quad\quad\quad\quad\quad\quad\quad\quad\quad\quad\quad\quad\quad\quad$ ↑ $\quad\quad\quad\quad\quad\quad\quad\quad$ 1 $\quad\quad\quad\quad\quad\quad\quad\quad\quad\quad\quad\quad\quad\quad\quad$ 1 $\quad\quad\quad\quad\quad\;$ βD-Gal$p$ $\quad\quad\quad\quad\quad\quad\quad$ αD-Glc$p$NAc (1 → 2) αL-Rha$p$	260
8	→4) βD-Glc$p$A (1 → 4) βD-Glc$p$ (1 → 4) αD-Glc$p$ (1 → 4) αD-Gal$p$ (1 →	261
9N	→4) αD-Glc$p$A (1 → 3) αD-Glc$p$ (1 → 3) βD-Man$p$NAc (1 → 4) βD-Glc$p$ (1 → 4) αD-Glc$p$NAc (1 → $\quad\quad\quad\quad\quad\quad\quad\quad\quad\quad\quad\quad\quad\quad\quad\quad\quad\quad\quad\quad\quad\quad$ 2 $\quad\quad\quad\quad\quad\quad\quad\quad\quad\quad\quad\quad\quad\quad\quad\quad\quad\quad\quad\quad\quad\quad$ ↑ $\quad\quad\quad\quad\quad\quad\quad\quad\quad\quad\quad\quad\quad\quad\quad\quad\quad\quad\quad\quad\quad$ OAc	262

9V	→4) αD-GlcpA (1 → 3) αD-Galp (1 → 3) βD-ManpNAc (1 → 4) βD-Glcp (1 → 4) αD-Glcp (1→ ↑ OAc	263
10A	βD-Galp 1 ↓ 6 →5) βD-Galf (1 → 3) βD-Galp (1 → 4) βD-GalpNAc (1 → 3) αD-Galp (1 → 2) D-Ribitol (5-OP0$_3$ → 3 ↑ 1 βD-Galf OAc ↓ 2/3	264
11A	→3) βD-Galf (1 → 4) βD-Glcp (1 → 6) αD-Glcp (1 → 4) αD-Galp (1→ 4-O PO$_3$-1-Glycerol	265
12F	→4) αL-FucpNAc (1 → 3) βD-GalpNAc (1 → 4) βD-ManpNAcA (1→ 3 ↑ 1 αD-Galp αD-Glcp (1 → 2) αD-Glcp	266

(Continued)

Table 4.4 Continued

Type	Structure	Reference
14	→4) βD-Glc$p$ (1 → 6) βD-Glc$p$NAc (1 → 3) βD-Gal$p$ (1→ 　　　　　　　　　　　　　　　4 　　　　　　　　　　　　　　　↑ 　　　　　　　　　　　　　　　1 　　　　　　　　　　　　　　βD-Gal$p$	267
15B	→6) βD-Glc$p$NAc (1 → 3) βD-Gal$p$ (1 → 4) βD-Glc$p$ (1→ 　　　　　　　　　　　　　4 2, 3, 4, 6 OAc　　　　↑ 　　│　　　　　　　　　1 αD-Gal$p$ (1 → 2) βD-Gal$p$ (-3-OPO$_3$) -2-Glycerol	268 269
17F	OAc 　　　　　　　　　　　　　　　　　　　　　↓ 　　　　　　　　　　　　　　　　　　　　　2 →3) βL-Rha$p$ (1 → 4) βD-Glc$p$ (1 → 3) αD-Gal$p$ (1 → 3) βL-Rha$p$ (1 → 4) αL-Rha$p$ (1 → 2) D-Arabinitol-1-OPO$_3$- 　　　　　　　　　　　　　　　　　　　4 　　　　　　　　　　　　　　　　　　　↑ 　　　　　　　　　　　　　　　　　　　1 　　　　　　　　　　　　　　　　　　βD-Gal$p$	270
18C	αD-Glc$p$ 　　　　　　　　1 　　　　　　　　↓ 　　　　　　　　2 →4) βD-Glc$p$ (1 → 4) βD-Gal$p$ (1 → 4) αD-Glc$p$ (1 → 3) αL-Rha$p$ (1 → 　　　　　　3-O 　　　　PO$_3$-1-Glycerol	271
19A	→4) βD-Man$p$NAc (1 → 4) αD-Glc$p$ (1 → 3) αL-Rha$p$-1-OPO$_3$-	272
19F	→4) βD-Man$p$NAc (1 → 4) αD-Glc$p$ (1 → 2) αL-Rha$p$-1-OPO$_3$-	273

20	→6) αD-Glcp (1 → 6) βD-Glcp (1 → 3) βD-Galf (1 → 3) βD-Glcp (1 → 3) αD-GlcpNAc-1-OPO$_3^-$
	                                                                                4
	                                                                                ↑
	                                                                                1
	                                                                                βD-Galf
		274

22F	αD-Glcp
	1
	↓
	3
	→4) βD-GlcpA (1 → 4) βL-Rhap (1 → 4) αD-Glcp (1 → 3) αD-Galf (1 → 2) αL-Rhap (1 →
	                                                              2
	                                                              ↑
	                                                              OAC
	                                                              αL-Rhap
		275

23F	1
	↓
	2
	→4) βD-Glcp (1 → 4) βD-Galp (1 → 4) βL-Rhap (1 →
	                                                3-O
	                                                PO$_3$-2-Glycerol
		276

33F	→3) βD-Galp (1 → 3) αD-Galp (1 → 3) βD-Galf (1 → 3) βD-Glcp (1 → 5) βD-Galf (1 →
	                                  2 2
	                                  ↑ |
	                                  1 OAc (0.5)
	                                  αD-Galp
		277

found in the Pn polysaccharide vaccine provided effective coverage for those Pn strains found predominantly in the United States. However, the prevailing thought at the time was that a 23-valent Pn conjugate vaccine was beyond reason. Instead, the strategy was to incorporate those serotypes most frequently associated with childhood disease into a combination vaccine.

Despite the use of polysaccharide vaccines for over 35 years, the immune correlates for protective efficacy remain poorly defined, which hampers the evaluation of the newer generation Pn conjugate vaccines. This is primarily due to a lack of a specific correlation between individual serotypes and efficacy, the varying epidemiological and regional differences of Pn worldwide, and the lack of suitable animal models for Pn. An analysis of pooled immunogenicity and efficacy data from all completed Pn polysaccharide trials to date has allowed the WHO to estimate the threshold for protective efficacy at 0.35 μg/ml serotype specific antibody [153, 154]. Furthermore, novel Pn conjugate vaccines should also demonstrate functional abilities such as opsonophagocytic activity as well as evidence for immunological memory (polysaccharide booster response; avidity maturation) [154].

The first and currently only licensed Pn conjugate vaccine is produced by Wyeth (Prevnar, 2000) and consists of 2 μg of each capsular polysaccharide from serotypes 4, 6B (4 μg), 9V, 14, 19F, 23F, and 18C oligosaccharide conjugated individually to the nontoxic diphtheria toxin mutant CRM_{197} (20 μg total) (PnC7-CRM) (Table 4.5).

This formulation accounts for those Pn serotypes associated with 86% of bacteremia, 83% of meningitis, and 65% of otitis media among children <6 years of age in the United States [155]. In addition, several other Pn conjugate vaccines are currently in later stages of evaluation (see Table 4.5) with either differences in the carrier protein employed (DT, TT, protein D) or increased serotype coverage (7-, 9-, 10-, 11-, and 13-valent conjugates) [156]. All the Pn conjugate vaccines evaluated to this point have demonstrated significant immunogenicity with serotype-specific differences in overall antibody levels with most studies reporting >0.35 μg/ml levels following a standard course of three doses [157]. Pn conjugate vaccines using sized polysaccharides in a pentavalent vaccine were found to be more immunogenic than comparable vaccines employing small oligosaccharides [158] similar to studies

Table 4.5 Pneumococcal Conjugate Vaccines

Conjugate	Serotypes	Carrier	Manufacturer
PnC7[a]	4, 6B, 9V, 14, 18C, 19F, 23F	CRM_{197}	Wyeth
	4, 6B, 9V, 14, 18C, 19F, 23F	OMP(NmB)	Merck
PnC9	1, 4, 5, 6B, 9V, 14, 18C, 19F, 23F	CRM_{197}	Wyeth
PnC11	1, 4, 5, 7F, 9V, 19F, 23F	TT+	Sanofi-Pasteur
	3, 6B, 14, 18C	DT	
PnC11	1, 3, 4, 5, 6B, 7F, 9V, 14, 18C, 19F, 23F	Protein D	GSK
PnC13	1, 3, 4, 5, 6A + B, 7F, 9V, 14, 18C, 19A + F, 23F	CRM_{197}	Wyeth

[a]Only licensed product to date.
Sources: Data taken from [145, 278, 279].

with Hib [87]. Few trials have assessed comparative immunogenicities of the candidate Pn conjugate vaccines, which help to illuminate differences in conjugate design (carrier protein, saccharide size, loading, kinetics) as they may relate to efficacy. A study comparing OMP versus CRM carrier proteins [159] revealed small differences in immunogenicity and avidity, but, overall, both possessed similar efficacies. Overall, more comparative studies are needed to concretely establish design differences that relate to superior immunogenicity and ultimately vaccine efficacy.

Pn conjugate vaccines, more than other bacterial conjugate vaccines, have the capacity to interfere or be interfered with by other concomitantly administered pediatric vaccines. This is due to their common scheduling regimes, the high multivalency of the vaccines, and therefore the concomitant need for significant amounts of carrier protein shared in common with other routine vaccines. A prime example of this phenomenon was observed with an 11-valent Pn mixed carrier protein conjugate vaccine where serotypes 1, 4, 5, 7F, 9V, and 19F were coupled to TT and 3, 6B, 14, and 18c were coupled to DT (Table 4.5). Due to the high TT loads, polysaccharide content required reduction to 1 μg per dose. The vaccine demonstrated carrier-induced epitope suppression (CIES) where those serotypes coupled to TT and administered concomitantly with a routine Diphtheria, Tetanus, acellular Pertussis, Hib (PRP-TT), and inactivated Polio combination vaccine (DTaP-Hib (PRP-TT)-IPV), had significantly depressed Pn serotype antibody responses [160, 161]. The Pn-DT serotype responses were negatively affected as well but only upon boosting. Furthermore, a single dose of the PnC11-D/T vaccine elicited similar levels of serotype-specific antibodies and functional activity as the same vaccine given as three doses [162] where anamnestic responses were expected. When serotype-specific antibody levels were compared between the established PnC7-CRM and the 11-valent mixed D/T vaccine, the former was found to induce much higher levels of antibody (with the exception of types 4 and 19F) [163, 164] compared to the 11-valent mixed carrier vaccine, which helps to explain the apparent discontinuation of the licensing process for this vaccine.

4.7.1 Impact on Invasive Pneumococcal Disease

PnC7-CRM was shown to be highly effective against invasive pneumococcal disease (IPD) in infants in the landmark Northern California Kaiser Permanente (NCKP) efficacy trial [165]. Over 90% of infants developed antibody to all 7 serotypes at levels >0.35 μg/ml and reduced IPD caused by vaccine serotypes by 97.4% and by 89% for all IPD. Similar to that observed for Hib and NmC, antibody titers dropped to slightly above baseline after year 1 but increased with memory kinetics following a boost with the conjugate vaccine or polysaccharide alone [166]. The vaccine was also found to be 77% effective in a population of American Indians where incidence rates approach 1800 cases/100,000 or ~10 times the normal rate [149] when given as a three-dose primary and one-dose booster series. In a similar high incidence setting, the PnC9-CRM vaccine in which serotypes 1 and 5 have been added for greater serotype coverage, provided 77% vaccine effectiveness against radiologically confirmed pneumonia for vaccine serotype disease [163]. Since the introduction of routine

PnC7-CRM immunization in the United States, vaccine serotype IPD in children <5 years has declined 94% and all IPD by 75% overall in a 3-year period of study [167].

4.7.2 Impact on Acute Otitis Media

Several large studies have addressed the efficacy of Pn conjugate vaccines on noninvasive otitis media [159, 165, 168], which represents a major health-related cost in industrialized countries. The protective effect on acute otitis media (AOM) is modest at 57–67% for vaccine serotypes or 34% for all pneumococcal-induced AOM disease or 6–8.9% against all AOM incidences. Despite the modest efficacy of the PnC7 vaccine and due to the high prevalence of AOM, the vaccine's impact has been significant by reducing AOM-related hospital visits per 1000 children [169] and on the spread of antibiotic resistance among pneumococci [146]. In follow-up surveillance, the impact of the PnC7 vaccine in the more serious forms of the disease (recurrent AOM, tympanocentesis) was also pronounced with up to 50% reductions over a 3-year follow-up period [170], due in part because Pn is associated with more severe AOM than other causative agents such as *H. influenzae* or *Moraxella catarrhalis* [171]. The vaccine was also found to be efficacious in the high-risk population of Native American Indians with 64% reductions for vaccine serotype OM [172]. The efficacy of the vaccine is mainly due to the reduction in serotype-specific carriage of Pn at mucosal surfaces [173, 174] presumably through IgG exudates [175]. The spin-off effect is the generation of herd immunity, which has the benefit of increased vaccine effectiveness. To emphasize the importance of this effect, the 94% decrease in IPD in children <5 years old was associated with 65% decrease in serotype-related disease in the nonvaccinated population comprising adults 20–39 (32% decrease) and the elderly >65 years of age (10% reduction) [167]. This also correlates with decreased carriage of vaccine serotype Pn from 28% of adults in the prevaccine era to 4.5% in 2004 [176]. A major event requiring close scrutiny is the phenomenon of *serotype replacement*, where Pn strains belonging to vaccine serotypes have been replaced with nonvaccine serotypes [177] that potentially have the capacity to reassert Pn disease. Although this has not yet been encountered to any great extent, the alarming recent rise in serotype replacement disease (serotypes 3, 6A, 7F, and 19A) in Alaskan Natives (140% increase), may be the harbinger of things yet to come in the general population [178].

4.8 GROUP B *Streptococcus*

Group B *Streptococcus* (GBS) or *Streptococcus agalactiae* is a leading cause of neonatal invasive disease associated with high morbidities and mortality [179, 180]. GBS disease in the neonate can be classified as either (1) early onset disease (EOD) in which symptoms occur 0–6 days and typically <24 h after birth or (2) late onset disease (LOD) in which GBS infection occurs after 7 days and up to 90 days after birth. GBS colonization of the neonate occurs primarily through maternal transfer either in utero or during passage through the birth canal from asymptomatic GBS colonized mothers; which represents ~5–40% of pregnant women [181]. In 1996 in the United States, routine administration of intrapartum antibiotics [182] to culture positive GBS mothers or mothers at high risk (early and prolonged membrane rupture, preterm delivery) had the effect of reducing the incidence of GBS EOD by 83% from 1.8/1000 births to 0.6/1000 births (1998), which has since decreased to 0.3/1000 births in 2004 as the practice has become widely adopted [183, 184]. Despite this successful intervention, 1100 U.S. cases of GBS with 50 deaths remain annually (5% case fatality) [185]. In addition, intrapartum prophylaxis has had little impact on LOD generally acquired horizontally from colonized mothers, which accounts for ~1400 cases annually with 3% fatality rates. Furthermore, GBS increasingly has emerged as an important contributor to nonpregnant adult disease, especially the elderly and those individuals with an underlying or chronic disease (rates in excess of 20 cases/ 100,000), with case fatality levels of 5% [185].

Group B streptococci are gram-positive organisms that are classified based on their capsular polysaccharides into at least nine different serotypes (Table 4.6) [9]. In the United States, EOD is associated primarily with serotypes III (38%), Ia (36%), V (13%), and II (11%) while LOD is primarily type III (60%) and Ia (23%), although this is not static since type V has only recently appeared as a contributor to disease [186, 187]. Screening of adult populations indicate type V as the predominant causative agent ranging from 24 to 31% of isolates [188–190]. It is presumed based on these serotype distributions that a pentavalent vaccine consisting of types Ia, Ib, II, III, and V could provide protection for 94–100% of invasive disease in neonates and pregnant women and 85–92% in nonpregnant adults [191].

The rationale behind active immunization of pregnant women as a means to combat invasive GBS disease in the neonate came from the early observation by Baker and Kasper [192] that infants with neonatal GBS disease were born from mothers with low levels of GBS-type specific antibodies—the rationale being the neonate does not receive placental transfer of type-specific antibody, which has been shown to be highly opsonic to homologous GBS organisms in vitro [193]. Since immunization of the neonate is not an effective option (immature immune system; 90% of EOD occurs within 48 h of birth), targeting the pregnant mother during the last trimester when placental transfer of antibody is highest [194] would be the most effective manner to provide protection for the neonate.

Capsular polysaccharides from the nine known serotypes have now been structurally and immunochemically characterized [186, 195–199], setting the basis for the production of a GBS combination vaccine. It has been reported that several GBS capsular serotypes contain varying degrees of Neu5Ac O-acetylation, which was found to be strain specific rather than type specific [200]. Early work with types Ia, II, and III established that the polysaccharides were safe but only modestly immunogenic in adults, unlike what is observed with pneumococcal capsular polysaccharides. Proof of concept, however, was established with the finding that type III immunization of pregnant women resulted in significant type-specific antibody transferred to the neonate at levels that promoted efficient in vitro opsonization and killing of GBS bacteria [201].

The variably poor immunogenicity of the GBS capsular polysaccharides can be related to structural homology with human tissue antigens as well as a lack of priming

Table 4.6 Structures of Capsular Polysaccharides of Group B *Streptococcus*

Type	Structure[a]	Reference
Ia	→4) βD-Glc*p* (1 → 4) βD-Gal*p* (1→ 3 ↑ 1 αD-Neu*p*NAc (2 → 3) βD-Gal*p* (1 → 4) βD-Glc*p*NAc	195
Ib	→4) βD-Glc*p* (1 → 4) βD-Gal*p* (1→ 3 ↑ 1 αD-Neu*p*NAc (2 → 3) βD-Gal*p* (1 → 4) βD-Glc*p*NAc	195
II	→4) βD-Glc*p*NAc (1 → 3) βD-Gal*p* (1 → 4) βD-Glc*p* (1 → 3) βD-Glc*p* (1 → 2) βD-Gal*p*-(1→ 6 3 ↑ ↑ 1 2 βD-Gal*p* αD-Neu*p*NAc	196
III	→4) βD-Glc*p* (1 → 6) βD-Glc*p*NAc (1 → 3) βD-Gal*p* (1→ 4 ↑ 1 αD-Neu*p*NAc (2 → 3) βD-Gal*p*	197
IV	→4) αD-Glc*p* (1 → 4) βD-Gal*p* (1 → 4) βD-Glc*p* (1→ 6 ↑ 1 αD-Neu*p*NAc (2 → 3) βD-Gal*p* (1 → 4) βD-Glc*p*NAc	198

```
V       →4) αD-Glcp (1 → 4) βD-Galp (1 → 4) βD-Glcp (1→                                    186
                            6
                            ↑
                            1
                         βD-Glcp
                            4
                            ↑
                            1
            αD-NeupNAc (2 → 3) βD-Galp (1 → 4) βD-GlcpNAcβD-Glcp

VI      →4) βD-Glcp (1 → 6) βD-Glcp (1 → 3) βD-Galp (1→                                    199
                                            4
                                            ↑
                                            1
                                 αD-NeupNAc (2 → 3) βD-Galp

VII     →4) βD-Glcp (1 → 4) αD-Galp (1 → 4) βD-Glcp (1→                                    280
                            6
                            ↑
                            1
            αD-NeupNAc (2 → 3) βD-Galp (1 → 4) βD-GlcpNAc

VIII    →4) βD-Glcp (1 → 4) βD-Galp (1 → 4) βL-Rhap (1→                                    281
                                            3
                                            ↑
                                            2
                                        αD-NeupNAc
```

[a]Strain-specific O-acetylation of Neu5Ac variably at positions 7, 8, or 9 has been reported [200].

from cross-reactive organisms [7, 30]. Interestingly, in a manner analogous to the NmB polysaccharide, the functional epitope of type III polysaccharide is conformational, requiring at minimum a decasaccharide for threshold binding [202, 203]—a phenomenon that may reflect deletion of B cells capable of recognizing shorter fragments present on tissue structures. The capsular polysaccharides of GBS all have pendant sialic acid branches that are amenable to simple periodate oxidation and reductive amination to protein [88], and this was the general strategy used to produce TT conjugates of all nine serotypes (see Chapter 2). Types Ia, Ib, II, III, and V-TT conjugates now have all undergone preliminary safety and dosing trials in healthy adults [191, 204–207]. All vaccines were well tolerated and all generated greater than fourfold rises in type-specific serum antibody in 80–93% of all individuals 28 days after immunization. Given that there are no immune correlates for protective efficacy, it was encouraging that immune sera uniformly killed type-specific GBS in in vitro opsonophagocytic assays. The importance of O-acetylation patterns [200] on GBS polysaccharide immunogenicity is not known at this time, however, this uniform in vitro killing of GBS suggests immune sera raised to de-O-acetylated GBS conjugates (see Chapter 2) is cross-reactive with potential O-acetylated structures on GBS as was observed with NmC conjugates [104]. Two candidate vaccines were assessed in humans for interference using a combination type II, III-TT vaccine that showed indistinguishable immunogenicity and killing of GBS relative to the individual vaccines alone [208]. With concern for carrier-induced epitope suppression, as had been observed with pneumococcal-TT conjugates [161], a second carrier protein (CRM_{197}) was evaluated for type V glycoconjugates [207]. Both the CRM and TT constructs possessed similar amounts of type V polysaccharide, and equivalent doses were administered to 35 healthy subjects. In terms of overall immunogenicity, both vaccines generated greater than four-fold rises in 93% of subjects, and both had equivalent capacity to kill GBS in opsonophagocytic assays indicating the interchangeability of the carrier protein. To assess the potential of the GBS conjugate vaccines for maternal transfer to neonates, 20 pregnant women were immunized with a type III-TT conjugate vaccine between 30 and 32 weeks of gestation [209]. No vaccine-related complications were observed. Greater than 50-fold increases of type III specific IgG were observed in maternal sera. There was a highly significant correlation between maternal levels and type-specific antibody found within cord blood samples of infants yielding a transfer ratio of 0.77 and persistence of antibody at 30% of cord blood levels 2 months after birth at 3 $\mu g/ml$ in the neonate. Infant sera containing >0.5 $\mu g/ml$ type-specific antibody uniformly killed GBS indicating functional activity.

The type V conjugate vaccine was also assessed in adults over the age of 65 [210] where the greatest burden of GBS disease occurs with ∼15% case fatality rates. Healthy men age 65–85 years old developed significant rises in type V specific IgG, IgM, and IgA antibodies with greater than fourfold increases in 68, 76, and 59% of recipients, respectively. Immune sera promoted opsonophagocytic killing of type V GBS at levels that were comparable to previous studies in younger adults.

In summary, employing conjugate technology has been effective in increasing type-specific and functional antibodies in all target groups—pregnant women, the elderly, and placentally transferred at high levels into neonates. A pentavalent vaccine, theoretically affording coverage to ~90% of all GBS-related disease in the United States is awaiting a large-scale phase III trial. Unfortunately, due to the great reductions in infections that prophylactic intrapartum antibiotic treatment has achieved, there is less of a sense of urgency in taking these vaccines to the next level, especially in light of the difficult issue of liability when immunizing pregnant women. This measure is short sighted, however, since antibiotic control is at best a temporary situation and prophylaxis is always the better option. Resistance to penicillin, the antibiotic of choice for GBS, has not yet surfaced, although GBS has already shown signs of resistance to clindamycin and erythromycin [211], which are used in penicillin allergic individuals. Furthermore, it is well known that intrapartum antibiotic treatment has little impact on LOD and there are currently no prophylactic strategies in place for the elderly or immunocompromised individuals.

4.9 Salmonella typhi

Typhoid fever is a major cause of morbidity and mortality worldwide with an estimated 22 million cases and 200,000 resultant deaths annually, mostly in South and East Asia, Australasia, Africa, and Central/ South America [212]. Endemic disease can reach as high as >100 cases/100,000 population in these areas; however, it is virtually unknown in industrialized countries with established sanitation systems. Typhoid fever strikes primarily schoolage children but has more recently been recognized as a serious enteric disease in younger children [213, 214] where it is spread via the fecal–oral route leading to mucosal invasion, high fever, and severe diarrhea.

The etiological agent, *Salmonella enterica* serovar typhi, possesses a linear capsular polysaccharide composed of repeating *N*-acetyl galacturonic acid groups variably *O*-acetylated at position 3, of which the latter has been shown to be immunodominant and a vaccine-required epitope [215, 216] (Table 4.1). The capsular polysaccharide known as the Vi antigen, forms the basis of one of two licensed *S. typhi* vaccines; the other one being an attenuated oral vaccine from the Ty21 strain [217]. Prophylaxis against typhoid fever is the desired goal since, although timely and judicious use of antibiotics is effective, increases in the antibiotic resistance patterns of *S. typhi* are troubling [218]. The Vi polysaccharide vaccine (TyphimVi, Typherix, or generic national program versions) [219] has been in use in persons >2 years old since 1994 and has been evaluated in numerous clinical trials for safety and efficacy. The subject of a recent meta-analysis involving the analysis of 100,000 participants in regions of high endemic disease, the Vi capsular polysaccharide vaccine has a demonstrated efficacy of 68% (year 1), 60% (year 2), and 50% (year 3) for a cumulative 3-year efficacy of 55% [217, 220]. This is in comparison to the oral Ty21a vaccine, which possesses a 3-year efficacy rate of 48%. Notably, the Vi polysaccharide vaccine behaves in the stereotypic manner of all capsular polysaccharides in that there is an age-dependant correlation with efficacy (increases with age), is uniformly poor in infants <2 years old, and does not elicit anamnestic responses to booster doses, although hyporesponsiveness does not appear to be a factor [221]. Although no immune correlates have been firmly established correlating Vi antibody levels to point efficacy, estimates have yielded levels of 1.33 µg/ml as an estimated protective correlate [222], which has been recently revised to 3.52 ELISA (enzyme-linked immunosorbent assay) units/ml [223].

To improve vaccine characteristics, the Vi polysaccharide has been conjugated to several carrier proteins of which a recombinant exoprotein of *Pseudomonas aeruginosa* (rEPA) has proven to be the best synthetic choice [224]. The merit of this strategy was shown when a Vi-rEPA conjugate vaccine was field tested in a large randomized trial for children 2–5 years of age in *S. typhi* elevated endemic conditions in Vietnam [225]. Of the children who had received two doses of the vaccine 91.1% were protected for over 27 months [225], which was subsequently updated to 89% over a 46-month period in follow-up surveillance [223]. On the basis of these promising results, the Vi-rEPA vaccine is now currently undergoing safety and immunogenicity studies in Vietnamese infants using a three-dose 2-, 4-, and 6-month regime [226] where expectations based on previous conjugate vaccines against other invasive bacteria are that this vaccine will prove to be efficacious in this vulnerable subset of the population.

A second and interesting approach to a Vi-like conjugate vaccine is also being developed by the Robbins group [227]. Using a structurally similar poly-α-(1-4)-D-Gal$p$A polysaccharide derived from plants (pectin), this pectin polysaccharide was converted into a Vi-like derivative by O-acetylating positions 2 and 3, with position 2 being the point of distinction between native Vi and the semisynthetic derivative (Table 4.1). The semisynthetic derivative is immunochemically indistinct from the Vi antigen [227]. The advantage of such a formulation is that pectin is abundant, it has no lipopolysaccharide (LPS) contaminants, and it requires only simple chemical modifications. Vi-like polysaccharide rEPA conjugates are currently undergoing phase I evaluation for safety in adults [228].

4.10 CONJUGATE VACCINES: FUTURE CONCERNS

The remarkable successes of the first-generation conjugate vaccines in humans and in particular with young children have set the stage for a plethora of new and improved vaccines in greater combinations and with wider disease coverage. For instance, a 13-valent Pn-CRM$_{197}$ conjugate vaccine has recently completed phase I studies [229], and GSK is seeking regulatory approval for a heptavalent combination vaccine (DTaP-Hib-IPV-NmA/C: Globorix) [230]. The introduction of these second-generation combination vaccines has raised a number of concerns that await further clinical trials to accurately gauge their impact.

Of prime importance is the effect of an increasing number of combination vaccines that share the same protein carrier, risking the creation of CIES where too much protein reduces the response against the hapten portion [160], or from vaccine interference where one component interferes with the normal response to another component of a combination vaccine. This latter case has been well documented with DTaP–Hib combinations [52, 231, 232] as well as with Pn and Nm combinations [42, 233]. Because of the already high reliance on bacterial toxoids as carrier proteins, there is a great need for an expanded pool of new carriers to limit CIES but at the same time to provide for functional diversity. An example of this latter case comes from a new unlicensed decavalent Pn conjugate vaccine using protein D from nontypeable *H. influenzae* (Hi) that has shown not only protection against Pn-induced AOM but also to Hi-induced AOM [229]. Furthermore, addressing concerns for a more functional memory response, conjugate vaccine constructs such as the Cuban NmC-*N. meningitidis* P64K protein conjugate [26] will elicit functional T-cell memory upon exposure to Nm (due to the shared P64K protein), which theoretically will enhance a quicker and more protective NmC memory B-cell response [19, 234].

The UK experience with Hib and NmC conjugates has illuminated the great need for more defined immunological correlates for vaccine protection, particularly in regard to what constitutes protective immunity. Initially, vaccine effectiveness in infants was beyond expectations (>90% for NmC [107]; 95% for Hib [235]) due in large part to high levels of circulating bactericidal antibody and subsequent carriage reductions and the establishment of herd immunity [236]. However, following 1 year postimmunization, the level of bactericidal antibody quickly declined to baseline in this age group; but in light of the robust immune memory response observed in these infants [97, 237], most believed functional protection was still in place [113]. Continued surveillance revealed, however, that vaccine effectiveness virtually disappeared after the first year in infants [111, 235] but not in older children, despite the fact that vaccine failure cases mounted strong memory responses [238]. Kinetic studies of the memory response identified a 7- to 8-day vulnerability window before a sufficiently strong memory B-cell response could be mounted, in which diseases caused by NmC, Hib, or other invasive agents could occur [24, 239]. It is now believed that the relatively low level of NmC disease in vaccinated infants was directly due to the reduction/elimination of NmC carriage and therefore spread of NmC to this vulnerable population.

The increasing routine use of conjugate vaccines has also identified vaccine pressure on microbial genetics as a future concern. While up to this point, Hib and NmC vaccines have not had the effect of increasing Hi or Nm disease burden from nonvaccine strains [61, 108, 240], the same cannot be said for Pn vaccination. Numerous studies have shown that carriage of vaccine serotype strains of Pn is suppressed in immunized persons, but this has been offset by the appearance of replacement strains such as serotypes 3, 6A, 7F, and 19A [178, 241, 242], which, fortunately, has not yet led to significant disease increases, although nonvaccine disease incidence is rising sharply, especially in native communities [243]. Perhaps more disconcerting is the possibility that Pn, Nm, and Hi will regain virulence through capsular switching [244] where, in a hypothetical example, the hypervirulent ST-11 clone, which is currently being held in check by NmC vaccination, acquires the B capsule through transformation leading to loss of vaccine control. Although not seen to this point in the United Kingdom [240, 245], circumstantial evidence for switch variants has been observed in Spain [246].

4.11 SUMMARY

Conjugate vaccines, as second-generation vaccines to bacterial polysaccharides, have proven most effective in controlling infant and childhood invasive bacterial disease and have led to tremendous impacts on related morbidity and mortality in the developed world. The present challenge is to make available these highly efficacious vaccines and/or transfer technology to developing countries where 350–700,000 annual child deaths are still due to Hib [247], ~900,000 children die annually from Pn-related disease [248], over 21 million annual cases of typhoid fever account for >200,000 deaths [212], and >500,000 cases of meningococcal disease result in 20% mortalities [75]. To emphasize the magnitude of the impact of conjugate vaccines, over a 3-year period in the United States (1998–2000), there were only 14 Hib-related deaths of which 70% of these occurred in non- or suboptimally immunized children [61]. The group A meningococcal vaccine initiative [77] is an example of what appears to be a successful model, and hopefully it may lend itself to other vaccines of dire importance for the developing world.

REFERENCES

1. Schiemann O, Casper W (1927) Sind die spezifisch pracipitablen Substanzen der 3 Pneumokokkentypen Haptene? *Med. Microbiol. Immunol.* 108:220–257.
2. Tillett WS, Francis T, Jr (1929) Cutaneous reactions to the polysaccharides and proteins of pneumococcus in lobar pneumonia. *J. Exp. Med.* 50:687–701.
3. MacLeod CM, Hodges RG, Heidelberger M, Bernhard WG (1945) Prevention of pneumococcal pneumonia by immunization with specific capsular polysaccharides. *J. Exp. Med.* 82:445–465.
4. Kelly DF, Moxon ER, Pollard AJ (2004) Haemophilus influenzae type b conjugate vaccines. *Immunology* 113:163–174.
5. Avery OT, Goebel WF (1929) Chemo-immunological studies on conjugated carbohydrate-proteins: II. Immunological specificity of synthetic sugar-protein antigens. *J. Exp. Med.* 50:533–550.
6. Jennings HJ, Sood RK (1994) Synthetic glycoconjugates as human vaccines. In Lee YC, Lee RT, Eds. *Neoglycoconjugates. Preparation and Applications.* Academic, London, pp. 325–371.
7. Jennings HJ, Pon RA (1996) Polysaccharides and glycoconjugates as human vaccines. In Dimitriu S, Ed. *Polysaccharides in Medicinal Applications.* Marcel Dekker, New York, pp. 473–479.
8. Robbins JB, Schneerson R, Szu SC, Pozsgay V (1999) Bacterial polysaccharide-protein conjugate vaccines. *Pure Appl. Chem.* 71:745–754.
9. Jennings HJ (1998) Polysaccharide vaccines against disease caused by Haemophilus influenzae, group B Streptococcus, and Salmonella typhi. *Carbohydrates Eur.* 21:17–23.
10. Stein KE (1992) Thymus-independent and thymus-dependent responses to polysaccharide antigens. *J. Infect. Dis.* 165(Suppl 1):S49–S52.
11. Dintzis RZ, Okajima M, Middleton MH, Greene G, Dintzis HM (1989) The immunogenicity of soluble haptenated polymers is determined by molecular mass and hapten valence. *J. Immunol.* 143:1239–1244.
12. Scott MG, Zachau HG, Nahm MH (1992) The human antibody V region repertoire to the type B capsular polysaccharide of Haemophilus influenzae. *Int. Rev. Immunol.* 9:45–55.

13. Jeurissen A, Wuyts G, Kasran A, Ramdien-Murli S, Blanckaert N, Boon L, Ceuppens JL, Bossuyt X (2004) The human antibody response to pneumococcal capsular polysaccharides is dependent on the CD40-CD40 ligand interaction. *Eur. J. Immunol.* 34:850–858.
14. Jeurissen A, Billiau AD, Moens L, Shengqiao L, Landuyt W, Wuyts G, Boon L, Waer M, Ceuppens JL, Bossuyt X (2006) CD4+ T lymphocytes expressing CD40 ligand help the IgM antibody response to soluble pneumococcal polysaccharides via an intermediate cell type. *J. Immunol.* 176:529–536.
15. Sen G, Khan AQ, Chen Q, Snapper CM (2005) In vivo humoral immune responses to isolated pneumococcal polysaccharides are dependent on the presence of associated TLR ligands. *J. Immunol.* 175:3084–3091.
16. Snapper CM, Mond JJ (1996) A model for induction of T cell-independent humoral immunity in response to polysaccharide antigens. *J. Immunol.* 157:2229–2233.
17. Heidelberger M, Dilapi MM, Siegel M, Walter AW (1950) Persistence of antibodies in human subjects injected with pneumococcal polysaccharides. *J.Immunol.* 65:535–541.
18. McHeyzer-Williams LJ, McHeyzer-Williams MG (2005) Antigen-specific memory B cell development. *Annu. Rev. Immunol.* 23:487–513.
19. Crotty S, Ahmed R (2004) Immunological memory in humans. *Semin. Immunol.* 16:197–203.
20. Insel RA, Anderson PW (1986) Oligosaccharide-protein conjugate vaccines induce and prime for oligoclonal IgG antibody responses to the Haemophilus influenzae b capsular polysaccharide in human infants. *J. Exp. Med.* 163:262–269.
21. Insel RA, Adderson EE, Carroll WL (1992) The repertoire of human antibody to the Haemophilus influenzae type b capsular polysaccharide. *Int. Rev. Immunol.* 9:25–43.
22. Hougs L, Juul L, Ditzel HJ, Heilmann C, Svejgaard A, Barington T (1999) The first dose of a Haemophilus influenzae type b conjugate vaccine reactivates memory B cells: Evidence for extensive clonal selection, intraclonal affinity maturation, and multiple isotype switches to IgA2. *J. Immunol.* 162:224–237.
23. Jakobsen H, Hannesdottir S, Bjarnarson SP, Schulz D, Trannoy E, Siegrist CA, Jonsdottir I (2006) Early life T cell responses to pneumococcal conjugates increase with age and determine the polysaccharide-specific antibody response and protective efficacy. *Eur. J. Immunol.* 36:287–295.
24. Clutterbuck EA, Salt P, Oh S, Marchant A, Beverley P, Pollard AJ (2006) The kinetics and phenotype of the human B-cell response following immunization with a heptavalent pneumococcal-CRM conjugate vaccine. *Immunology* 119:328–337.
25. Goldblatt D, Richmond P, Millard E, Thornton C, Miller E (1999) The induction of immunologic memory after vaccination with Haemophilus influenzae type b conjugate and acellular pertussis-containing diphtheria, tetanus, and pertussis vaccine combination. *J. Infect. Dis.* 180:538–541.
26. Guirola M, Urquiza D, Alvarez A, Cannan-Haden L, Caballero E, Guillen G (2006) Immunologic memory response induced by a meningococcal serogroup C conjugate vaccine using the P64k recombinant protein as carrier. *FEMS Immunol. Med. Microbiol.* 46:169–179.
27. Anon (1989) Pneumococcal polysaccharide vaccine. *MMWR Morb. Mortal. Wkly. Rep.* 38:64–66.
28. Lee CJ (1996) Bacterial capsular polysaccharides. In Dimitriu S, Ed. *Polysaccharides in Medicinal Applications*. Marcel Dekker, New York, pp. 411–441.

29. Trollfors B, Lagergard T, Claesson BA, Thornberg E, Martinell J, Schneerson R (1992) Characterization of the serum antibody response to the capsular polysaccharide of Haemophilus influenzae type b in children with invasive infections. *J. Infect. Dis.* 166:1335–1339.

30. Schneerson R, Robbins JB (1975) Induction of serum Haemophilus influenzae type B capsular antibodies in adult volunteers fed cross-reacting Escherichia coli 075:K100:H5. *N. Engl. J. Med.* 292:1093–1096.

31. Guinamard R, Okigaki M, Schlessinger J, Ravetch JV (2000) Absence of marginal zone B cells in Pyk-2-deficient mice defines their role in the humoral response. *Nat. Immunol.* 1:31–36.

32. Jeurissen A, Ceuppens JL, Bossuyt X (2004) T lymphocyte dependence of the antibody response to T lymphocyte independent type 2' antigens. *Immunology* 111:1–7.

33. Weller S, Faili A, Aoufouchi S, Gueranger Q, Braun M, Reynaud CA, Weill JC (2003) Hypermutation in human B cells in vivo and in vitro. *Ann. N.Y. Acad. Sci.* 987:158–165.

34. Gold R, Lepow ML, Goldschneider I, Draper TF, Gotshlich EC (1979) Kinetics of antibody production to group A and group C meningococcal polysaccharide vaccines administered during the first six years of life: Prospects for routine immunization of infants and children. *J. Infect. Dis.* 140:690–697.

35. Makela PH, Sibakov M, Herva E, Henrichsen J, Luotonen J, Timonen M, Leinonen M, Koskela M, Pukander J, Pontynen S, Gronroos P, Karma P (1980) Pneumococcal vaccine and otitis media. *Lancet* 2:547–551.

36. Makela PH, Peltola H, Kayhty H, Jousimies K, Pettay O (1977) Polysaccharide vaccines of group A *Neisseria meningitidis* and *Haemophilus influenzae* type b: A field trial in Finland. *J. Infect. Dis.* 136(Suppl):43–50.

37. Wessels MR, Kasper DL, Johnson KD, Harrison LH (1998) Antibody responses in invasive group B streptococcal infection in adults. *J. Infect. Dis.* 178:569–572.

38. Butcher S, Chahel H, Lord JM (2000) Review article: Ageing and the neutrophil: No appetite for killing? *Immunology* 100:411–416.

39. Hedlund J, Ortqvist A, Konradsen HB, Kalin M (2000) Recurrence of pneumonia in relation to the antibody response after pneumococcal vaccination in middle-aged and elderly adults. *Scand. J. Infect. Dis.* 32:281–286.

40. Overturf GD (2002) Pneumococcal vaccination of children. *Semin. Pediatr. Infect. Dis.* 13:155–164.

41. Singleton RJ, Butler JC, Bulkow LR, Hurlburt D, O'Brien KL, Doan W, Parkinson AJ, Hennessy TW (2007) Invasive pneumococcal disease epidemiology and effectiveness of 23-valent pneumococcal polysaccharide vaccine in Alaska native adults. *Vaccine* 25:2288–2295.

42. Bar-Zeev N, Buttery JP (2006) Combination conjugate vaccines. *Expert. Opin. Drug Saf.* 5:351–360.

43. Anon (1991) Haemophilus b conjugate vaccines for prevention of Haemophilus influenzae type b disease among infants and children two months of age and older. Recommendations of the immunization practices advisory committee (ACIP). *MMWR Recomm. Rep.* 40:1–7.

44. Anderson P, Peter G, Johnston R-BJ, Wetterlow LH, Smith DH (1972) Immunization of humans with polyribophosphate, the capsular antigen of Hemophilus influenzae, type b. *J. Clin. Invest.* 51:39–44.

45. Parke J-CJ, Schneerson R, Robbins JB, Schlesselman JJ (1977) Interim report of a controlled field trial of immunization with capsular polysaccharides of Haemophilus influenzae type b and group C Neisseria meningitidis in Mecklenburg county, North Carolina (March 1974–March 1976). *J. Infect. Dis.* 136(Suppl):S51–S56.
46. Smith DH, Peter G, Ingram DL, Harding AL, Anderson P (1973) Responses of children immunized with the capsular polysaccharide of Hemophilus influenzae, type b. *Pediatrics* 52:637–644.
47. Kayhty H, Karanko V, Peltola H, Makela PH (1984) Serum antibodies after vaccination with Haemophilus influenzae type b capsular polysaccharide and responses to reimmunization: No evidence of immunologic tolerance or memory. *Pediatrics* 74:857–865.
48. Schneerson R, Barrera O, Sutton A, Robbins JB (1980) Preparation, characterization, and immunogenicity of Haemophilus influenzae type b polysaccharide-protein conjugates. *J. Exp. Med.* 152:361–376.
49. Verez-Bencomo V, Fernandez-Santana V, Hardy E, Toledo ME, Rodriguez MC, Heynngnezz L, Rodriguez A, Baly A, Herrera L, Izquierdo M, Villar A, Valdes Y, Cosme K, Deler ML, Montane M, Garcia E, Ramos A, Aguilar A, Medina E, Torano G, Sosa I, Hernandez I, Martinez R, Muzachio A, Carmenates A, Costa L, Cardoso F, Campa C, Diaz M, Roy R (2004) A synthetic conjugate polysaccharide vaccine against Haemophilus influenzae type b. *Science* 305:522–525.
50. Anon (2006) Recommended childhood and adolescent immunization schedule—United States, 2006. *MMWR Morb. Mortal. Wkly. Rep.* 54:Q1–Q4.
51. Claesson BA, Trollfors B, Lagergard T, Knutsson N, Schneerson R, Robbins JB (2005) Antibodies against Haemophilus influenzae type b capsular polysaccharide and tetanus toxoid before and after a booster dose of the carrier protein nine years after primary vaccination with a protein conjugate vaccine. *Pediatr. Infect. Dis. J.* 24:463–464.
52. Eskola J, Ward J, Dagan R, Goldblatt D, Zepp F, Siegrist CA (1999) Combined vaccination of Haemophilus influenzae type b conjugate and diphtheria-tetanus-pertussis containing acellular pertussis. *Lancet* 354:2063–2068.
53. Mulholland K, Hilton S, Adegbola R, Usen S, Oparaugo A, Omosigho C, Weber M, Palmer A, Schneider G, Jobe K, Lahai G, Jaffar S, Secka O, Lin K, Ethevenaux C, Greenwood B (1997) Randomised trial of Haemophilus influenzae type-b tetanus protein conjugate vaccine [corrected] for prevention of pneumonia and meningitis in Gambian infants. *Lancet* 349:1191–1197.
54. Booy R, Hodgson S, Carpenter L, Mayon-White RT, Slack MP, Macfarlane JA, Haworth EA, Kiddle M, Shribman S, Roberts JS, Moxon ER (1994) Efficacy of Haemophilus influenzae type b conjugate vaccine PRP-T. *Lancet* 344:362–366.
55. Vadheim CM, Greenberg DP, Partridge S, Jing J, Ward JI (1993) Effectiveness and safety of an Haemophilus influenzae type b conjugate vaccine (PRP-T) in young infants. Kaiser-UCLA Vaccine Study Group. *Pediatrics* 92:272–279.
56. Obonyo CO, Lau J (2006) Efficacy of Haemophilus influenzae type b vaccination of children: A meta-analysis. *Eur. J. Clin. Microbiol. Infect. Dis.* 25:90–97.
57. Black SB, Shinefield HR, Fireman B, Hiatt R, Polen M, Vittinghoff E (1991) Efficacy in infancy of oligosaccharide conjugate Haemophilus influenzae type b (HbOC) vaccine in a United States population of 61,080 children. The Northern California Kaiser Permanente Vaccine Study Center Pediatrics Group. *Pediatr. Infect. Dis. J.* 10:97–104.

58. Ward J, Brenneman G, Letson GW, Heyward WL (1990) Limited efficacy of a Haemophilus influenzae type b conjugate vaccine in Alaska Native infants. The Alaska H. influenzae Vaccine Study Group. *N. Engl. J. Med.* 323:1393–1401.
59. Chandran A, Watt JP, Santosham M (2005) Prevention of Haemophilus influenzae type b disease: Past success and future challenges. *Expert. Rev. Vaccines* 4:819–827.
60. Granoff DM, Holmes SJ, Osterholm MT, McHugh JE, Lucas AH, Anderson EL, Belshe RB, Jacobs JL, Medley F, Murphy TV (1993) Induction of immunologic memory in infants primed with Haemophilus influenzae type b conjugate vaccines. *J. Infect. Dis.* 168:663–671.
61. Anon (2002) Progress toward elimination of Haemophilus influenzae type b invasive disease among infants and children—United States, 1998–2000. *MMWR Morb. Mortal. Wkly. Rep.* 51:234–237.
62. Barbour ML, Mayon-White RT, Coles C, Crook DW, Moxon ER (1995) The impact of conjugate vaccine on carriage of Haemophilus influenzae type b. *J. Infect. Dis.* 171:93–98.
63. Takala AK, Eskola J, Leinonen M, Kayhty H, Nissinen A, Pekkanen E, Makela PH (1991) Reduction of oropharyngeal carriage of Haemophilus influenzae type b (Hib) in children immunized with an Hib conjugate vaccine. *J. Infect. Dis.* 164:982–986.
64. Rosenstein NE, Perkins BA, Stephens DS, Lefkowitz L, Cartter ML, Danila R, Cieslak P, Shutt KA, Popovic T, Schuchat A, Harrison LH, Reingold AL (1999) The changing epidemiology of meningococcal disease in the United States, 1992–1996. *J. Infect. Dis.* 180:1894–1901.
65. Anon (2003) Summary of notifiable diseases—United States, 2001. *MMWR Morb. Mortal. Wkly. Rep.* 50:i–108.
66. Rosenstein NE, Perkins BA, Stephens DS, Popovic T, Hughes JM (2001) Meningococcal disease. *N. Engl. J. Med.* 344:1378–1388.
67. Goldschneider I, Gotschlich EC, Artenstein MS (1969) Human immunity to the meningococcus. I. The role of humoral antibodies. *J. Exp. Med.* 129:1307–1326.
68. Wahdan MH, Rizk F, El Akkad AM, El Ghoroury A, Hablas R, Girgis NI, Amer A, Boctar W, Sippel JE, Gotschlich EC, Triau R, Sanborn WR, Cvjetanovic B (1973) A controlled field of a serogroup A meningococcal polysaccharide vaccine. *Bull. World Health Organ.* 48:667–673.
69. Soriano-Gabarro M, Toe L, Tiendrebeogo SR, Nelson CB, Dabal M, Djingarey MH, Plikaytis B, Rosenstein N (2007) Effectiveness of a trivalent serogroup A/C/W135 meningococcal polysaccharide vaccine in Burkina Faso, 2003. *Vaccine* 25(Suppl 1): A92–A96.
70. Anon (1985) Recommendation of the Immunization Practices Advisory Committee Meningococcal Vaccines. *MMWR Morb. Mortal. Wkly. Rep.* 34:255–259.
71. Soriano-Gabarro M, Toe L, Tiendrebeogo SR, Nelson CB, Dabal M, Djingarey MH, Plikaytis B, Rosenstein N (2007) Effectiveness of a trivalent serogroup A/C/W135 meningococcal polysaccharide vaccine in Burkina Faso, 2003. *Vaccine* 25(Suppl 1): A92–A96.
72. Reingold AL, Broome CV, Hightower AW, Ajello GW, Bolan GA, Adamsbaum C, Jones EE, Phillips C, Tiendrebeogo H, Yada A (1985) Age-specific differences in duration of clinical protection after vaccination with meningococcal polysaccharide A vaccine. *Lancet* 2:114–118.

73. MacLennan J, Obaro S, Deeks J, Williams D, Pais L, Carlone G, Moxon R, Greenwood B (1999) Immune response to revaccination with meningococcal A and C polysaccharides in Gambian children following repeated immunisation during early childhood. *Vaccine* 17:3086–3093.
74. Berry DS, Lynn F, Lee CH, Frasch CE, Bash MC (2002) Effect of O acetylation of Neisseria meningitidis serogroup A capsular polysaccharide on development of functional immune responses. *Infect. Immun.* 70:3707–3713.
75. Girard MP, Preziosi MP, Aguado MT, Kieny MP (2006) A review of vaccine research and development: Meningococcal disease. *Vaccine* 24:4692–4700.
76. Jodar L, Feavers IM, Salisbury D, Granoff DM (2002) Development of vaccines against meningococcal disease. *Lancet* 359:1499–1508.
77. LaForce FM, Konde K, Viviani S, Preziosi MP (2007) The Meningitis Vaccine Project. *Vaccine* 25(Suppl 1):A97–100.
78. Robbins JB, Schneerson R, Gotschlich EC (2000) A rebuttal: Epidemic and endemic meningococcal meningitis in sub-Saharan Africa can be prevented now by routine immunization with group A meningococcal capsular polysaccharide vaccine. *Pediatr. Infect. Dis. J.* 19:945–953.
79. Patel M, Lee CK (2005) Polysaccharide vaccines for preventing serogroup A meningococcal meningitis. *Cochrane. Database Syst. Rev.* CD001093.
80. Lieberman JM, Chiu SS, Wong VK, Partidge S, Chang SJ, Chiu CY, Gheesling LL, Carlone GM, Ward JI (1996) Safety and immunogenicity of a serogroups A/C Neisseria meningitidis oligosaccharide-protein conjugate vaccine in young children. A randomized controlled trial. *JAMA* 275:1499–1503.
81. Fairley CK, Begg N, Borrow R, Fox AJ, Jones DM, Cartwright K (1996) Conjugate meningococcal serogroup A and C vaccine: Reactogenicity and immunogenicity in United Kingdom infants. *J. Infect. Dis.* 174:1360–1363.
82. Leach A, Twumasi PA, Kumah S, Banya WS, Jaffar S, Forrest BD, Granoff DM, LiButti DE, Carlone GM, Pais LB, Broome CV, Greenwood BM (1997) Induction of immunologic memory in Gambian children by vaccination in infancy with a group A plus group C meningococcal polysaccharide-protein conjugate vaccine. *J. Infect. Dis.* 175:200–204.
83. MacLennan J, Obaro S, Deeks J, Lake D, Elie C, Carlone G, Moxon ER, Greenwood B (2001) Immunologic memory 5 years after meningococcal A/C conjugate vaccination in infancy. *J. Infect. Dis.* 183:97–104.
84. Campagne G, Garba A, Fabre P, Schuchat A, Ryall R, Boulanger D, Bybel M, Carlone G, Briantais P, Ivanoff B, Xerri B, Chippaux JP (2000) Safety and immunogenicity of three doses of a Neisseria meningitidis A + C diphtheria conjugate vaccine in infants from Niger. *Pediatr. Infect. Dis. J.* 19:144–150.
85. Rennels M, King JJ, Ryall R, Papa T, Froeschle J (2004) Dosage escalation, safety and immunogenicity study of four dosages of a tetravalent meningococcal polysaccharide diphtheria toxoid conjugate vaccine in infants. *Pediatr. Infect. Dis. J.* 23:429–435.
86. Joseph H, Ryall R, Bybel M, Papa T, MacLennan J, Buttery J, Borrow R (2003) Immunogenicity and immunological priming of the serogroup a portion of a bivalent meningococcal A/C conjugate vaccine in 2-year-old children. *J. Infect. Dis.* 187: 1142–1146.
87. Anderson P, Pichichero M, Edwards K, Porch CR, Insel R (1987) Priming and induction of Haemophilus influenzae type b capsular antibodies in early infancy by Dpo20, an oligosaccharide–protein conjugate vaccine. *J. Pediatr.* 111:644–650.

88. Jennings HJ, Lugowski C (1981) Immunochemistry of groups A, B, and C meningococcal polysaccharide-tetanus toxoid conjugates. *J. Immunol.* 127:1011–1018.
89. Jessouroun E, Da Silveira IAF, Bastos RC, Frasch CE, Lee CJ (2005) Process for preparing polysaccharide-protein conjugate vaccines Patent No. WO/2005/037320.
90. Kshirsagar N, Mur N, Thatte U, Gogtay N, Viviani S, Preziosi MP, Elie C, Findlow H, Carlone G, Borrow R, Parulekar V, Plikaytis B, Kulkarni P, Imbault N, LaForce FM (2007) Safety, immunogenicity, and antibody persistence of a new meningococcal group A conjugate vaccine in healthy Indian adults. *Vaccine* 25(Suppl 1):A101–A107.
91. Gray SJ, Trotter CL, Ramsay ME, Guiver M, Fox AJ, Borrow R, Mallard RH, Kaczmarski EB (2006) Epidemiology of meningococcal disease in England and Wales 1993/94 to 2003/04: Contribution and experiences of the Meningococcal Reference Unit. *J. Med. Microbiol.* 55:887–896.
92. Ashton FE, Ryan JA, Borczyk A, Caugant DA, Mancino L, Huang D (1991) Emergence of a virulent clone of Neisseria meningitidis serotype 2a that is associated with meningococcal group C disease in Canada. *J. Clin. Microbiol.* 29:2489–2493.
93. Borrow R, Andrews N, Goldblatt D, Miller E (2001) Serological basis for use of meningococcal serogroup C conjugate vaccines in the United Kingdom: Reevaluation of correlates of protection. *Infect. Immun.* 69:1568–1573.
94. Richmond P, Goldblatt D, Fusco PC, Fusco JD, Heron I, Clark S, Borrow R, Michon F (1999) Safety and immunogenicity of a new Neisseria meningitidis serogroup C-tetanus toxoid conjugate vaccine in healthy adults. *Vaccine* 18:641–646.
95. Borrow R, Southern J, Andrews N, Peake N, Rahim R, Acuna M, Martin S, Miller E, Kaczmarski E (2001) Comparison of antibody kinetics following meningococcal serogroup C conjugate vaccine between healthy adults previously vaccinated with meningococcal A/C polysaccharide vaccine and vaccine-naive controls. *Vaccine* 19:3043–3050.
96. Richmond P, Kaczmarski E, Borrow R, Findlow J, Clark S, McCann R, Hill J, Barker M, Miller E (2000) Meningococcal C polysaccharide vaccine induces immunologic hyporesponsiveness in adults that is overcome by meningococcal C conjugate vaccine. *J. Infect. Dis.* 181:761–764.
97. Richmond P, Borrow R, Goldblatt D, Findlow J, Martin S, Morris R, Cartwright K, Miller E (2001) Ability of 3 different meningococcal C conjugate vaccines to induce immunologic memory after a single dose in UK toddlers. *J. Infect. Dis.* 183:160–163.
98. MacLennan JM, Shackley F, Heath PT, Deeks JJ, Flamank C, Herbert M, Griffiths H, Hatzmann E, Goilav C, Moxon ER (2000) Safety, immunogenicity, and induction of immunologic memory by a serogroup C meningococcal conjugate vaccine in infants: A randomized controlled trial. *JAMA* 283:2795–2801.
99. Bramley JC, Hall T, Finn A, Buttery RB, Elliman D, Lockhart S, Borrow R, Jones IG (2001) Safety and immunogenicity of three lots of meningococcal serogroup C conjugate vaccine administered at 2, 3 and 4 months of age. *Vaccine* 19:2924–2931.
100. Richmond P, Borrow R, Miller E, Clark S, Sadler F, Fox A, Begg N, Morris R, Cartwright K (1999) Meningococcal serogroup C conjugate vaccine is immunogenic in infancy and primes for memory. *J. Infect. Dis.* 179:1569–1572.
101. Borrow R, Longworth E, Gray SJ, Kaczmarski EB (2000) Prevalence of de-O-acetylated serogroup C meningococci before the introduction of meningococcal serogroup C conjugate vaccines in the United Kingdom. *FEMS Immunol. Med. Microbiol.* 28:189–191.

102. Glode MP, Lewin EB, Sutton A, Le CT, Gotschlich EC, Robbins JB (1979) Comparative immunogenicity of vaccines prepared from capsular polysaccharides of group C Neisseria meningitidis O-acetyl-positive and O-acetyl-negative variants and Escherichia coli K92 in adult volunteers. *J. Infect. Dis.* 139:52–59.
103. Steinhoff MC, Lewin EB, Gotschlich EC, Robbins JB (1981) Group C Neisseria meningitidis variant polysaccharide vaccines in children. *Infect. Immun.* 34:144–146.
104. Richmond P, Borrow R, Findlow J, Martin S, Thornton C, Cartwright K, Miller E (2001) Evaluation of De-O-acetylated meningococcal C polysaccharide-tetanus toxoid conjugate vaccine in infancy: Reactogenicity, immunogenicity, immunologic priming, and bactericidal activity against O-acetylated and De-O-acetylated serogroup C strains. *Infect. Immun.* 69:2378–2382.
105. Rubinstein LJ, Garcia-Ojeda PA, Michon F, Jennings HJ, Stein KE (1998) Murine immune responses to Neisseria meningitidis group C capsular polysaccharide and a thymus-dependent toxoid conjugate vaccine. *Infect. Immun.* 66:5450–5456.
106. Michon F, Huang CH, Farley EK, Hronowski L, Di J, Fusco PC (2000) Structure activity studies on group C meningococcal polysaccharide-protein conjugate vaccines: Effect of O-acetylation on the nature of the protective epitope. *Dev. Biol. (Basel)* 103:151–160.
107. Miller E, Salisbury D, Ramsay M (2001) Planning, registration, and implementation of an immunisation campaign against meningococcal serogroup C disease in the UK: A success story. *Vaccine* 20(Suppl 1):S58–S67.
108. Balmer P, Borrow R, Miller E (2002) Impact of meningococcal C conjugate vaccine in the UK. *J. Med. Microbiol.* 51:717–722.
109. Larrauri A, Cano R, Garcia M, Mateo S (2005) Impact and effectiveness of meningococcal C conjugate vaccine following its introduction in Spain. *Vaccine* 23:4097–4100.
110. De Wals P, Deceuninck G, Boulianne N, De Serres G (2004) Effectiveness of a mass immunization campaign using serogroup C meningococcal conjugate vaccine. *JAMA* 292:2491–2494.
111. Trotter CL, Andrews NJ, Kaczmarski EB, Miller E, Ramsay ME (2004) Effectiveness of meningococcal serogroup C conjugate vaccine 4 years after introduction. *Lancet* 364:365–367.
112. Trotter CL, McVernon J, Andrews NJ, Burrage M, Ramsay ME (2003) Antibody to Haemophilus influenzae type b after routine and catch-up vaccination. *Lancet* 361:1523–1524.
113. Borrow R, Goldblatt D, Finn A, Southern J, Ashton L, Andrews N, Lal G, Riley C, Rahim R, Cartwright K, Allan G, Miller E (2003) Immunogenicity of, and immunologic memory to, a reduced primary schedule of meningococcal C-tetanus toxoid conjugate vaccine in infants in the United Kingdom. *Infect. Immun.* 71:5549–5555.
114. Sexton K, Lennon D, Oster P, Crengle S, Martin D, Mulholland K, Percival T, Reid S, Stewart J, O'Hallahan J (2004) The New Zealand Meningococcal Vaccine Strategy: A tailor-made vaccine to combat a devastating epidemic. *N. Z. Med. J.* 117:U1015.
115. Tondella ML, Popovic T, Rosenstein NE, Lake DB, Carlone GM, Mayer LW, Perkins BA (2000) Distribution of Neisseria meningitidis serogroup B serosubtypes and serotypes circulating in the United States. The Active Bacterial Core Surveillance Team. *J. Clin. Microbiol.* 38:3323–3328.

116. Finne J, Leinonen M, Makela PH (1983) Antigenic similarities between brain components and bacteria causing meningitis. Implications for vaccine development and pathogenesis. *Lancet* 2:355–357.
117. Devi SJN, Robbins JB, Schneerson R (1991) Antibodies to poly[(2-8)-{alpha}-N-acetylneuraminic acid] and poly[(2-9)-{alpha}-N-acetylneuraminic acid] are elicited by immunization of mice with Escherichia coli K92 conjugates: Potential vaccines for Groups B and C meningococci and E. coli K1. *Proc. Natl. Acad. Sci. U.S.A.* 88:7175–7179.
118. Stein DM, Robbins J, Miller MA, Lin FY, Schneerson R (2006) Are antibodies to the capsular polysaccharide of Neisseria meningitidis group B and Escherichia coli K1 associated with immunopathology? *Vaccine* 24:221–228.
119. Jennings HJ, Roy R, Gamian A (1986) Induction of meningococcal group B polysaccharide-specific IgG antibodies in mice by using an N-propionylated B polysaccharide-tetanus toxoid conjugate vaccine. *J. Immunol.* 137:1708–1713.
120. Jennings HJ, Gamian A, Ashton FE (1987) N-propionylated group B meningococcal polysaccharide mimics a unique epitope on group B Neisseria meningitidis. *J. Exp. Med.* 165:1207–1211.
121. Jennings HJ (1988) Chemically modified capsular polysaccharides as vaccines. *Adv. Exp. Med. Biol.* 228:495–550.
122. Granoff DM, Bartoloni A, Ricci S, Gallo E, Rosa D, Ravenscroft N, Guarnieri V, Seid RC, Shan A, Usinger WR, Tan S, McHugh YE, Moe GR (1998) Bactericidal monoclonal antibodies that define unique meningococcal B polysaccharide epitopes that do not cross-react with human polysialic acid. *J. Immunol.* 160:5028–5036.
123. Coquillat D, Bruge J, Danve B, Latour M, Hurpin C, Schulz D, Durbec P, Rougon G (2001) Activity and cross-reactivity of antibodies induced in mice by immunization with a group B meningococcal conjugate. *Infect. Immun.* 69:7130–7139.
124. Moe GR, Dave A, Granoff DM (2005) Epitopes recognized by a nonautoreactive murine anti-N-propionyl meningococcal group B polysaccharide monoclonal antibody. *Infect. Immun.* 73:2123–2128.
125. Pon RA, Lussier M, Yang QL, Jennings HJ (1997) N-Propionylated group B meningococcal polysaccharide mimics a unique bactericidal capsular epitope in group B Neisseria meningitidis. *J. Exp. Med.* 185:1929–1938.
126. Fusco PC, Michon F, Tai JY, Blake MS (1997) Preclinical evaluation of a novel group B meningococcal conjugate vaccine that elicits bactericidal activity in both mice and non-human primates. *J. Infect. Dis.* 175:364–372.
127. Moore S, Farley EK, Fusco PC, Michon F (2007) Specificity of the immune response to a modified group B meningococcal polysaccharide conjugate vaccine. *Clin. Vaccine Immunol.* 14:106–109.
128. Bruge J, Bouveret-Le-Cam N, Danve B, Rougon G, Schulz D (2004) Clinical evaluation of a group B meningococcal *N*-propionylated polysaccharide conjugate vaccine in adult, male volunteers. *Vaccine* 22:1087–1096.
129. Plested JS, Harris SL, Wright JC, Coull PA, Makepeace K, Gidney MA, Brisson JR, Richards JC, Granoff DM, Moxon ER (2003) Highly conserved Neisseria meningitidis inner-core lipopolysaccharide epitope confers protection against experimental meningococcal bacteremia. *J. Infect. Dis.* 187:1223–1234.
130. Plested JS, Makepeace K, Jennings MP, Gidney MA, Lacelle S, Brisson J, Cox AD, Martin A, Bird AG, Tang CM, Mackinnon FM, Richards JC, Moxon ER (1999)

Conservation and accessibility of an inner core lipopolysaccharide epitope of Neisseria meningitidis. *Infect. Immun.* 67:5417–5426.
131. Verheul AF, Boons GJ, Van der Marel GA, Van Boom JH, Jennings HJ, Snippe H, Verhoef J, Hoogerhout P, Poolman JT (1991) Minimal oligosaccharide structures required for induction of immune responses against meningococcal immunotype L1, L2, and L3,7,9 lipopolysaccharides determined by using synthetic oligosaccharide-protein conjugates. *Infect. Immun.* 59:3566–3573.
132. Perrett KP, Pollard AJ (2005) Towards an improved serogroup B Neisseria meningitidis vaccine. *Expert. Opin. Biol. Ther.* 5:1611–1625.
133. Harrison LH (2006) Prospects for vaccine prevention of meningococcal infection. *Clin. Microbiol. Rev.* 19:142–164.
134. Sierra GV, Campa HC, Varcacel NM, Garcia IL, Izquierdo PL, Sotolongo PF, Casanueva GV, Rico CO, Rodriguez CR, Terry MH (1991) Vaccine against group B Neisseria meningitidis: Protection trial and mass vaccination results in Cuba. *NIPH Ann.* 14:195–207.
135. Oster P, Lennon D, O'Hallahan J, Mulholland K, Reid S, Martin D (2005) MeNZB: A safe and highly immunogenic tailor-made vaccine against the New Zealand Neisseria meningitidis serogroup B disease epidemic strain. *Vaccine* 23:2191–2196.
136. Bjune G, Hoiby EA, Gronnesby JK, Arnesen O, Fredriksen JH, Halstensen A, Holten E, Lindbak AK, Nokleby H, Rosenqvist E, Solberg LK, Closs O, Eng J, Froholm LO, Lystad A, Bakketeig LS, Hareide B (1991) Effect of outer membrane vesicle vaccine against group B meningococcal disease in Norway. *Lancet* 338:1093–1096.
137. Bilukha OO, Rosenstein N (2005) Prevention and control of meningococcal disease. Recommendations of the Advisory Committee on Immunization Practices (ACIP). *MMWR Recomm. Rep.* 54:1–21.
138. Keyserling H, Papa T, Koranyi K, Ryall R, Bassily E, Bybel MJ, Sullivan K, Gilmet G, Reinhardt A (2005) Safety, immunogenicity, and immune memory of a novel meningococcal (groups A, C, Y, and W-135) polysaccharide diphtheria toxoid conjugate vaccine (MCV-4) in healthy adolescents. *Arch. Pediatr. Adolesc. Med.* 159:907–913.
139. Rennels M, King JJ, Ryall R, Papa T, Froeschle J (2004) Dosage escalation, safety and immunogenicity study of four dosages of a tetravalent meninogococcal polysaccharide diphtheria toxoid conjugate vaccine in infants. *Pediatr. Infect. Dis. J.* 23:429–435.
140. Pichichero M, Casey J, Blatter M, Rothstein E, Ryall R, Bybel M, Gilmet G, Papa T (2005) Comparative trial of the safety and immunogenicity of quadrivalent (A, C, Y, W-135) meningococcal polysaccharide-diphtheria conjugate vaccine versus quadrivalent polysaccharide vaccine in two- to ten-year-old children. *Pediatr. Infect. Dis. J.* 24:57–62.
141. Sanofi Pasteur (2007) Menactra meningococcal conjugate vaccine approved for use in children 2 years through to 10 years of age. Available at: http://www.vaccineplace.com/docs/MENACTRA_FDA_APPROVAL_181007_ENG.pdf.
142. Snape MD, Perrett KP, Ford KJ, John TM, Pace D, Yu LM, Langley JM, McNeil S, Dull PM, Ceddia F, Anemona A, Halperin SA, Dobson S, Pollard AJ (2008) Immunogenicity of a tetravalent meningococcal glycoconjugate vaccine in infants—A randomized controlled trial. *JAMA* 299:173–184.
143. Anon (2008) Novartis Menveo® vaccine shows superior immune response against four types of meningitis disease in pivotal phase III trial. Available at: http://www.novartis.com/newsroom/media-releases/en/2008/1180471.shtml.

144. Trotter CL, Ramsay ME (2007) Vaccination against meningococcal disease in Europe: Review and recommendations for the use of conjugate vaccines. *FEMS Microbiol. Rev.* 31:101–107.
145. Anon (2007) Pneumococcal conjugate vaccine for childhood immunization—WHO position paper. *Wkly. Epidemiol. Rec.* 82:93–104.
146. Low DE (2005) Changing trends in antimicrobial-resistant pneumococci: It's not all bad news. *Clin. Infect. Dis.* 41(Suppl 4):S228–S233.
147. Lee CJ (1999) Advances in pneumococcal vaccines. *Infect. Med.* 16:596–612.
148. Anon (2007) Pneumococcal disease. Available at: www.cdc.gov/vaccines/pubs/pinkbook/downloads/pneumo.pdf.
149. O'Brien KL, Moulton LH, Reid R, Weatherholtz R, Oski J, Brown L, Kumar G, Parkinson A, Hu D, Hackell J, Chang I, Kohberger R, Siber G, Santosham M (2003) Efficacy and safety of seven-valent conjugate pneumococcal vaccine in American Indian children: Group randomised trial. *Lancet* 362:355–361.
150. Robbins JB, Austrian R, Lee CJ, Rastogi SC, Schiffman G, Henrichsen J, Makela PH, Broome CV, Facklam RR, Tiesjema RH, Parke JC (1983) Considerations for formulating the second-generation pneumococcal capsular polysaccharide vaccine with emphasis on the cross-reactive types within groups. *J. Infect. Dis.* 148:1136–1159.
151. Shapiro ED, Berg AT, Austrian R, Schroeder D, Parcells V, Margolis A, Adair RK, Clemens JD (1991) The protective efficacy of polyvalent pneumococcal polysaccharide vaccine. *N. Engl. J. Med.* 325:1453–1460.
152. Melegaro A, Edmunds WJ (2004) The 23-valent pneumococcal polysaccharide vaccine. Part I. Efficacy of PPV in the elderly: A comparison of meta-analyses. *Eur. J. Epidemiol.* 19:353–363.
153. Jodar L, Butler J, Carlone G, Dagan R, Goldblatt D, Kayhty H, Klugman K, Plikaytis B, Siber G, Kohberger R, Chang I, Cherian T (2003) Serological criteria for evaluation and licensure of new pneumococcal conjugate vaccine formulations for use in infants. *Vaccine* 21:3265–3272.
154. World Health Organization (2005) Recommendations for the production and control of pneumococcal conjugate vaccines. WHO Technical Report Series. 927:64–98.
155. Butler JC (1997) Epidemiology of pneumococcal serotypes and conjugate vaccine formulations. *Microb. Drug Resist.* 3:125–129.
156. Farrell DJ, Jenkins SG, Reinert RR (2004) Global distribution of Streptococcus pneumoniae serotypes isolated from paediatric patients during 1999–2000 and the in vitro efficacy of telithromycin and comparators. *J. Med. Microbiol.* 53:1109–1117.
157. Oosterhuis-Kafeja F, Beutels P, Van Damme P (2007) Immunogenicity, efficacy, safety and effectiveness of pneumococcal conjugate vaccines (1998–2006). *Vaccine* 25:2194–2212.
158. Daum RS, Hogerman D, Rennels MB, Bewley K, Malinoski F, Rothstein E, Reisinger K, Block S, Keyserling H, Steinhoff M (1997) Infant immunization with pneumococcal CRM197 vaccines: Effect of saccharide size on immunogenicity and interactions with simultaneously administered vaccines. *J. Infect. Dis.* 176:445–455.
159. Eskola J, Kilpi T, Palmu A, Jokinen J, Haapakoski J, Herva E, Takala A, Kayhty H, Karma P, Kohberger R, Siber G, Makela PH (2001) Efficacy of a pneumococcal conjugate vaccine against acute otitis media. *N. Engl. J. Med.* 344:403–409.

160. Dagan R, Eskola J, Leclerc C, Leroy O (1998) Reduced response to multiple vaccines sharing common protein epitopes that are administered simultaneously to infants. *Infect. Immun.* 66:2093–2098.
161. Dagan R, Goldblatt D, Maleckar JR, Yaich M, Eskola J (2004) Reduction of antibody response to an 11-valent pneumococcal vaccine coadministered with a vaccine containing acellular pertussis components. *Infect. Immun.* 72:5383–5391.
162. Lucero MG, Puumalainen T, Ugpo JM, Williams G, Kayhty H, Nohynek H (2004) Similar antibody concentrations in Filipino infants at age 9 months, after 1 or 3 doses of an adjuvanted, 11-valent pneumococcal diphtheria/tetanus-conjugated vaccine: A randomized controlled trial. *J. Infect. Dis.* 189:2077–2084.
163. Cutts FT, Zaman SM, Enwere G, Jaffar S, Levine OS, Okoko JB, Oluwalana C, Vaughan A, Obaro SK, Leach A, McAdam KP, Biney E, Saaka M, Onwuchekwa U, Yallop F, Pierce NF, Greenwood BM, Adegbola RA (2005) Efficacy of nine-valent pneumococcal conjugate vaccine against pneumonia and invasive pneumococcal disease in the Gambia: Randomised, double-blind, placebo-controlled trial. *Lancet* 365:1139–1146.
164. Dagan R, Kayhty H, Wuorimaa T, Yaich M, Bailleux F, Zamir O, Eskola J (2004) Tolerability and immunogenicity of an eleven valent mixed carrier Streptococcus pneumoniae capsular polysaccharide-diphtheria toxoid or tetanus protein conjugate vaccine in Finnish and Israeli infants. *Pediatr. Infect. Dis. J.* 23:91–98.
165. Black S, Shinefield H, Fireman B, Lewis E, Ray P, Hansen JR, Elvin L, Ensor KM, Hackell J, Siber G, Malinoski F, Madore D, Chang I, Kohberger R, Watson W, Austrian R, Edwards K (2000) Efficacy, safety and immunogenicity of heptavalent pneumococcal conjugate vaccine in children. Northern California Kaiser Permanente Vaccine Study Center Group. *Pediatr. Infect. Dis. J.* 19:187–195.
166. Choo S, Seymour L, Morris R, Quataert S, Lockhart S, Cartwright K, Finn A (2000) Immunogenicity and reactogenicity of a pneumococcal conjugate vaccine administered combined with a haemophilus influenzae type B conjugate vaccine in United Kingdom infants. *Pediatr. Infect. Dis. J.* 19:854–862.
167. Anon (2005) Direct and indirect effects of routine vaccination of children with 7-valent pneumococcal conjugate vaccine on incidence of invasive pneumococcal disease—United States, 1998–2003. *MMWR Morb. Mortal. Wkly. Rep.* 54:893–897.
168. Kilpi T, Ahman H, Jokinen J, Lankinen KS, Palmu A, Savolainen H, Gronholm M, Leinonen M, Hovi T, Eskola J, Kayhty H, Bohidar N, Sadoff JC, Makela PH (2003) Protective efficacy of a second pneumococcal conjugate vaccine against pneumococcal acute otitis media in infants and children: Randomized, controlled trial of a 7-valent pneumococcal polysaccharide-meningococcal outer membrane protein complex conjugate vaccine in 1666 children. *Clin. Infect. Dis.* 37:1155–1164.
169. Fletcher MA, Fritzell B (2007) Brief review of the clinical effectiveness of PREVENAR against otitis media. *Vaccine* 25:2507–2512.
170. Palmu AA, Verho J, Jokinen J, Karma P, Kilpi TM (2004) The seven-valent pneumococcal conjugate vaccine reduces tympanostomy tube placement in children. *Pediatr. Infect. Dis. J.* 23:732–738.
171. Palmu AA, Jokinen JT, Kaijalainen T, Leinonen M, Karma P, Kilpi TM (2005) Association of clinical signs and symptoms with pneumococcal acute otitis media by serotype—implications for vaccine effect. *Clin. Infect. Dis.* 40:52–57.

172. O'Brien KL, David AB, Chandran A, Moulton LH, Reid R, Weatherholtz R, Santosham M (2008) Randomized, controlled trial efficacy of pneumococcal conjugate vaccine against otitis media among Navajo and White Mountain Apache infants. *Pediatr. Infect. Dis. J.* 27:71–73.

173. Dagan R, Givon-Lavi N, Zamir O, Fraser D (2003) Effect of a nonavalent conjugate vaccine on carriage of antibiotic-resistant Streptococcus pneumoniae in day-care centers. *Pediatr. Infect. Dis. J.* 22:532–540.

174. Choo S, Zhang Q, Seymour L, Akhtar S, Finn A (2000) Primary and booster salivary antibody responses to a 7-valent pneumococcal conjugate vaccine in infants. *J. Infect. Dis.* 182:1260–1263.

175. Bogaert D, Veenhoven RH, Ramdin R, Luijendijk IH, Rijkers GT, Sanders EA, de Groot R, Hermans PW (2005) Pneumococcal conjugate vaccination does not induce a persisting mucosal IgA response in children with recurrent acute otitis media. *Vaccine* 23:2607–2613.

176. Hammitt LL, Bruden DL, Butler JC, Baggett HC, Hurlburt DA, Reasonover A, Hennessy TW (2006) Indirect effect of conjugate vaccine on adult carriage of Streptococcus pneumoniae: An explanation of trends in invasive pneumococcal disease. *J. Infect. Dis.* 193:1487–1494.

177. Byington CL, Samore MH, Stoddard GJ, Barlow S, Daly J, Korgenski K, Firth S, Glover D, Jensen J, Mason EO, Shutt CK, Pavia AT (2005) Temporal trends of invasive disease due to Streptococcus pneumoniae among children in the intermountain west: Emergence of nonvaccine serogroups. *Clin. Infect. Dis.* 41:21–29.

178. Peters TR, Poehling KA (2007) Invasive pneumococcal disease: The target is moving. *JAMA* 297:1825–1826.

179. Law MR, Palomaki G, Alfirevic Z, Gilbert R, Heath P, McCartney C, Reid T, Schrag S (2005) The prevention of neonatal group B streptococcal disease: A report by a working group of the Medical Screening Society. *J. Med. Screen.* 12:60–68.

180. Baker CJ, Edwards MS (2003) Group B streptococcal conjugate vaccines. *Arch. Dis. Child* 88:375–378.

181. Schuchat A, Wenger JD (1994) Epidemiology of group B streptococcal disease. Risk factors, prevention strategies, and vaccine development. *Epidemiol. Rev.* 16:374–402.

182. Schrag S, Gorwitz R, Fultz-Butts K, Schuchat A (2002) Prevention of perinatal group B streptococcal disease. Revised guidelines from CDC. *MMWR Recomm. Rep.* 51:1–22.

183. Schuchat A, Oxtoby M, Cochi S, Sikes RK, Hightower A, Plikaytis B, Broome CV (1990) Population-based risk factors for neonatal group B streptococcal disease: Results of a cohort study in metropolitan Atlanta. *J. Infect. Dis.* 162:672–677.

184. Anon (2005) Early-onset and late-onset neonatal group B streptococcal disease—United States, 1996–2004. *MMWR Morb. Mortal. Wkly. Rep.* 54:1205–1208.

185. Schrag SJ, Zywicki S, Farley MM, Reingold AL, Harrison LH, Lefkowitz LB, Hadler JL, Danila R, Cieslak PR, Schuchat A (2000) Group B streptococcal disease in the era of intrapartum antibiotic prophylaxis. *N. Engl. J. Med.* 342:15–20.

186. Wessels MR, DiFabio JL, Benedi VJ, Kasper DL, Michon F, Brisson JR, Jelinkova J, Jennings HJ (1991) Structural determination and immunochemical characterization of the type V group B Streptococcus capsular polysaccharide. *J. Biol. Chem.* 266:6714–6719.

187. Hervas JA, Benedi VJ (1993) Neonatal sepsis caused by a new group B streptococcal serotype (type V). *J. Pediatr.* 123:839.
188. Farley MM (2001) Group B streptococcal disease in nonpregnant adults. *Clin. Infect. Dis.* 33:556–561.
189. Harrison LH, Elliott JA, Dwyer DM, Libonati JP, Ferrieri P, Billmann L, Schuchat A (1998) Serotype distribution of invasive group B streptococcal isolates in Maryland: Implications for vaccine formulation. Maryland Emerging Infections Program. *J. Infect. Dis.* 177:998–1002.
190. Zaleznik DF, Rench MA, Hillier S, Krohn MA, Platt R, Lee ML, Flores AE, Ferrieri P, Baker CJ (2000) Invasive disease due to group B Streptococcus in pregnant women and neonates from diverse population groups. *Clin. Infect. Dis.* 30:276–281.
191. Baker CJ, Rench MA, Paoletti LC, Edwards MS (2007) Dose-response to type V group B streptococcal polysaccharide-tetanus toxoid conjugate vaccine in healthy adults. *Vaccine* 25:55–63.
192. Baker CJ, Kasper DL (1976) Correlation of maternal antibody deficiency with susceptibility to neonatal group B streptococcal infection. *N. Engl. J. Med.* 294:753–756.
193. Wessels MR, Paoletti LC, Kasper DL, DiFabio JL, Michon F, Holme K, Jennings HJ (1990) Immunogenicity in animals of a polysaccharide-protein conjugate vaccine against type III group B Streptococcus. *J. Clin. Invest* 86:1428–1433.
194. Englund JA, Glezen WP (2003) Maternal immunization with Haemophilus influenzae type b vaccines in different populations. *Vaccine* 21:3455–3459.
195. Jennings HJ, Katzenellenbogen E, Lugowski C, Kasper DL (1983) Structure of native polysaccharide antigens of type Ia and type Ib group B Streptococcus. *Biochemistry* 22:1258–1264.
196. Jennings HJ, Rosell KG, Katzenellenbogen E, Kasper DL (1983) Structural determination of the capsular polysaccharide antigen of type II group B Streptococcus. *J. Biol. Chem.* 258:1793–1798.
197. Wessels MR, Pozsgay V, Kasper DL, Jennings HJ (1987) Structure and immunochemistry of an oligosaccharide repeating unit of the capsular polysaccharide of type III group B Streptococcus. A revised structure for the type III group B streptococcal polysaccharide antigen. *J. Biol. Chem.* 262:8262–8267.
198. Di Fabio JL, Michon F, Brisson JR, Benedi VJ, Wessels MR, Kaspar DL, Jennings HJ (1989) Structure of the capsular polysaccharide antigen of Type IV group B Streptococcus. *Can. J. Chem.* 67:877–882.
199. Kogan G, Uhrin D, Brisson JR, Paoletti LC, Kaspar DL, von Hunolstein C, Orefici G, Jennings HJ (1994) Structure of the type VI group B *Streptococcus* capsular polysaccharide determined by high resolution NMR spectroscopy. *J. Carbohydr. Chem.* 13:1071–1078.
200. Lewis AL, Nizet V, Varki A (2004) Discovery and characterization of sialic acid O-acetylation in group B Streptococcus. *Proc. Natl. Acad. Sci. U.S.A.* 101:11123–11128.
201. Baker CJ, Rench MA, Edwards MS, Carpenter RJ, Hays BM, Kasper DL (1988) Immunization of pregnant women with a polysaccharide vaccine of group B streptococcus. *N. Engl. J. Med.* 319:1180–1185.
202. Wessels MR, Munoz A, Kasper DL (1987) A model of high-affinity antibody binding to type III group B Streptococcus capsular polysaccharide. *Proc. Natl. Acad. Sci. U.S.A.* 84:9170–9174.

203. Brisson JR, Uhrinova S, Woods RJ, vanderZwan M, Jarrell HC, Paoletti LC, Kasper DL, Jennings HJ (1997) NMR and molecular dynamics studies of the conformational epitope of the type III group B Streptococcus capsular polysaccharide and derivatives. *Biochemistry* 36:3278–3292.

204. Kasper DL, Paoletti LC, Wessels MR, Guttormsen HK, Carey VJ, Jennings HJ, Baker CJ (1996) Immune response to type III group B streptococcal polysaccharide-tetanus toxoid conjugate vaccine. *J. Clin. Invest.* 98:2308–2314.

205. Baker CJ, Paoletti LC, Wessels MR, Guttormsen HK, Rench MA, Hickman ME, Kasper DL (1999) Safety and immunogenicity of capsular polysaccharide-tetanus toxoid conjugate vaccines for group B streptococcal types Ia and Ib. *J. Infect. Dis.* 179:142–150.

206. Baker CJ, Paoletti LC, Rench MA, Guttormsen HK, Carey VJ, Hickman ME, Kasper DL (2000) Use of capsular polysaccharide-tetanus toxoid conjugate vaccine for type II group B Streptococcus in healthy women. *J. Infect. Dis.* 182:1129–1138.

207. Baker CJ, Paoletti LC, Rench MA, Guttormsen HK, Edwards MS, Kasper DL (2004) Immune response of healthy women to 2 different group B streptococcal type V capsular polysaccharide-protein conjugate vaccines. *J. Infect. Dis.* 189:1103–1112.

208. Baker CJ, Rench MA, Fernandez M, Paoletti LC, Kasper DL, Edwards MS (2003) Safety and immunogenicity of a bivalent group B streptococcal conjugate vaccine for serotypes II and III. *J. Infect. Dis.* 188:66–73.

209. Baker CJ, Rench MA, McInnes P (2003) Immunization of pregnant women with group B streptococcal type III capsular polysaccharide-tetanus toxoid conjugate vaccine. *Vaccine* 21:3468–3472.

210. Palazzi DL, Rench MA, Edwards MS, Baker CJ (2004) Use of type V group B streptococcal conjugate vaccine in adults 65–85 years old. *J. Infect. Dis.* 190:558–564.

211. Edwards MS (2006) Issues of antimicrobial resistance in group B streptococcus in the era of intrapartum antibiotic prophylaxis. *Semin. Pediatr. Infect. Dis.* 17:149–152.

212. Crump JA, Luby SP, Mintz ED (2004) The global burden of typhoid fever. *Bull. World Health Organ.* 82:346–353.

213. Sinha A, Sazawal S, Kumar R, Sood S, Reddaiah VP, Singh B, Rao M, Naficy A, Clemens JD, Bhan MK (1999) Typhoid fever in children aged less than 5 years. *Lancet* 354:734–737.

214. Lin FY, Vo AH, Phan VB, Nguyen TT, Bryla D, Tran CT, Ha BK, Dang DT, Robbins JB (2000) The epidemiology of typhoid fever in the Dong Thap Province, Mekong Delta region of Vietnam. *Am. J. Trop. Med. Hyg.* 62:644–648.

215. Szu SC, Li XR, Stone AL, Robbins JB (1991) Relation between structure and immunologic properties of the Vi capsular polysaccharide. *Infect. Immun.* 59:4555–4561.

216. Robbins JD, Robbins JB (1984) Reexamination of the protective role of the capsular polysaccharide (Vi antigen) of Salmonella typhi. *J. Infect. Dis.* 150:436–449.

217. Fraser A, Goldberg E, Acosta CJ, Paul M, Leibovici L (2007) Vaccines for preventing typhoid fever. *Cochrane. Database Syst. Rev.* CD001261.

218. Rowe B, Ward LR, Threlfall EJ (1997) Multidrug-resistant Salmonella typhi: A worldwide epidemic. *Clin. Infect. Dis.* 24(Suppl 1):S106–S109.

219. Lebacq E (2001) Comparative tolerability and immunogenicity of Typherix or Typhim Vi in healthy adults: 0, 12-month and 0, 24-month administration. *BioDrugs* 15(Suppl 1):5–12.

220. Fraser A, Paul M, Goldberg E, Acosta CJ, Leibovici L (2007) Typhoid fever vaccines: Systematic review and meta-analysis of randomised controlled trials. *Vaccine* 25: 7848–7857.
221. Keitel WA, Bond NL, Zahradnik JM, Cramton TA, Robbins JB (1994) Clinical and serological responses following primary and booster immunization with Salmonella typhi Vi capsular polysaccharide vaccines. *Vaccine* 12:195–199.
222. Klugman KP, Koornhof HJ, Robbins JB, Le Cam NN (1996) Immunogenicity, efficacy and serological correlate of protection of Salmonella typhi Vi capsular polysaccharide vaccine three years after immunization. *Vaccine* 14:435–438.
223. Mai NL, Phan VB, Vo AH, Tran CT, Lin FY, Bryla DA, Chu C, Schiloach J, Robbins JB, Schneerson R, Szu SC (2003) Persistent efficacy of Vi conjugate vaccine against typhoid fever in young children. *N. Engl. J. Med.* 349:1390–1391.
224. Szu SC, Taylor DN, Trofa AC, Clements JD, Shiloach J, Sadoff JC, Bryla DA, Robbins JB (1994) Laboratory and preliminary clinical characterization of Vi capsular polysaccharide-protein conjugate vaccines. *Infect. Immun.* 62:4440–4444.
225. Lin FY, Ho VA, Khiem HB, Trach DD, Bay PV, Thanh TC, Kossaczka Z, Bryla DA, Shiloach J, Robbins JB, Schneerson R, Szu SC (2001) The efficacy of a Salmonella typhi Vi conjugate vaccine in two-to-five-year-old children. *N. Engl. J. Med.* 344: 1263–1269.
226. Anon (2006) Evaluation of the safety, immunogenicity, and compatibility with DTP of an investigational Vi-rEPA(2) conjugate vaccine for typhoid fever when administered to infants in Vietnam concurrently with DTP. Available at: http://clinicaltrials.gov/ct2/show/nct00342628.
227. Kossaczka Z, Bystricky S, Bryla DA, Shiloach J, Robbins JB, Szu SC (1997) Synthesis and immunological properties of Vi and di-*O*-acetyl pectin protein conjugates with adipic acid dihydrazide as the linker. *Infect. Immun.* 65:2088–2093.
228. Anon (2006) Salmonella typhi Vi *O*-acetyl pectin-rEPA conjugate vaccine. Available at: http://clinicaltrials.gov/ct2/show/NCT00277147.
229. Scott DA, Komjathy SF, Hu BT, Baker S, Supan LA, Monahan CA, Gruber W, Siber GR, Lockhart SP (2007) Phase 1 trial of a 13-valent pneumococcal conjugate vaccine in healthy adults. *Vaccine* 25:6164–6166.
230. Anon (2007) GlaxoSmithKline files meningococcal conjugate vaccine. Available at: http://www.gsk.com/media/pressreleases/2007/2007_03_30_GSK1006.htm.
231. Poolman J, Kaufhold A, De Grave D, Goldblatt D (2001) Clinical relevance of lower Hib response in DTPa-based combination vaccines. *Vaccine* 19:2280–2285.
232. Greenberg DP, Wong VK, Partridge S, Chang SJ, Jing J, Howe BJ, Ward JI (2000) Immunogenicity of a Haemophilus influenzae type b-tetanus toxoid conjugate vaccine when mixed with a diphtheria-tetanus-acellular pertussis-hepatitis B combination vaccine. *Pediatr. Infect. Dis. J.* 19:1135–1140.
233. O'Brien KL, Hochman M, Goldblatt D (2007) Combined schedules of pneumococcal conjugate and polysaccharide vaccines: Is hyporesponsiveness an issue? *Lancet Infect. Dis.* 7:597–606.
234. McVernon J, Mitchison NA, Moxon ER (2004) T helper cells and efficacy of Haemophilus influenzae type b conjugate vaccination. *Lancet Infect. Dis.* 4:40–43.
235. Ramsay ME, McVernon J, Andrews NJ, Heath PT, Slack MP (2003) Estimating Haemophilus influenzae type b vaccine effectiveness in England and Wales by use of the screening method. *J. Infect. Dis.* 188:481–485.

236. Maiden MC, Stuart JM (2002) Carriage of serogroup C meningococci 1 year after meningococcal C conjugate polysaccharide vaccination. *Lancet* 359:1829–1831.
237. Johnson NG, Ruggeberg JU, Balfour GF, Lee YC, Liddy H, Irving D, Sheldon J, Slack MP, Pollard AJ, Heath PT (2006) Haemophilus influenzae type b reemergence after combination immunization. *Emerg. Infect. Dis.* 12:937–941.
238. Auckland C, Gray S, Borrow R, Andrews N, Goldblatt D, Ramsay M, Miller E (2006) Clinical and immunologic risk factors for meningococcal C conjugate vaccine failure in the United Kingdom. *J. Infect. Dis.* 194:1745–1752.
239. Kelly DF, Pollard AJ, Moxon ER (2005) Immunological memory: The role of B cells in long-term protection against invasive bacterial pathogens. *JAMA* 294:3019–3023.
240. Diggle MA, Clarke SC (2005) Increased genetic diversity of Neisseria meningitidis isolates after the introduction of meningococcal serogroup C polysaccharide conjugate vaccines. *J. Clin. Microbiol.* 43:4649–4653.
241. Pichichero ME, Casey JR (2007) Emergence of a multiresistant serotype 19A pneumococcal strain not included in the 7-valent conjugate vaccine as an otopathogen in children. *JAMA* 298:1772–1778.
242. Hanage WP, Huang SS, Lipsitch M, Bishop CJ, Godoy D, Pelton SI, Goldstein R, Huot H, Finkelstein JA (2007) Diversity and antibiotic resistance among nonvaccine serotypes of Streptococcus pneumoniae carriage isolates in the post-heptavalent conjugate vaccine era. *J. Infect. Dis.* 195:347–352.
243. Singleton RJ, Hennessy TW, Bulkow LR, Hammitt LL, Zulz T, Hurlburt DA, Butler JC, Rudolph K, Parkinson A (2007) Invasive pneumococcal disease caused by nonvaccine serotypes among Alaska native children with high levels of 7-valent pneumococcal conjugate vaccine coverage. *JAMA* 297:1784–1792.
244. Beall B (2007) Vaccination with the pneumococcal 7-valent conjugate: A successful experiment but the species is adapting. *Expert. Rev. Vaccines* 6:297–300.
245. Trotter CL, Ramsay ME, Gray S, Fox A, Kaczmarski E (2006) No evidence for capsule replacement following mass immunisation with meningococcal serogroup C conjugate vaccines in England and Wales. *Lancet Infect. Dis.* 6:616–617.
246. Alcala B, Arreaza L, Salcedo C, Uria MJ, De La Fuente L, Vazquez JA (2002) Capsule switching among C : 2b : P1.2,5 meningococcal epidemic strains after mass immunization campaign, Spain. *Emerg. Infect. Dis.* 8:1512–1514.
247. Feikin DR, Levine O, Nelson C, Mohsnie E, Watt J, Wenger J, Kou U (2001) Estimating the local burden of Haemophilus influenzae type b (Hib) disease preventable by vaccination, a rapid assessment tool. Available at: www.who.int/vaccines-documents/DocsPDF01/www625.pdf.
248. Williams BG, Gouws E, Boschi-Pinto C, Bryce J, Dye C (2002) Estimates of world-wide distribution of child deaths from acute respiratory infections. *Lancet Infect. Dis.* 2:25–32.
249. Crisel RM, Baker RS, Dorman DE (1975) Capsular polymer of Haemophilus influenzae, type b. I. Structural characterization of the capsular polymer of strain Eagan. *J. Biol. Chem.* 250:4926–4930.
250. Bundle DR, Smith ICP, Jennings HJ (1974) Determination of the structure and conformation of bacterial polysaccharides by carbon 13 nuclear magnetic resonance. Studies on the group-specific antigens of *Neisseria meningitidis* serogroups A and X. *J. Biol. Chem.* 249:2275–2281.
251. Bhattacharjee AK, Jennings HJ, Kenny CP, Martin A, Smith IC (1975) Structural determination of the sialic acid polysaccharide antigens of Neisseria meningitidis

serogroups B and C with carbon 13 nuclear magnetic resonance. *J. Biol. Chem.* 250:1926–1932.

252. Bhattacharjee AK, Jennings HJ, Kenny CP, Martin A, Smith IC (1976) Structural determination of the polysaccharide antigens of Neisseria meningitidis serogroups Y, W-135, and BO1. *Can. J. Biochem.* 54:1–8.

253. Heyns K, Kiessling G, Lindenberg W, Paulsen H, Webster ME (1959) D-Galaktosaminuronsäure (2-Amino-2-desoxy-2-galakturonsäure) als baustein des Vi-Antigens. *Chemische. Berichte.* 92:2435–2438.

254. Lindberg B, Lindqvist B, Lonngren J, Powell DA (1980) Structural studies of the capsular polysaccharide from Streptococcus pneumoniae type 1. *Carbohydr. Res.* 78:111–117.

255. Jansson PE, Lindberg B, Anderson M, Lindquist U, Henrichsen J (1988) Structural studies of the capsular polysaccharide from Streptococcus pneumoniae type 2, a reinvestigation. *Carbohydr. Res.* 182:111–117.

256. Reeves RE, Goebel WF (1941) Chemoimmunological studies on the soluble specific substance of pneumococcus. V. The structure of the Type III polysaccharide. *J. Biol. Chem.* 139:511–519.

257. Jansson PE, Lindberg B, Lindquist U (1981) Structural studies of the capsular polysaccharide from Streptococcus pneumoniae type 4. *Carbohydr. Res.* 95:73–80.

258. Jansson PE, Lindberg B, Lindquist U (1985) Structural studies of the capsular polysaccharide from Streptococcus pneumoniae type 5. *Carbohydr. Res.* 140:101–110.

259. Kenne L, Lindberg B, Madden JK (1979) Structural studies of the capsular antigen from Streptococcus pneumoniae type 26. *Carbohydr. Res.* 73:175–182.

260. Moreau M, Richards JC, Perry MB, Kniskern PJ (1988) Application of high-resolution n.m.r. spectroscopy to the elucidation of the structure of the specific capsular polysaccharide of Streptococcus pneumoniae type 7F. *Carbohydr. Res.* 182:79–99.

261. Jones JKN, Perry MB (1957) The structure of the Type VIII pneumococcus specific polysaccharide. *J. Am. Chem. Soc.* 79:2787–2793.

262. Jones C, Mulloy B, Wilson A, Dell A, Oates JE (1985) Structure of the capsular polysaccharide from *Streptococcus pneumoniae* type 9. *J. Chem. Soc. Perkins Trans.* 1: 1665–1673.

263. Perry MB, Daoust V, Carlo DJ (1981) The specific capsular polysaccharide of Streptococcus pneumoniae type 9V. *Can. J. Biochem.* 59:524–533.

264. Jones C (1995) Full assignment of the NMR spectrum of the capsular polysaccharide from Streptococcus pneumoniae serotype 10A. *Carbohydr. Res.* 269:175–181.

265. Richards JC, Perry MB, Moreau M (1988) Elucidation and comparison of the specific capsular polysaccharides of *Streptococcus pneumoniae* group 11 (11F, 11B, 11C, and 11A). *Adv. Exp. Med. Biol.* 228:595.

266. Leontein K, Lindberg B, Lonngren J (1981) Structural studies of the capsular polysaccharide from *Streptococcus pneumoniae* Type 12F. *Can. J. Chem.* 59:2081–2085.

267. Lindberg B, Lonngren J, Powell DA (1977) Structural studies on the specific type-14 pneumococcal polysaccharide. *Carbohydr. Res.* 58:177–186.

268. Jansson PE, Lindberg B, Lindquist U, Ljungberg J (1987) Structural studies of the capsular polysaccharide from Streptococcus pneumoniae types 15B and 15C. *Carbohydr. Res.* 162:111–116.

269. Jones C, Lemercinier X (2005) Full NMR assignment and revised structure for the capsular polysaccharide from Streptococcus pneumoniae type 15B. *Carbohydr. Res.* 340:403–409.
270. Jones C, Whitley C, Lemercinier X (2000) Full assignment of the proton and carbon NMR spectra and revised structure for the capsular polysaccharide from Streptococcus pneumoniae Type 17F. *Carbohydr. Res.* 325:192–201.
271. Lugowski C, Jennings HJ (1984) Structural determination of the capsular polysaccharide of Streptococcus pneumoniae type 18C (56). *Carbohydr. Res.* 131:119–129.
272. Katzenellenbogen E, Jennings HJ (1983) Structural determination of the capsular polysaccharide of Streptococcus pneumoniae type 19A (57). *Carbohydr. Res.* 124:235–245.
273. Jennings HJ, Rosell KG, Carlo DJ (1980) Structural determination of the capsular polysaccharide of *Streptococcus pneumoniae* type-19 (19F). *Can. J. Chem.* 58:1069–1074.
274. Richards JC, Perry MB, Carlo DJ (1983) The specific capsular polysaccharide of *Streptococcus pneumoniae* type 20. *Biochem. Cell Biol.* 61:178–190.
275. Richards JC, Perry MB, Kniskern PJ (1989) Structural analysis of the specific capsular polysaccharide of Streptococcus pneumoniae type 22F. *Can. J. Chem.* 67:1038–1050.
276. Richards JC, Perry MB (1988) Structure of the specific capsular polysaccharide of Streptococcus pneumoniae type 23F (American type 23). *Biochem. Cell Biol.* 66:758–771.
277. Lemercinier X, Jones C (2006) Full assignment of the 1H and 13C spectra and revision of the O-acetylation site of the capsular polysaccharide of Streptococcus pneumoniae Type 33F, a component of the current pneumococcal polysaccharide vaccine. *Carbohydr. Res.* 341:68–74.
278. Wuorimaa T, Kayhty H (2002) Current state of pneumococcal vaccines. *Scand. J. Immunol.* 56:111–129.
279. Schuerman L, Prymula R, Henckaerts I, Poolman J (2007) ELISA IgG concentrations and opsonophagocytic activity following pneumococcal protein D conjugate vaccination and relationship to efficacy against acute otitis media. *Vaccine* 25:1962–1968.
280. Kogan G, Brisson JR, Kasper DL, von Hunolstein C, Orefici G, Jennings HJ (1995) Structural elucidation of the novel type VII group B Streptococcus capsular polysaccharide by high resolution NMR spectroscopy. *Carbohydr. Res.* 277:1–9.
281. Kogan G, Uhrin D, Brisson JR, Paoletti LC, Blodgett AE, Kasper DL, Jennings HJ (1996) Structural and immunochemical characterization of the type VIII group B Streptococcus capsular polysaccharide. *J. Biol. Chem.* 271:8786–8790.

5

CARBOHYDRATE-BASED ANTIVIRAL VACCINES

Benjamin M. Swarts and Zhongwu Guo
Department of Chemistry, Wayne State University, Detroit, Michigan 48202

5.1 INTRODUCTION

Viral infections cause a myriad of diseases ranging from the common cold, flu, and chicken pox to the more life-threatening illnesses such as acquired immunodeficiency syndrome (AIDS), Ebola hemorrhagic fever, and severe acute respiratory syndrome (SARS). Throughout history, humankind has been devastated by viral epidemics. For example, an estimated 70% of the Native American population was killed by contagious diseases, most notably smallpox, during the European colonization of North America [1]. The 1918 Spanish flu pandemic, caused by a deadly strain of influenza A virus, claimed between 20 and 100 million lives [2]. At present, the AIDS pandemic has killed over 25 million people [3]. Meanwhile, some viruses have been implicated in the onset of cancer in humans. For instance, hepatitis B and C can induce a chronic viral infection that is an established cause of hepatic cancer [4], while the human papillomavirus can lead to cervical cancer in women [5]. Furthermore, viral infections, such as AIDS caused by the human immunodeficiency virus (HIV), can result in immunocompromisation, in which patients display increased susceptibility to opportunistic infections and tumors.

The prevalence and severity of many diseases caused by viral infections justify tremendous efforts aimed at the development of antiviral drugs and vaccines. In recent history, antiviral vaccines have played a key role in the control of viral morbidity and mortality. For example, vaccines were instrumental in reducing the transmission

Carbohydrate-Based Vaccines and Immunotherapies. Edited by Zhongwu Guo and Geert-Jan Boons
Copyright © 2009 John Wiley & Sons, Inc.

of—or eradicating, in the case of smallpox—polio, measles, mumps, and other viral diseases.

Many effective antiviral vaccines are composed of killed or attenuated viruses. However, recent advances in the understanding of biochemistry and viral structure at the molecular level have enabled the development of subunit vaccines, which contain purified antigenic portions of a specific virus. Because subunit vaccines are not infectious, they are generally considered safer, especially for immunocompromised patients. In addition, subunit vaccines cause fewer side effects because undesired immune responses are limited.

Recent studies have shown that many viruses express glycans and that protein glycosylation is crucial for their survival and virulence [6]. Since carbohydrates are generally known to be relatively exposed and structurally conserved, viral glycans have become attractive molecular targets in the development of antiviral subunit vaccines. However, viruses usually make use of host cell glycosylation pathways to construct their glycans. Therefore, the similarity between endogenous and viral glycans may help the virus evade immune system detection and make the identification of carbohydrate epitopes for subunit vaccine development quite challenging. Despite this hurdle, significant progress has been made in characterizing viral glycosylation and in the development of carbohydrate-based antiviral vaccines. While the genome of many viruses encodes for envelope glycoproteins that can be useful targets, this chapter will mainly focus on the structures, functions, and potential therapeutic applications of HIV, influenza, and hepatitis C virus (HCV) glycans, as most recent studies on carbohydrate-based vaccine and drug development have focused on these viruses.

5.2 VIRAL GLYCOSYLATION

5.2.1 Viral *N*-glycosylation

Recently, much attention has been focused on studying *N*-glycosylation of viral proteins, and with good reason. Some viruses are shielded by a dense layer of carbohydrates that serve critical functions such as assisting protein folding, aiding entry into host cells, and evading detection by the immune system. Since viruses use host glycosylation pathways for protein *N*-glycosylation, it would be helpful to briefly review the general *N*-glycosylation process.

Asparagine *N*-glycosylation (Fig. 5.1) is a ubiquitous co-translational protein modification that occurs in eukaryotic and some prokaryotic systems [7], and the functions of *N*-glycosylation in eukaryotes have been extensively studied [8]. *N*-glycosylation is initiated by the transfer of a precursor oligosaccharide ($Glc_3Man_9GlcNAc_2$) to the amide nitrogen of asparagine in the Asn-Xaa-Ser/Thr (where X is any amino acid except proline) sequence during protein synthesis. This oligosaccharide then undergoes enzymatic modification in the endoplasmic reticulum (ER) and the Golgi, including glucose and mannose trimming and subsequent glycosylation with various monosaccharides such as glucose, galactose, fucose, sialic acid, and others. The resultant *N*-glycans may have different forms, broadly divided into high-mannose, hybrid,

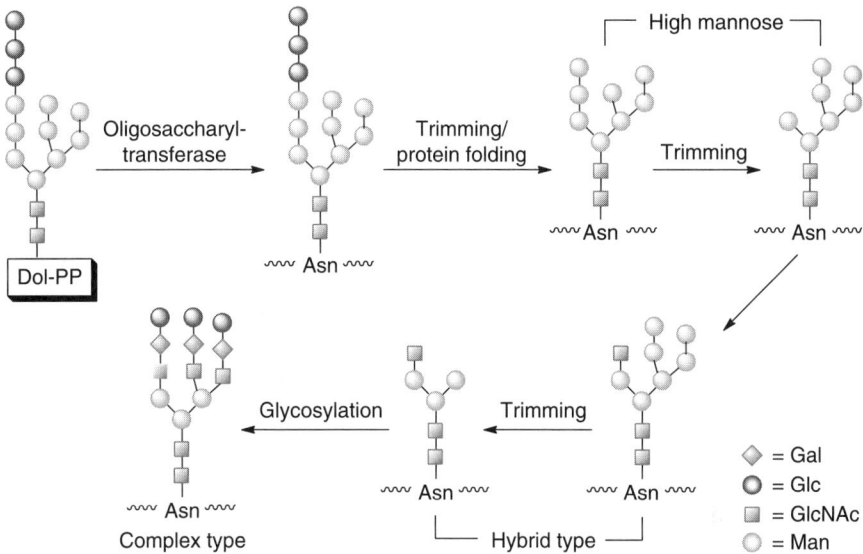

Figure 5.1 Biosynthetic pathway for N-glycans. (See color insert.)

and complex-type glycans according to their structures. For a more in-depth discussion of N-glycosylation, please refer to Chapter 1 Section 1.2.3.2 or one of many reviews on the subject [9, 10].

Viruses can be considered "cell hijackers" that co-opt the host cell's biosynthetic machinery, including the protein glycosylation pathways, to translate their genetic codes and facilitate their reproduction. Thus, the type and extent of N-glycosylation in a viral protein is determined by the biosynthetic environment that the virus has infected. Accordingly, viral protein N-glycosylation occurs at asparagine residues in the consensus peptide sequence Asn-Xaa-Ser/Thr. However, the precise location and number of N-glycosylation sites are encoded in the viral genome and thus are determined by the virus itself. Virus evolution via genome mutation can result in addition, deletion, or alteration of glycosylation sites, increasing the potential diversity of a virus. Predictably, such changes can dramatically alter viral protein structure and function.

The cellular functions of N-glycans in viral proteins generally parallel those characteristic of the host cell [6]. For instance, of great importance to viral biosynthesis is the ability of N-glycans to facilitate proper protein trafficking and correct protein folding, which may involve the host's quality control chaperones calnexin and calreticulin. The same structural attributes that allow viruses to take advantage of the host cell's quality control machinery, namely host-synthesized glycans, also enable viruses to escape detection by the immune system. N-glycans are also instrumental in viral infection, replication, tropism, and other processes. Thus, carbohydrates play a critical role in the persistence of many viruses. The following sections will provide an overview of some specific viral glycosylation motifs and their biological functions.

5.2.2 Carbohydrates of HIV

HIV-1 is a retrovirus that can lead to AIDS, a disease in which immune system failure results in life-threatening opportunistic infections and/or cancer. The surface glycoproteins of HIV-1 are responsible for many important functions, including virus attachment and fusion to the host cell at the initial stage of infection [11]. The HIV-1 envelope gene encodes for glycoprotein 160 (gp160), which is enzymatically cleaved to form two glycoproteins, gp120 and gp41, that assemble to create the viral envelope [12]. Gp41 is buried in the envelope unless assisting cell fusion, and it is non-covalently attached to gp120 [13]. Gp120 is exposed on the envelope surface to mediate HIV attachment to host cells via interaction with CD4 [14]. It is one of the most heavily glycosylated proteins in nature [15], with carbohydrates accounting for approximately half of its mass [16]. The N-glycans of gp120 are presently the primary molecular templates for the development of carbohydrate-based HIV vaccines.

5.2.2.1 Biosynthesis of gp120 The biosynthesis of gp120 starts with its parent glycoprotein, gp160. After the nascent polypeptide of gp160 is biosynthesized by the host cell, it is transported to the ER, where N-glycosylation begins. First, precursor oligosaccharide $Glc_3Man_9GlcNAc_2$ is covalently linked to the appropriate asparagine residues. Next, deglucosylation occurs to give gp160 bearing $GlcMan_9GlcNAc_2$, which is then bound by calnexin/calreticulin to undergo proper folding and assembly. After removal of the last glucose residue from the N-glycan, gp160 is no longer a substrate for calnexin/calreticulin and is released into the Golgi, where it is cleaved by furin in the host cell to give gp41 and gp120 [17]. Further enzymatic modification of the N-glycans in gp120 leads to a mixture of high-mannose, hybrid, and complex-type oligosaccharides [18–21].

5.2.2.2 Structure of gp120 There are approximately 25 N-glycosylation sites in gp120, and although the HIV-1 genome is highly variable, there is no net tendency for glycosylation sites to increase or decrease over time [22]. There is, however, evidence that glycosylation sites may shift in location over the course of HIV-1 infection, which can result in enhanced immune evasion referred to as "dynamic glycan shielding" [23]. Additional variability is imparted by glycan microheterogeneity. In fact, an average of five major glycoforms for each N-glycosylation site would mean that a maximum of 5^{25} gp120 glycoforms are possible for an HIV clone [20, 24]. While such structural diversity raises questions as to whether a carbohydrate-based vaccine could be operative against a broad set of HIV clones, gp120 does contain several highly conserved N-glycosylation sites, most of which display high-mannose glycans [20].

Interestingly, high-mannose and complex-type N-glycans in gp120 are located in different regions on the glycoprotein surface. While most complex-type glycans are present in a region where CD4 binding occurs (the "neutralizing face"), high-mannose glycans are concentrated on the "immunosilent face" of gp120 (Fig. 5.2) [20, 25]. An additional feature of gp120 glycans is that they lack the conformational flexibility typically observed in N-glycans [26] because they are compacted into abnormally tight

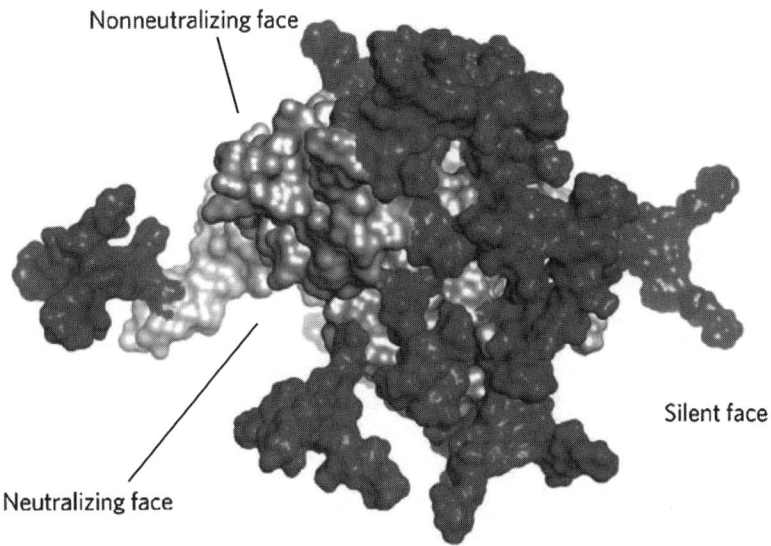

Figure 5.2 Antigenic map of monomeric gp120, with *N*-glycans shown in blue. The neutralizing face contains receptor-binding sites; the nonneutralizing face is hidden when gp120 and gp41 assemble into their active trimeric form; the silent face is composed mostly of clustered high-mannose glycans. (Image reproduced, with permission, from Ref. [24].) (See color insert.)

clusters, as observed in the crystal structure of $gp120_{SIV}$—the envelope glycoprotein of the related simian immunodeficiency virus [27, 28]. The highly conserved and abnormally tight oligomannose clusters on the gp120 surface are immunologically distinct and represent a promising target for vaccine development (see Section 5.3.1 for more details).

5.2.2.3 Functional Roles of N-glycans in gp120
The *N*-glycans in gp120 are instrumental in many of the biological roles mentioned above, including proper protein folding/transport, host cell entry, tropism, and immune evasion. For instance, deletion of certain *N*-glycosylation sites in gp160 prevented its enzymatic cleavage to gp120/gp41 and impaired its transport to the Golgi [29–31]. Early studies suggested that although glycosylation was necessary for creating the functional conformation of the viral glycoproteins and for their intracellular trafficking, it was not critical for CD4 binding, host infection, or virus replication. Recent studies revealed that although loss of glycans in gp120 moderately decreased binding between the virus and CD4, it did not deter host infection or virus replication [31–35].

However, gp120 glycans are crucial for the virus to escape detection by the immune system, as the HIV-1 virus is coated with immunologically "self" carbohydrates [23, 36]. Moreover, the variability of the HIV virus imparted by genome mutation, recombination, and glycan reordering [37] can be considered a mechanism to provide a "dynamic glycan shield" that effectively stays one step ahead of the immune system. For instance, one recent study showed that over the course of an infection, the envelope protein variable loops 1 and 2 (V1 and V2) expand in size and add glycosylation sites that increase diversity and reduce sensitivity to antibody neutralization [38, 39]. These observations lead to the conclusion that this evolution is driven by the need not only to protect the underlying protein epitopes but also to constantly redefine the immunological landscape of the virus by eradicating "non-self" patterns. These properties of HIV render the development of HIV vaccines particularly challenging.

5.2.3 Carbohydrates of Influenza A Virus

Influenza A is a highly variable virus that, although most commonly causes relatively mild illness, has the potential to be fatal to high-risk groups, including the elderly and those with pulmonary or cardiac disease. Occasionally, influenza viruses are transmitted from wild water foul to domestic poultry, potentially causing an outbreak that could lead to human influenza pandemic. The influenza A viral envelope contains two extensively glycosylated surface proteins, hemagglutinin (HA) and neuraminidase (NA). Both glycoproteins function by recognizing sialic acid molecules on the host cell. While HA binds to sialylated receptors on the host to initiate virus infection, NA hydrolyzes sialic acid residues from host cell receptors to assist in the release of progeny virus that go on to infect neighboring cells [40]. The functional interplay between HA and NA in virus attachment/release is reliant on the glycans of HA [41]. Overall, the carbohydrates in HA and NA serve a variety of purposes, including assistance in protein folding, intracellular transport, receptor binding, virus release, infection, immune evasion, and neurovirulence [42–50].

Hemagglutinin assembles as a trimer, and each monomer is composed of two distinct regions: a globular head, which is involved in receptor binding, and a stalk, which anchors HA to the viral envelope [51]. HA biosynthesis occurs in a manner similar to that discussed for the HIV envelope protein gp120. During translation, N-glycosylation at multiple asparagine residues occurs, and glycans assist in the folding and protection of the nascent peptide chain throughout the maturation process, which also involves aid from calnexin and calreticulin [42]. After proper assembly, the N-glycans of HA undergo further modification by host enzymes to generate a mixture of high-mannose, hybrid, and complex-type oligosaccharides [52, 53]. While the complex-type carbohydrates display a considerable degree of heterogeneity, sites containing high-mannose glycans are quite homogeneous [53, 54].

The number of glycosylation sites in HA ranges from 5 to 11, and several glycosylation sites are conserved among an array of animal and human influenza A viruses [55–57]. However, in many virus subtypes, selection has resulted in the introduction of additional glycosylation sites over time [58]. For example, most of the human H3N2 viruses currently exhibit six to seven N-glycosylation sites in the globular head, whereas the prototype H3N2 strain, isolated in 1968, contained only two [58]. Although an evolutionary increase in glycosylation sites is not general for all influenza strains, it is a common occurrence [22]. As the glycans of the HA have been clearly implicated in immune evasion [47, 50, 59], this global increase/variation in glycosylation sites is believed to generate antigenic variants, and thus the concept of an evolving or dynamic glycan shield applies here. On the other hand, HA also contains several regions that have always lacked carbohydrates, and site-specific mutagenesis experiments demonstrated that incorporating glycans into these regions resulted in the disruption of HA transport to the cell surface [60].

The role HA carbohydrates play in host cell attachment and release has been studied in detail. The first step in influenza cell entry involves proteolytic activation of HA, which can be hindered when glycans near the HA cleavage site reduce the accessibility of cellular proteases [61]. Conversely, glycans in close proximity to the receptor binding site appear to be essential to viral replication and release, as they regulate receptor binding affinity [43, 62–64]. Optimal virus replication relies on a delicate functional balance between the surface proteins HA and NA, and the glycosylation state of HA has a direct impact on this balance [40]. Thus, the carbohydrates associated with influenza virus surface proteins are highly involved in many viral functions, and their presence can result in positive or damaging effects depending on location.

5.2.4 Carbohydrates in Hepatitis C Virus

Hepatitis C virus is a major cause of liver diseases, including liver inflammation, fibrosis, cirrhosis, and in some cases liver cancer or failure. HCV is a small envelope virus that contains two transmembrane surface glycoproteins, E1 and E2, which assemble noncovalently as heterodimers and are located primarily in the ER [65]. E1 and E2 display 4–5 and 11 potential N-glycosylation sites, respectively [66]. Most of the sites are highly conserved, suggesting their importance in protein structure and activity [22, 67].

Early studies to probe the effects of glycosylation on E1 function were mostly focused on viral biosynthesis, and the importance of N-glycans during protein folding, assembly, and transportation was confirmed [68–70]. Recent studies suggest that at least two E2 glycosylation sites are crucial for proper folding and assembly of the E1E2 heterodimer, while other E2 glycans are implicated in virus entry into host cells [67, 71]. Several reports have suggested that, like HIV and the influenza virus, carbohydrates surrounding HCV—primarily located in the E2 glycoprotein—form a glycan shield that serves to protect underlying protein epitopes [67, 72, 73]. However, the highly conserved nature of the N-glycosylation sites implies that, unlike HIV and influenza, HCV does not have an "evolving glycan shield." This likely is a result of the E2 glycans serving a dual role in which a compromise between facilitating virus entry and escape from antibody neutralization must be reached [67].

In addition, the E2 glycoprotein binds with high affinity to DC-SIGN (dendritic cell–specific intercellular adhesion molecule–grabbing nonintegrin, also known as CD209) and L-SIGN (liver/lymph node-specific intercellular adhesion molecule–grabbing integrin, also known as CD209L), which are mannose-binding C-type lectins present on macrophages and dendritic cells [74, 75]. These lectins may mediate trans infection of liver cells with HCV [76, 77]. The ability of E2 to interact with mannose-specific lectins is consistent with a recent structural analysis of the E2 glycans, which showed that the majority of carbohydrates are high-mannose-type glycans [78]. The presence of conserved high-mannose glycans suggests that evaluation of carbohydrate-binding agents (CBA) is a logical step toward the identification of anti-HCV drugs [79, 80].

5.2.5 Carbohydrates in Other Viruses

The West Nile virus (WNV) is a mosquito-borne flavivirus that can lead to relatively mild fever and occasionally lethal encephalitis, a neuroinvasive disease. The flavivirus envelope (E) glycoprotein is the major target for immune response [81]. Many WNV strains have one to two E protein N-glycosylation sites, which are not strictly conserved, while some strains do not contain any glycosylation sites [82]. Glycosylation has been suggested to be linked to neurovirulence in several strains [82–85]. Additionally, various studies showed that N-glycans affect WNV protein folding and expression [86], as well as virus infection, release, replication, and tropism [87–90].

Ebola is a member of the Filoviridae family of viruses, and infection can lead to potentially fatal hemorrhagic fever. The envelope glycoprotein of ebola, GP, is composed of two subunits (G1 and G2) that are connected by a disulfide bond to form the heterodimer $GP_{1,2}$ [91]. $GP_{1,2}$ is heavily glycosylated, with N- and O-glycans making up approximately half of its mass [92]. Recent studies have established that the N-glycans of GP1 and GP2 are involved in infectivity and immune evasion [93, 94]. Cyanovirin-N (CV-N), a high-mannose binding agent, was found to associate with $GP_{1,2}$ and inhibit viral entry [95, 96].

The SARS coronavirus (SARS-CoV) causes severe acute respiratory syndrome, a recently emergent human disease that affects the respiratory system and has a relatively high mortality rate, especially among the elderly. Several proteins associated with SARS-CoV exhibit N-glycosylation sites, including the spike (S) protein, which is located in the viral envelope [97]. The N-glycans present on the S protein are critical for DC-SIGN- and L-SIGN-mediated entry into host cells [98, 99]. As these C-type lectins operate by binding high-mannose glycans, the use of mannose-binding CV-N and other carbohydrate-binding agents as anti-SARS-CoV drugs were tested, and several were found to inhibit viral entry at a post-host cell attachment stage [100, 101].

In addition to those discussed above, many other viruses, such as Hendra, Hantaan, Nipah, metapneumoniavirus, cytomegalovirus, Bunyamwera, Newcastle, and Lassa, also exhibit N-glycosylation carrying carbohydrates that serve a variety of functions [102–110]. However, these viruses will not be covered in detail here.

It is clear that the elucidation of viral structure at the molecular level has illuminated the importance of N-glycosylation. As viral glycosylation becomes better understood in specific cases, the development of antiviral vaccines and drugs based on or targeting special virus carbohydrate motifs has emerged as a promising strategy.

5.3 VACCINE AND DRUG DEVELOPMENT

5.3.1 Human Immune Deficiency Virus

HIV-1 gp120 is a primary target for the development of HIV vaccines. Immune response to the protein portion of gp120 in infected individuals is not typically observed due to extensive camouflaging of the conserved peptide epitopes by carbohydrates. Therefore, developing HIV vaccines based on the carbohydrate epitopes of gp120 has become a natural choice. As mentioned above, however, when considering

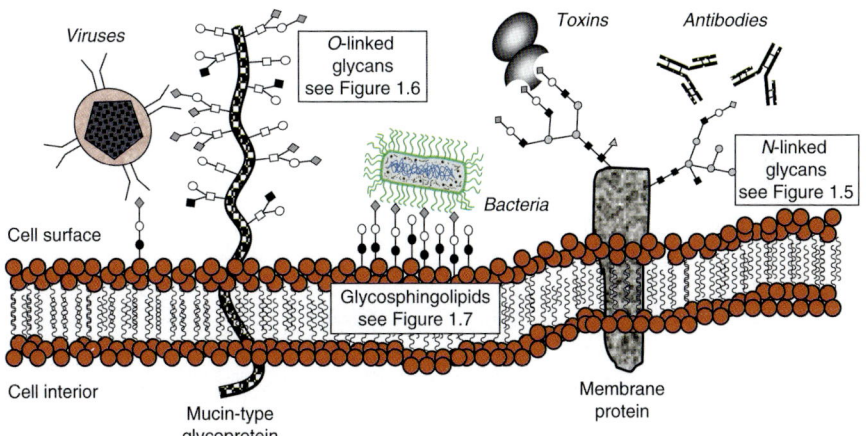

Figure 1.1 Landscape of the cell is dominated by carbohydrates. The surface of a mammalian cell is decorated by various classes of complex carbohydrates; these oligosaccharides are often referred to as *glycans* and collectively as the *glyocalyx*. An overview of glycan biosynthesis is provided in Figures 1.2–1.4 and more information on each class of these molecules is given in Figures 1.5–1.7 and the accompanying text. This graphic is not drawn to scale (as is evident from the relative sizes of toxins, antibodies, viruses, and bacteria, which are all pathogens that exploit surface sugars for entry into a cell). It is notable that even though the glycocalyx comprises only about 8–10% of the mass of the plasma membrane, in a typical mammalian cell it forms a continuous (albeit not uniform) layer ∼8 nm thick, occupying roughly the same volume as the lipid and protein constituents of the membrane.

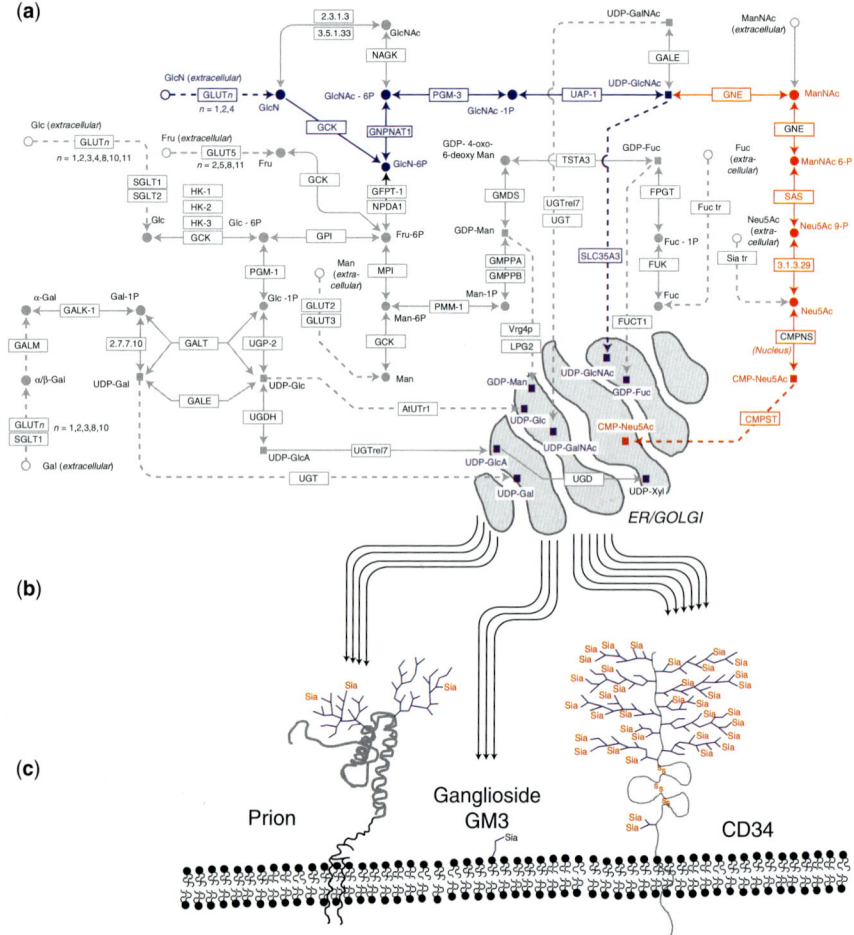

Figure 1.2 Overview of glycan biosynthesis. (a) Monosaccharide uptake and processing (see Fig. 1.3 for structures). Sugars obtained exogenously by cells such as Gal, Glc, and GlcN (full names and chemical structures of common mammalian monosaccharides are shown in Fig. 1.3a) are taken up by families of membrane transporters [32, 33] and converted to nucleotide sugar donors such as UDP-GlcNAc and CMP-Neu5Ac (Fig. 1.3c) by mostly cytosolic enzymatic reactions. Additional information on the enzymes is provided in other review articles [34] or from online search engines or databases such as Pubmed [35], KEGG [36], or HUGO [37]. (b) Glycan assembly (see Fig. 1.4 for details of sialylation). The nucleotide sugars are assembled by a suite of glycosyltransferases (illustrated in detail for sialyltransferases in Fig. 1.4) into structurally complex surface-displayed glycans. (c) Representative glycans include GPI-anchored prions (see Fig. 1.7b, which in turn bear *N*-linked glycans, Fig. 1.5), the glycosphingolipid ganglioside GM3 (see Figs. 1.7a and 1.10e), and CD34, a mucin-type glycoprotein that bears numerous *O*-linked glycans that often include TACAs (see Figs. 1.6a and 1.6b).

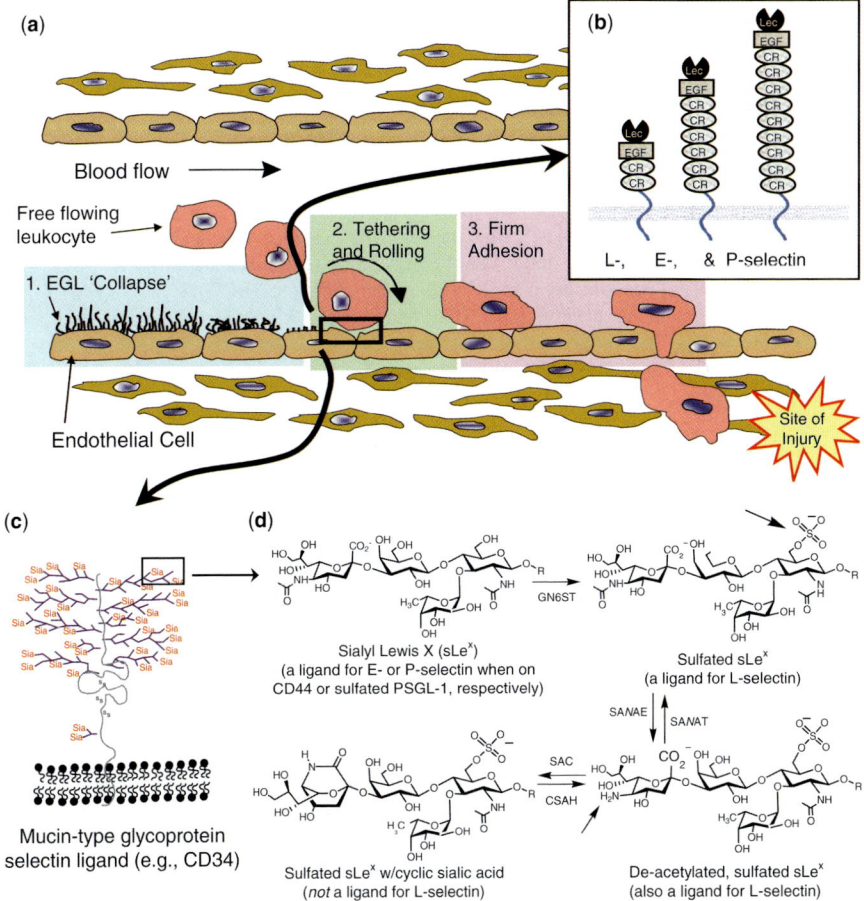

Figure 1.8 Role of glycans in leukocyte extravasation. (a) Free-flowing leukocyte is shown in a cross section of a venule (not to scale; the leukocytes are typically 6–12 μm in diameter, the vessels where extravasation takes place have a diameter of 30–80 μm, and the EGL is approximately 0.2–0.5 μm thick) where extravasation to sites of injury or homing in the lymph node occurs. (step 1) The EGL collapses to a height of less than 50 μm allowing (step 2) selectin-mediated tethering and rolling to take place followed (step 3) by integrin-mediated firm adhesion and extravasation. (b) Graphic of selectin structures where CR represents the cysteine-rich consensus repeat domains, EGF represents the EGF-like domain, and Lec represents the carbohydrate recognition (lectin) domain. (c) CD34 exemplifies mucin-type counter-receptor for selectins where the peptide forms a scaffold for multivalent glycan display; up to 80% of the mass of CD34 can be carbohydrate [53]. (d) Chemical modification of sLe^x determines physiological binding specificity between various selectins and counter-receptors. Of note, the reverse situation, shown where selectins are found on the epithelial substrate (e.g., P-selectin) and interact with ligands (e.g., PSGL-1) on the incoming cell, can also occur.

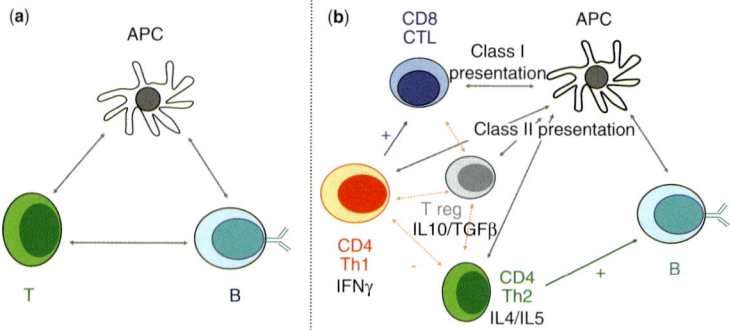

Figure 3.1 Immunological *ménage à trois*. (a) Simplified view of T, B, and APCs interactions. (b) Extended view, showing the major orientations of Th responses triggered by initial APC activation. APC class II presentation to CD4 cells will stimulate Th1, Th2, or Treg responses according to the cytokine environment. Th2 responses will further help B-cell responses predominantly while Th1 responses will support CTL C8 responses, should Ag also be presented by class I antigens to these latter cells. Tregs can regulate all different types of responses.

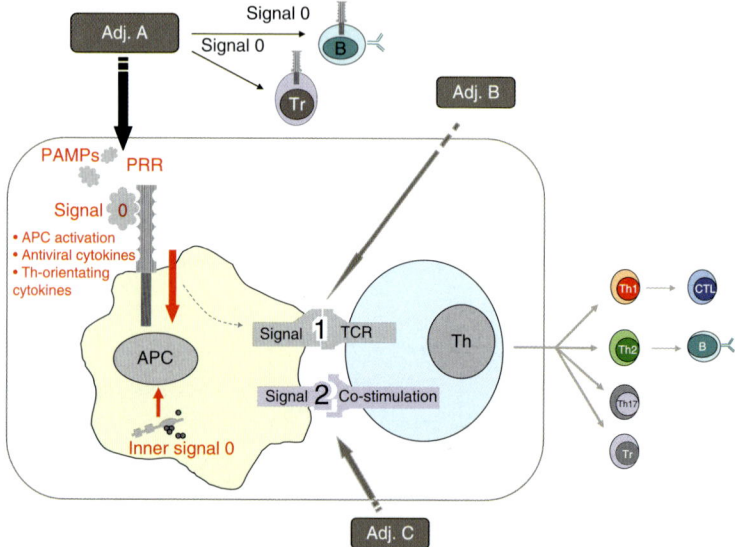

Figure 3.2 Where can adjuvants act? Initiation of T-helper responses needs three signals, as signals 0, 1, and 2. Adjuvants/formulations can in theory act, alone or in combination, on each of these three signals and can be referred as adjuvants A, B, and C, respectively. Most specific adjuvants developed recently, such as TLR agonists, can be considered as type A adjuvants: They act on signal 0, and indirectly on signal 2, by activating antigen-presenting cells (APCs) and triggering the secretion of cytokines such as IL12 to orientate subsequent Th1 responses, for instance. Some TLR agonist can also directly act on B cells and Tregs. In addition, TLR agonists can act on signal 1 by favoring an efficient presentation of the co-administered antigen (Ag). Adjuvants/formulations targeting APCs or favoring Ag capture can be viewed as type B adjuvants, acting on signal 1, as their effect is eventually mediated by enhanced Ag presentation to T cells. Liposomes, microspheres, and some emulsions can be found in this category. As stressed in this review, targeting signal 1 is not sufficient and an immunostimulant signal should be co-delivered for an optimal response. Finally, some specific ligands of co-stimulatory molecules can directly enhance signal 2, acting as type C adjuvants; such compounds must, however, be used with some caution, most likely in therapeutic applications. (Modified from Ref. [1].)

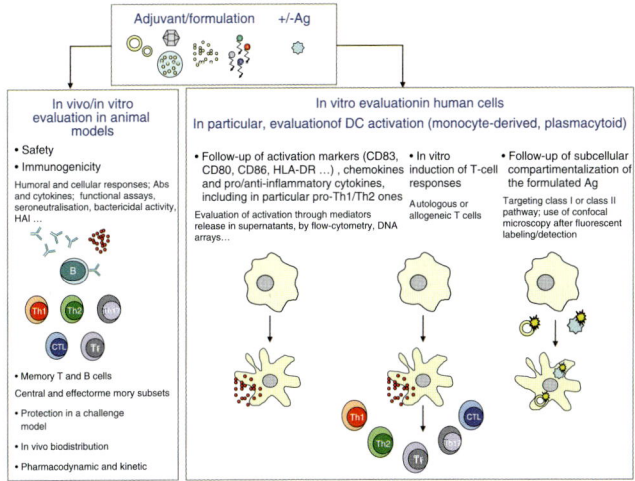

Figure 3.3 In vivo and in vitro preclinical evaluation of adjuvants. In addition to preclinical in vivo evaluation in animal models, it can be of interest to investigate in parallel adjuvants/formulations in vitro in human cells, in particular in APCs such as dendritic cells (DCs). The combination of information obtained in vivo and in vitro may increase our knowledge on both immunogenicity and safety of these formulations before going into clinics. (Modified from Ref. [1].)

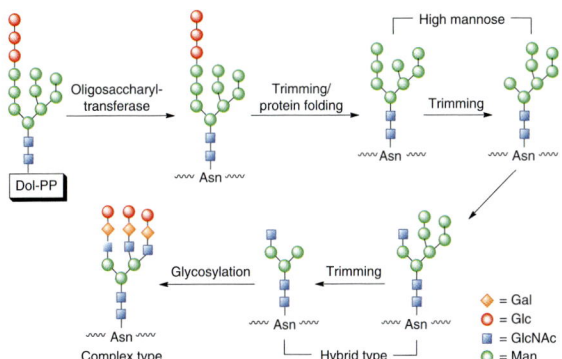

Figure 5.1 Biosynthetic pathway for *N*-glycans.

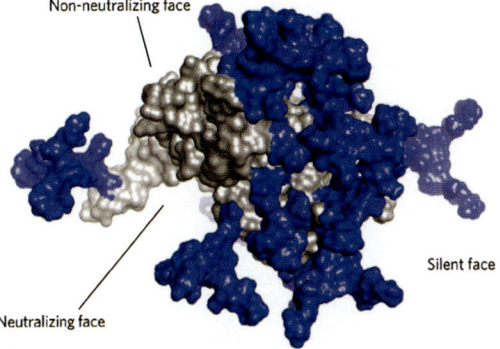

Figure 5.2 Antigenic map of monomeric gp120, with *N*-glycans shown in blue. The neutralizing face contains receptor-binding sites; the nonneutralizing face is hidden when gp120 and gp41 assemble into their active trimeric form; the silent face is composed mostly of clustered high-mannose glycans. (Image reproduced, with permission, from Ref. [24].)

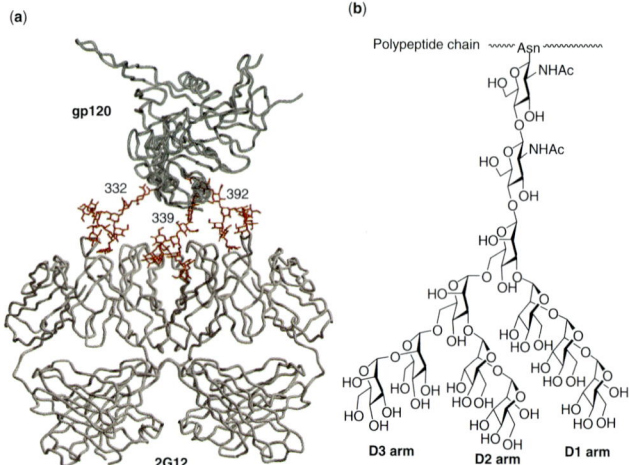

Figure 5.3 (a) Model structure of gp120 in complex with mAb 2G12; glycans involved in the binding are highlited in red color. (Reproduced, with permission, from Ref. [25].) (b) Structure of N-linked high-mannose glycan with the three oligomannose branches defined as D1, D2, and D3 arms.

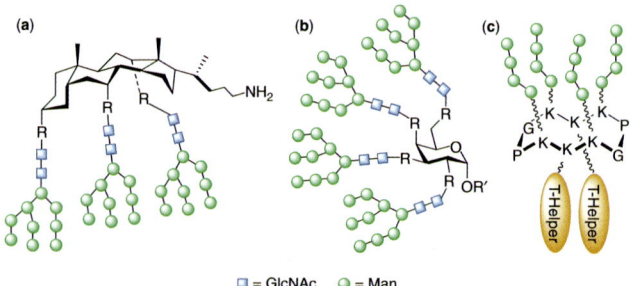

Figure 5.4 Oligomannose clusters built with (a) cholic acid, (b) D-galactose, and (c) RAFT as scaffolds. R = linker; R′ = OEt, KLH, or T helper peptide.

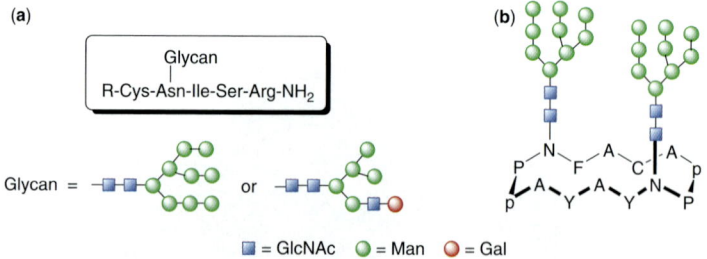

Figure 5.5 Multivalent glycan clusters for anti-HIV vaccine development. (a) High-mannose and hybrid-type glycans are linked to 5mer and 20mer peptides of the gp120 sequence. R = NH$_2$ or appropriate gp120 15mer peptide. (b) Cyclic peptide scaffold was used to construct mono-, di-, and trivalent clusters.

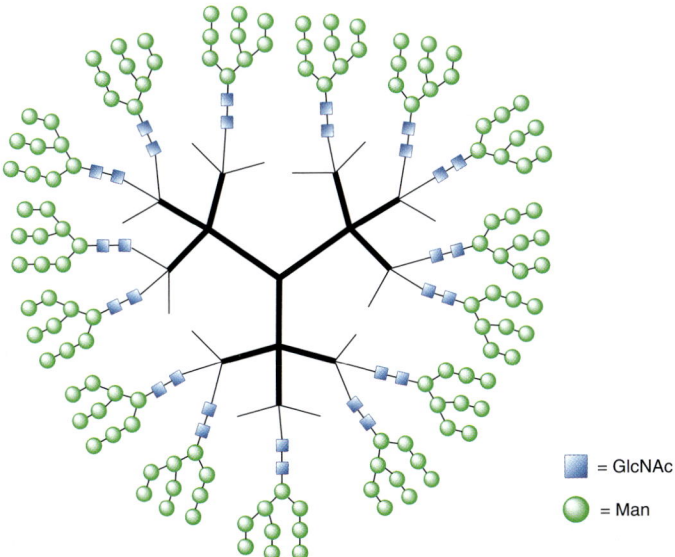

Figure 5.6 Schematic representation of multivalent Man$_9$ on a dendrimeric scaffold.

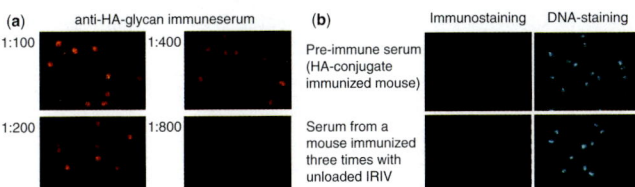

Figure 6.8 Immunization with *Leishmania* cap glycan loaded IRIV generated *L. donovani* cross-reactive IgG. (a) Immunofluorescence staining of parasites by serial dilution (1:100 to 1:800) of serum from mice immunized three times with HA-glycan. (b) Lack of immunofluorescence staining (left panel) of parasites by preimmune sera (dilution 1:100) and by serum (dilution 1:100) of a mouse immunized three times with unloaded IRIV. Presence of parasites on the microscopic slides was demonstrated by staining of parasite DNA with Hoechst dye 33258 (right panel).

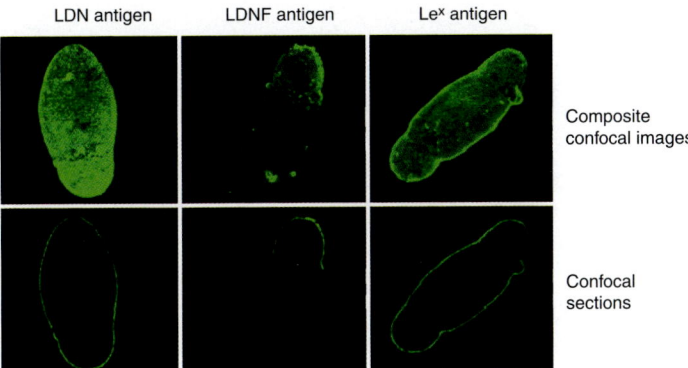

Figure 6.10 Antibodies to LDN, LDNF, and Le^X bind these structures on the surface to the immuno-vulnerable 3-h-old *S. mansoni* schistosomula.

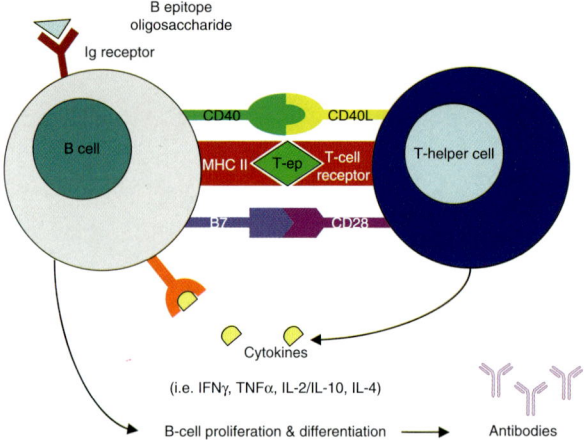

Figure 9.2 Schematic presentation of the interaction between B cells and helper T cells. Activation of B cells is initiated by specific recognition of antigens by the Ig receptors, which will lead to production of low-affinity IgM antibodies. To achieve a class-switch to specific IgG antibodies, the B cell needs activation by an activated helper T cell. In this context, B cells display processed protein antigens on their cell surface in the form of a peptide–MHC II complex and express the co-stimulator B7. Helper T cells that have been activated with the same peptide epitope by an APC, recognize the peptide antigen and B7 presented by the B cell. The T-helper cell is stimulated to express CD40 ligand (CD40L) and to secret cytokines. CD40L then binds to CD40 on the B cells and initiates B-cell proliferation and differentiation into an antibody secreting cell. T-ep = T-helper epitope.

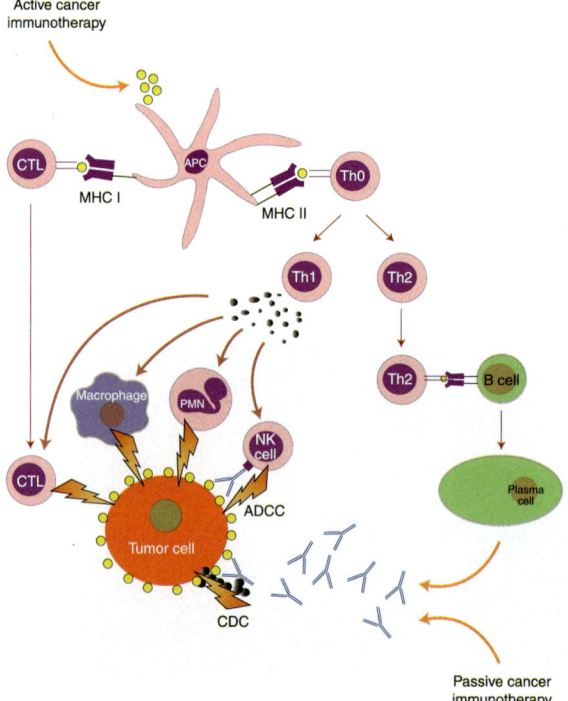

Figure 11.1 Principles of active and passive cancer immunotherapy.

the use of carbohydrates in the development of HIV vaccines, two problematic issues arise. First, the glycans exposed on the HIV envelope surface are immunologically "self" molecules. Second, the combination of variability in the viral genome and microheterogeneity of *N*-glycans means that an extraordinarily high degree of HIV structural diversity exists. Therefore, to develop a potentially useful HIV vaccine, the important, albeit challenging, first step is to identify defined, highly conserved, and immunologically active carbohydrate-containing epitopes in HIV gp120 as potential antigenic targets.

5.3.1.1 The 2G12 Epitope Although there is limited immune response to HIV-1 in vivo, a few neutralizing monoclonal antibodies (mAb) have been isolated that recognize conserved regions of the viral envelope [111–116]. Of great interest is the highly conserved and carbohydrate-rich 2G12 epitope, which elicits a broadly neutralizing mAb 2G12 [117]. This epitope is located on the so-called immunosilent face of gp120 and is composed of a dense cluster of primarily high-mannose glycans. Two studies using a combination of site-directed mutagenesis and enzymatic digestion elucidated some structural elements of the 2G12 epitope, finding that the high-mannose and/or hybrid-type glycans linked to residues 295, 332, and 392 are central to binding to mAb 2G12, while glycans linked to residues 386 and 448 contribute to binding peripherally [118, 119]. More specifically, it was reported that the 2G12 antibody recognizes Man $\alpha 1 \rightarrow$ 2Man moieties that are present at the nonreducing termini of oligomannose glycans. The gp120–mAb 2G12 interaction (Fig. 5.3a) was characterized with the help of crystal structures, which highlight the importance of oligomannose D1 and D3 arms (Fig. 5.3b) in binding [25, 120].

The facts that the 2G12 epitope is able to elicit an immune response and bind strongly to the broadly neutralizing mAb 2G12 suggest that it is a promising target for HIV vaccine development. Despite the fact that the carbohydrates adorning gp120 are host-synthesized and can be considered self, the abnormally dense clustering of oligomannoses may indicate that the self carbohydrates are arranged in a "non-self" pattern, which in this case is a distinguishing immunogenic trait [24]. In addition, many of the *N*-glycosylation sites on the "silent" face, including those pivotal for mAb 2G12 binding, are highly conserved and exhibit a minimal amount of glycan microheterogeneity [20]. This reduced variability allows mAb 2G12 to be broadly neutralizing. In fact, 41% of 93 HIV viruses with diverse backgrounds were neutralized with mAb 2G12 [121]. The intensive studies outlined here indicate that HIV vaccine development based on the oligomannose-rich 2G12 epitope is a promising direction.

5.3.1.2 Synthetic HIV Vaccines The use of chemical synthesis to construct mimics of carbohydrate-based antigenic epitopes is a valuable strategy for the developmet of subunit vaccines. Recent advances in carbohydrate synthesis have enabled the preparation of significant quantities of homogenous and structurally defined oligosaccharides and glycoconjugates. In the context of HIV vaccine development with the 2G12 epitope as the target, the overall design is to construct a glycoconjugate that contains dense clusters of oligomannoses, which can hopefully

176 CARBOHYDRATE-BASED ANTIVIRAL VACCINES

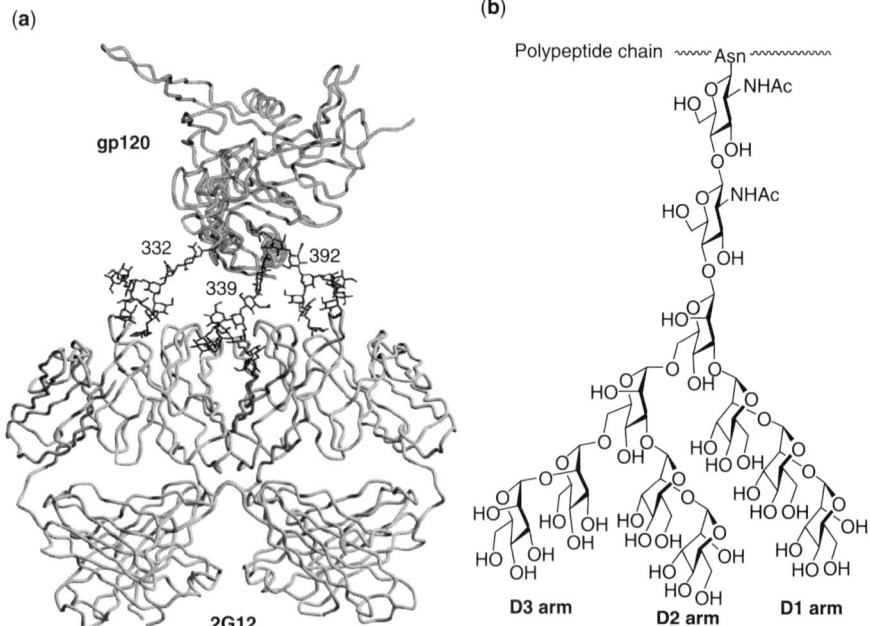

Figure 5.3 (a) Model structure of gp120 in complex with mAb 2G12; glycans involved in the binding are highlited in red color. (Reproduced, with permission, from Ref. [25].) (b) Structure of N-linked high-mannose glycan with the three oligomannose branches defined as D1, D2, and D3 arms. (See color insert.)

induce a robust immune response and raise neutralizing antibodies against a broad spectrum of HIV viruses.

The Wang lab [122, 123] was among a few pioneering groups that set out to explore the possibility of synthetic HIV vaccines based on the 2G12 epitope. Their strategy was to assemble oligomannose clusters on molecular scaffolds to mimic the 2G12 epitope. After analyzing the reported crystal structures [18, 124] of gp120 to determine the distances between the β-amide nitrogen atoms of asparagine residues 332, 339, and 392 (deemed essential for interaction with mAb 2G12), they designed properly functionalized/spaced templates for derivatization by high-mannose glycans. In one report, they used cholic acid as the scaffold [122]. Three unprotected $Man_9GlcNAc_2Asn$ moieties, prepared from soybean agglutinin according to a reported procedure [125], were attached to the template, resulting in the desired trivalent oligomannose cluster (Fig. 5.4a). The template also contained an amino group to facilitate its conjugation to a carrier protein. In a competitive inhibition assay, the synthetic oligomannose cluster Half Maximal Inhibitory Concentration ($IC_{50} = 21$ μM) showed a 46-fold increase in affinity for mAb 2G12 over $Man_9GlcNAc_2Asn$ ($IC_{50} = 960$ μM). However, gp120 binds mAb 2G12 in the nanomolar range. In another study, Wang et al. used D-galactose as a molecular scaffold to construct di-, tri-, and tetravalent oligomannose clusters (Fig. 5.4b) [123]. While slightly increased mAb 2G12 affinity was

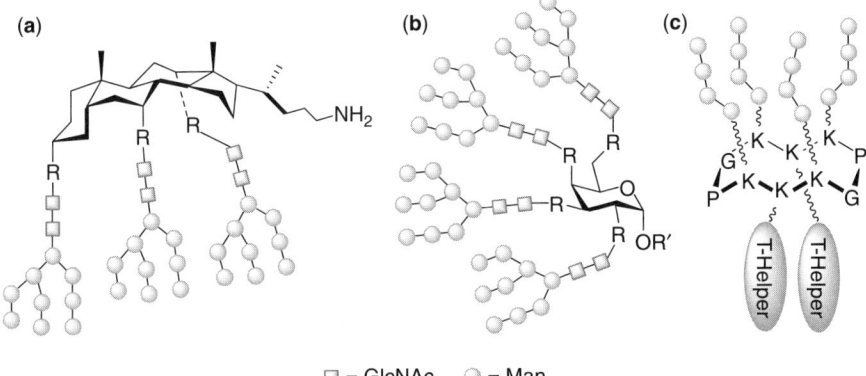

Figure 5.4 Oligomannose clusters built with (a) cholic acid, (b) D-galactose, and (c) RAFT as scaffolds. R = linker; R′ = OEt, KLH, or T helper peptide. (See color insert.)

observed for the tetravalent cluster (IC_{50} = 13 μM, a 73-fold increase over Man$_9$GlcNAc$_2$Asn), it still could not reach the nanomolar range observed for gp120. Evidently, the synthetic oligomannose cluster required optimization in glycan spacing or flexibility.

Wang et al. also synthesized oligomannose conjugates that contained either a carrier protein or the universal T-cell helper peptide epitope and studied their immunological properties [126]. The synthetic glycoconjugates did raise IgG-type antibodies, but the majority of antibodies were directed toward the linkers rather than the clustered carbohydrates. Moreover, during animal immunization studies, carbohydrate-specific antibodies generated in antisera showed weak cross-reactivity for HIV-1 gp120 but did not achieve a neutralizing effect. In their most recent report, Wang et al. altered their molecular scaffold/linker in an attempt to eliminate linker-directed immune response, which suppressed the antibody response to carbohydrate-based epitopes [127]. For this purpose, a conformationally stabilized cyclic decapeptide, RAFT (regioselectively addressable functionalized template), was chosen as the scaffold. To one face of the RAFT was added multiple oligomannose glycans and to the other was added two T-helper peptide epitopes (Fig. 5.4c). Rather than using Man$_9$GlcNAc$_2$Asn, chemically synthesized azidomannoses, including a Man$_4$ designed according to the important D1 arm of the core oligosaccharide Man$_9$GlcNAc$_2$, were employed in the preparation of glycoconjugates. The Man$_4$ moiety was previously shown to exhibit mAb 2G12 binding comparable to its parent glycan Man$_9$GlcNAc$_2$ [128], and thus it can be used as a functional yet simplified analog in HIV vaccine development. The glycans were introduced to RAFT via copper(I)-catalyzed alkyne-azide 1,3-diploar cycloaddition. A tetravalent Man$_4$ cluster containing an extended linker displayed the highest affinity for mAb 2G12, whereas the monomeric glycoconjugates predictably showed no measurable affinity for mAb 2G12. Currently, animal immunization studies using T-helper-modified RAFT-based oligomannose clusters are underway in the Wang lab to determine whether a strong carbohydrate-specific immune response can be generated.

The Danishefsky lab [129, 130] is also pursuing HIV vaccines based on the high-mannose and hybrid-type glycans of gp120. Besides the carbohydrate epitopes, their synthetic targets also included short peptide sequences of gp120 containing one of the key glycosylated asparagine residues (Asn-332). It was expected that the immune system would recognize not only the sugar but also the unique peptide sequence of gp120. They employed total chemical synthesis in their construction of the gp120 fragments that contain the high-mannose (Man$_9$GlcNAc$_2$) and hybrid-type (Man$_5$GalGlcNAc$_3$) glycans [129, 130]. En route to the high-mannose gp120 fragment (Fig. 5.5a) [129], two synthetic strategies, namely the "layer approach" and the "block approach," which make use of linear and convergent carbohydrate assembly, respectively, were used to efficiently construct a fully protected Man$_9$GlcNAc$_2$ building block. A strategy similar to the convergent block approach was applied in the total synthesis of gp120 fragments bearing mature hybrid-type glycans (Fig. 5.5a) [130].

Danishefsky and co-workers found that a dimer resulting from disulfide bond formation of the synthetic high-mannose glycopentapeptide exhibited substantial binding to mAb 2G12, while the monomer showed significantly reduced binding, thus offering additional evidence that multivalent binding is operative in the gp120–mAb 2G12 interaction [131]. In the same study, the hybrid-type glycopentapeptide, whether in monomeric or dimeric form, did not bind to mAb 2G12. This result demonstrates that hybrid glycans, which lack the crucial D1 arm present in high-mannose glycans, are not recognized by mAb 2G12.

With these structural criteria in mind, the Danishefsky group constructed multivalent oligomannose motifs on a cyclic peptide scaffold [132] similar to the one used by Wang [127] (Fig. 5.5b). As expected, the monovalent glycopeptide did not bind to mAb 2G12, but strong binding was observed for the di- and trivalent high-mannose glycopeptides. The cyclic glycopeptides were then coupled to a carrier lipoprotein that can be used to boost carbohydrate immunogenicity. These conjugates showed significant binding to mAb 2G12, although their affinity for mAb 2G12

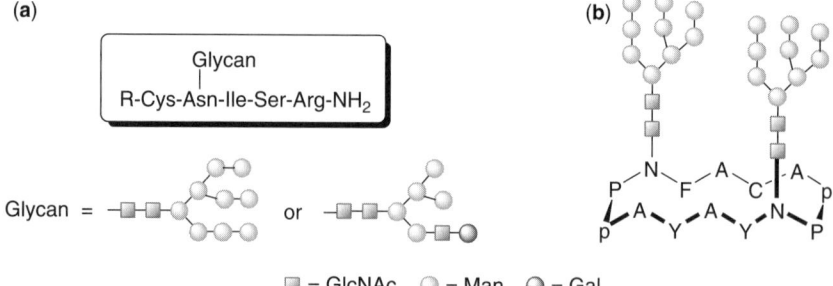

Figure 5.5 Multivalent glycan clusters for anti-HIV vaccine development. (a) High-mannose and hybrid-type glycans are linked to 5mer and 20mer peptides of the gp120 sequence. R = NH$_2$ or appropriate gp120 15mer peptide. (b) Cyclic peptide scaffold was used to construct mono-, di-, and trivalent clusters. (See color insert.)

remained weaker than gp120. Animal vaccination studies aimed at evaluating the synthetic conjugates as HIV vaccines are presently underway.

Early work by the Wong group was focused on development of synthetic methodology for and identification of functional oligomannose derivatives. For example, the "programmable reactivity-based one-pot oligosaccharide synthesis" technique [133] was employed in the preparation of a variety of oligomannose mimitopes [128]. Interestingly, simplified synthetic oligomannoses based on the D1 and/or D3 arms of Man$_9$GlcNAc$_2$ were able to disrupt gp120–mAb 2G12 interaction better than Man$_9$GlcNAc$_2$. Specifically, a Man$_4$ oligosaccharide, which does not contain GlcNAc residues or the difficult-to-achieve β-mannose linkage, successfully inhibited gp120–mAb 2G12 binding. Using simpler but effective carbohydrate building blocks may significantly reduce the complexity and cost of synthetic HIV vaccines.

More recently, the synthesis and evaluation of glycodendrons bearing Man$_4$ and Man$_9$ glycan clusters (Fig. 5.6) was reported by the Wong lab [134]. This strategy is conceptually similar to those used by Wang and Danishefsky, in that high-mannose glycan clusters are presented in a dense, multivalent fashion on a suitable molecular scaffold. In Wong's glycodendrons, variably branched/sized dendrimers were decorated with Man$_4$ and Man$_9$ via a copper(I)-catalyzed cycloaddition between the terminal alkynes of the dendrimer and Man$_4$- and Man$_9$-azides. Binding studies indicated that a second-generation dendron functionalized with Man$_9$, displaying an average

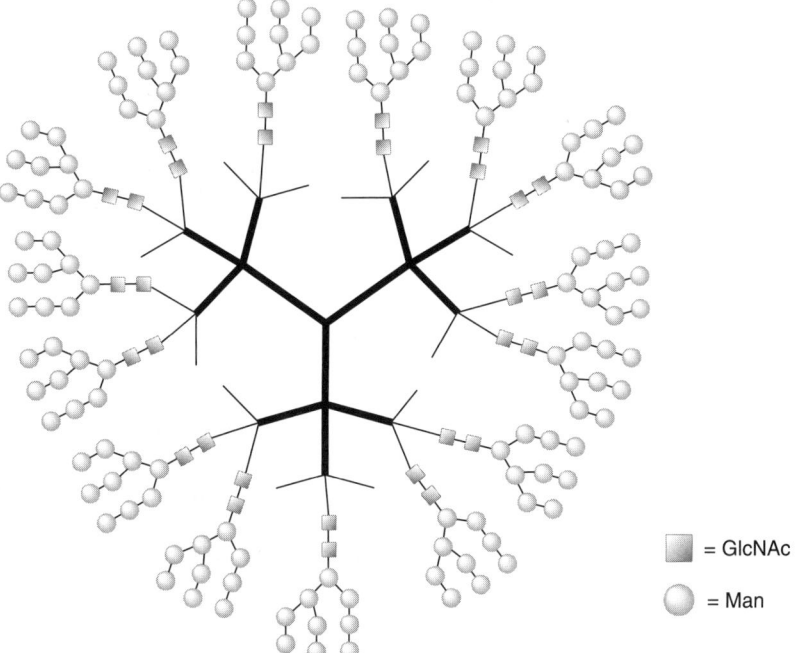

Figure 5.6 Schematic representation of multivalent Man$_9$ on a dendrimeric scaffold. (See color insert.)

of nine glycans, showed the most promising properties, as it exhibited significant inhibition of both the gp120–mAb 2G12 and gp120–DC-SIGN [135] interactions. The ability of a synthetic mimitope to inhibit dendritic cell-mediated HIV infection could lead to the development of an antiviral agent that prevents trans infection. Thus, these glycodendrons may be explored as both carbohydrate vaccine candidates and antiviral therapeutic agents.

5.3.1.3 Biosynthetic HIV Vaccines Luallen et al. have shown the potential of genetically engineering yeast for anti-HIV vaccine and drug development [136]. The *Saccharomyces cerevisiae* genome encodes for numerous proteins that have a large number of potential N-glycosylation sites, a trait that may lead to glycoproteins with densely packed high-mannose glycans, such as gp120. To achieve this, Luallen et al. created a mutant strain of *S. cerevisiae*, which had a crippled N-glycosylation pathway that generated—almost exclusively—unprocessed $Man_8GlcNAc_2$ glycans. Promisingly, mAb 2G12 had significant affinity for the mutant strain of *S. cerevisiae*, but not the wild type, and for four glycoproteins synthesized by the genetically engineered yeast, each of which exhibited densely clustered oligomannose glycans. Most significantly, rabbit immunization with whole cells of the mutant strain of *S. cerevisiae* produced sera that showed carbohydrate-dependent cross-reactivity with multiple HIV-1/SIV envelope glycoproteins. These results, along with the facts that *S. cerevisiae* is inexpensive and genetically malleable, represent a promising route to potential HIV vaccine candidates.

5.3.1.4 Carbohydrate-Binding Agents The heavily glycosylated surface glycoproteins of HIV are certainly a focus for vaccine development, as discussed above. On the other hand, because much of the viral envelope is covered by high-mannose glycans, another carbohydrate-based approach to AIDS therapy that has been explored is the use of mannose-specific binding agents to target HIV. In this context, lectins innate to the immune system, specifically the C-type mannose-binding lectins (MBL), have been explored [137]. While MBL is able to bind gp120 in vitro and can neutralize some lab-adapted HIV strains [138], MBL-mediated neutralization of HIV primary isolates is significantly reduced [139]. Nevertheless, treatment of cells with mannosidase I inhibitor 1-deoxymannojirimycin (DMJ) increased neutralization of primary isolates by MBL, which is a result of a higher terminal mannose content in the N-glycans [140]. An interesting consideration arose from this study, namely that if MBL-mediated HIV neutralization is increased by treatment with mannosidase inhibitors such as DMJ, a selective pressure for loss of glycans in gp120 could be evoked, thus resulting in a breakdown of the glycan shield and increased susceptibility of the virus to antibody neutralization [137].

Despite the potential ability of MBL to neutralize the virus in vivo, it appears that HIV actually uses some endogenous lectins as a means for survival and infection. For example, DC-SIGN, a C-type lectins on dendritic cells that bind mannose, has been shown to bind to HIV and enhance infection of T cells, a process known as trans infection [135, 141]. Therefore, MBL may also be operative in preventing trans infection, as it has been shown to block DC-SIGN-mediated trans infection of T cells in vitro [142].

Many exogenous lectins have also been evaluated for their efficacy against HIV. Of particular interest are lectins that show specificity for mannose or even N-acetylglucosamine, as many lectins in the latter category show anti-HIV activity as well [79, 143]. The mechanisms of anti-HIV action for these lectins are based on inhibition of the gp120–DC-SIGN interaction [144], as discussed above, or disruption of conformational changes/membrane fusion that result in abortion of the initial HIV infection process [145]. Cyanovirin-N, which operates by the latter mechanism, is an 11-kDa bacterial lectin that specifically binds to the $\alpha 1,2$-linked mannose residues present in the high-mannose glycans of gp120 [146]. CV-N was recently shown to inhibit HIV infection in vivo in a vaginal transmission model, suggesting its potential as a topical anti-HIV therapeutic [147]. While CV-N and other related exogenous lectins are promising tools for the exploration of HIV treatment and prevention, their systemic use could be problematic in terms of cross-reactivity with host glycans and unwanted immune response [143].

5.3.2 Influenza A Virus

Vaccination to combat seasonal influenza is often recommended for high-risk populations, such as the elderly. Influenza A vaccines are comprised of killed or attenuated whole virus or viral components (subunit vaccines). Although presently applied strategies for influenza A vaccination are quite effective for many individuals, improvement is needed. For example, some studies have shown that vaccination of the elderly reduced influenza contraction by only 50% [148]. While significant progress is being made in influenza vaccine improvement and development in a broader sense, the focus of this section will be a carbohydrate-based approach toward this goal. In this respect, the extensively glycosylated envelope protein HA, discussed in detail in Section 5.2.2, is the major target.

The Galili lab [149] has described a carbohydrate-based strategy for improving the immunogenicity of influenza vaccine by exploiting the anti-Gal antibody, which is naturally existing and the most abundant antibody in humans [150]. Anti-Gal is also the only human antibody that can target antigens to antigen-presenting cells (APCs). Consequently, any antigen that contains the α-Gal epitope (Gal1α1-3Galβ1-4GlcNAc-R) will form immune complexes with anti-Gal and be marked for uptake by APCs, and a positive immune response should follow. The modified influenza vaccines explored by Galili et al. harness these special properties of the α-Gal epitope and anti-Gal [151]. They employed enzymatic engineering to incorporate the α-Gal epitope into the N-glycans of HA [152]. It was expected that the anti-Gal antibody would direct the modified HA to APCs, and effective uptake of influenza vaccines by APCs is a prerequisite for efficacious vaccinations.

This strategy was tested with $\alpha 1,3$-galactosyltransferase ($\alpha 1,3$GT) knockout mice, which effectively mimic the human immune system in this context, using the PR8 influenza virus strain [152]. Modification of the PR8 N-glycans by recombinant $\alpha 1,3$GT produced the α-Gal epitope-containing influenza virus strain, PR8α_{Gal}, which was used as a modified influenza vaccine. Mice immunized with PR8α$_{Gal}$ produced higher quantities of PR8-specific CD8$^+$ and CD4$^+$ T cells and higher titers of anti-PR8 antibodies than mice immunized with PR8 lacking α-Gal epitopes. Furthermore, when treated with a lethal dose of live PR8 virus, 89% of PR8α$_{Gal}$-immunized mice survived, while only 11% of PR8-immunized mice lived over a window of 30 days. This work represents an attractive strategy for enhancing immune response to influenza virus vaccines that can also be applied to the improvement of other vaccines. In fact, expression of α-Gal epitopes on surface glycoproteins has also been used for improving immunogenicity of HIV gp120 [153] and tumor vaccines [154].

5.3.3 Hepatitis C Virus

Currently, no vaccine to prevent HCV infection is available. While the HCV envelope glycoproteins (E1 and E2) do contain highly conserved N-glycosylation sites that may be potential vaccine targets (as discussed in Section 5.2.4), most research has been focused on carbohydrate-based drug development by screening CBAs for anti-HCV activity. Because the E1E2 heterodimer contains high-mannose N-glycans, a variety of mannose-specific CBAs were able to bind to and inhibit host cell entry of HCV [79]. In addition, CBAs could also inhibit capture of HCV by DC-SIGN, which may reduce trans infection. These results, coupled with the ability of cyanovirin-N to inhibit HCV infection [80], show that CBAs are promising lead drugs for deactivation of HCV at the stage of host cell entry.

5.4 CONCLUSIONS

In recent years, the importance of host-cell-mediated N-glycosylation of viral envelope proteins has been documented in growing detail. The biological roles of these glycans, as outlined in this chapter, seem to be integral to every phase of the viral life cycle. At the outset of biosynthesis, N-glycans are directly involved in the folding and transport of many viral proteins. The carbohydrates of mature envelope glycoproteins assist in functions specific to the virus, often including host cell attachment and release, infectivity, replication, and determination of tropism. Meanwhile, viruses also elegantly combine the immunologically inactive property of host-cell-biosynthesized self carbohydrates with genetic variability, which adds another layer of carbohydrate diversity in addition to natural microheterogeneity, to construct an evolving glycan shield that counters the attempts by the immune system to generate protective reactions against them.

On the other hand, for some viruses, it is these functionally necessary glycans, which play a dual function as infection mediators and protective shields, that may prove to be their Achilles' heel. A delicate balance must be reached between highly conserved glycans that are essential to virus infectivity and highly variable glycans that are essential to virus survival, which means that despite the extraordinary diversity possible for a viral glycoprotein, some carbohydrate epitopes must be well conserved, therefore exposing the targets for vaccine and drug development. For example, the same highly conserved oligomannose glycans in HIV gp120 that appear to be essential for trans infection also elicited one of few broadly neutralizing antibodies—mAb 2G12. This discovery has spurred the development of HIV vaccine candidates. Similarly, the highly conserved oligomannose glycans in HCV glycoproteins, which have been implicated in HCV trans infection as well, have been identified as the binding receptors for CBAs to inhibit HCV infection. Thus, as a virus is forced to strike a compromise between infectivity and immune evasion, its immunological security can be breached.

Although much work remains to be done before successful carbohydrate-based antiviral vaccines and drugs are a reality, the foundation for their development has been set forth. Further elucidation of viral glycan structures and functions will facilitate the identification of clearly defined and conserved carbohydrate epitopes, which may prove paramount to combating viral infections in the future.

ACKNOWLEDGMENT

The authors thank the National Institutes of Health (NIH) for financial support (R01 CA095142).

REFERENCES

1. Patterson K, Runge T (2002) Smallpox and the Native American. *Am. J. Med. Sci.* 323:216–222.
2. Johnson NPAS, Mueller J (2002) Updating the accounts: Global mortality of the 1918–1920 "Spanish" influenza pandemic. *B. Hist. Med.* 76:105–115.
3. UNAIDS (2007) Overview of the global aids epidemic. Available at: http://www.unaids.org/en/KnowledgeCentre/HIVData/.
4. Idilman R, Maria ND, Colantoni A, Thiel DHV (1998) Pathogenesis of hepatitis B and C-induced hepatocellular carcinoma. *J. Viral Hepat.* 5:285–299.
5. Schiffman M, Castle PE, Jeronimo J, Rodriguez AC, Wacholder S (2007) Human papillomavirus and cervical cancer. *Lancet* 370:890–907.
6. Vigerust DJ, Shepherd VL (2007) Virus glycosylation: Role in virulence and immune interactions. *Trends Microbiol.* 15:211–218.
7. Weerapana E, Imperiali B (2006) Asparagine-linked protein glycosylation: From eukaryotic to prokaryotic systems. *Glycobiology* 16:91R–101R.
8. Varki A (1993) Biological roles of oligosaccharides: All of the theories are correct. *Glycobiology* 3:97–130 and references cited therein.
9. Lehle L, Strahl S, Tanner W (2006) Protein glycosylation, conserved from yeast to man: A model organism helps elucidate congenital human diseases. *Angew. Chem. Int. Ed. Engl.* 45:6801–6818.
10. Helenius A, Aebi M (2004) Roles of N-linked glycans in the endoplasmic reticulum. *Annu. Rev. Biochem.* 73:1019–1049.
11. Wyatt R, Sodroski J (1998) The HIV-1 envelope glycoproteins: Fusogens, antigens, and immunogens. *Science* 280:1884–1888.

12. Center RJ, Leapman RD, Lebowitz J, Arthur LO, Earl PL, Moss B (2002) Oligomeric structure of the human immunodeficiency virus type 1 envelope protein on the virion surface. *J. Virol.* 76:7863–7867.
13. Chan DC, Fass D, Berger JM, Kim PS (1997) Core structure of gp41 from the HIV envelope glycoprotein. *Cell* 89:263–273.
14. McDougal JS, Kennedy MS, Sligh JM, Cort SP, Mawle A, Nicholson JK (1986) Binding of HTLV-III/LAV to T4+ T cells by a complex of the 110K viral protein and the T4 molecule. *Science* 231:382–385.
15. Myers G, MacInnes K, Korber B (1992) The emergence of simian/human immunodeficiency viruses. *AIDS Res. Hum. Retroviruses* 8:373–386.
16. Geyer H, Holschbach C, Hunsmann G, Schneider J (1988) Carbohydrates of human immunodeficiency virus. Structures of oligosaccharides linked to the envelope glycoprotein 120. *J. Biol. Chem.* 263:11760–11767.
17. Veronese FD, DeVico AL (1985) Characterization of gp41 as the transmembrane protein coded by the HTLV-III/LAV envelope gene. *Science* 229:1402–1405.
18. Wyatt R, Kwong PD, Desjardins E, Sweet RW, Robinson J, Hendrickson WA, Sodroski JG (1998) The antigenic structure of the HIV gp120 envelope glycoprotein. *Nature* 393:705–711.
19. Cutalo JM, Deterding LJ, Tomer KB (2004) Characterization of glycopeptides from HIV-ISF2 gp120 by liquid chromatography mass spectrometry. *J. Am. Chem. Mass Spectrom.* 15:1545–1555.
20. Zhu X, Borchers C, Bienstock RJ, Tomer KB (2000) Mass spectrometric characterization of the glycosylation pattern of HIV-gp120 expressed in CHO cells. *Biochemistry* 39:11194–11204.
21. Leonard CK, Spellman MW, Riddle L, Harris RJ, Thomas JN, Gregory TJ (1990) Assignment of intrachain disulfide bonds and characterization of potential glycosylation sites of the type 1 recombinant human immunodeficiency virus envelope glycoprotein (gp120) expressed in Chinese hamster ovary cells. *J. Biol. Chem.* 265:10373–10382.
22. Zhang M, Gaschen B, Blay W, Foley B, Haigwood N, Kuiken C, Korber B (2004) Tracking global patterns of N-linked glycosylation site variation in highly variable viral glycoproteins: HIV, SIV, and HCV envelopes and influenza hemagglutinin. *Glycobiology* 14:1229–1246.
23. Wei X, Decker JM, Wang S, Hui H, Kappes JC, Wu X, Salazar-Gonzalez JF, Salazar MG, Kilby JM, Saag MS, Komarova NL, Nowak MA, Hahn BH, Kwong PD, Shaw GM (2004) Antibody neutralization and escape by HIV-1. *Nature* 422:307–312.
24. Scanlan CN, Offer J, Zitzmann N, Dwek RA (2007) Exploiting the defensive sugars of HIV-1 for drug and vaccine design. *Nature* 446:1038–1045.
25. Calarese DA, Scanlan CN, Zwick MB, Deechongkit S, Mimura Y, Kunert R, Zhu P, Wormald MR, Stanfield RL, Roux KH, Kelly JW, Rudd PM, Dwek RA, Katinger H, Burton DR, Wilson IA (2003) Antibody domain exchange is an immunological solution to carbohydrate cluster recognition. *Science* 300:2065–2071.
26. Woods RJ, Pathiaseril A, Wormald MR, Edge CJ, Dwek RA (1998) The high degree of internal flexibility observed for an oligomannose oligosaccharide does not alter the overall topology of the molecule. *Eur. J. Biochem.* 258:372–386.

27. Chen B, Vogan EM, Gong H, Skehel JJ, Wiley DC, Harrison SC (2005) Determining the structure of an unliganded and fully glycosylated SIV gp120 envelope glycoprotein. *Structure* 13:197–211.
28. Chen B, Vogan EM, Gong H, Skehel JJ, Wiley DC, Harrison SC (2005) Structure of an unliganded simian immunodeficiency virus gp120 core. *Nature* 433:834–841.
29. Dash B, McIntosh A, Barrett W, Daniels R (1994) Deletion of a single N-linked glycosylation site from the transmembrane envelope protein of human immunodeficiency virus type 1 stops cleavage and transport of gp160 preventing env-mediated fusion. *J. Gen. Virol.* 75:1389–1397.
30. Fenouillet E, Jones IM (1995) The glycosylation of human immunodeficiency virus type 1 transmembrane glycoprotein (gp41) is important for the efficient intracellular transport of the envelope precursor gp160. *J. Gen. Virol.* 76:1509–1514.
31. Li Y, Luo L, Rasool N, Kang CY (1993) Glycosylation is necessary for the correct folding of human immunodeficiency virus gp120 in CD4 binding. *J. Virol.* 67:584–588.
32. Montefiori DC, Robinson WE, Mitchell WM (1988) Role of protein N-glycosylation in pathogenesis of human immunodeficiency virus type 1. *Proc. Natl. Acad. Sci. U. S. A.* 85:9248–9252.
33. Fenouillet E, Gluckman JC, Bahraoui E (1990) Role of N-linked glycans of envelope glycoproteins in infectivity of human immunodeficiency virus type 1. *J. Virol.* 64:2841–2848.
34. Gruters RA, Neefjes JJ, Tersmette M, de Goede REY, Tulp A, Huisman HG, Miedema F, Ploegh HL (1987) Interference with HIV-induced syncytium formation and viral infectivity by inhibitors of trimming glucosidase. *Nature* 330:74–77.
35. Pal R, Hoke GM, Sarngadharan MG (1989) Role of oligosaccharides in the processing and maturation of envelope glycoproteins of human immunodeficiency virus type 1. *Proc. Natl. Acad. Sci. U.S.A.* 86:3384–3388.
36. Reitter JN, Means RE, Desrosiers RC (1998) A role for carbohydrates in immune evasion in AIDS. *Nature Med.* 4:679–684.
37. Korber B, Gaschen B, Yusim K, Thakallapally R, Kesmir C, Detours V (2001) Evolutionary and immunological implications of contemporary HIV-1 variation. *Br. Med. Bull.* 58:19–42.
38. Sagar M, Wu X, Lee S, Overbaugh J (2006) Human immunodeficiency virus type 1 V1-V2 envelope loop sequences expand and add glycosylation sites over the course of infection, and these modifications affect antibody neutralization sensitivity. *J. Virol.* 80:9586–9598.
39. Wolk T, Schreiber M (2006) N-glycans in the gp120 V1/V2 domain of the HIV-1 strain NL4-3 are indispensable for viral infectivity and resistance against antibody neutralization. *Med. Microbiol. Immunol. (Berl.)* 195:165–172.
40. Wagner R, Matrosovich M, Klenk H (2002) Functional balance between haemagglutinin and neuraminidase in influenza virus infections. *Rev. Med. Virol.* 12:159–166.
41. Klenk H-D, Wagner R, Heuer D, Wolff T (2002) Importance of hemagglutinin glycosylation for the biological functions of influenza virus. *Virus Res.* 82:73–75.
42. Daniels R, Kurowski B, Hebert AEJN (2003) N-linked glycans direct the cotranslational folding pathway of influenza hemagglutinin. *Mol. Cell* 11:79–90.

43. Wagner R, Wolff T, Herwig A, Pleschka S, Klenk H-D (2000) Interdependence of hemagglutinin glycosylation and neuraminidase as regulators of influenza virus growth: A study by reverse genetics. *J. Virol.* 74:6316–6323.
44. Baignet SJ, McCauley JW (2001) Glycosylation of haemagglutinin and stalk-length of neuraminidase combine to regulate the growth of avian influenza viruses in tissue culture. *Virus Res.* 79:177–185.
45. Tsuchiya E, Sugawara K, Hongo S, Matsuzaki Y, Muraki Y, Nakamura K (2002) Role of overlapping glycosylation sequons in antigenic properties, intracellular transport and biological activities of influenza A/H2N2 virus haemagglutinin. *J. Gen. Virol.* 83:3067–3074.
46. Tsuchiya E, Sugawara K, Hongo S, Matsuzaki Y, Muraki Y, Li Z-N, Nakamura K (2002) Effect of addition of new oligosaccharide chains to the globular head of influenza A/H2N2 virus haemagglutinin on the intracellular transport and biological activities of the molecule. *J. Gen. Virol.* 83:1137–1146.
47. Abe Y, Takashita E, Sugawara K, Matsuzaki Y, Hongo YM (2004) Effect of the addition of oligosaccharides on the biological activities and antigenicity of influenza A/H3N2 virus hemagglutinin. *J. Virol.* 78:9605–9611.
48. Kaverin NV, Rudneva IA, Ilyushina NA, Varich NL, Lipatov AS, Smirnov YA, Govorkova EA, Gitelman AK, Lvov DK, Webster RG (2002) Structure of antigenic sites on the haemagglutinin molecule of H5 avian influenza virus and phenotypic variation of escape mutants. *J. Gen. Virol.* 83:2497–2505.
49. Segawa H, Inakawa A, Yamashita T, Taira H (2003) Functional analysis of individual oligosaccharide chains of Sendai virus hemagglutinin-neuraminidase protein. *Biosci. Biotechnol. Biochem.* 67:592–598.
50. Skehel JJ, Stevens DJ, Daniels RS, Douglas AR, Knossow M, Wilson IA, Wiley DC (1984) A carbohydrate side chain on hemagglutinins of Hong Kong influenza viruses inhibits recognition by a monoclonal antibody. *Proc. Natl. Acad. Sci. U.S.A.* 81:1779–1783.
51. Wilson IA, Wiley JJSC (1981) Structure of the haemagglutinin membrane glycoprotein of influenza virus at 3 Å resolution. *Nature* 289:366–373.
52. Basak S, Pritchard DG, Bhown AS, Compans RW (1981) Glycosylation sites of influenza viral glycoproteins: Characterization of tryptic glycopeptides from the A/USSR(H1N1) hemagglutinin glycoprotein. *J. Virol.* 37:549–558.
53. Ward CW, Dopheide TA (1981) Amino acid sequence and oligosaccharide distribution of the haemagglutinin from an early Hong Kong influenza virus variant A/Aichi/2/68 (X-31). *Biochem. J.* 193:953–962.
54. Mir-Shekari SY, Ashford DA, Harvey DJ, Dwek RA, Schulze IT (1997) The glycosylation of the influenza A virus hemagglutinin by mammalian cells. A site-specific study. *J. Biol. Chem.* 272:4027–4036.
55. Gallagher PJ, Henneberry JM, Sambrook JF, Gething MJ (1992) Glycosylation requirements for intracellular transport and function of the hemagglutinin of influenza virus. *J. Virol.* 66:7136–7145.
56. Roberts PC, Garten W, Klenk HD (1993) Role of conserved glycosylation sites in maturation and transport of influenza A virus hemagglutinin. *J. Virol.* 67:3048–3060.
57. Nobusawa E, Aoyama T, Kato H, Suzuki Y, Tateno Y, Nakajima K (1991) Comparison of complete amino acid sequences and receptor-binding properties among 13 serotypes of hemagglutinins of influenza A viruses. *Virology* 182:475–485.

58. Igarashi M, Ito K, Kida H, Takada A (2008) Genetically destined potentials for N-linked glycosylation of influenza virus hemagglutinin. *Virology* 376:323–329.
59. Vigerust D, Ulett K, Boyd K, Madsen J, Hawgood S, McCullers J (2007) N-linked glycosylation attenuates H3N2 influenza viruses. *J. Virol.* 81:8593–8600.
60. Gallagher P, Henneberry J, Wilson I, Sambrook J, Gething M (1988) Addition of carbohydrate side chains at novel sites on influenza virus hemagglutinin can modulate the folding, transport, and activity of the molecule. *J. Cell Biol.* 107:2059–2073.
61. Deshpande KL, Fried VA, Ando M, Webster RG (1987) Glycosylation affects cleavage of an H5N2 influenza virus hemagglutinin and regulates virulence. *PNAS* 84:36–40.
62. Ohuchi M, Ohuchi R, Feldmann A, Klenk H (1997) Regulation of receptor binding affinity of influenza virus hemagglutinin by its carbohydrate moiety. *J. Virol.* 71:8377–8384.
63. Wagner R, Heuer D, Wolff T, Herwig A, Klenk H-D (2002) N-glycans attached to the stem domain of haemagglutinin efficiently regulate influenza A virus replication. *J. Gen. Virol.* 83:601–609.
64. Ohuchi R, Ohuchi M, Garten W, Klenk H (1997) Oligosaccharides in the stem region maintain the influenza virus hemagglutinin in the metastable form required for fusion activity. *J. Virol.* 71:3719–3725.
65. Op De Beeck A, Cocquerel L, Dubuisson J (2001) Biogenesis of hepatitis C virus envelope glycoproteins. *J. Gen. Virol.* 82:2589–2595.
66. Goffard A (2003) Glycosylation of hepatits C virus envelope proteins. *Biochemie* 85:295–301.
67. Helle F, Goffard A, Morel V, Duverlie G, McKeating J, Keck Z-Y, Foung S, Penin F, Dubuisson J, Voisset C (2007) The neutralizing activity of anti-hepatitis C virus antibodies is modulated by specific glycans on the E2 envelope protein. *J. Virol.* 81:8101–8111.
68. Meunier J-C, Fournillier A, Choukhi A, Cahour A, Cocquerel L, Dubuisson J, Wychowski C (1999) Analysis of the glycosylation sites of hepatitis C virus (HCV) glycoprotein E1 and the influence of E1 glycans on the formation of the HCV glycoprotein complex. *J. Gen. Virol.* 80:887–896.
69. Nakano I, Fukuda Y, Katano Y, Hayakawa T (1998) Conformational epitopes detected by cross-reactive antibodies to envelope 2 glycoprotein of the hepatitis C virus. *J. Infect. Dis.* 180:1328–1333.
70. Slater-Handshy T, Droll DA, Fan X, Bisceglie AMD, Chambers TJ (2004) HCV E2 glycoprotein: Mutagenesis of N-linked glycosylation sites and its effects on E2 expression and processing. *Virology* 319:36–48.
71. Goffard A, Callens N, Bartosch B, Wychowski C, Cosset F-L, Montpellier C, Dubuisson J (2005) Role of N-linked glycans in the functions of hepatitis C virus envelope glycoproteins. *J. Virol.* 79:8400–8409.
72. Li P, Wan Q, Feng Y, Liu M, Wu J, Chen X, Zhang X-L (2006) Engineering of N-glycosylation of hepatitis C virus envelope protein E2 enhances T cell responses for DNA immunization. *Vaccine* 25:1544–1551.
73. Liu M, Chen H, Luo F, Li P, Pan Q, Xia B, Qi Z, Ho W-Z, Zhang X-L (2007) Deletion of N-glycosylation sites of hepatitis C virus envelope protein E1 enhances specific cellular and humoral immune responses. *Vaccine* 25:6572–6580.
74. Lozach P-Y, Lortat-Jacob H, de Lavalette DA, Staropoli I, Foung S, Amara A, Houlès C, Fieschi F, Schwartz O, Virelizier J-L, Arenzana-Seisdedos F, Altmeyer R (2003)

DC-SIGN and L-SIGN are high affinity binding receptors for hepatitis C virus glycoprotein E2. *J. Biol. Chem.* 278:20358–20366.
75. Wang Q, Feng Z, Nie Q, Zhou Y (2004) DC-SIGN: Binding receptors for hepatitis C virus. *Chin. Med. J.* 117:1395–1400.
76. Cormier EG, Durso RJ, Tsamis F, Boussemart L, Manix C, Olson WC, Gardner JP, Dragic T (2004) L-SIGN (CD209L) and DC-SIGN (CD209) mediate transinfection of liver cells by hepatitis C virus. *Proc. Natl. Acad. Sci. U.S.A.* 101:14067–14072.
77. Lozach P-Y, Amara A, Bartosch B, Virelizier J-L, Arenzana-Seisdedos F, Cosset F-L, Altmeyer R (2004) C-type lectins L-SIGN and DC-SIGN capture and transmit infectious hepatitis C virus pseudotype particles. *J. Biol. Chem.* 279:32035–32045.
78. Iacob RE, Perdivara I, Przybylski M, Tomer KB (2008) Mass spectrometric characterization of glycosylation of hepatitis C virus E2 envelope glycoprotein reveals extended microheterogeneity of N-glycans. *J. Am. Soc. Mass Spectrom.* 19:428–444.
79. Bertaux C, Daelemans D, Meertens L, Cormier EG, Reinus JF, Peumans WJ, Van Damme EJM, Igarashi Y, Oki T, Schols D, Dragic T, Balzarini J (2007) Entry of hepatitis C virus and human immunodeficiency virus is selectively inhibited by carbohydrate-binding agents but not by polyanions. *Virology* 366:40–50.
80. Helle F, Wychowski C, Vu-Dac N, Gustafson KR, Voisset C, Dubuisson J (2006) Cyanovirin-N inhibits hepatitis C virus entry by binding to envelope protein glycans. *J. Biol. Chem.* 281:25177–25183.
81. Wang T, Anderson J, Magnarelli L, Bushmich S, Wong S, Koski R, Fikrig E (2001) West Nile virus envelope protein: Role in diagnosis and immunity. *Ann. N. Y. Acad. Sci.* 951:325–327.
82. Chambers TJ, Hahn CS, Galler R, Rice CM (1990) Flavivirus genome organization, expression, and replication. *Annu. Rev. Microbiol.* 44:649–688.
83. Shirato K, Miyoshi H, Goto A, Ako Y, Ueki T, Kariwa H, Takashima I (2004) Viral envelope protein glycosylation is a molecular determinant of the neuroinvasiveness of the New York strain of West Nile virus. *J. Gen. Virol.* 85:3637–3645.
84. Beasley DWC, Whiteman MC, Zhang S, Huang CY-H, Schneider BS, Smith DR, Gromowski GD, Higgs S, Kinney RM, Barrett ADT (2005) Envelope protein glycosylation status influences mouse neuroinvasion phenotype of genetic lineage 1 West Nile virus strains. *J. Virol.* 79:8339–8347.
85. Halevy M, Akov Y, Ben-Nathan D, Kobiler D, Lachmi B, Lustig S (1994) Loss of active neuroinvasiveness in attenuated strains of West Nile virus: Pathogenicity in immunocompetent and SCID mice. *Arch. Virol.* 137:355–370.
86. Lad V, Shende V, Gupta A, Koshy A, Roy A (2000) Effect of tunicamycin on expression of epitopes on Japanese encephalitis virus glycoprotein E in porcine kidney cells. *Acta Virol.* 44:359–364.
87. Hanna SL, Pierson TC, Sanchez MD, Ahmed AA, Murtadha MM, Doms RW (2005) N-linked glycosylation of West Nile virus envelope proteins influences particle assembly and infectivity. *J. Virol.* 79:13262–13274.
88. Li J, Bhuvanakantham R, Howe J, Ng M-L (2006) The glycosylation site in the envelope protein of West Nile virus (Sarafend) plays an important role in replication and maturation processes. *J. Gen. Virol.* 87:613–622.
89. Davis CW, Mattei LM, Nguyen H-Y, Ansarah-Sobrinho C, Doms RW, Pierson TC (2006) The location of asparagine-linked glycans on West Nile virions controls their Interactions

with CD209 (dendritic cell-specific ICAM-3 grabbing nonintegrin). *J. Biol. Chem.* 281:37183–37194.

90. Davis CW, Nguyen H-Y, Hanna SL, Sánchez MD, Doms RW, Pierson TC (2006) West Nile virus discriminates between DC-SIGN and DC-SIGNR for cellular attachment and infection. *J. Virol.* 80:1290–1301.
91. Sanchez A, Yang Z-Y, Xu L, Nabel GJ, Crews T, Peters CJ (1998) Biochemical analysis of the secreted and virion glycoproteins of ebola virus. *J. Virol.* 72:6442–6447.
92. Feldmann H, Nichol ST, Klenk H-D, Peters CJ, Sanchez A (1994) Characterization of filoviruses based on differences in structure and antigenicity of the virion glycoprotein. *Virology* 199:469–473.
93. Dowling W, Thompson E, Badger C, Mellquist J, Garrison A, Smith J, Paragas J, Hogan R, Schmaljohn C (2007) Influences of glycosylation on antigenicity, immunogenicity, and protective efficacy of ebola virus GP DNA vaccines. *J. Virol.* 81:1821–1837.
94. Lin G, Simmons G, Pöhlmann S, Baribaud F, Ni H, Leslie GJ, Haggarty BS, Bates P, Weissman D, Hoxie JA, Doms RW (2003) Differential N-linked glycosylation of human immunodeficiency virus and ebola virus envelope glycoproteins modulates interactions with DC-SIGN and DC-SIGNR. *J. Virol.* 77:1337–1346.
95. Barrientos L, Gronenborn A (2005) The highly specific carbohydrate-binding protein cyanovirin-N: Structure, anti-HIV/Ebola activity and possibilities for therapy. *Mini. Rev. Med. Chem.* 5:21–31.
96. Barrientos LG, O'Keefe BR, Bray M, Sanchez A, Gronenborn AM, Boyd MR (2003) Cyanovirin-N binds to the viral surface glycoprotein, $GP_{1,2}$ and inhibits infectivity of Ebola virus. *Antiviral Res.* 58:47–56.
97. Rota P, Oberste M, Monroe S, Nix W, Campagnoli R, Icenogle J, Peñaranda S, Bankamp B, Maher K, Chen M, Tong S, Tamin A, Lowe L, Frace M, DeRisi J, Chen Q, Wang D, Erdman D, Peret T, Burns C, Ksiazek T, Rollin P, Sanchez A, Liffick S, Holloway B, Limor J, McCaustland K, Olsen-Rasmussen M, Fouchier R, Günther S, Osterhaus A, Drosten C, Pallansch M, Anderson L, Bellini W (2003) Characterisation of a novel coronavirus associated with severe acute respiratory syndrome. *Science* 300:1394–1399.
98. Shih Y, Chen C, Liu S, Chen K, Lee Y, Chao Y, Chen Y (2006) Identifying epitopes responsible for neutralizing antibody and DC-SIGN binding on the spike glycoprotein of the severe acute respiratory syndrome coronavirus. *J. Virol.* 80:10315–10324.
99. Han DP, Lohani M, Cho MW (2007) Specific asparagine-linked glycosylation sites are critical for DC-SIGN- and L-SIGN-mediated severe acute respiratory syndrome coronavirus entry. *J. Virol.* 81:12029–12039.
100. van der Meer FJUM, de Haan CAM, Schuurman NMP, Haijema BJ, Peumans WJ, Van Damme EJM, Delputte PL, Balzarini J, Egberink HF (2007) Antiviral activity of carbohydrate-binding agents against Nidovirales in cell culture. *Antiviral Res.* 76:21–29.
101. van der Meer FJUM, de Haan CAM, Schuurman NMP, Haijema BJ, Verheije MH, Bosch BJ, Balzarini J, Egberink HF (2007) The carbohydrate-binding plant lectins and the non-peptidic antibiotic pradimicin A target the glycans of the coronavirus envelope glycoproteins. *J. Antimicro. Chemother.* 60:741–749.
102. Bossart KN, Crameri G, Dimitrov AS, Mungall BA, Feng Y-R, Patch JR, Choudhary A, Wang L-F, Eaton BT, Broder CC (2005) Receptor binding, fusion inhibition, and induction of cross-reactive neutralizing antibodies by a soluble G glycoprotein of *Hendra virus*. *J. Virol.* 79:6690–6702.

103. Zheng F, Ma L, Shao L, Wang G, Chen F, Zhang Y, Yang S (2007) Defining the N-linked glycosylation site of Hantaan virus envelope glycoproteins essential for cell fusion. *J. Microbiol.* 45:41–47.

104. Moll M, Kaufmann A, Maisner A (2004) Influence of N-glycans on processing and biological activity of the Nipah virus fusion protein. *J. Virol.* 78:7274–7278.

105. Schowalter RM, Smith SE, Dutch RE (2006) Characterization of human metapneumovirus F protein-promoted membrane fusion: Critical roles for proteolytic processing and low pH. *J. Virol.* 80:10931–10941.

106. Griffin C, Wang ECY, McSharry BP, Rickards C, Browne H, Wilkinson GWG, Tomasec P (2005) Characterization of a highly glycosylated form of the human cytomegalovirus HLA class I homologue gpUL18. *J. Gen. Virol.* 86:2999–3008.

107. Novoa RR, Calderita G, Cabezas P, Elliott RM, Risco C (2005) Key golgi factors for structural and functional maturation of Bunyamwera virus. *J. Virol.* 79:10852–10863.

108. Melanson VR, Iorio RM (2006) Addition of N-glycans in the stalk of the Newcastle disease virus HN protein blocks. Its interaction with the F protein and prevents fusion. *J. Virol.* 80:623–633.

109. Kunz S, Rojek JM, Kanagawa M, Spiropoulou CF, Barresi R, Campbell KP, Oldstone MBA (2005) Posttranslational modification of α-dystroglycan, the cellular receptor for arenaviruses, by the glycosyltransferase LARGE is critical for virus binding. *J. Virol.* 79:14282–14296.

110. Eichler R, Lenz O, Garten W, Strecker T (2006) The role of single N-glycans in proteolytic processing and cell surface transport of the Lassa virus glycoprotein GP-C. *Virology J.* 3:41–47.

111. Burton DR, Stanfield RL, Wilson IA (2005) Antibody vs. HIV in a clash of evolutionary titans. *Proc. Natl. Acad. Sci. U. S. A.* 102:14943–14948.

112. Pantophlet R, Burton DR (2006) Gp120: Target for neutralizing HIV-1 antibodies. *Annu. Rev. Immunol.* 24:739–769.

113. Conley AJ, Kessler JA, Boots LJ, Tung JS, Arnold BA, Keller PM, Shaw AR, Emini EA (1994) Neutralization of divergent human immunodeficiency virus type 1 variants and primary isolates by IAM-41-2F5, an anti-gp41 human monoclonal antibody. *Proc. Natl. Acad. Sci. U.S.A.* 91:3348–3352.

114. Zwick MB, Labrijn AF, Wang M, Spenlehauer C, Saphire EO, Binley JM, Moore JP, Stiegler G, Katinger H, Burton DR, Parren PWHI (2001) Broadly neutralizing antibodies targeted to the membrane-proximal external region of human immunodeficiency virus type 1 glycoprotein gp41. *J. Virol.* 75:10892–10905.

115. Burton D, Pyati J, Koduri R, Sharp S, Thornton G, Parren P, Sawyer L, Hendry R, Dunlop N, Nara P, Lamacchia M, Garratty E, Stiehm E, Bryson Y, Cao Y, Moore J, Ho D, Barbas C (1994) Efficient neutralization of primary isolates of HIV-1 by a recombinant human monoclonal antibody. *Science* 266:1024–1027.

116. Trkola A, Pomales A, Yuan H, Korber B, Maddon P, Allaway G, Katinger H, Barbas C, Burton D, Ho D (1995) Cross-clade neutralization of primary isolates of human immunodeficiency virus type 1 by human monoclonal antibodies and tetrameric CD4-IgG. *J. Virol.* 69:6609–6617.

117. Trkola A, Purtscher M, Muster T, Ballaun C, Buchacher A, Sullivan N, Srinivasan K, Sodroski J, Moore J, Katinger H (1996) Human monoclonal antibody 2G12 defines a

distinctive neutralization epitope on the gp120 glycoprotein of human immunodeficiency virus type 1. *J. Virol.* 70:1100–1108.

118. Sanders RW, Venturi M, Schiffner L, Kalyanaraman R, Katinger H, Lloyd KO, Kwong PD, Moore JP (2002) The mannose-dependent epitope for neutralizing antibody 2G12 on human immunodeficiency virus type 1 glycoprotein gp120. *J. Virol.* 76:7293–7305.

119. Scanlan CN, Pantophlet R, Wormald MR, Saphire EO, Stanfield R, Wilson IA, Katinger H, Dwek RA, Rudd PM, Burton DR (2002) The broadly neutralizing anti-human immunodeficiency virus type 1 antibody 2G12 recognizes a cluster of 12 mannose residues on the outer face of gp120. *J. Virol.* 76:7306–7321.

120. Calarese DA, Lee H-K, Huang C-Y, Best MD, Astronomo RD, Stanfield RL, Katinger H, Burton DR, Wong C-H, Wilson IA (2005) Dissection of the carbohydrate specificity of the broadly neutralizing anti-HIV-1 antibody 2G12. *Proc. Natl. Acad. Sci. U.S.A.* 102:13372–13377.

121. Binley JM, Wrin T, Korber B, Zwick MB, Wang M, Chappey C, Stiegler G, Kunert R, Zolla-Pazner S, Katinger H, Petropoulos CJ, Burton DR (2004) Comprehensive cross-clade neutralization analysis of a panel of anti-human immunodeficiency virus type 1 monoclonal antibodies. *J. Virol.* 78:13232–13252.

122. Li H, Wang L-X (2004) Design and synthesis of a template-assembled oligomannose cluster as an epitope mimic for human HIV-neutralizing antibody 2G12. *Org. Biomol. Chem.* 2:483–488.

123. Wang L-X, Ni J, Singh S, Li H (2004) Binding of high-mannose-type oligosaccharides and synthetic oligomannose clusters to human antibody 2G12: Implications for HIV-1 vaccine design. *Chem. Biol.* 11:127–134.

124. Kwong PD, Wyatt R, Robinson J, Sweet RW, Sodroski J, Hendrickson WA (1998) Structure of an HIV gp120 envelope glycoprotein in complex with the CD4 receptor and a neutralizing human antibody. *Nature* 393:648–659.

125. Lis H, Sharon N (1978) Soybean agglutinin—a plant glycoprotein. Structure of the carboxydrate unit. *J. Biol. Chem.* 253:3468–3476.

126. Ni J, Song H, Wang Y, Stamatos NM, Wang L-X (2006) Toward a carbohydrate-based HIV-1 vaccine: Synthesis and immunological studies of oligomannose-containing glycoconjugates. *Bioconjug. Chem.* 17:493–500.

127. Wang J, Li H, Zou G, Wang L-X (2007) Novel template-assembled oligosaccharide clusters as epitope mimics for HIV-neutralizing antibody 2G12. Design, synthesis, and antibody binding study. *Org. Biomol. Chem.* 5:1529–1540.

128. Lee H-K, Scanlan CN, Huang C-Y, Chang AY, Calarese DA, Dwek RA, Rudd PM, Burton DR, Wilson IA, Wong C-H (2004) Reactivity-based one-pot synthesis of oligomannoses: Defining antigens recognized by 2G12, a broadly neutralizing anti-HIV-1 antibody. *Angew. Chem. Int. Ed. Engl.* 43:1000–1003.

129. Geng X, Dudkin VY, Mandal M, Danishefsky SJ (2004) In pursuit of carbohydrate-based HIV vaccines, part 2: The total synthesis of high-mannose-type gp120 fragments—evaluation of strategies directed to maximal convergence. *Angew. Chem. Int. Ed. Engl.* 43:2562–2565.

130. Mandal M, Dudkin VY, Geng X, Danishefsky SJ (2004) In pursuit of carbohydrate-based HIV vaccines, part 1: The total synthesis of hybrid-type gp120 fragments. *Angew. Chem. Int. Ed. Engl.* 43:2557–2561.

131. Dudkin VY, Orlova M, Geng X, Mandal M, Olson WC, Danishefsky SJ (2004) Toward fully synthetic carbohydrate-based HIV antigen design: On the critical role of bivalency. *J. Am. Chem. Soc.* 126:9560–9562.

132. Krauss I, Joyce J, Finnefrock A, Song H, Dudkin V, Geng X, Warren J, Chastain M, Shiver J, Danishefsky S (2007) Fully synthetic carbohydrate HIV antigens designed on the logic of the 2G12 antibody. *J. Am. Chem. Soc.* 129:11042–11044.

133. Zhang ZY, Ollmann IR, Ye X-S, Wischnat R, Baasov T, Wong C-H (1999) Programmable one-pot oligosaccharide synthesis. *J. Am. Chem. Soc.* 121:734–753.

134. Wang S-K, Liang P-H, Astronomo RD, Hsu T-L, Hsieh S-L, Burton DR, Wong C-H (2008) Targeting the carbohydrates on HIV-1: Interaction of oligomannose dendrons with human monoclonal antibody 2G12 and DC-SIGN. *Proc. Natl. Acad. Sci. U.S.A.* 105:3690–3695.

135. Geijtenbeek TBH, Kwon DS, Torensma R, van Vliet SJ, van Duijnhoven GCF, Middel J, Cornelissen ILMHA, Nottet HSLM, KewalRamani VN, Littman DR, Figdor CG, van Kooyk Y (2000) DC-SIGN, a dendritic cell–specific HIV-1-binding protein that enhances trans-infection of T cells. *Cell* 100:587–597.

136. Luallen RJ, Lin J, Fu H, Cai KK, Agrawal C, Mboudjeka I, Lee F-H, Montefiori D, Smith DF, Doms RW, Geng Y (2008) An engineered saccharomyces cerevisiae strain binds the broadly neutralizing human immunodeficiency virus type 1 antibody 2G12 and elicits mannose-specific gp120-binding antibodies. *J. Virol.* 82:6447–6457.

137. Ji X, Gewurz H, Spear GT (2005) Mannose binding lectin (MBL) and HIV. *Mol. Immunol.* 42:145–152.

138. Ezekowitz RA, Kuhlman M, Groopman JE, Byrn RA (1989) A human serum mannose-binding protein inhibits in vitro infection by the human immunodeficiency virus. *J. Exp. Med.* 169:185–196.

139. Ying H, Ji X, Hart ML, Gupta K, Saifuddin M, Zariffard MR, Spear GT (2004) Interaction of mannose-binding lectin with HIV type 1. Is sufficient for virus opsonization but not neutralization. *AIDS Res. Hum. Retroviruses* 20:327–335.

140. Hart ML, Saifuddin M, Spear GT (2003) Glycosylation inhibitors and neuraminidase enhance human immunodeficiency virus type 1 binding and neutralization by mannose-binding lectin. *J. Gen. Virol.* 84:353–360.

141. Geijtenbeek T, van Kooyk Y (2003) DC-SIGN: A novel HIV receptor on DCs that mediates HIV-1 transmission. *Curr. Top. Microbiol. Immunol.* 276:31–54.

142. Spear GT, Zariffard MR, Xin J, Saifuddin M (2003) Inhibition of DC-SIGN-mediated trans infection of T cells by mannose-binding lectin. *Immunology* 110:80–85.

143. Balzarini J (2006) Inhibition of HIV entry by carbohydrate-binding proteins. *Antiviral Res.* 71:237–247.

144. Balzarini J, Van Herrewege Y, Vermeire K, Vanham G, Schols D (2007) Carbohydrate-binding agents efficiently prevent dendritic cell-specific intercellular adhesion molecule-3-grabbing nonintegrin (DC-SIGN)-directed HIV-1 transmission to T lymphocytes. *Mol. Pharmacol.* 71:3–11.

145. Esser MT, Mori T, Mondor I, Sattentau QJ, Dey B, Berger EA, Boyd MR, Lifson JD (1999) Cyanovirin-N binds to gp120 to interfere with CD4-dependent human immunodeficiency virus type 1 virion binding, fusion, and infectivity but does not affect the CD4 binding site on gp120 or soluble CD4-induced conformational changes in gp120. *J. Virol.* 73:4360–4371.

146. Bewley CA (2001) Solution structure of a cyanovirin-N: Manα1-2Manα complex: Structural basis for high-affinity carbohydrate-mediated binding to gp120. *Structure* 9:931–940.
147. Tsai CC, Emau P, Jiang Y, Agy M, Shattock RJ, Schmidt A, Morton WR, Gustafson KR, Boyd MR (2004) Cyanovirin-N inhibits AIDS virus infections in vaginal transmission models. *AIDS Res. Hum. Retroviruses* 20:11–18.
148. Webster RG (2000) Immunity to influenza in the elderly. *Vaccine* 18:1686–1689.
149. Abdel-Motal UM, Guay HM, Wigglesworth K, Welsh RM, Galili U (2007) Immunogenicity of influenza virus vaccine is increased by anti-Gal-mediated targeting to antigen-presenting cells. *J. Virol.* 81:9131–9141.
150. Galili U, Rachmilewitz E, Peleg A, Flechner I (1984) A unique natural human IgG antibody with anti-alpha-galactosyl specificity. *J. Exp. Med.* 160:1519–1531.
151. Galili U (1993) Evolution and pathophysiology of the human natural anti-α-galactosyl IgG (anti-Gal) antibody. *Springer Semin. Immunopathol.* 15:155–171.
152. Henion TR, Gerhard W, Anaraki F, Galili U (1997) Synthesis of α-gal epitopes on influenza virus vaccines, by recombinant α1,3galactosyltransferase, enables the formation of immune complexes with the natural anti-Gal antibody. *Vaccine* 15:1174–1182.
153. Abdel-Motal U, Wang S, Lu S, Wigglesworth K, Galili U (2006) Increased immunogenicity of human immunodeficiency virus gp120 engineered to express Gal1-3Galß1-4GlcNAc-R epitopes. *J. Virol.* 80:6943–6951.
154. LaTemple D, Abrams J, Zhang S, Galili U (1999) Increased immunogenicity of tumor vaccines complexed with anti-Gal: Studies in knockout mice for alpha1,3galactosyltransferase. *Cancer Res.* 59:3417–3423.

6

CARBOHYDRATE-BASED ANTIPARASITIC VACCINES

Faustin Kamena, Xinyu Liu, and Peter H. Seeberger
Laboratory for Organic Chemistry, Swiss Federal Institute of Technology (ETH) Zurich, Zurich, Switzerland

6.1 INTRODUCTION

Parasitic infections are among the most devastating and prevalent diseases in humans and other mammals. Notable examples include malaria, leishmaniasis, African trypanosomiasis, Chagas' disease, and schistosomiasis, which pose a risk to 3 billion people worldwide and claim more than 2 million lives annually. The term *parasite* in the broader sense represents a microorganism living at the expense of its host, in the vernacular, however, the term is restricted to pathogenic protozoan and helminths. The parasitic diseases occur worldwide but pose rather more serious health threats in the developing country than in western countries. Factors related to climate but also to the improved living conditions are among the reasons why the spread of parasites has dropped to an almost insignificant level in western countries in contrast to most developing countries. In addition, the rise of infections such as HIV/AIDS (human immunodeficiency virus/acquired immune deficiency syndrome) has contributed to worsen the stand of parasitic diseases in developing countries as they are also opportunistic diseases that appear when the host immune system is weakened.

The treatment of parasitic infections in both humans and livestock relies extensively on the use of antiparasitic drugs [1]. However, the limitation of such an approach is obvious, as the reinfection that occurs rapidly in endemic regions is not prevented. In addition, the widespread appearance of drug-resistant parasites in

---

Carbohydrate-Based Vaccines and Immunotherapies. Edited by Zhongwu Guo and Geert-Jan Boons
Copyright © 2009 John Wiley & Sons, Inc.

animals and humans calls for the development of alternative methods of control [2]. An antiparasitic vaccine thus becomes an obvious target.

Historically, vaccines were primarily whole agent vaccines containing inactivated or attenuated organisms. The modern approach is to exploit the molecular and biochemical understanding of pathogens to generate subunit vaccines, where the subunits may be antigenic peptide, protein, or carbohydrate fragments derived from the pathogen. Remarkable success has been achieved in developing vaccines for many different human pathogens (bacteria and virus), including smallpox, polio, influenza, measles, diphtheria, tetanus, pertussis, varicella, mumps, rubella, hepatitis B, *Pneumococcus*, and *Haemophilus influenzae* type b [3, 4]. However, there is no commercial vaccine against a human parasite to date, while only a few antiparasitic vaccines for livestock have reached the point of being marketed [5].

The ineffectiveness of the strategies explored for antiparasitic vaccine development is probably due to a combination of multiple factors. First of all, our understanding of the molecular interactions that occur between parasites and their (intermediate) hosts is still in its infancy, as is our understanding of the alterations in expression levels of antigens during parasite development in the infected host. Compared to other pathogenic microorganisms such as bacteria or viruses, parasites often undergo several morphological changes as part of their complex life cycle, and these changes are in line with alterations in the expression levels of antigens. As a result the host immune system is constantly involved in a new battle against a new set of antigenic structures. One example is illustrated with the malaria parasite, where many proteins are encoded by the *var* genes, which makes sure that only one copy of more than 60 different possible copies is expressed at a given time [6, 7]. Most parasites also accomplish multiple steps of their life cycles in different hosts, which increases their chances of surviving. With respect to the major problem of antiparasitic vaccine, the difficulty for the host immune system to cope with the constant variation of the protein antigen epitopes, carbohydrate-based vaccination strategy could prove decisive [8]. Unlike protein and nucleic acids, carbohydrates are not primary gene products; hence carbohydrate epitopes are less susceptible to variation after punctual mutation in the genome than are protein epitopes.

In recent years, great success has been witnessed in utilizing bacterial cell surface capsular polysaccharides (CSP) as primary antigens for conjugate vaccine development. Several vaccines based on this strategy are now commercially available against *Neisseria meningitides* groups A, C, W-135, and Y, *Steptococcus pneumoniae* (23 serotypes), *Haemophilus influenza* type b, and *Salmonella typhi* [9–12].

However, a primary hurdle exists in the adaptation of bacterial CSP-based vaccine development strategies to that of antiparasitic vaccines. In contrast to bacterial CSP that can be isolated from bacteria in bulk quantities, it is virtually impossible in the case of parasites. Therefore, chemical synthesis of defined parasite-specific carbohydrate antigens becomes an indispensable approach for antiparasitic carbohydrate-based vaccine development. In the following section we will present a few examples of promising carbohydrate-based vaccine candidates where carbohydrate chemical synthesis has played or may play a central role in the production of the antigen.

6.2 GPI-BASED ANTIMALARIAL VACCINE

6.2.1 GPI as a Malaria Toxin

Malaria is a devastating parasitic disease that threatens 30% of the world's population and claims more than one million lives each year, mostly young children in developing countries. Transmitted by female *Anopheles* mosquitoes, the disease is caused by protozoan parasite of the genus *Plasmodium*. Of the four known species of *Plasmodium* to infect human, *Plasmodium falciparum* accounts for the majority of infections and leads to the most severe forms of the disease.

The cell surface of *P. falciparum* expresses abundant amounts of glycosylphosphatidylinositols (GPIs). GPIs are a class of glycolipids that link proteins covalently to the plasma membrane of the cells (Fig. 6.1a). Following the initial structural elucidations of GPIs of *Trypnosoma brucei* variable surface glycoprotein (VSG) and rat brain Thy-1 in 1988 [13, 14], hundreds of GPI-anchored proteins have been identified in various organisms, including protozoa, fungi, yeast, plants, slime-modes, mollusks, insects, and vertebrates, indicating that this mode of protein modification is ubiquitous in eukaryotes [15]. Virtually all protein-linked GPI anchors identified to date share a common core structure, a pseudopentasaccharide consisting of *myo*-inositol, glucosamine, and three mannoses (Fig. 6.1a). Depending on the nature of the proteins attached to GPIs, structural variations including additional glycan residues, extra ethanolamine phosphate moieties, and inositol lipidation may decorate the core pentasaccharide backbone (Fig. 6.1a).

The GPIs constitute more than 95% of the total carbohydrate modification of *P. falciparum* schizonts and reflect the virtual absence of *N*- and *O*-linked glycosylation in these parasites [16]. Early studies indicated that the administration of purified *P. falciparum* GPI to mice induced syndromes similar to accute malaria, including the release of proinflammatory cytokines, pyrexia, hypoglycaemia, and eventually caused deaths [17]. These observations led to the speculation that GPI is a potential malaria toxin and anti-GPI antibodies may provide protection against malaria in human.

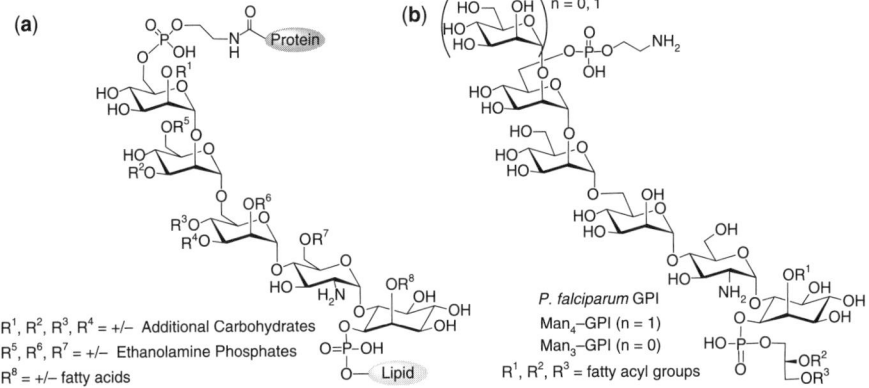

Figure 6.1 GPI structural diversities and *P. falciparum* GPI structures.

Plasmodium falciparum presents two mature GPI species with slightly different glycan compositions [16, 18]. One consists of the evolutionarily conserved core GPI glycan with three mannosyl residues (Fig. 6.1b, Man_3-GPI), whereas the other carries a terminal fourth mannose residue in an $\alpha 1-2$ linkage to the third mannose in the core glycan (Fig. 6.1b, Man_4-GPI). Interestingly, parasite proteins were found to be anchored exclusively by the Man_4-GPIs, indicating a high degree of selectivity toward the presence of the fourth, terminal mannose residue in the GPI anchor moiety; however, it is not clear whether protein-free GPI present in *P. falciparum* is dominantly Man_3-GPI or mixed with Man_4-GPI.

There is no recombinant approach for the production of GPI from natural sources as it is often the case for DNA/RNA (deoxyribonucleic acid/ribonucleic acid) and proteins. Therefore, the GPI is purified from bulk parasite cultures [16, 19]. Obtaining adequate amounts of parasites free of host cell components for detailed biochemical studies and vaccine trials proved difficult. Furthermore, standardized methods to purify native malaria GPI to compositional homogeneity are not clearly established [19]. Indeed, several biological studies using native GPIs prepared by single high-performance liquid chromatography (HPLC) purification have provided contradictory findings [20–25]. To precisely define the structural requirements for malarial GPI bioactivity and evaluate the potential of *P. falciparum* GPI as malaria vaccine antigen lead, chemical synthesis became the only option.

6.2.2 Synthetic GPI as Antitoxic Malaria Vaccine Candidate

Based on the hypothesis that the *P. falciparum* GPI is a malarial toxin, a consensus *P. falciparum* GPI glycan, derived from chemical and enzymatic hydrolysis of the native GPI and nontoxic itself, was synthesized (Fig. 6.2). The initial synthesis was carried out in solution phase [26], and the synthetic process was further accelerated using an automated oligosaccharide synthesizer (Fig. 6.2) [27]. The synthetic GPI glycan was modified with 2-iminothiolane and conjugated to maleimide-activated carrier protein, keyhole limpet hemocyanin (KLH) to give the vaccine construct (Fig. 6.3a) [26]. Cysteine-conjugated KLH was prepared in a similar manner as the control (Fig. 6.3b). The GPI glycan–KLH conjugate was immunogenic in rodents.

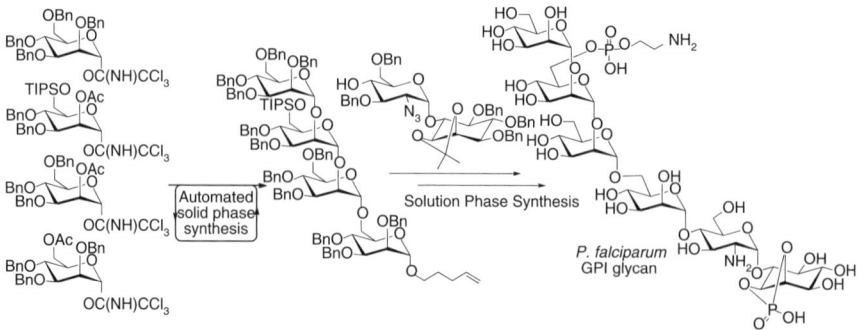

Figure 6.2 Semiautomated synthesis of a *P. falciparum* GPI glycan as immunogen.

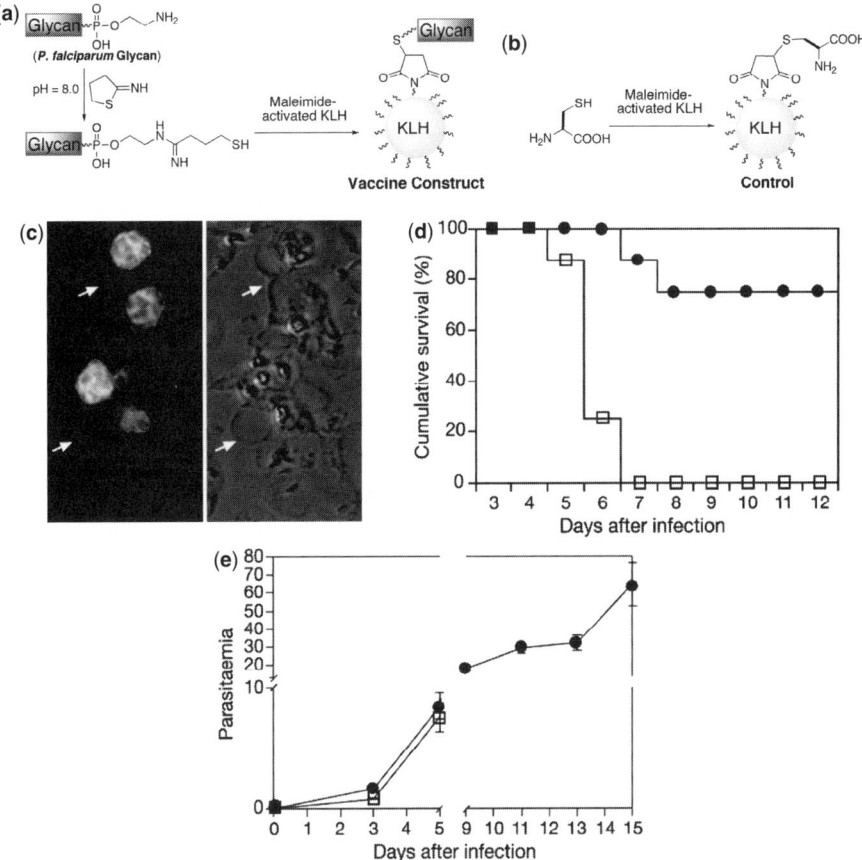

Figure 6.3 Synthetic GPI glycan–KLH conjugate protects against murine cerebral malaria. (a) Preparation of *P. falciparum* GPI glycan–KLH conjugate; (b) preparation of a cysteine–KLH conjugate as control; (c) left panel: reactivity of antiglycan IgG antibodies with *P. falciparum* trophozoites and schizonts. Right panel: same field under white light (d) Kaplan–Meier survival plots, and (e) parasitaemias, to 2 weeks postinfection, of KLH-glycan-immunized (closed circles) and sham-immunized (open squares) mice challenged with malaria parasite.

Antibodies from KLH–glycan immunized animals gave positive IgG titers and selectively recognized intact trophozoites and schizonts, but not uninfected erythrocytes, due to the GPI structural differences between human and parasites (Fig. 6.3c). Furthermore, mice treated with the synthetic GPI–KLH conjugate were substantially protected from death by malaria. Between 60 and 75% of the vaccinated mice survived for 2 weeks after the parasitic challenge, whereas the sham-immunized mice all died within 7 days (Fig. 6.3d).

More interestingly, immunization of mice did not alter infection rates of the animals and overall parasitaemia, indicating that the antibody to the GPI neutralized toxicity without killing the parasites (Fig. 6.3e) [26]. This study thus provided convincing evidence for the hypothesis that GPI is the dominant toxin of *P. falciparum* origin.

Further development of this vaccine candidate to the clinical level is currently being pursued by Ancora Pharmaceuticals (Medford, Cambridge, MA). Synthetic strategies to access multigram quantities of the GPI glycan have been devised and preclinical studies are currently underway.

A central question concerning the association of altered risk of malaria in human with anti-GPI antibodies remains to be clarified. There has been interest in investigating acquired antibody responses to GPI in humans. In several recent studies, Gowda and co-workers have used native *P. falciparum* GPI prepared by a one-step HPLC fractionation of total organic solvent- and water-extractable parasite material, where the GPI-containing peak represents over 20% of the total fractionation gradient [19]. Studies using this material in enzyme-linked immunsorbent assay (ELISA) have provided contradictory findings, some reporting a statistically significant association of anti-GPI IgG response with protection against symptoms of severe malaria, and others finding no such relationship [19, 28–33]. One study found the lipid and two studies found the glycan as the dominant epitope within the GPI. To unequivocally address this question, a novel approach based on microarrays of defined synthetic GPI–glycan structures related to the *P. falciparum* GPIs was developed and has shed more light on the anti-GPI immune response from the host.

6.2.3 Synthetic GPI Microarray to Define Antimalarial Antibody Response

Microarray technology has been used successfully in genomics and proteomics research and has recently begun to impact glycomics [34]. This chip-based format offers many advantages over conventional methods. These include the ability to screen several thousand binding events in parallel, and the fact that a minimal amount of analyte and ligand are required for study.

Briefly, a series of oligosaccharides related to *P. falciparum* GPI (**1–7**) with a terminal thiol group were chemically synthesized and covalently attached to maleimide-activated bovine serum albumin (BSA)-coated glass slides (Fig. 6.4). Seven synthetic GPIs were arrayed as quadruplicates in a square. Seven squares, each representing one of the GPI epitopes **1–7** and one square containing the buffer control made up the 32 spots that were printed in close proximity. Placement of a 64-well adhesive gasket to the surface of the microarray slides following printing created separated wells, each containing the 7 GPIs as well as a control (Fig. 6.4). This setup has helped to define both the minimal length of the glycan structure susceptible to induce immune response as well as the specific anti-GPI response strictly related to malaria infection [35].

The results obtained with this setup show interesting features that would have been almost impossible to obtain using natural materials. First, it appears that only the largest (i.e., Man_3- and Man_4-containing) GPI structures bind anti-GPI antibodies. In addition, the study provided evidence that the fine specificities of anti-GPI antibodies are individually by dependent. Finally, it was possible to establish that sera from people without any malaria history have substantial amounts of anti-GPI antibodies. These malaria nonspecific antibodies appeared to bind the Man_4-GPI, whereas

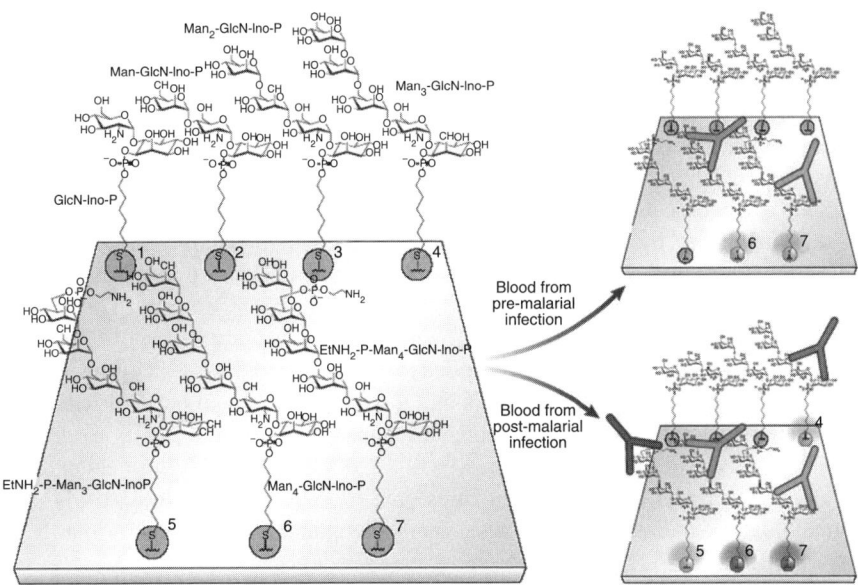

Figure 6.4 Covalently immobilized synthetic GPI glycan structures are used as targets to bind anti-GPI antibodies in normal and malarial infection sera. Fluorescent spots indicate the positive glycan targets, and the patterns of these spots distinguish normal from malarial sera.

antibodies to the Man$_3$-GPI could only be found from those individuals with previous contact with the malaria parasite. A possible explanation of the presence of anti-Man$_4$-GPI response in malaria-unexposed subjects is that it could be the result of infection with nonmalarial parasites or even fungi that typically express Man$_4$ GPIs.

6.3 LPG-BASED ANTILEISHMANIAL VACCINE

6.3.1 LPG in Leishmaniasis Pathogenesis

Leishmaniasis is a class of vector-borne infectious disease in humans and other mammals. The disease is caused by variety of *Leishmania* species and transmitted by sand fly bites. Although leishmaniasis is not a household name like malaria, it still threatens 350 million people in 88 countries around the world with an annual death toll of around 60,000. Mainly, three forms of the disease have been classified, including cutaneous, mucocutaneous, and visceral leishmaniasis [36]. Cutaneous leishmaniasis represents more than 50% of all new cases and is mainly caused by *L. major*, *L. tropica*, *L. aethiopia*, and *L. mexicana*. Visceral leishmaniasis, also known as kala-azar and black fever, is the most severe form and caused by *L. donovani*.

Clinical treatment of leishmaniasis still relies on half-century-old drugs (antimonials), which remain expensive and have significant side effects. Efforts toward the development of the cost-effective vaccinations have been initiated in various ways, although there is no effective vaccine for any form of leishmaniasis yet [36].

The cell surface of leishamania parasites are coated with the glycocalyx, of which the major component is lipophosphonoglycans (LPGs) [37]. LPGs, present primarily on the surface of the insect stage of *Leishmania* spp., is an important virulence factor and is crucial for the survival of the parasite in the insect vector and for the infectivity of the parasite when it enters the mammalian host [38]. It is a highly complex glycolipid, containing -6Galβ1-4Manα1-PO_4- repeat units, and can be substituted by monosaccharides or oligosaccharides in a species-specific manner (Fig. 6.5) [39]. It is linked to cell membrane by a GPI anchor and capped with a neutral oligosaccharide (Fig. 6.5). The same phosphodisaccharide repeat unit is also found in proteophosphoglycans (PPGs) secreted by the intracellular amastigote stages, as well as the promastigote stages, of the parasites [40]. The protective effect of purified LPG on mice challenged with *L. major* was first described in the late 1980s [41, 42]. However, these studies were later questioned by others that it was the protein co-purified with LPG contributed to this effect [43]. To unambiguously address this question, defined synthetic carbohydrate antigens based on the LPG substructures thus emerged as the only option. Various partial structures of LPG have been synthesized [44–54]. Immunological studies based on these synthetic compounds have been primarily on the cap oligosaccharides and the repeating phosphodisaccharides.

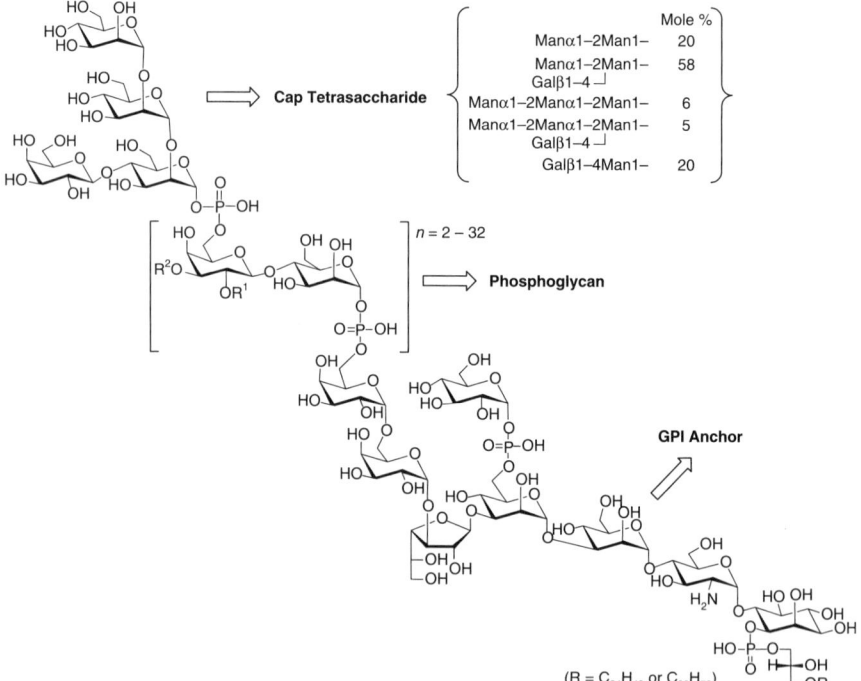

Figure 6.5 *L. donovani* LPG structure ($R^1 = R^2 = H$). Cap tetrassacharide bracket shows the compositional distribution of cap oligosaccharides in *L. donovani* LPG. R^1 and R^2 can be various oligosaccharides in other *Leishmania* species.

6.3.2 Synthetic Phosphoglycan Repeating Unit as Potential Antileishmanial Vaccine

A large array of different oligophosphosaccharides, representing LPG substructures from and *L. donovani*, *L. mexicana*, and *L. major*, were synthesized by Nikolaev's group at the University of Dundee (Scotland) (Fig. 6.6) [44, 48, 52, 54–56]. The synthetic oligosaccharides were equipped with a dec-9-enyl linker and allowed for the subsequent transformation into an aldehyde group to couple to a carrier protein tetanus toxoid via reductive amination [57].

Detailed vaccination studies using these chemically defined synthetic glycovaccines were performed in a BALB/c mouse/*L. mexicana* model [58]. Immunization with the synthetic glycovaccine containing the glycans found in *L. mexicana* LPG provided significant protection against challenge by the bite of infected sand flies. Only the glycan from *L. mexicana* was protective with a 94% reduction in the parasite burden. The result is particularly significant as the protection was achieved without using any adjuvant in the mice. Those from other species did not protect against *L. mexicana* infection, suggesting this glycovaccine may induce the species- and glycan-specific protection, although these findings will have to be confirmed with further challenge studies.

A more intriguing point disclosed during this study was that the glycovaccine did not protect against the artificial needle challenge, which is traditionally used in

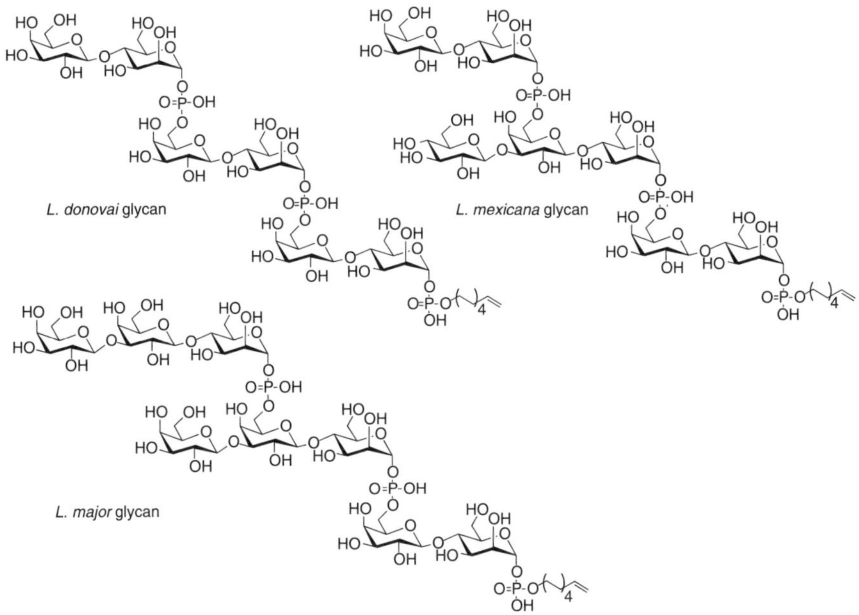

Figure 6.6 Chemical structures of the synthetic LPG phosphoglycan fragments from different *Leishmania* species. The alkene function in the linker was converted to aldehyde via ozonolysis and conjugated to tetanus toxoid via reductive amination to give the corresponding glycovaccine conjugates.

antileishmanial vaccine development. In contrast, an antigen preparation that was shown effective against needle challenge did not offer any protection against sand fly bites. This clearly demonstrates that the method of challenge is crucial in vaccine development and the critical role that the vector plays in the evaluation of candidate vaccines for leishmaniasis and other vector-borne diseases.

6.3.3 Synthetic LPG Cap Oligosaccharide as Antileishmanial Vaccine Candidate

Another approach toward LPG-fragment-based vaccine against leishmaniasis explored the unique, structurally well-defined capping oligosaccharides that terminate the LPGs of leishmania parasites and have been shown to be crucial for the invasion of parasite into macrophages (Fig. 6.7). The tetrasaccharide epitope, representing a portion of *L. donovani* LPG, has been synthesized by several groups [46, 47, 50, 53]. Several conjugate vaccine constructs have been prepared, including the conjugates with carrier protein KLH and immunostimulator tripalmitoyl-*S*-glycerylcysteine [47].

A more recent immunization study in the murine models utilized the *Leishmania* cap tetrasaccharide conjugated virosome as the vaccine candidate (Fig. 6.8) [59]. Immunostimulating reconstituted influenza virosomes (IRIVs) are an antigen delivery platform with integrated carrier and adjuvant properties with a proven safety record, as two IRIV-based vaccines against hepatitis A and influenza are currently on the market [60]. As discussed in the previous chapters, carbohydrate antigen alone typically elicits only T-cell-independent short-lived and low-affinity IgM antibody responses. The synthetic carbohydrate vaccine design critically depends on the development of an antigen delivery platform that promotes the generation of T-cell-dependent immune responses against oligosaccharides. This is generally achieved by conjugating carbohydrate antigens to a carrier protein. In addition, the immunogenicity has to be enhanced by delivery of the conjugates with an immunological adjuvant. Alum has remained the dominant adjuvant for human vaccines since many other candidate

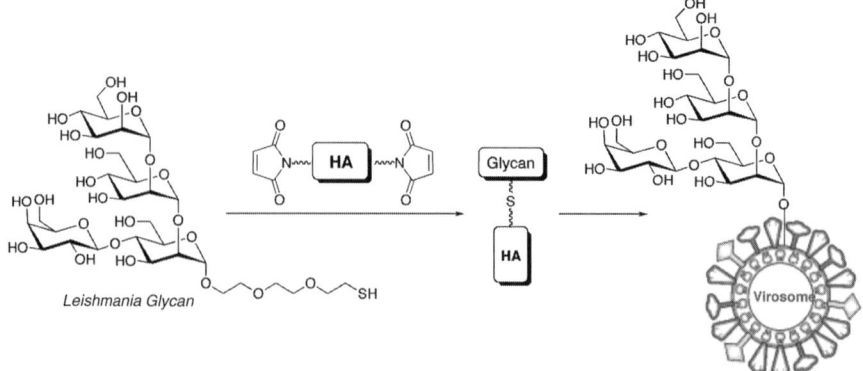

Figure 6.7 Synthetic glycovaccine based on *L. donovani* cap tetrasaccharide. Formulation of the synthetic glycan into virosome via conjugation to solubilized influenza hemagglutinin.

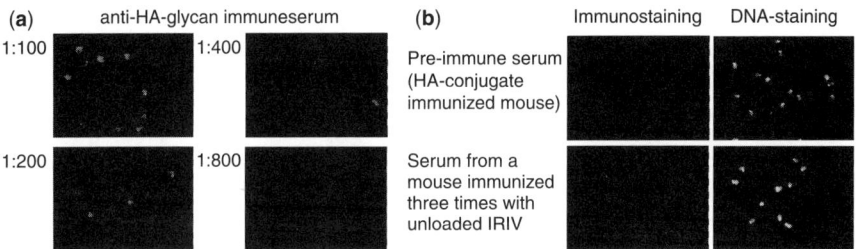

Figure 6.8 Immunization with *Leishmania* cap glycan loaded IRIV generated *L. donovani* cross-reactive IgG. (a) Immunofluorescence staining of parasites by serial dilution (1 : 100 to 1 : 800) of serum from mice immunized three times with HA-glycan. (b) Lack of immunofluorescence staining (left panel) of parasites by preimmune sera (dilution 1 : 100) and by serum (dilution 1 : 100) of a mouse immunized three times with unloaded IRIV. Presence of parasites on the microscopic slides was demonstrated by staining of parasite DNA with Hoechst dye 33258 (right panel). (See color insert.)

adjuvants have shown unsuitable properties, including reactogenicity, toxicity, instability, or high costs.

Synthetic *Leishmania* cap tetrasaccharide antigen was formulated into the virosomal carrier by conjugating to the succinimide-activated solubilized influenza hemagglutinin (HA) [59]. The resulting glycan–HA conjugate was resuspended in octaethylenglycol and combined with egg phosphatidylcholine, surface glycoproteins, and phospholipids of inactivated influenza to give the HA–glycan-loaded IRIV construct (Fig. 6.7). Mice immunized with glycan–HA-loaded IRIV generated glycan-specific IgG and IgM antibody responses as analyzed by ELISA. Immunofluorescence staining using the mice sera immunized with glycan-loaded IRIV showed cross-reactive IgG with natural carbohydrate antigens expressed by the *L. donovani* parasite (Fig. 6.8). No staining was observed with preimmune sera and after three immunizations of mice with IRIV that were not loaded with the synthetic glycan (Fig. 6.8). The antiglycan antisera stained also parasites in liver sections from *Leishmania*-infected hamsters. These assays clearly demonstrated that the vaccine candidate is highly immunogenic and elicits IgG antibodies that recognize *Leishmania* parasites. The question whether these antibodies are protective will have to be addressed in future challenge studies. Overall, this study demonstrated that IRIV represents an attractive alternative to currently used carbohydrate vaccine formulations suitable to induce T-cell-dependent antibody responses against oligosaccharide antigens, and the *L. donovani* LPG cap tetrasaccharide constitutes a viable antigen for future antileishmanial vaccine development.

6.4 OTHER EXAMPLES

6.4.1 Fucosylated *N*-glycan as Potential Vaccine Lead against Schistosomiasis

An equally interesting candidate antigen based on carbohydrate comes from the case of schistosomiasis. Schistosomiasis, also known as bilharzia, is a major parasitic infection in humans and is caused by trematode worms of the genus *Schistosoma*.

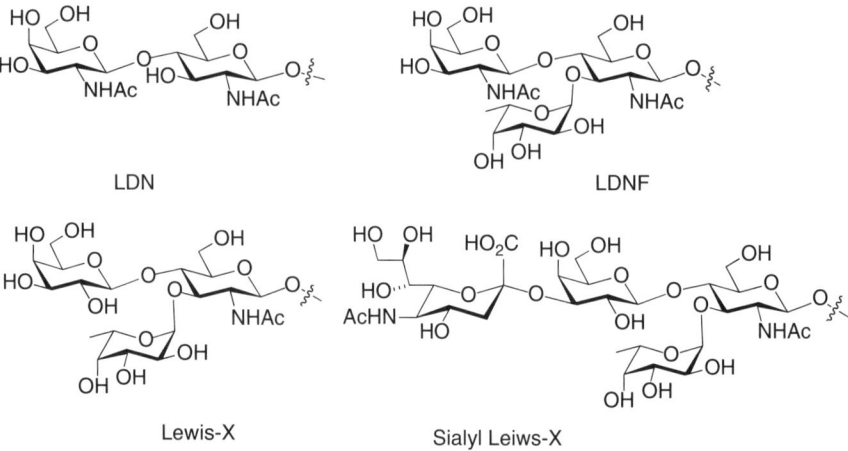

Figure 6.9 Structures of specific immunogenic glycans from *Schistosoma* genus.

The disease is almost exclusively restricted to tropical and subtropical areas due to the climatic requirement of the intermediate hosts to the parasite. It has been known for more than a quarter century that glycosylated antigens play a majror role in the development of the humorale response in schistosomiasis [61, 62]. More specifically a correlation between blood levels of antibody (both IgMs and IgGs) against egg polysaccharides and the succeptibility of reinfection has been established [63].

The major humoral response to *Schistosoma mansoni* has appeared to be directed against complex-type *N*-glycans expressing LDN (GalNAcβ1-4GlcNAc), LDNF (GalNAc [Fucα1-3]β1-4GlcNAc), and polymeric Lewis x (Le^x) epitopes (Fig. 6.9) [64–67]. Antibodies against these epitopes bind specifically to the immunovulnerable 3-h-old *S. mansoni* schistosomula (Fig. 6.10).

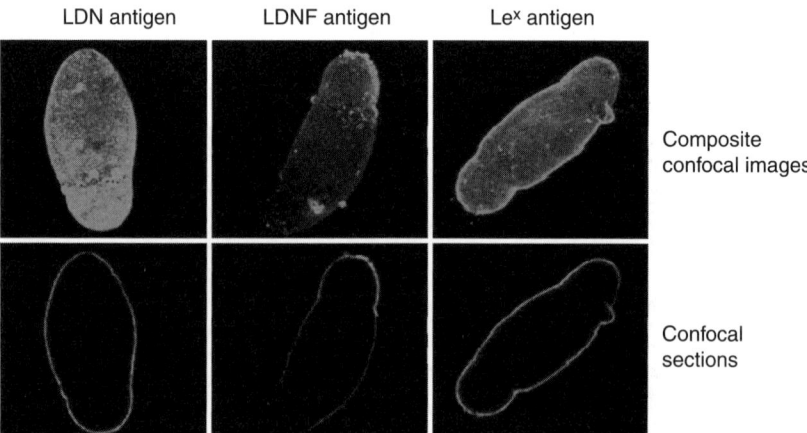

Figure 6.10 Antibodies to LDN, LDNF, and Le^x bind these structures on the surface to the immuno-vulnerable 3-h-old *S. mansoni* schistosomula. [From A.K. Nyame et al. (2000) *Archives of Biochemistry and Biophysics* 426:182–200.] (See color insert.)

Strikingly, in the case of rhesus monkeys specific antibodies to these carbohydrate determinants were able to stimulate complement-mediated cytolysis of cells expressing the corresponding antigens [67, 68], raising the possibility of the use of these structures as schistosomiasis vaccine candidates. In another study Naus et al. analyzed sera of individuals migrating into *S. mansoni* endemic regions using Surface Plasmon Resonance (SPR) on immobilized synthetic LDNF, and LDN-DF (GalNAcβ1-4[Fucα1-2Fucα1-3]GlcNAc) chips. They showed that the acute response against schistosome glycoconjugate antigens in young children was predominantly directed toward the LDN-DF epitope [69]. However, the observed high levels of IgG1 immediately after migration to the endemic decreased subsequently to reach the level observed in resident controls within 2 years. In contrast IgM responses constantly increased during the entire time of the study. Although the biological significance of these observations needs further elucidation, these results provide a useful basis for the development of a carbohydrate-based vaccine approach against *S. mansoni*.

6.4.2 GPIs as Potential Vaccine Lead against Toxoplasmosis and Chagas' Disease

Many parasitic protozoa express abundant amounts of GPIs in comparison to mammals. In these organisms, besides providing membrane anchorage, GPIs are packed densely to form a *glycocalyx*, the protecting "sugar" coat on the cell surface that is critical for their survival within the mammalian host [37]. As discussed in the previous paragraphs, GPI constitutes the primary toxin of *P. faciparium*, and immunization against *P. falciparum* GPI offered protection from malaria in a rodent model. The role of parasitic GPI anchors as virulence factors and in modulating the host immune response to infection are also observed in other parasites. Notable examples include the GPIs from *Toxoplasma gondii* and *Trypanosoma cruzi*, the causative agents of toxoplasmosis and Chargas' disease.

Toxoplasma gondii is a ubiquitous parasitic protozoan that causes congenital infectious diseases as well as severe encephalitis, a major cause of death among immune-deficient persons, such as AIDS patients. Schwarz and co-workers have shown that highly purified GPIs from *T. gondii* tachyzoites (Fig. 6.11a), as well as their core glycan, can induce Tumor Necrosis Factor-alpha (TNF-α) production in macrophages [70], which implicates the participation of *T. gondii* GPIs as bioactive factors in the production of TNF-α during toxoplasmal pathogenesis.

Chagas' disease, also refered as *American trypanosomiasis*, is a human tropical parasitic disease that occurs primarily in South America and is caused by the flagellate protozoan *T. cruzi*. *T. cruzi* have toxinlike proinflammatory activities that may be responsible for the pathology of Chagas' disease. Ferguson and co-workers showed that a highly purified and characterized GPI from *T. cruzi* trypomastigote mucins exhibits the potent TNF-α, interleukin-12, and nitric-oxide-inducing activities that are comparable to the activities of bacterial endotoxin and *Mycoplasma* lipopeptide and constitute one of the most potent microbial proinflammatory agents known to date (Fig. 6.11b) [71].

Synthetic strategies to access the GPIs of *T. gondii* and *T. cruzi* origins have been developed [72–75]. It would be worthwhile to investigate to what extent a vaccine prepared on the basis of these GPIs can be suitable to prevent these infections.

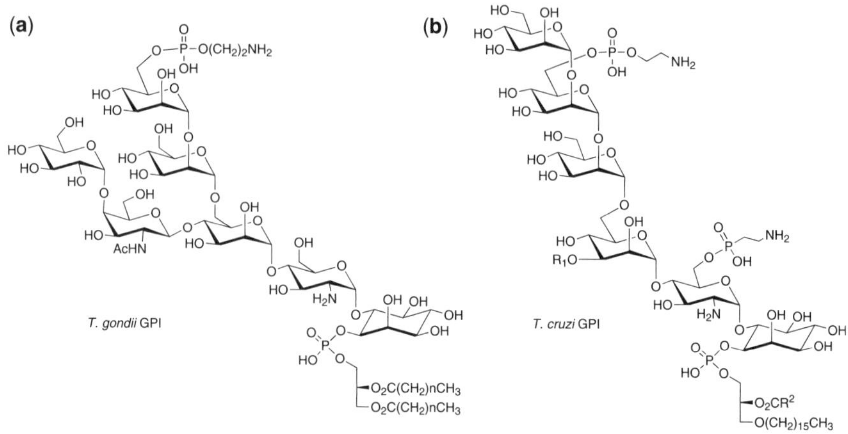

Figure 6.11 Structures of *T. gondii* and *T. cruzi* GPIs that were found exhibit toxinlike properties. (a) *T. gondii* GPI (*n* is not specified). (b) *T. cruzi* GPI (R^1 can be oligogalactose, R^2 is derived from palmitic, or oleic, or linoleic acid).

6.5 PERSPECTIVES AND FUTURE CHALLENGE

Parasitic diseases including malaria are major health problems affecting mainly individuals in developing countries. The increasing prevalence of parasitic infections and the emergence of resistance against conventional drugs have led to a concerted commitment of scientific and economic communities to establish efficient health care structures. For both the immense success rate and the generally affordable cost protection through vaccination are still the ultimate choices to fight infectious diseases. The quest for vaccines against human parasites has been nothing but a succession of frustrating experiences as none of the candidates has reached the market. It is striking to observe that none of the candidate vaccines is designed around a carbohydrate epitope.

Carbohydrates have a long-lasting reputation of being useless immunogenic molecules, but that theory is loosing ground, as recent approaches using conjugate vaccines have proven protective in many cases involving bacterial epitopes. In the specific case of parasitic diseases carbohydrate-based vaccination will certainly be more robust as the probability for the parasite to change the epitope is far lower than with peptide antigens. One major difficulty, however, remains the significantly weak number of specific carbohydrate epitopes described to date. If a carbohydrate-based approach for vaccination is to become a method of choice, the identification of novel specific epitopes is required. In this respect already existing analytical tools such as carbohydrate microarrays will have to be optimized for a better application and new tools will be most welcome. The involvement of synthetic carbohydrate chemistry at this level is invaluable as it is required for both the generation of oligosaccharide libraries to perform the screening and the production of strictly defined vaccine conjugates.

ACKNOWLEDGMENT

Research in the Seeberger laboratory on carbohydrate-based vaccines has been supported by ETH Zurich, the Swiss National Science Foundation, Fondation Bay, and a Koerber Award.

REFERENCES

1. Pink R, Hudson A, Mouriès M-A, Bendig M (2005) Opportunities and challenges in antiparasitic drug discovery. *Nat. Rev. Drug Discov.* 4:727–740.
2. Kaplan RM (2004) Drug resistance in nematodes of veterinary importance: A status report. *Trends in Parasitol.* 20:477–481.
3. Rappuoli R (2005) *Vaccination of Humans, Encyclopedia of Life Sciences.* John Wiley & Sons, New York.
4. Bachmann MF, Dyer MR (2004) Therapeutic vaccination for chronic diseases: A new class of drugs in sight. *Nat. Rev. Drug Discov.* 3:81–88.
5. Knox DP, Redmond DL (2006) Parasite vaccines—recent progress and problems associated with their development. *Parasitology* 133:S1–S8.
6. Su XZ, Heatwole VM, Wertheimer SP, Guinet F, Herrfeldt JA, Peterson DS, Ravetch JA, Wellems TE (1995) The large diverse gene family var encodes proteins involved in cytoadherence and antigenic variation of Plasmodium falciparum-infected erythrocytes. *Cell* 82:89–100.
7. Smith JD, Chitnis CE, Craig AG, Roberts DJ, Hudson-Taylor DE, Peterson DS, Pinches R, Newbold CI, Miller LH (1995) Switches in expression of *Plasmodium falciparum* var genes correlate with changes in antigenic and cytoadherent phenotypes of infected erythrocytes. *Cell* 82:101–110.
8. Crampton A, Vanniasinkam T (2007) Parasite vaccine: The new generation. *Infect. Genet. Evol.* 7:664–673.
9. Jennings HJ, Pon RA (1996) Polysaccharides and glycoconjugates as human vaccines. *Polysaccharides Med. Appl.* 443–479.
10. Moreau M, Schulz D (2000) Polysaccharide based vaccines for the prevention of pneumococcal infections. *J. Carbohydr. Chem.* 19:419–434.
11. Robbins JB, Schneerson R, Szu SC, Pozsgay V (1999) Bacterial polysaccharide-protein conjugate vaccines. *Pure Applied Chem.* 71:745–754.
12. Weintraub A (2003) Immunology of bacterial polysaccharide antigens. *Carbohydr. Res.* 338:2539–2547.
13. Ferguson MAJ, Homans SW, Dwek RA, Rademacher TW (1988) Glycosyl-phosphatidylinositol moiety that anchors Trypnosoma brucei variant surface glycoprotein to the membrane. *Science* 239:753–759.
14. Homans SW, Ferguson MA, Dwek RA, Rademacher TW, Anand R, Williams AF (1988) Complete structure of the glycosyl phosphatidylinositol membrane anchor of rat brain Thy-1 glycoprotein. *Nature* 333:269–272.
15. Ikezawa H (2002) Glycosylphosphatidylinositol (GPI)-anchored proteins. *Biol. Pharm. Bull.* 25:409–417.

16. Gerold P, Schofield L, Blackman MJ, Holder AA, Schwarz RT (1996) Structural analysis of the glycosyl-phosphatidylinositol membrane anchor of the merozoite surface proteins-1 and -2 of *Plasmodium falciparum*. *Mol. Biochem. Parasitol.* 75:131–143.
17. Schofield L, Hackett F (1993) Signal transduction in host cells by a glycosylphosphatidylinositol toxin of malaria parasites. *J. Exp. Med.* 177:145–153.
18. Naik RS, Branch OH, Woods AS, Vijaykumar M, Perkins DJ, Nahlen BL, Lal AA, Cotter RJ, Costello CE, Ockenhouse CF, Davidson EA, Gowda DC (2000) Glycosylphosphatidylinositol anchors of *Plasmodium falciparum*: Molecular characterization and naturally elicited antibody response that may provide immunity to malaria pathogenesis. *J. Exp. Med.* 192:1563–1576.
19. Naik RS, Branch OH, Woods AS, Vijaykumar M, Perkins DJ, Nahlen BL, Lal AA, Cotter RJ, Costello CE, Ockenhouse CF, Davidson EA, Gowda DC (2000) Glycosylphosphatidylinositol anchors of *Plasmodium falciparum*: Molecular characterization and naturally elicited antibody response that may provide immunity to malaria pathogenesis. *J. Exp. Med.* 192:1563–1575.
20. Boutlis CS, Gowda DC, Naik RS, Maguire GP, Mgone CS, Bockarie MJ, Lagog M, Ibam E, Lorry K, Anstey NM (2002) Antibodies to *Plasmodium falciparum* glycosylphosphatidylinositols: Inverse association with tolerance of parasitemia in Papua New Guinean children and adults. *Infect. Immun.* 70:5052–5057.
21. de Souza JB, Todd J, Krishegowda G, Gowda DC, Kwiatkowski D, Riley EM (2002) Prevalence and boosting of antibodies to *Plasmodium falciparum* glycosylphosphatidylinositols and evaluation of their association with protection from mild and severe clinical malaria. *Infect. Immun.* 70:5045–5051.
22. Boutlis CS, Fagan PK, Gowda DC, Lagog M, Mgone CS, Bockarie MJ, Anstey NM (2003) Immunoglobulin G (IgG) responses to Plasmodium falciparum glycosylphosphatidylinositols are short-lived and predominantly of the IgG3 subclass. *J. Infect. Dis.* 187:862–865.
23. Hudson Keenihan SN, Ratiwayanto S, Soebianto S, Krisin, Marwoto H, Krishnegowda G, Gowda DC, Bangs MJ, Fryauff DJ, Richie TL, Kumar S, Baird JK (2003) Age-dependent impairment of IgG responses to glycosylphosphatidylinositol with equal exposure to *Plasmodium falciparum* among Javanese migrants to Papua, Indonesia. *Am. J. Trop. Med. Hyg.* 69:36–41.
24. Perraut R, Diatta B, Marrama L, Garraud O, Jambou R, Longacre S, Krishnegowda G, Dieye A, Gowda DC (2005) Differential antibody responses to Plasmodium falciparum glycosylphosphatidylinositol anchors in patients with cerebral and mild malaria. *Microbes Infect.* 7:682–687.
25. Naik RS, Krishnegowda G, Ockenhouse CF, Gowda DC (2006) Naturally elicited antibodies to glycosylphosphatidylinositols (GPIs) of Plasmodium falciparum require intact GPI structures for binding and are directed primarily against the conserved glycan moiety. *Infect. Immun.* 74:1412–1415.
26. Schofield L, Hewitt MC, Evans K, Siomos MA, Seeberger PH (2002) Synthetic GPI as a candidate anti-toxic vaccine in a model of malaria. *Nature* 418:785–789.
27. Hewitt MC, Snyder DA, Seeberger PH (2002) Rapid synthesis of a glycosylphosphatidylinositol-based malaria vaccine using automated solid-phase oligosaccharide synthesis. *J. Am. Chem. Soc.* 124:13434–13436.

28. Boutlis CS, Fagan PK, Gowda DC, Lagog M, Mgone CS, Bockarie MJ, Anstey NM (2003) Immunoglobulin G (IgG) responses to *Plasmodium falciparum* glycosylphosphatidylinositols are short-lived and predominantly of the IgG3 subclass. *J. Infect. Dis.* 187:862–865.

29. Boutlis CS, Gowda DC, Naik RS, Maguire GP, Mgone CS, Bockarie MJ, Lagog M, Ibam E, Lorry K, Anstey NM (2002) Antibodies to *Plasmodium falciparum* glycosylphosphatidylinositols: Inverse association with tolerance of parasitemia in Papua New Guinean children and adults. *Infect. Immun.* 70:5052–5057.

30. de Souza JB, Todd J, Krishegowda G, Gowda DC, Kwiatkowski D, Riley EM (2002) Prevalence and boosting of antibodies to *Plasmodium falciparum* glycosylphosphatidylinositols and evaluation of their association with protection from mild and severe clinical malaria. *Infect. Immun.* 70:5045–5051.

31. Hudson Keenihan SN, Ratiwayanto S, Soebianto S, Krisin, Marwoto H, Krishnegowda G, Gowda DC, Bangs MJ, Fryauff DJ, Richie TL, Kumar S, Baird JK (2003) Age-dependent impairment of IgG responses to glycosylphosphatidylinositol with equal exposure to *Plasmodium falciparum* among Javanese migrants to Papua, Indonesia. *Am. J. Trop. Med. Hyg.* 69:36–41.

32. Perraut R, Diatta B, Marrama L, Garraud O, Jambou R, Longacre S, Krishnegowda G, Dieye A, Gowda DC (2005) Differential antibody responses to *Plasmodium falciparum* glycosylphosphatidylinositol anchors in patients with cerebral and mild malaria. *Microbes Infect.* 7:682–687.

33. Suguitan AL, Jr, Gowda DC, Fouda G, Thuita L, Zhou A, Djokam R, Metenou S, Leke RG, Taylor DW (2004) Lack of an association between antibodies to *Plasmodium falciparum* glycosylphosphatidylinositols and malaria-associated placental changes in Cameroonian women with preterm and full-term deliveries. *Infect. Immun.* 72:5267–5273.

34. Ratner DM, Adams EW, Disney MD, Seeberger PH (2004) Tools for glycomics: Mapping interactions of carbohydrates in biological systems. *ChemBioChem.* 5:1375–1383.

35. Kamena F, Tamborrini M, Liu XY, Kwon YU, Thompson F, Pluschke G, Seeberger PH (2008) Synthetic GPI array to study antitoxic malaria response. *Nat. Chem. Biol.* 4:238–240.

36. Kedzierski L, Zhu Y, Handman E (2006) Leishmania vaccines: Progress and problems. *Parasitology* 133:S87–112.

37. Guha-Niyogi A, Sullivan DR, Turco SJ (2001) Glycoconjugate structures of parasitic protozoa. *Glycobiology* 11:45R–59R.

38. Mahoney A, Sacks D, Saraiva E, Modi G, Turco S (1999) Intraspecies and stage-specific polymorphisms in lipophosphoglycan structure control Leishmania donovani-sand fly interactions. *Biochemistry* 38:9813–9823.

39. McConville M, Schnur L, Jaffe C, Schneider P (1995) Structure of Leishmania lipophosphoglycan: Inter- and intra-specific polymorphism in Old World species. *Biochem. J.* 310:807–818.

40. Ilg T, Handman A, Stierhof Y (1999) Proteophosphoglycans from Leishmania promastigotes and amastigotes. *Biochem. Soc. Trans.* 27:518–525.

41. Russell D, Alexander J (1988) Effective immunization against cutaneous leishmaniasis with defined membrane antigens reconstituted into liposomes. *J. Immunol.* 140:1274–1279.

42. McConville M, Bacic A, Mitchell G, Handman E (1987) Lipophosphoglycan of Leishmania major that vaccinates against cutaneous leishmaniasis contains an alkylglycerophosphoinositol lipid anchor. *Proc. Natl. Acad. Sci. U.S.A.* 84:8491–9495.
43. Jardim A, Tolson D, Turco S, Pearson T, Olafson R (1991) The Leishmania donovani lipophosphoglycan T lymphocyte-reactive component is a tightly associated protein complex. *J. Immunol.* 147:3538–3544.
44. Higson AP, Ross AJ, Tsvetkov YE, Routier FH, Sizova OV, Ferguson MAJ, Nikolaev AV (2005) Synthetic fragments of antigenic lipophosphoglycans from Leishmania major and Leishmania mexicana and their use for characterisation of the Leishmania elongating alpha-D-mannopyranosylphosphate transferase. *Chem. Eur. J.* 11:2019–2030.
45. Ruhela D, Vishwakarma RA (2004) A facile and novel route to the antigenic branched phosphoglycan of the protozoan Leishmania major parasite. *Tetrahedron Lett.* 45:2589–2592.
46. Hewitt MC, Seeberger PH (2001) Automated solid-phase synthesis of a branched Leishmania cap tetrasaccharide. *Org. Lett.* 3:3699–3702.
47. Hewitt MC, Seeberger PH (2001) Solution and solid-support synthesis of a potential leishmaniasis carbohydrate vaccine. *J. Org. Chem.* 66:4233–4243.
48. Ross AJ, Ivanova IA, Ferguson MAJ, Nikolaev AV (2001) Parasite glycoconjugates. Part 11. Preparation of phosphodisaccharide synthetic probes, substrate analogues for the elongating alpha-D-mannopyranosylphosphate transferase in the Leishmania. *J. Chem. Soc. Perkin Trans.* 1:72–81.
49. Ruhela D, Vishwakarma RA (2001) Efficient synthesis of the antigenic phosphoglycans of the Leishmania parasite. *Chem. Commun.* 2024–2025.
50. Upreti M, Ruhela D, Vishwakarma RA (2000) Synthesis of the tetrasaccharide cap domain of the antigenic lipophosphoglycan of Leishmania donovani parasite. *Tetrahedron* 56:6577–6584.
51. Ruda K, Lindberg J, Garegg PJ, Oscarson S, Konradsson P (2000) Synthesis of the Leishmania LPG core heptasaccharyl myo-inositol. *J. Am. Chem. Soc.* 122:11067–11072.
52. Nikolaev AV, Watt GM, Ferguson MAJ, Brimacombe JS (1997) Parasite glycoconjugates. 6. Chemical synthesis of phosphorylated, penta- and hepta-saccharide fragments of Leishmania major antigenic lipophosphoglycan. *J. Chem. Soc. Perkin Trans.* 1:969–979.
53. Arasappan A, Fraser-Reid B (1996) n-Pentenyl glycoside methodology in the stereoselective construction of the tetrasaccharyl cap portion of Leishmania lipophosphoglycan. *J. Org. Chem.* 61:2401–2406.
54. Nikolaev AV, Rutherford TJ, Ferguson MAJ, Brimacombe JS (1995) Parasite glycoconjugates. 4. Chemical synthesis of disaccharide and phosphorylated oligosaccharide fragments of Leishmania donovani antigenic lipophosphoglycan. *J. Chem. Soc. Perkin Trans.* 1:1977–1987.
55. Ivanova IA, Ross AJ, Ferguson MAJ, Nikolaev AV (1999) Parasite glycoconjugates. Part 9. Synthesis of dec-9-enyl beta-D-galactopyranosyl-(1->4)-alpha-D-mannopyranosyl phosphate and its epimers at the D-galactose moiety, substrate analogues for the elongating alpha-D-mannopyranosylphosphate transferase in the Leishmania. *J. Chem. Soc. Perkin Trans.* 1:1743–1753.
56. Higson AP, Tsvetkov YE, Ferguson MAJ, Nikolaev AV (1998) Parasite glycoconjugates. Part 8. Chemical synthesis of a heptaglycosyl triphosphate fragment of Leishmania

mexicana lipo- and proteo-phosphoglycan and of a phosphorylated trisaccharide fragment of Leishmania donovani surface lipophosphoglycan. *J. Chem. Soc. Perkin Trans.* 1:2587–2595.

57. Routier FH, Nikolaev AV, Ferguson MAJ (1999) The preparation of neoglycoconjugates containing inter-saccharide phosphodiester linkages as potential anti-Leishmania vaccines. *Glycoconjugate J.* 16:773–780.

58. Rogers ME, Sizova OV, Ferguson MAJ, Nikolaev AV, Bates PA (2006) Synthetic glycovaccine protects against the bite of Leishmania-infected sand flies. *J. Infect. Dis.* 194:512–518.

59. Liu X, Siegrist S, Amacker M, Zurbriggen R, Pluschke G, Seeberger PH (2006) Enhancement of the immunogenicity of synthetic carbohydrates by conjugation to virosomes: A leishmaniasis vaccine candidate. *ACS Chem. Biol.* 1:161–164.

60. Zurbriggen R (2003) Immunostimulating reconstituted influenza virosomes. *Vaccine* 21:921–924.

61. Ali PO, Hagan P, Wilkins HA, Simpson AJ (1989) Antibody to schistosomulum surface carbohydrate epitopes in subjects infected with Schistosoma haematobium. *Trans. R. Soc. Trop. Med. Hyg.* 83:358–361.

62. Nash TE, Lunde MN, Cheever AW (1981) Analysis and antigenic activity of a carbohydrate fraction derived from adult Schistosoma mansoni. *J. Immunol.* 126:805–810.

63. Butterworth A, Dunne D, Fulford A, Capron M, Khalife J, Capron A, Koech D, Ouma J, Sturrock R (1988) Immunity in human schistosomiasis mansoni: Cross-reactive IgM and IgG2 anti-carbohydrate antibodies block the expression of immunity. *Biochimie* 70:1053–1063.

64. Thomas PG, Harn DA, Jr (2004) Immune biasing by helminth glycans. *Cell Microbiol.* 6:13–22.

65. Hokke CH, Deelder AM (2001) Schistosome glycoconjugates in host-parasite interplay. *Glycoconj. J.* 18:573–587.

66. Cummings RD, Nyame AK (1996) Glycobiology of schistosomiasis. *FASEB J.* 10:838–848.

67. Nyame AK, Pilcher JB, Tsang VC, Cummings RD (1996) Schistosoma mansoni infection in humans and primates induces cytolytic antibodies to surface Le(x) determinants on myeloid cells. *Exp. Parasitol.* 82:191–200.

68. van Dam GJ, Claas FH, Yazdanbakhsh M, Kruize YC, van Keulen AC, Ferreira ST, Rotmans JP, Deelder AM (1996) Schistosoma mansoni excretory circulating cathodic antigen shares Lewis-x epitopes with a human granulocyte surface antigen and evokes host antibodies mediating complement-dependent lysis of granulocytes. *Blood* 88:4246–4251.

69. Naus CWA, Remoortere Av, Ouma JH, Kimani G, Dunne DW, Kamerling JP, Deelder AM, Hokke CH (2003) Specific antibody responses to three schistosome-related carbohydrate structures in recently exposed immigrants and established residents in an area of schistosoma mansoni endemicity. *Infect. Immun.* 71:5676–5681.

70. Debierre-Grockiego F, Azzouz N, Schmidt J, Dubremetz JF, Geyer H, Geyer R, Weingart R, Schmidt RR, Schwarz RT (2003) Roles of glycosylphosphatidylinositols of Toxoplasma gondii—Induction of tumor necrosis factor-alpha production in macrophages. *J. Biol. Chem.* 278:32987–32993.

71. Almeida IC, Camargo MM, Procopio DO, Silva LS, Mehlert A, Travassos LR, Gazzinelli RT, Ferguson MAJ (2000) Highly purified glycosylphosphatidylinositols from Trypanosoma cruzi are potent proinflammatory agents. *EMBO J.* 19:1476–1485.
72. Kwon Y-U, Liu X, Seeberger PH (2005) Total syntheses of fully lipid glycosylphosphatidylinositol anchors of Toxoplasma gondii. *Chem. Commun.* 2280–2282.
73. Pekari K, Schmidt RR (2002) A variable concept for the preparation of branched glycosyl phosphatidyl inositol anchors. *J. Org. Chem.* 68:1295–1308.
74. Yashunsky DV, Borodkin VS, Ferguson MAJ, Nikolaev AV (2006) The chemical synthesis of bioactive glycosylphosphatidylinositols from Trypanosoma cruzi containing an unsaturated fatty acid in the lipid. *Angew. Chem. Int. Ed.* 45:468–474.
75. Ali A, Gowda DC, Vishwakarma RA (2005) A new approach to construct full-length glycosylphosphatidylinositols of parasitic protozoa and [4-deoxy-Man-III]-GPI analogues. *Chem. Commun.* 519–521.

7

CARBOHYDRATE-BASED ANTIFUNGAL VACCINES

Magdia De Jesus, Liise-anne Pirofski, and Arturo Casadevall

Department of Microbiology and Immunology, Albert Einstein College of Medicine, Bronx, New York

7.1 INTRODUCTION

In the past few decades there has been a sharp increase in the number of immunocompromised patients for a variety of reasons that include increased use of immunosupressive medications to treat patients for neoplasms and autoimmune diseases and to prevent organ and bone marrow transplant rejection [1, 2]. Additionally, the widespread use of broad-spectrum antibacterial drugs, predominantly in hospitals, has resulted in an increase in the prevalence of patients at risk for fungal infections [3]. The human immunodeficiency virus (HIV) epidemic has also contributed to the increased number of patients with immune impairment or immunosuppression [1, 2, 4]. Many microbes, including fungi that were previously thought to be nonpathogenic, have emerged as significant pathogens in immunosuppressed populations. This has led to an interest in vaccine development for pathogens that affect patients with impaired immunity [2]. For example, both *Candida albicans* and *Cryptococcus neoformans* were relatively rare pathogens at the beginning of the twentieth century, but in recent years both candidiasis and cryptococcosis are distressingly common diseases.

There are currently no licensed vaccines for medically important fungi [4, 5]. The current interest in fungal vaccines arises from the fact that fungal diseases have emerged as a major cause for morbidity and mortality in patients with underlying

---

Carbohydrate-Based Vaccines and Immunotherapies. Edited by Zhongwu Guo and Geert-Jan Boons
Copyright © 2009 John Wiley & Sons, Inc.

immune impairment and current antifungal drug regimens are often unable to cure them. Many of these fungi are ubiquitous soil inhabitants such as *Aspergillus* spp. and *C. neoformans*. For these organisms, infection is thought to be acquired by inhalation of fungal spores, and the lung is the most common site of primary infection [5, 6]. For example, *C. neoformans* and *gattii* are found in pigeon feces and eucalyptus trees, respectively. These yeasts are unique among the pathogenic fungi because they possess a polysaccharide capsule that is antiphagocytic and has many deleterious effects on host immunity. Cryptococcosis in the immunocompromised host has a predilection for the nervous system [7, 8]. In contrast to the pathogenic fungi found in the soil, nearly all *Candida* spp. are only found in mammalian hosts. This is best exemplified by the most common *Candida* species, *C. albicans*, which is a commensal microbe that resides in the human mouth, gastrointestinal tract, and vaginal tract. *C. albicans* has become an increasingly important pathogen since the introduction of broad-spectrum antibiotic therapy and the increased number of patients with impaired immunity. In the immunocompromised state, *C. albicans* has the potential to be a pathogen, rather than a commensal microbe, though not every individual at risk will develop a fungal infection [5].

Here we review the experience with the development of glycoconjugate vaccines focusing primarily on the three fungal pathogens for which the most information is available: *Cryptococcus neoformans*, *Candida albicans*, and *Aspergillus fumigatus*.

7.2 TERMINOLOGY

7.2.1 Vaccination versus Immunization

The terms vaccination and immunization are often loosely used to mean the same thing. However, they are different. Vaccination refers to the administration of a preparation of live or inactivated organisms or its antigens (a vaccine), which have been formulated to stimulate specific immunity. Immunization is the process by which immunity to a pathogen or a vaccine is elicited. The reason that these two terms are used interchangeably is that immunization encompasses the act of vaccination to acquire a specific protective immunity [2].

7.2.2 Toxoids

Toxoids are a modified preparation of a toxin that inactivates the native toxin but retains its antigenicity. Toxoids can elicit immune responses that can neutralize the native toxin. Certain toxoids have been used as protein carriers in the development of glycoconjugates to make the carbohydrate antigens more immunogenic.

7.2.3 Glycoconjugates

A glycoconjugate vaccine is a vaccine consisting of a carbohydrate antigen linked to a protein carrier, such as a toxoid, that enhances the immune response to the antigen. In the next section we will explore how some of these glycoconjugates have been produced with fungal antigens and the type of immune responses that they elicit.

7.3 ANTIFUNGAL GLYCOCONJUGATE VACCINES

7.3.1 *C. neoformans* Polysaccharide–Protein Conjugates

Cryptococcus neoformans has a polysaccharide capsule with functional similarities to those of encapsulated bacteria such as *Streptococcus pneumoniae* and *Haemophilus influenzae* [9, 10]. Cryptococcal polysaccharide shares antigenic determinants with certain pneumococcal polysaccharides [11, 12]. The *C. neoformans* capsule is composed of glucuronoxylomannan (GXM), galactoxylomannan (GalXM), and mannoproteins (MP) (Fig. 7.1). GXM is a copolymer of up to six different repeating units that consist mostly of a linear $\alpha(1\rightarrow 3)$-mannan trisaccharide with side groups consisting of a $\beta(1\rightarrow 2)$-glucopyranosyluronic acid and $\beta(1\rightarrow 2)$ and $\beta(1\rightarrow 4)$-xylopyranosyl [13]. The mannan backbone of GXM is modified by acetyl groups, although specific patterns of acetylation are unknown [14–16] (Fig. 7.2). GXM can interfere with many aspects of immune function, including leukocyte migration and phagocytosis [17–20]. The ability of GXM to interfere with immune function is believed to contribute to *C. neoformans* virulence by allowing the fungal cells to evade the immune system. In addition, GXM synthesis is different from that of other capsular polysaccharides in that it is a heteropolymer that in theory could display an almost infinite number of structural combinations, that could in turn translate into antigenic changes and enhance its structural complexity [16]. The rationale for GXM-based conjugate vaccines is that they elicit antibodies that are opsonic and bind soluble polysaccharide, promoting its clearance.

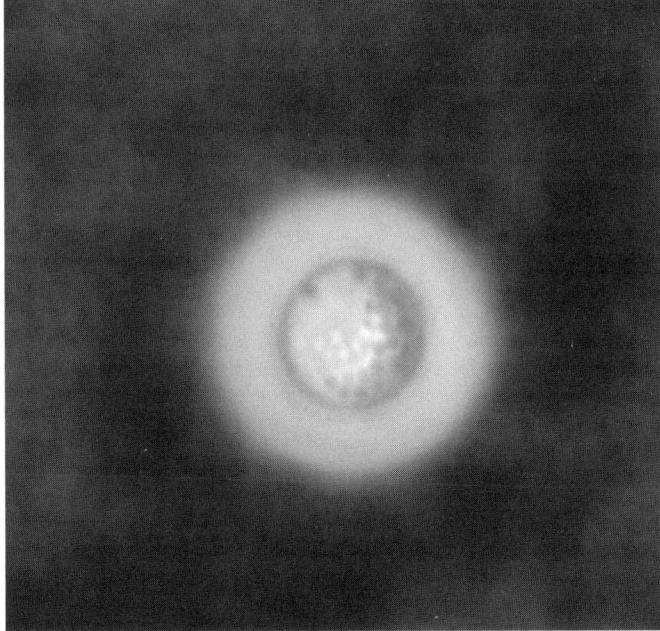

Figure 7.1 India ink stain of *C. neoformans*: The capsular polysaccharide can be easily visualized by India ink stain.

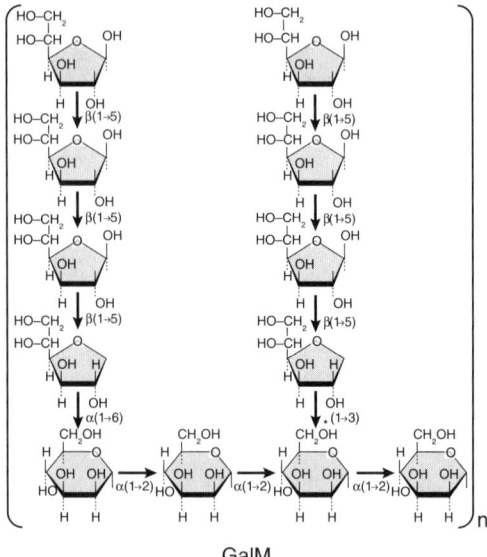
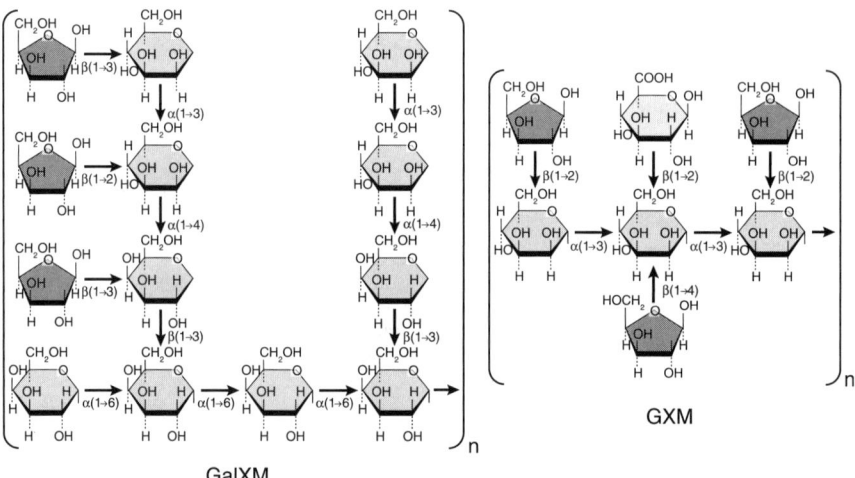

Figure 7.2 Structures of GXM, GalXM, and GalM: Comparison of the structures of the major repeating units of GalM from *Aspergillus* spp. and GalXM and GXM from *C. neoformans*. The asterisk near the 1→3 galactofuranosylmannopyranose bond indicates that there is currently uncertainty as to whether this linkage is α or β in configuration.

The first immunization studies with GXM were done by injecting the polysaccharide with an ion exchange resin or in an emulsion with Freund's adjuvant; however, this strategy led to the production of a weak antibody response mostly of the IgM isotype [21–23]. In the late 1960s Goren and Middlebrook generated the first GXM (or cryptococcal polysaccharide)-based glycoconjugate vaccine by coupling unfractionated capsular polysaccharide (which most likely contained all three capsular components) to a bovine γ-globulin by nitrocarbanilation and diazotization [24]. This vaccine was highly immunogenic but did not elicit a protective antibody response. In the 1990s Devi et al. [25] developed a second glycoconjugate vaccine based on fractionating the GXM component and linking it to tetanus-toxoid (TT). Initial attempts at conjugating the native GXM resulted in precipitation. Next, the polysaccharide was depolymerized by ultrasonic irradiation. The sonicated GXM was subjected to gel filtration through a Sepharose 2B-Cl column. The GXM fractions eluted from the middle of the column were collected and assigned as GXM. These investigators tried two different strategies to link the GXM to a protein carrier. In one strategy, GXM was derivatized and a bifunctional spacer linker was added that linked the polysaccharide and the toxoid protein. Adipic acid dihydrazide (ADH), the linker was introduced into GXM by two different methods. In one method, carboxyl groups on GXM were activated using N-(3-Dimethylaminopropyl)-N-ethylcarboiimide hydrochloride (EDAC) in which they called this preparation 1 or GXM-TT1. In the second method, GXM hydroxyl groups were randomly activated using cyanogen bromide, to produce a second conjugate known as GXM-TT2. To synthesize the final conjugate, the derivatized GXM-AH, produced by either method, was incubated with equal concentrations of the toxoid for 3 h with continuous monitoring of pH to avoid GXM de-O-acetylation. Both conjugates were verified by nuclear magnetic resonance (NMR) spectroscopy.

When both conjugates were tested in mice, the initial antibody response contained predominantly IgM. However, after a second immunization only GXM-TT2 elicited the highest IgG levels. GXM-TT1 yielded an antibody response similar to that induced by GXM alone. This is important because the antibody response to vaccination becomes a key issue since the ability of an antibody to promote fungal elimination may be dependent on isotype [5]. Glucuronyl and O-acetyl groups are recognized to be the major immunodeterminants of GXM. Serum antibodies induced by GXM-TT2 reacted with both chemically unmodified and de-O-acetylated forms of GXM, whereas serum antibodies elicted by GXM-TT1 reacted with only the chemically unmodified form. These results suggested that GXM-TT1 was less immunogenic because it retained only O-acetyl determinant, whereas GXM-TT2 was more immunogenic since boosting immunization yielded IgG antibodies, and it retained both the glucuronyl and the O-acetyl determinants.

We have also investigated the development of a glycoconjugate against the minor polysaccharide component of the *C. neoformans* capsule, galactoxylomannan. GalXM constitutes about 7% of the capsular mass and has an $\alpha(1\to 6)$ galactan backbone containing four potential short oligosaccharide branch structures. The branches are 3-O-linked to the backbone and consist of an $\alpha(1\to 3)$-Man, $\alpha(1\to 4)$-Man, β-galactosidase trisaccharide with variable amounts of $\beta(1\to 2)$- or $\beta(1\to 3)$-xylose side groups [26]. Using the same methodology described by Devi et al. [25], GalXM was conjugated to the protective antigen (PA) of *Bacillus anthracis* using the bifunctional linker ADH. This conjugate was immunogenic in both mice and rats. [Parts of this work were presented at the 106th General Meeting of the American Society for Microbiology, Orlando, FL, May 2006 (abstr. F-013).]

7.3.2 Development of Alternative Vaccines for *C. neoformans*

An alternative approach in *C. neoformans* as a vaccine antigen that is conceptually related to glycoconjugate vaccines in the type of immunity elicited was the development of a peptide that mimics a GXM epitope. A mimetic is a peptide epitope that inhibits the binding of an antibody to its native antigen [4]. The peptide can elicit a protective antibody response by stimulating the production of protective antibodies that bind to polysaccharides [4]. The approach used to develop the peptide mimetope vaccine employed a decapeptide phage display library in which peptide epitopes were expressed on the surface protein of the M13 bacteriophage, and these were screened by using human IgM monoclonal antibody to *C. neoformans* GXM, 2E9. This MAb was produced from immortalized lymphocytes from a volunteer who received the GXM-TT vaccine [27]. A group of these phage peptide epitopes were isolated and were considered GXM mimetics.

The peptide mimetic, P13 was conjugated to TT or bovine serum albumin (BSA) and used to vaccinate mice. The vaccine elicited both IgM and IgG antibodies to both GXM and P13, suggesting that P13 is a GXM mimetope. A mimetope is a mimetic that induces an immune response to the native antigen [4]. The P13-TT conjugate induced protection in mice that is probably antibody mediated. Despite promising results, mimetope-mediated protection and vaccine efficacy is not fully understood.

7.3.3 *C. albicans* Mannan–Protein Conjugates

In the 1990s Han and co-workers [28] developed a vaccine for *C. albicans* based on a mannan conjugate. Mannans are responsible for the attachment of the fungus to the splenic and lymph node macrophages in mice during systemic infection.

The experimental system used a liposome-encapsulated yeast cell L-adhesin fraction. To do this, a *C. albicans* cell surface mannan-enriched fraction was isolated by a β-mercaptoethanol treatment of yeast cells and encapsulated into multilamellar liposomes [28, 29]. Control liposomes L-Diluent Phosphate Buffered Saline (L-DPBS) were prepared by adding diluent buffer (DPBS) instead of mannan during the preparation. Vaccination with liposome-encapsulated *C. albicans* surface mannan (L-mann) elicited a protective antibody response that protected against disseminated disease. Monoclonal antibodies were also generated in mice immunized with the vaccine. The monoclonal antibodies described were B6 and B6.1, both specific for the *Candida* mannan, but recognizing different epitopes. The antibody B6.1 enhanced resistance of normal and Severe Combined Immunodeficiency (SCID) mice against disseminated candidiasis and was also shown to be protective in neutropenic mice. The second MAb, B6, had no protective activity. Both MAbs were IgM and both agglutinated yeast cells [28].

In a later study the group tested the ability of the L-mann vaccine and the MAbs to enhance the resistance of mice to *Candida* vaginal infection. The group used active (the vaccine) and passive (the MAb) immunization together and these conferred protective effects. Although the mechanisms by which the liposome vaccine and the MAbs protected against *Candida* vaginal infections are unknown, MAb B6.1 promoted in vitro neutrophil candidacidal activity [28].

Liposomal encapsulation of the mannan extract in the vaccine was essential because the extract alone was poorly immunogenic. By 1999 these investigators observed that although this liposome vaccine induced protective responses, there were limitations to this approach. The liposomal part of the formula was unstable after 2 weeks of storage, and many booster immunizations with the vaccine were required to produce a protective antibody response. Consequently, they sought to improve the vaccine by conjugating the *Candida* mannan extract to BSA. The glycoconjugate was then tested for protective antibodies in mice.

The BSA conjugate was produced by extracting the yeast cell surface mannan by β-mercaptoethanol. Chemical conjugation of the mannan to BSA was done by activation with cyanogen bromide (CNBr). The resulting mannan hydrazide was added to an equal weight of BSA, by reacting the mixture in the presence of 0.1 M 1-ethyl-3 (3-dimethylaminoprophyl) carbodiimide-HCl (EDC) [30].

Mice immunized with the conjugate fraction became more resistant to disseminated candidiasis, and immune sera transferred protection against disseminated disease and against a *Candida* vaginal infection. These results were similar to those obtained by use of the mannan extract-liposomal formulation in that protection was associated with the development of serum antibodies specific for the *Candida* mannan. Unlike the liposomal vaccine, which required intravenous (IV) priming and four boosters, priming and a single intraperitoneal (IP) booster of the conjugate–adjuvant vaccine generated protective antibody levels [30].

7.3.4 β-Glucan–Protein Conjugates

Torosantucci et al. [31] reported that a glycoconjugate vaccine protected against two fungal pathogens, *C. albicans* and *A. fumigatus*. The vaccine works by eliciting cross-reactive antibodies. This strategy was novel because it defied the dogma that vaccine efficacy is dependent on the induction of specific antibody responses [31, 32]. The rationale for the vaccine was that β-glucans, which are present on all human pathogenic fungi, are critical for fungal viability. Many β-glucans are found on fungal cell walls. To avoid any contamination with other fungal antigens the authors used β-glucans from a nonfungal source, namely, laminarin, a β(1→3) glucan from brown alga *Laminaria digitata*. Since laminarin is a poor immunogen, it was conjugated to the diptheria toxoid protein Cross Reactive Material 197 (CRM 197). To accomplish this, terminal reductive groups on laminarin were subjected to reductive amination. The aminated polysaccharides were then reacted with an *N*-hydroxysuccinimide diester of adipic acid. The conjugation was done by incubating a 20-fold excess of the laminarin-linker with CRM197 toxoid, and the conjugate was verified by NMR [31].

In mice, the Lam–CRM conjugate was immunogenic and protective. Mice were infected with *C. albicans* followed by vaccination with Lam–CRM in Freund's adjuvant. The results showed a dramatic reduction in fungal burden. Additionally, the conjugate elicited IgG antibodies. Given that systemic and mucosal host defense against *C. albicans* are different, the vaccine was also tested in a vaginal candidiasis model in rats. Rats that were intravaginally immunized with Lam–CRM exhibited an accelerated rate of *Candida* clearance. The finding that the vaccine could protect against both forms of the disease suggests that its efficacy could be dependent on antibody-related antimicrobial effects [32].

Consistent with this notion, Lam–CRM immune serum was able to inhibit the growth of *A. fumigatus* in its conidial and hyphal forms. Furthermore, a monoclonal antibody elicited by immunization with the Lam–CRM conjugate, 2G8 was tested for its efficacy effect upon *C. neoformans*. Cell wall β-glucans are present beneath the capsular polysaccharide of *C. neoformans*. The results revealed that the capsule does not prevent the MAb 2G8 from binding to β-glucan. The study also showed several positive features of the MAb 2G8: (1) It bound to the cell wall of both acapsular and encapsulated *C. neoformans*; (2) it inhibited the in vitro growth of *C. neoformans*; (3) it reduced the capsule size of the fungus; (4) it increased the amount of and killing of acapsular, but not encapsulated, strains of *C. neoformans*; and (5) it was protective against *C. neoformans* in vivo [33]. The laminarin-CRM197 vaccine is intriguing because it suggests that a vaccine can be developed against a common cross-reactive antigen shared by many pathogenic fungi. Although antibodies to β-glucans inhibit fungal growth in vitro, their effect in vivo may be immunomodulatory since their activity prevents the development of inflammation [34].

7.4 ANTIFUNGAL VACCINES AND THE IMMUNE SYSTEM

In this chapter we have discussed that fungal glycoconjugate vaccines elicit protective antibodies; however, we have not discussed how they elicit such antibodies. Fungal antigens, which contain a high number of polysaccharide repeats, are recognized by the B-cell receptor, but interactions with T cells are less vigorous. This type of interaction produces antibodies that are mainly IgM. These types of antigens are referred to as the T-cell-independent antigens. However, T-cell-independent antigens can be rendered more immunogenic when they are presented to the immune system as glycoconjugates. A polysaccharide coupled to a protein carrier will elicit the involvement of T-helper (T_H) cells that secrete cytokines; these are called T-cell-dependent antigens. Depending on the type of cytokine that is produced, the antibody response can shift from IgM, IgG (and subclasses), IgA, and IgE. Cytokines that are produced by glycoconjugates can elicit B-cell clonal expansion resulting in immunological memory if there is another encounter with the T-cell-dependent antigen. For example, there is a hierarchical efficacy in the clearance of *C. neoformans* in the host that is dictated by the type of antibody produced IgG1 > IgG3 > IgM > IgA [5].

So far we have mentioned that the role of a glycoconjugate is to shift the immune response from a T-cell-independent to a T-cell-dependent response that activates and involves T-helper cells in the response. Where do these T-helper cells come from? $CD4^+$ and $CD8^+$ T cells are crucial for the elimination of fungal infections. There are two distinct subclasses of $CD4^+$ cells T helper 1 (T_H1) and T helper 2 (T_H2) in mice. In primary fungal infection the T_H1 response is dominant, and these cells secrete cytokines such as Interferon-γ (INF-γ), Tissue Necrosis Factor-α (TNF-α), and Interleukin-12 (IL-12). In progressive disease such as occurs in an immunocompromised host, there is a shift from a (T_H1) to a (T_H2) response that is marked by an increase in IL-4, IL-5, and Il-10 cytokines. T_H1 responses promote the synthesis of

IgG2a and IgG3, whereas T_H2 cells promote the synthesis of IgM, IgG1, IgA, and IgE. Vaccine design is geared toward promoting a T_H1 response since the cytokines secreted by this response are associated with protective immunity.

Antibodies are thought to mediate host protection against fungi by mechanisms that are similar to those that protect against bacterial agents. Possible mechanisms of antibody efficacy involve their ability to serve as opsonins for the enhancement of phagocytosis and killing by macrophages, enhance antigen presentation, neutralize fungal toxins, inhibit fungal growth, modulate the host inflammatory response, and enhance the organization of T cells [35, 36]. The Ab dose appears to be a critical parameter for protection since high doses can cause dramatic prozone-like effects whereby larger doses are less protective than smaller doses. In *C. neoformans* dose–response studies have been carried out testing both IgM and IgG MAbs. In vivo studies using IgM, mAb 12A1 reduced survival observed in mice receiving large doses and resulted in enhanced brain infection. In vitro studies suggested that prozone effects were observed because large amounts of Ab in the capsule reduce opsonic efficacy and interfered with the microbicidal activity of nitrogen-derived oxidants [37]. In a second study IgG and subclasses MAbs were tested and large differences in cytokine responses were measured at high and low antibody dose. Whether prozone-like phenomena occur as a consequence of high titer response to vaccines is unknown. However, it is increasingly clear that antibody-mediated protection is observed only for certain antibody doses. Thus, a better understanding of the prozone effect may be essential for evaluating vaccine efficacy and designing better vaccines and for developing Ab-based therapies [38].

7.5 SUMMARY

Current vaccines are effective because they have the ability to prevent disease not infection [39]. The efficacy of a vaccine depends on inducing an immune response that can reduce host damage caused by host–microbe interactions during an infection [3]. The normal human host can successfully limit the spread of medically important fungi such as *Histoplama capsulatum*, *C. neoformans*, *C. albicans*, *A. fumigatus*, and *Coccidioides immitis* within tissues. The fact that infection with these organisms is common yet disease is rare provides considerable optimism that the immune system can be mobilized to prevent disease in susceptible hosts. Despite the relative efficacy of host defense mechanism, these fungi may remain as reservoirs or in a dormant state awaiting reactivation if host immunity becomes impaired. Hence, in addition to protecting against diseases from acute infection, the ideal fungal vaccine will also protect against reactivation. A vaccine of this type would be important for the immunocompromised because it would prevent the pathogenic fungus from reactivating. Another alternative is to design a vaccine that can limit latent fungal infection [5, 32].

One of the major challenges in the development of fungal vaccines is the type of immune response it will elicit. A vaccine that elicits strong immune responses can be deleterious due to an exuberant inflammatory response. Alternatively, a vaccine could also be deleterious by causing the inability to stimulate weak immune responses [3]. There is also concern that vaccines may not be effective in individuals who are immunocompromised and thus at high risk for fungal diseases.

Although there are currently no vaccines against fungal pathogens, many approaches are being developed to specifically target them and more recently to target more than one fungus with a single vaccine. Glycoconjugates are promising reagents in that they provide immunogenicity for a poor immunogenic antigen. These vaccines have also been used as tools to develop antibodies as reagents that can be used to study fungal organisms. An existing challenge to the development of fungal vaccines is that these have to be efficacious in immune-compromised patients. Highlighting this challenge, HIV-infected individuals have a deficit in the expression of the heavy-chain variable region of the antibodies encoded by the V_H3 immunoglobulin family. This dysregulation causes a reduction in the repertoire of antibodies to which capsular polysaccharides are generated. Furthermore, this deficit in expression not only translates to increased susceptibility to encapsulated pathogens but also impaired antibody responses to capsular polysaccharide vaccines in development [4]. Given the importance of the V_H3 antibody repertoire in the response to encapsulated microbes, it is hopeful that the status of this repertoire was improved by antiretroviral therapy [40].

In addition to glycoconjugate vaccines, live viable cells, cellular components, and mimetics have been used to elicit immune responses against fungi. Although many have elicited protection in the host, the fungi can remain viable at the site of vaccination or that cell antigens induce variable responses in different models, and these are dependent by the route of vaccination. In the case of peptide mimetics the mechanisms of how they work are not fully understood.

Although the understanding of fungal vaccines is incomplete, significant progress has been made, and there is considerable evidence that a conjugate vaccine can be developed. Nevertheless, the efficacy of such a vaccine in immunocompromised individuals remains uncertain, either because such patients respond poorly or the antibodies elicited are not as protective in the presence of impaired immunity [5]. However, the identification of new vaccine adjuvants could lead to the development of strategies to improve the response of immunocompromised patients.

ACKNOWLEDGMENT

We thank the American Society for Microbiology Journals Department for permission to reproduce Figure 7.2 from a work [26] published in the *Journal of Clinical Vaccine Immunology*.

REFERENCES

1. Dan JM, Levitz SM (2006) Prospects for development of vaccines against fungal diseases. *Drug Resist. Update* 9(3):105.
2. Pirofski LA, Casadevall A (1998) Use of licensed vaccines for active immunization of the immunocompromised host. *Clin. Microbiol. Rev.* 11(1):1.
3. Datta K, Pirofski LA (2006) Towards a vaccine for Cryptococcus neoformans: Principles and caveats. *FEMS Yeast Res.* 6(4):525.
4. Pirofski LA (2001) Polysaccharides, mimotopes and vaccines for fungal and encapsulated pathogens. *Trends Microbiol.* 9(9):445.

5. Deepe GS Jr (1997) Prospects for the development of fungal vaccines. *Clin. Microbiol. Rev.* 10(4):585.
6. Feldmesser M (2005) Prospects of vaccines for invasive aspergillosis. *Med. Mycol.* 43(7): 571.
7. Goldman DL, Lee SC, Casadevall A (1995) Tissue localization of Cryptococcus neoformans glucuronoxylomannan in the presence and absence of specific antibody. *Infect. Immun.* 63(9):3448.
8. Monari C, Bistoni F, Vecchiarelli A (2006) Glucuronoxylomannan exhibits potent immunosuppressive properties. *FEMS Yeast Res.* 6(4):537.
9. Kang, YS, Kim JY, Bruening SA, Pack M, Charalambous A, Pritsker A, Moran TM, Loeffler JM, Steinman RM, Park CG (2004) The C-type lectin SIGN-R1 mediates uptake of the capsular polysaccharide of Streptococcus pneumoniae in the marginal zone of mouse spleen. *Proc. Natl. Acad. Sci. U.S.A.* 101(1):215–220.
10. Baker PJ (1992) T cell regulation of the antibody response to bacterial polysaccharide antigens: An examination of some general characteristics and their implications. *J. Infect. Dis.* 165 (Suppl 1):S44.
11. Pirofski LA, Casadevall A (1996) Cryptococcus neoformans: Paradigm for the role of antibody immunity against fungi? *Zentralbl. Bakteriol.* 284(4):475.
12. Maitta RW, Datta K, Lees A, Belouski SS, Pirofski LA (2004) Immunogenicity and efficacy of Cryptococcus neoformans capsular polysaccharide glucuronoxylomannan peptide mimotope-protein conjugates in human immunoglobulin transgenic mice. *Infect. Immun.* 72(1):196.
13. Cherniak R, Valafar H, Morris LC, Valafar F (1998) *Cryptococcus neoformans* chemotyping by quantitative analysis of ^{1}H nuclear magnetic resonance spectra of glucuronoxylomannans with a computer-simulated artificial neural network. *Clin. Diagn. Lab. Immunol.* 5(2):146.
14. Cherniak R, Jones RG, Reiss E (1988) *Cryptococcus neoformans* serotype A-variant glucuronoxylomannan by 13C-n.m.r. spectroscopy. *Carbohydr. Res.* 172(1):113.
15. McFadden DC, De Jesus M, Casadevall A (2006) The physical properties of the capsular polysaccharides from Cryptococcus neoformans suggest features for capsule construction. *J. Biol. Chem.* 281(4):1868.
16. McFadden DC, Fries BC, Wang F, Casadevall A (2007) Capsule structural heterogeneity and antigenic variation in Cryptococcus neoformans. *Eukaryot. Cell.* 6(8):1464.
17. Murphy JW, Cozad GC (1972) Immunological unresponsiveness induced by cryptococcal capsular polysaccharide assayed by the hemolytic plaque technique. *Infect. Immun.* 5(6):896.
18. Lendvai N, Casadevall A, Liang Z, Goldman DL, Mukherjee J, Zuckier L (1998) Effect of immune mechanisms on the pharmacokinetics and organ distribution of cryptococcal polysaccharide. *J. Infect. Dis.* 177(6):1647–1659.
19. Kozel TR, Gotschlich EC (1982) The capsule of cryptococcus neoformans passively inhibits phagocytosis of the yeast by macrophages. *J. Immunol.* 129(4):1675.
20. Monari C, Pericolini E, Bistoni G, Casadevall A, Kozel TR, Vecchiarelli A (2005) Cryptococcus neoformans capsular glucuronoxylomannan induces expression of fas ligand in macrophages. *J. Immunol.* 174(6):3461–3468.
21. Cauley LK, Murphy JW (1979) Response of congenitally athymic (nude) and phenotypically normal mice to Cryptococcus neoformans infection. *Infect. Immun.* 23(3):644.
22. Gadelbusch HH (1958) Active immunization against Cryptococcus neoformans. *J. Infec. Dis.* 102:219.

23. Kozel TR, Gulley WF, Cazin J Jr (1977) Immune response to Cryptococcus neoformans soluble polysaccahride immunological unresponsiveness. *Infect. Immun.* 18(3):701.
24. Goren MB, Middlebrook GM (1967) Protein conjugates of polysaccharide from Cryptococcus neoformans. *J. Immunol.* 98(5):901.
25. Devi SJ, Schneerson R, Egan W, Ulrich TJ, Bryla D, Robbins JB, Bennett JE (1991) Cryptococcus neoformans serotype A glucuronoxylomannan-protein conjugate vaccines: Synthesis, characterization, and immunogenicity. *Infect. Immun.* 59(10):3700–3707.
26. De Jesus M, Hackett E, Durkin M, Connolly P, Casadevall A, Petraitiene R, Walsh TJ, Wheat LJ (2007) Galactoxylomannan does not exhibit cross-reactivity in the platelia Aspergillus enzyme immunoassay. *Clin. Vaccine. Immunol.* 14(5):624–627.
27. Valadon P, Nussbaum G, Boyd LF, Margulies DH, Scharff MD (1996) Peptide libraries define the fine specificity of anti-polysaccharide antibodies to Cryptococcus neoformans. *J. Mol. Biol.* 261(1):11.
28. Han Y, Morrison RP, Cutler JE (1998) A vaccine and monoclonal antibodies that enhance mouse resistance to Candida albicans vaginal infection. *Infect. Immun.* 66(12):5771.
29. Han Y, Cutler JE (1995) Antibody response that protects against disseminated candidiasis. *Infect. Immun.* 63(7):2714.
30. Han Y, Ulrich MA, Cutler JE (1999) Candida albicans mannan extract-protein conjugates induce a protective immune response against experimental candidiasis. *J. Infect. Dis.* 179(6):1477.
31. Torosantucci A, Bromuro C, Chiani P, De Bernardis F, Berti F, Galli C, Norelli F, Bellucci C, Polonelli L, Costantino P, Rappuoli R, Cassone A (2005) A novel glyco-conjugate vaccine against fungal pathogens. *J. Exp. Med.* 202(5):597–606.
32. Casadevall A, Pirofski LA (2006) Polysaccharide-containing conjugate vaccines for fungal diseases. *Trends Mol. Med.* 12(1):6.
33. Rachini A, Pietrella D, Lupo P, Torosantucci A, Chiani P, Bromuro C, Proietti C, Bistoni F, Cassone A, Vecchiarelli A (2007) An anti-beta-glucan monoclonal antibody inhibits growth and capsule formation of Cryptococcus neoformans in vitro and exerts therapeutic, anticryptococcal activity in vivo. *Infect. Immun.* 75(11):5085.
34. Casadevall A, Pirofski LA (2007) Antibody-mediated protection through cross-reactivity introduces a fungal heresy into immunological dogma. *Infect. Immun.* 75(11):5074.
35. Cutler JE, Deepe GS Jr, Klein BS (2007) Advances in combating fungal diseases: Vaccines on the threshold. *Nat. Rev. Microbiol.* 5(1):13.
36. Feldmesser M, Casadevall A (1998) Mechanism of action of antibody to capsular polysaccharide in Cryptococcus neoformans infection. *Front. Biosci.* 3:d136.
37. Taborda CP, Casadevall A (2001) Immunoglobulin M efficacy against Cryptococcus neoformans: Mechanism, dose dependence, and prozone-like effects in passive protection experiments. *J. Immunol.* 166(3):2100.
38. Taborda CP, Rivera J, Zaragoza O, Casadevall A (2003) More is not necessarily better: Prozone-like effects in passive immunization with IgG. *J. Immunol.* 170(7):3621.
39. Casadevall A, Pirofski LA (2005) Feasibility and prospects for a vaccine to prevent cryptococcosis. *Med. Mycol.* 43(8):667.
40. Subramaniam KS, Segal R, Lyles RH, Rodriguez-Barradas MC, Pirofski LA (2003) Qualitative change in antibody responses of human immunodeficiency virus-infected individuals to pneumococcal capsular polysaccharide vaccination associated with highly active antiretroviral therapy. *J. Infect. Dis.* 187(5):758.

8

CANCER-ASSOCIATED AND RELATED GLYCOSPHINGOLIPID ANTIGENS

Steven B. Levery

Department of Cellular and Molecular Medicine, Faculty of Health Sciences, University of Copenhagen, Copenhagen N, Denmark

8.1 INTRODUCTION

A considerable body of literature documents the expression of so-called tumor-associated, cancer-associated, carcinoembryonic, and onco-developmental glycosphingolipid (GSL) antigens [1–7]. However, the initial optimistic view that there could be one or more GSL antigens uniquely expressed in specific cancers, constituting by their mere appearance reliable indicators of pathology, has not been borne out, for example, by widespread commercial development of diagnostic methods based on detection of such antigens. Furthermore, incorporation of such GSL antigens into vaccine strategies has not yet contributed to development of a clinically successful therapeutic. Nevertheless, it is still abundantly clear that the overall pattern of GSL expression in a tissue can change dramatically during the process of carcinogenesis, just as it does during early development and differentiation events. At the same time, there are observable changes in the organization of membrane lipids, including GSLs, as well as increased shedding of cancer-associated GSL antigens into

Carbohydrate-Based Vaccines and Immunotherapies. Edited by Zhongwu Guo and Geert-Jan Boons
Copyright © 2009 John Wiley & Sons, Inc.

extracellular media, and concomitant increases in titers of circulating anti-GSL antibodies. Taken together, these changes could yet provide a basis for detection of cancer pathology and, with increased understanding of the immunological processes involved, future use in a successful vaccine is not out of the question.

The following discussion is an attempt to survey the structure-based literature on the expression of GSLs in cancer tissues, with some discussion of their immunological properties, emphasizing antigens that have been studied (or perhaps ignored) as potential targets of immunotherapeutic treatments. Detailed evaluations of their relative suitability, as well as the results and status of synthetic programs and clinical trials, have mainly been left for other chapters. Discussion of the normal expression and cancer-associated changes of the ABO histo-blood groups has not been attempted (for a review, see, e.g., [7]). Extensive discussion of the functional roles of GSLs in tumor pathobiology is also beyond the scope of this review. For the most part, the focus has been confined to GSLs that are primarily of interest as human antigens; and structural overlap with other glycoconjugates, while extensive in some areas, such as poly-LacNAc-based determinants, is only hinted at. The discussion as such is far from comprehensive; the volume of available literature has been daunting, and apologies are offered in advance for any issues or authors whose work has been left out.

8.2 STRUCTURAL CLASSIFICATION OF ANTIGENS

It might appear reasonable to list antigens according to tissue expression or pathological association, but it is far less complicated to associate them by structural criteria. The antigens listed herein (Table 8.1) are grouped according to their core sequences, with some groups further divided into neutral and sialic-acid-containing structures (gangliosides). These are the globo- (Gb; Galα4Galβ4Glcβ1Cer = Gb$_3$Cer), ganglio- (Gg; Galβ4Galβ4Glcβ1Cer = Gg$_3$Cer), and lacto-/neolacto-series (Lc and nLc, respectively; Galβ3GlcNAcβ3Galβ4Glcβ1Cer = Lc$_4$Cer; Galβ4GlcNAcβ3Galβ4Glcβ1Cer = nLc$_4$Cer), biosynthesized via divergent pathways from the common precursor, lactosylceramide (LacCer, Galβ4Glcβ1Cer) or, in the case of Lc/nLc, from the common triglycosylceramide intermediate, GlcNAcβ3Galβ4Glcβ1Cer (Lc$_3$Cer). An additional pathway from LacCer exists (iGb$_3$) in mammals but will not be discussed further here. Hybrid core structures also occur, and some of interest are included. This system, which is already in common usage for those mammalian GSLs based on processing of Glcβ1Cer (GlcCer), inherently encodes the connection between structures and processing pathways based on glycosyltransferases known to direct GSL biosynthesis.

8.3 "ABNORMAL" EXPRESSION OF GLYCOSPHINGOLIPID (GSL) GLYCAN STRUCTURES IN CANCER TISSUES

It is understood that GSL expression is a complex phenomenon, and that pathological or abnormal expression must be defined carefully with respect to the normal pattern of

Table 8.1 Structure Key

A. Globo-series

1. Gb₄: GalNAcβ3Galα4Galβ4Glcβ1Cer
2. Gb₅: Galβ3GalNAcβ3Galα4Galβ4Glcβ1Cer
3. Globo-H: Fucα2Galβ3GalNAcβ3Galα4Galβ4Glcβ1Cer
4. Sialosyl-Gb₅: NeuAcα3Galβ3GalNAcβ3Galα4Galβ4Glcβ1Cer
   NeuAcα6
5. Disialosyl-Gb₅: NeuAcα3Galβ3GalNAcβ3Galα4Galβ4Glcβ1Cer

B. Ganglio-series gangliosides

6. G_{M3}: NeuAcα3Galβ4Glcβ1Cer
7. G_{M2}: GalNAcβ4Galβ4Glcβ1Cer
   NeuAcα3
8. G_{D2}: GalNAcβ4Galβ4Glcβ1Cer
   NeuAcα8NeuAcα3
9. G_{D3}: NeuAcα8NeuAcα3Galβ4Glcβ1Cer
10. G_{T3}: NeuAcα8NeuAcα8NeuAcα3Galβ4Glcβ1Cer
11. 9-O-Acetyl-G_{D3}: 9-O-Ac-NeuAcα8NeuAcα3Galβ4Glcβ1Cer
12. Fucosyl-G_{M1}: Fucα2Galβ3GalNAcβ4Galβ4Glcβ1Cer
    NeuAcα3

C. Lacto-series (Type 1)

13. Leᵃ: Galβ3GlcNAcβ3Galβ4Glcβ1Cer
    Fucα4
14. Leᵇ: Fucα2Galβ3GlcNAcβ3Galβ4Glcβ1Cer
    Fucα4
15. Leᵃ-Leᵃ: Galβ3GlcNAcβ3Galβ3GalNAcβ3Galβ4Glcβ1Cer
    Fucα4   Fucα4
16. Leᵇ-Leᵃ: Fucα2Galβ3GlcNAcβ3Galβ3GalNAcβ3Galβ4Glcβ1Cer
    Fucα4   Fucα4

D. Lacto-series (Type 1) gangliosides

17. Monosialosyl-Lc₄: NeuAcα3Galβ3GlcNAcβ3Galβ4Glcβ1Cer
    NeuAcα6
18. Disialosyl-Lc₄/FH9: NeuAcα3Galβ3GlcNAcβ3Galβ4Glcβ1Cer
19. SLeᵃ: NeuAcα3Galβ3GlcNAcβ3Galβ4Glcβ1Cer
    Fucα4

(Continued)

Table 8.1 Continued

20. Disialosyl-Le^a/FH7:	NeuAcα6 NeuAcα3Galβ3GlcNAcβ3Galβ4Glcβ1Cer 　　　　　　　　Fucα4
E. Neolacto-series (Type 2)	
21. Le^x:	Galβ4GlcNAcβ3Galβ4Glcβ1Cer 　　Fucα3
22. Le^x-Le^x/FH4:	Galβ4GlcNAcβ3Galβ4GlcNAcβ3Galβ4Glcβ1Cer 　Fucα3　　　　　　Fucα3
23. "Y2":	Galβ4GlcNAcβ3Galβ4GlcNAcβ3Galβ4Glcβ1Cer 　　　　　　　　Fucα3
24. Tri-Le^x:	Galβ4GlcNAcβ3Galβ4GlcNAcβ3Galβ4GlcNAcβ3Galβ4Glcβ1Cer 　Fucα3　　　　　Fucα3　　　　　Fucα3
25. "Z2a":	Galβ4GlcNAcβ3Galβ4GlcNAcβ3Galβ4GlcNAcβ3Galβ4Glcβ1Cer 　Fucα3　　　　　Fucα3
26. "Z1":	Galβ4GlcNAcβ3Galβ4GlcNAcβ3Galβ4GlcNAcβ3Galβ4Glcβ1Cer 　　　　　　　　Fucα3
27. "Z3":	Galβ4GlcNAcβ3Galβ4GlcNAcβ3Galβ4GlcNAcβ3Galβ4Glcβ1Cer 　Fucα3
28. Le^y:	Fucα2Galβ4GlcNAcβ3Galβ4Glcβ1Cer 　　　　　Fucα3
29. "extended Le^y":	Fucα2Galβ4GlcNAcβ3Galβ4GlcNAcβ3Galβ4Glcβ1Cer 　　　　　Fucα3
30. Le^y-Le^x:	Fucα2Galβ4GlcNAcβ3Galβ4GlcNAcβ3Galβ4Glcβ1Cer 　　　　　Fucα3　　　　　Fucα3
F. Neolacto-series (Type 2) gangliosides	
(I) Miscellaneous Short-Chain and Branched Poly-LacNac	
31. IV^{3}NeuAc-nLC$_4$:	NeuAcα3Galβ4GlcNAcβ3Galβ4Glcβ1Cer
32. IV^{6}NeuAc-nLC$_4$:	NeuAcα6Galβ4GlcNAcβ3Galβ4Glcβ1Cer
33. VI^{3}NeuAc-nLC$_6$:	NeuAcα3Galβ4GlcNAcβ3Galβ4GlcNAcβ3Galβ4Glcβ1Cer
34. VI^{6}NeuAc-nLC$_6$:	NeuAcα6Galβ4GlcNAcβ3Galβ4GlcNAcβ3Galβ4Glcβ1Cer
35. 'G-8':	Galβ4GlcNAcβ6 NeuAcα3Galβ4GlcNAcβ3Galβ4GlcNAcβ3Galβ4Glcβ1Cer

36. 'G-10'/NUH-2:

NeuAcα3Galβ4GlcNAcβ6
NeuAcα3Galβ4GlcNAcβ3Galβ4GlcNAcβ3Galβ4Glcβ1Cer

37. SLe^x:

NeuAcα3Galβ4GlcNAcβ3Galβ4Glcβ1Cer
               Fucα3

38. G14-1:

NeuAcα3Galβ4GlcNAcβ3Galβ4GlcNAcβ3Galβ4Glcβ1Cer
               Fucα3

39. VIM-2/ACFH-18/G14-2:

NeuAcα3Galβ4GlcNAcβ3Galβ4GlcNAcβ3Galβ4Glcβ1Cer
                                 Fucα3

40. SLe^x-Le^y/6B/FH6/G17-1:

NeuAcα3Galβ4GlcNAcβ3Galβ4GlcNAcβ3Galβ4Glcβ1Cer
               Fucα3   Fucα3

41. G10-1:

NeuAcα6Galβ4GlcNAcβ3Galβ4Glcβ1Cer
               Fucα2

42. 6C/G15:

NeuAcα6Galβ4GlcNAcβ3Galβ4GlcNAcβ3Galβ4Glcβ1Cer
                                 Fucα3

43. G17-2:

NeuAcα6Galβ4GlcNAcβ3Galβ4GlcNAcβ3Galβ4Glcβ1Cer
               Fucα2   Fucα3

(II) Long-Chain Poly-LacNac "Myeloglycans" and Related/Isomeric Structures

44.

NeuAcα3Galβ4GlcNAcβ3Galβ4GlcNAcβ3Galβ4GlcNAcβ3Galβ4Glcβ1Cer
               Fucα3

45.

NeuAcα3Galβ4GlcNAcβ3Galβ4GlcNAcβ3Galβ4GlcNAcβ3Galβ4Glcβ1Cer
                                 Fucα3

46.

NeuAcα3Galβ4GlcNAcβ3Galβ4GlcNAcβ3Galβ4GlcNAcβ3Galβ4Glcβ1Cer
                                                 Fucα3

47.

NeuAcα3Galβ4GlcNAcβ3Galβ4GlcNAcβ3Galβ4GlcNAcβ3Galβ4Glcβ1Cer
               Fucα3

48.

NeuAcα3Galβ4GlcNAcβ3Galβ4GlcNAcβ3Galβ4GlcNAcβ3Galβ4Glcβ1Cer
               Fucα3

49.

NeuAcα3Galβ4GlcNAcβ3Galβ4GlcNAcβ3Galβ4GlcNAcβ3Galβ4Glcβ1Cer
               Fucα3   Fucα3

50.

NeuAcα3Galβ4GlcNAcβ3Galβ4GlcNAcβ3Galβ4GlcNAcβ3Galβ4Glcβ1Cer
               Fucα3   Fucα3

51.

NeuAcα3Galβ4GlcNAcβ3Galβ4GlcNAcβ3Galβ4GlcNAcβ3Galβ4Glcβ1Cer
               Fucα3   Fucα3

(Continued)

Table 8.1 Continued

52.	NeuAcα3Galβ4GlcNAcβ3Galβ4GlcNAcβ3Galβ4GlcNAcβ3Galβ4Glcβ1Cer
	Fucα3 Fucα3
53.	NeuAcα3Galβ4GlcNAcβ3Galβ4GlcNAcβ3Galβ4GlcNAcβ3Galβ4GlcNAcβ3Galβ4Glcβ1Cer
	Fucα3
54.	NeuAcα3Galβ4GlcNAcβ3Galβ4GlcNAcβ3Galβ4GlcNAcβ3Galβ4Glcβ1Cer
	Fucα3 Fucα3
55.	NeuAcα3Galβ4GlcNAcβ3Galβ4GlcNAcβ3Galβ4GlcNAcβ3Galβ4Glcβ1Cer
	Fucα3
56.	NeuAcα3Galβ4GlcNAcβ3Galβ4GlcNAcβ3Galβ4GlcNAcβ3Galβ4Glcβ1Cer
	Fucα3 Fucα3

G. Lacto/neolacto and (neo)lacto/ganglio hybrid cores

57. Le^a-Le^x:	Galβ3GlcNAcβ3Galβ4GlcNAcβ3Galβ4Glcβ1Cer
	Fucα4 Fucα3
58. GalNAcβ4-Monosialosyl-Lc$_4$:	GalNAcβ4Galβ3GlcNAcβ3Galβ4Glcβ1Cer
	NeuAcα3
	NeuAcα6
59. GalNAcβ4-Disialosyl-Lc$_4$:	GalNAcβ4Galβ3GlcNAcβ3Galβ4Glcβ1Cer
	NeuAcα3

expression within a particular tissue, organism, state of development and differentiation, and the like. Among mammalian GSLs, one rarely finds a completely unique structure observed only in a particular cancer tissue or cell line and nowhere else. However, (1) it may be expressed in a cancer tissue at a level many times that found in the nonpathological state of the same tissue; (2) it may have a structure normally expressed by the organism but in a different tissue or stage of development; (3) it may be normally expressed in the same tissue of a different species; or (4) there may be a change in the organization of the cancer cell surface that permits immune recognition of a normally cryptic GSL component. The latter circumstance could result from a variety of factors, for example, from quantitative or qualitative changes in the overall pattern of cell surface GSL expression, or from changes in quantity or structure of other molecules found in noncovalent association with GSLs.

Because GSLs are not direct gene products, alterations in their expression patterns need to be interpreted with respect to induction, amplification, or suppression of glycosyltransferase activities. Hakomori [6] defined two descriptive mechanisms for aberrant GSL glycosylation: "(i) incomplete synthesis of certain GSLs with associated accumulation of precursor, or (ii) enhanced 'neosynthesis' of GSLs which are minimal or absent in normal cells or tissues." (p. 633). It is worth noting that (ii) could result directly from expression or up-regulation of one or more glycosyltransferases normally unexpressed or minimally active in the relevant GSL pathway; or it could result from (i), that is, a precursor normally used as the substrate for a deleted or suppressed glycosyltransferase activity could be utilized as the substrate for another glycosyltransferase, normally expressed but poorly competitive with the first. In this latter case, precursor accumulation may be less apparent due to shunting to another available pathway. Either case of neosynthesis could be observed as a "pathway-switching" event.

In this light it is important to be aware that the key transferase(s) involved may not necessarily be the most obvious [3]. For example, it has been proposed that the elevated enzyme activity most responsible for accumulation of $(Le^x)_n$, Le^y-$(Le^x)_n$, NeuAcα3-$(Le^x)_n$, Le^a-Le^x, Le^a-Le^a, or NeuAcα3-$(Le^a)_2$ structures in some colonic cancers is not β1,4 or β1,3Gal-T, α2,3NeuAc-T, α1,4 or α1,3Fuc-T but the β1,3GlcNAc-T, which is normally rate limiting for the production of $[Gal\beta1 \to 4/3GlcNAc\beta1 \to 3]_n$ substrates [8]. On the other hand, enhanced expression of terminal NeuAcα2 $\to$ 6Galβ1 $\to$ 4GlcNAcβ1 $\to$ 3 in gangliosides of human colonic adenocarcinoma [9, 10], and of Fucα1 $\to$ 2G_{M1} and Fucα1 $\to$ 2(Galα1 $\to$ 3)G_{M1} gangliosides in rat hepatoma and precancerous liver tissue [11], do correlate with elevated activities of α2,6NeuAc-T [12, 13] and a specific α1,2Fuc-T [14, 15], respectively, in these tissues. It is also possible that changes in the physical organization of the biosynthetic pathway may bring glycosyltransferase in contact with a GSL substrate that it normally never "sees." Some circumstantial evidence for this has been found in observation of abnormal ceramide as well as glycan distributions in GSLs of cancer tissues. A partial list of GSLs proposed to be cancer-associated antigens is appended.

As mentioned above, qualitative and quantitative changes in cell surface distribution of GSLs may be accompanied by changes of membrane organization [1, 3], as well as increased shedding of GSLs into the extracellular space [16–21]. These could result in (1) abnormal interactions of GSLs with intrinsic membrane proteins, such as those involved in transmembrane signal transduction; (2) activation of immune response; (3) suppression of immune response [18, 22–25]; (4) increased accessibility of certain GSLs to (a) antibodies, or other molecules with an affinity for GSL glycans, such as (b) adhesion receptors; and (5) suppression of intercellular communications responsible for contact inhibition and proper tissue differentiation. All of these [except 2 and 4(a)] could serve to mediate tumor progression and metastasis [3].

8.4 DISCUSSION OF SELECTED ANTIGENS

8.4.1 Globo-Series and Related Antigens

8.4.1.1 Globo-H A putative globo-H ($V^2Fuc\alpha$-Gb_5Cer; **3**) GSL antigen was originally identified, although incompletely characterized, as a component of human meconium [26] (the aglyconic hydroxyl of the Galβ $\rightarrow$ GalNAc linkage was not specified). Thereafter, it was identified and unambiguously characterized among a series of "extended globo-series" GSLs (i.e., containing the Gb_5 = Galβ3GalNAcβ3Galα4Galβ4Glcβ1Cer core; **2**; see further discussion below) expressed by a human teratocarcinoma cell line, 2102Ep, distinguished by their reactivity with a monoclonal antibody (mAb) directed to SSEA-3 (SSEA = stage-specific embryonic antigen) [27–29]. Globo-H was subsequently characterized as a major component of the breast cancer cell line MCF-7 [30], having been identified by its strong reaction with mAb MBr1 [31, 32]. It was observed to be strongly expressed in breast cancer and a variety of epithelial neoplasms. MBr1 staining was also observed with normal mammary gland ductal epithelia and apocrine sweat gland, among a limited variety of tissues tested [31]. Not long after, it was identified as an MBr1-reactive minor component of human O erythrocytes, and later as a significant component of normal human kidney tissue [33]. Zhang et al. [4], in their immunochemical study of the tissue expression of a variety of tumor antigens as possible targets for anticancer immunotherapy, observed MBr1 staining of normal human lung, breast, stomach, pancreas, uterus, ovary, and prostate epithelia (it is worth noting that significant cross-reactivity of MBr1 was observed with Fucα2-asialoG_{M1} [30]). Despite the wide spectrum of MBr1 cross-reactivity, it was considered a good target for vaccine development [4] and has been incorporated into a wide variety of synthetic strategies and clinical trials, as discussed in Chapter 11.

8.4.1.2 Gb_5Cer, Monosialosyl-Gb_5Cer, and Disialosyl-Gb_5Cer A putative Galβ3-extended globopentaosylceramide antigen was originally identified, although not unambiguously characterized, as a component of a green monkey kidney cell line, GMK AH-1 [34]. The authors suggested that this structure, together with the proposed globo-H GSL from human meconium [26], pointed to a previously unsuspected complexity in the globo-series, whose only pentaglycosylceramide had been thought to be the Forssmann antigen (GalNAcα3GalNAcβ3Galα4Galβ4Glcβ1Cer) [35]. Thereafter, a Galβ3-extended globopentaosylceramide antigen was similarly identified among the anti-SSEA-3 [27, 28] reactive globo-series

GSLs from 2102Ep cells and unambiguously characterized as Galβ3GalNAc β3Galα4Galβ4Glcβ1Cer (Gb$_5$, **2**) [29].

Monosialosyl-Gb$_5$Cer (also referred to as monosialosyl-galactosyl-globoside, MSGG) was also identified as an SSEA-3 antigen in 2102Ep cells and unambiguously characterized as V^{3}NeuAcα-Gb$_5$Cer (**4**) [29]; it was also identified as the ganglioside component reactive with another mAb defining a stage-specific embryonic antigen, SSEA-4 [36]. The same year, MSGG was identified as a ganglioside of normal chicken skeletal muscle [37]. It was subsequently identified as a component of renal carcinoma tissue reactive with mAb RM1 established by immunization of mice with cells of a renal carcinoma cell line (from lung metastatic deposit) TOS-1 [38]. MSGG has also been established as a component of normal human kidney, which is a preferred binding site for uropathogenic *Escherichia coli* [39].

The structure and immunochemistry of disialosyl-Gb$_5$Cer (also referred to as disialosyl-galactosyl-globoside, DSGG) have been the subject of several misunderstandings, which are all the more significant since it has been regarded as a promising marker of metastatic potential in renal cell carcinoma [40, 41]. DSGG was originally identified as a disialosyl-ganglioside of human erythrocytes by Kundu et al. [42], who proposed a structure with both NeuAc residues attached to the terminal Gal residue of Gb$_5$, that is, V^{3}NeuAcαV^{6}NeuAcα-Gb$_5$Cer. It was later reported as a second globo-series ganglioside of chicken skeletal muscle and assigned the same structure [43]. Its subsequent identification, along with MSGG, as a component of renal carcinoma tissue led to a thorough reinvestigation of the structure from all three sources. The renal carcinoma antigen was identified as V^{3}NeuAcα,IV^{6}NeuAcα-Gb$_5$Cer (**5**); in other words, the NeuAcα2 → 6 residue was linked to GalNAcβ3, not Galβ3 [38]. At the same time, the structures of both human erythrocyte and chicken muscle DSGG were proven as well to be V^{3}NeuAcα,IV^{6}NeuAcα-Gb$_5$Cer [44].

The DSGG of renal carcinoma had initially been identified by its reactivity with RM2, a second mAb established by immunization with TOS-1 cells [38]. While MSGG and DSGG are both found in normal kidney, their expression appeared to be restricted to cells and tissues of mesodermal origin [38]. By immunohistochemical staining using mAbs RM1 and RM2, directed to MSGG and DSGG, respectively, all renal carcinomas were found to express high levels of MSGG, while DSGG was observed in some cases and not others (6 out of 12 tested). It was proposed that this difference could have potential implications for the pathobiology of renal carcinoma, related mechanistically to, as well as providing a means for distinguishing, their relative malignancy, that is, metastatic potential and invasiveness [38, 45]. Further work suggested that metastasis of renal cell carcinoma to lung could be mediated by a specific, non E/P (endothelial/platelet)-selectin, receptor for DSGG expressed by lung tissue [40].

The terminal tetrasaccharide sequence of DSGG is identical to that of GD1α (IV^{3}NeuAcα,III^{6}NeuAcα-Gg$_4$Cer) and to the disialosyl-T antigen, except for the configuration of the GalNAc anomeric center. However, mAb RM2 was observed to react only with DSGG, among the three disialosyl antigens tested, suggesting the importance of the carrier structure to the antigenicity of otherwise similar determinants, especially when sampled in situ at the cell surface, for example, immunohistologically, or as the ligand for a complementary carbohydrate-binding protein (CBP) receptor [38]. The differences could result from a combination of differences in (1) the structure of the carrier itself, (2) the relative orientation of the determinant with respect to the carrier, and (3) the spatial presentation of the determinant at the cell membrane surface. Molecular modeling using a simple hard-sphere exo-anomeric (HSEA) force-field mechanics program helped to visualize these differences [38].

All of the points brought forward in this work appear to be valid, even though they were based on incomplete data and several faulty assumptions. As it turned out, TOS-1 cells, the immunogen employed to generate mAbs RM1 and RM2, do not express DSGG significantly [41, 46]. Instead, their major disialoganglioside components were found to be $IV^3NeuAc\alpha,III^6NeuAc\alpha$-$Lc_4Cer$ (**18**; disialosyl-Lc_4, previously observed as a colonic cancer-associated antigen recognized by mAb FH9 [47]; see later section) and a novel hybrid core antigen, $IV^4GalNAc\beta,IV^3NeuAc\alpha,$ $III^6NeuAc\alpha$-Lc_4Cer (**59**; "G1") [46]. The terminal trisaccharide structure of this novel GSL is identical to that of G_{M2} ganglioside (**7**), while the internal hexaglycosyl-ceramide without the GalNAc residue is identical to disialosyl-Lc_4Cer (**18**), the FH9 antigen (a related monosialosylated antigen, $IV^4GalNAc\beta,IV^3NeuAc\alpha$-$Lc_4Cer$ (**58**), has been detected in human meconium [48]). It further turned out that mAb RM2 reacts strongly with the novel TOS-1 heptaglycosylceramide, G1, but not significantly with DSGG, which was purified for this study from another metastatic renal carcinoma cell line, ACHN [46]. ACHN cells, which express very little of the Lc_4-core-based antigens, were in turn used as an immunogen to generate another mAb, 5F3, which was then with great care shown to be specific for DSGG [41] and not cross-reactive with the newly discovered RM2 antigen G1.

Together, mAbs 5F3, anti-SSEA-3, MK1-8 (directed to G_{M2} ganglioside), and RM2 were used to clarify the ganglioside expression of renal carcinoma tissue, as well as explain the earlier mistaken assignment of the specificity of the latter mAb. Ito et al. [41, 46] suggested that the confusion could be explained by the fact that (1) G1 and DSGG have similar, overlapping relative retention factor (Rf) values in high performance thin-layer chromatography (HPTLC), which was used for the earlier purification of the putative RM2 antigen, and (2) the proportion of globo-series and ganglio/lacto-series antigens varies widely among renal carcinoma tissue samples, from which the putative RM2 antigen was purified. Some express predominantly one type or the other; while some cases might express both significantly, it is clear that pooling of samples would inevitably result in obtaining a mixture. A sample thus prepared for the purpose of structure elucidation could appear to be essentially pure DSGG, while a spectroscopically undetectable trace of G1 could have given the appearance that the isolate was indeed the RM2-reactive antigen.

Immunohistochemical staining of human renal carcinoma and normal kidney tissues with 5F3 and RM2 showed some interesting patterns: In normal kidney, the two mAbs showed mutually exclusive staining of the luminal epithelia of renal tubules, 5F3 staining the proximal tubules and RM2 staining the distal [41]. About half the primary renal carcinoma samples (19/41) exhibited positive staining with either one or the other mAb, and a quarter (10/41) exhibited staining with both. Of these, 8 cases were also positive with RM1, directed to MSGG, with metastasis observed in 6 of these (unpublished data reported in Ito et al. [41]). Both mAbs also stained a variety of normal tissues, notably gastrointestinal (GI) and colonic goblet cells and stroma, and prostate stroma, but not epithelial cells from these tissues; smooth muscle from bladder, ureter, and GI tract were stained with 5F3 but not RM2; lung, thyroid, aorta, liver, pancreas, and spleen were all negative with both mAbs, although the number of samples studied was relatively small [41]).

8.4.2 Ganglio-Series Antigens

Some terminal epitopes of ganglio-series antigens are shared with glycoprotein glycans.

8.4.2.1 G_{M3}, G_{M2}, G_{D3}, G_{D2}, 9-O-acetyl-G_{D3}, and 9-O-acetyl-G_{D2}

Along with other gangliosides, G_{M3} (**6**) is overexpressed in some cancer tissues, notably melanoma [49, 50]. It has also been found in bladder tumors, although high expression was associated with superficial rather than invasive tumors [51]. Based on its widespread expression in normal cells, such as neural tissues, Zhang et al. [4] concluded that, as with $G_{M1(a)}$ and G_{D1a}, it would not be worthwhile to pursue as a target for immunotherapy. On the other hand, Hakomori and co-workers [52] have stressed that its cell surface presentation and organization, and consequent immune recognition, may be density dependent, and have proposed it as a therapeutic target. This hypothesis has been supported by studies using a murine B16 melanoma model [53]. A G_{M3}-conjugate vaccine was described [54] and studies in the B16 melanoma model supported the role of this ganglioside in induced immune response [55]. The current status of this vaccine is discussed in Chapters 9 and 11. Ohyama and co-workers [51, 56, 57] have proposed G_{M3} as both a negative indicator of potential invasiveness in bladder tumors and a target for therapeutic intervention via administration or induced overexpression.

Another line of study is suggested by the reported accumulation of gangliosides and other glycoconjugates incorporating *N*-glycolyl neuraminic acid (NeuGc), which cannot be synthesized by *Homo sapiens* [58, 59], in a variety of human cancer tissues [60–65]. The NeuGc accumulating in these tissues is hypothesized to be dietary in origin [66]. Based on the detection of NeuGc-G_{M3} in some human cancers, for example, of colon [60], breast [67, 68], and in retinoblastoma [69] and melanoma [61, 68], it has been suggested as a potential immunotherapeutic target [70, 71].

G_{M3}, G_{M2} (**7**), and G_{D3} (**9**) were observed long ago as characteristic antigens of human melanoma, for example, by Portoukalian et al. [49, 72]. G_{D3} has been suggested as an immunotherapeutic target for melanoma [73], despite its known pattern of distribution in normal tissues, notably the central nervous system (CNS) and connective tissue of various organs [4]. It was noted that G_{D3} in plasma was elevated about four times normal levels in patients with malignant melanoma [74]. Related to this, Bernhard et al. [75] studied the shedding of gangliosides into media by cultured human melanoma cells in vitro. They found that in general the profiles of gangliosides in the media matched those of the cell lines grown in them, G_{D3} and G_{M3} being observed as prominent components in most cases.

A few years after the initial observation of G_{M3}, G_{M2}, and G_{D3} in melanoma, G_{D2} (**8**) [49] and an alkali-labile ganglioside, which ultimately proved to be 9-O-acetyl-G_{D3} (**11**), were also reported; the latter, which could be missed by commonly applied analytical protocols using strong base at some stage before detection, was strongly recognized by an mAb, D1.1 [76, 77]. The recognition by D1.1 was abolished with base treatment, demonstrating the 9-O-acetyl moiety as a crucial component of the epitope. A later survey of melanoma samples by Hamilton et al. [50] confirmed the consistent presence of these components, along with an additional ganglioside, G_{T3} (**10**). Tsuchida et al. [78] compared ganglioside expression between melanoma biopsy samples and melanoma-derived cell lines and found that, against a background of considerable heterogeneity of expression, G_{M2} and G_{D2} were major components of many melanoma cell lines, but minor components of biopsied melanomas. The alkali-labile 9-O-acetyl-G_{D3} was found to be a consistent component of the melanoma biopsy samples, confirming the importance of this ganglioside as a melanoma marker. In a broad immunohistochemical study of normal and cancer tissues, Zhang et al. [4] confirmed that the expression of 9-O-acetyl-G_{D3} is almost completely restricted to melanoma, although connective tissues of a few organs (prostate, stomach, ovary) appeared to be stained by mAb D1.1 in their study. It was also detected in 1/5 breast tumor samples, consistent with the detection and mass spectrometric confirmation of an O-acetylated G_{D3} present in gangliosides extracted from breast tumor tissue [67]. Melanoma cells may also express small amounts of 9-O-acetyl-G_{D2} [50, 79], and O-acetylated G_{D2} has been detected as well in neuroblastoma and other tumors [80].

The potential lability of 9-O-acetyl-G_{D3} might appear to render it less attractive as a vaccine candidate, but it has been considered [4, 81–83]. It was observed by Ravindranath et al. [81] that a "melanoma cell vaccine" (MCV) administered to melanoma patients induced serum antibodies cross-reacting to both 9-O-acetyl-G_{D3} and G_{D3}, whereas a purified G_{D3} vaccine had not induced an effective immune response in human patients [84]. It was hypothesized that the 9-O-acetyl modification of a sialic acid residue could render the molecule carrying it more antigenic in humans, analogous to the effect of N-glycolylation [61], even if the 9-O-acetyl group was not itself an essential component of the natural epitope [81]. Their results do not conflict with the general lack of human immune response to G_{D3} compared with, for example, G_{M2}, observed earlier by Tai et al. [85] and more or less confirmed by subsequent experiences [86]. On the other hand, they do not explain why Tai et al. [85] failed to induce an anti-G_{D3} response, when they had also immunized patients with a "tumor cell vaccine" (TCV) that contained a mixture of gangliosides extracted from the same cultured melanoma cell lines as used later by Ravindranath et al. [81].

However, a close comparison of the protocols shows that the MCV used in the later study consisted of whole cells killed by irradiation, while the TCV used in the earlier study was a purified acidic GSL extract that had been treated by strong base as a final step before administration. The results could not have reflected, therefore, any effect of 9-O-acetyl-G_{D3} since that had been destroyed. Nevertheless, a later study by Ritter et al. [83] failed to show that a general response to G_{D3}/9-O-acetyl-G_{D3} could be elicited by immunization with purified 9-O-acetyl-G_{D3} from various sources. They found that immunization of patients with more highly O-acetylated G_{D3} components was more effective in that it did induce an IgM/IgG response—however, the serological reaction appeared to be to 7-O-acetylation of G_{D3} and did not extend to 9-O-acetyl-G_{D3} or G_{D3}. The same group also observed that immunization with other modified G_{D3} derivatives (lactones, amides, carboxyl-reduced) induced specific responses to the respective modified immunogens, but cross-reactivity to G_{D3} itself was absent [87]. It should be noted that none of the conflicting studies employed a whole cell vaccine as used by Ravindranath et al. [81]; whether this is a significant difference is unclear, but it is doubtful that an MCV of this nature in the long run represents a practical therapeutic. In the meantime, approaches to boosting the immune response to G_{D3} are still being investigated (e.g., [88]).

By now, considerable attention has been given over to G_{M2} as a target for immunotherapy of melanoma and other tumors of neuroectodermal origin, particularly by Livingston and co-workers [4, 86, 89–96]. Tai et al. [85] considered the relative immunogenicity of melanoma gangliosides, as determined by titer of circulating antiganglioside antibodies in sera of 26 melanoma patients who were immunized with a tumor-cell vaccine prepared from cultured melanoma cells. None of the patients developed anti-G_{M3} or anti-G_{D3} titers, while about 40% developed IgM anti-G_{M2}; 2 patients developed IgG anti-G_{M2} and 2 developed IgM anti-G_{D2} [85]. From this it could be concluded that G_{M2} is the most immunogenic ganglioside expressed on human melanoma cells in humans. It was pointed out that G_{M3} and G_{D3} are present in human sera and in human normal tissues, which might establish immunologic tolerance, or circulating G_{M3} and G_{D3} could neutralize anti-G_{M3} and anti-G_{D3} antibodies if these had been produced. In this connection it is worth noting that Kundu et al. [97] reported isolation of 10 distinct neutral GSLs and 15 gangliosides from human plasma. Of the gangliosides, G_{M3} was reported to be by far the most abundant, followed by G_{D3}, G_{D1a}, G_{T1b}, and G_{Q1b} [97]. The concentration of G_{M3} in plasma was estimated by Lin et al. [98] to be $\sim$8 nmol/ml in sera of both nonpregnant and pregnant women. This is roughly equivalent to 10 μg/ml, or 2.5 μg/ml G_{M3}-bound NeuAc. Kundu et al. [97] estimated plasma levels of lipid-bound NeuAc to be $\sim$2 μg/ml. Sela et al. [99] monitored levels of G_{D2}, G_{D3}, and 9-O-acetyl-G_{D3} immunochemically in sera of patients with melanoma and healthy adults. In their assay, G_{D2} was detectable in normal sera at 2 ng/ml; serum levels of G_{D2} and G_{D3} were increased approximately six-fold and five-fold, respectively, in patients with disseminated melanoma, compared with those of healthy adults. Interestingly, the key melanoma-specific antigen 9-O-acetyl-G_{D3} was not detected in serum in this study.

G_{M2} is expressed in normal brain and luminal epithelial cells of glandular tissues [4, 50]; it is also characteristically expressed or overexpressed in a wide variety of tumor tissues, including melanoma and other tumors of neuroectodermal origin [50, 100]. Higashi et al. [101] detected G_{M2} in sera and tumor tissues of hepatoma patients. The G_{M2} in sera from these patients was detected at levels 20–100 times higher than in sera from normal adults. Zhang et al. [4] detected G_{M2} by immunohistochemical staining with an anti-G_{M2} mAb, 696, in all tumor tissues tested; these included tumors of both neuroectodermal and epithelial origins. G_{M2} is the major ganglioside component of a human medulloblastoma cell line, Daoy, along with lesser amounts of G_{M3} and G_{D1a} [102]; moreover, Daoy and two other human medulloblastoma cells were found to shed these gangliosides profusely into culture media [21].

The related disialosyl-ganglioside G_{D2} is also expressed in a variety of cancers and normal tissues [4]. Similar to G_{D3}, it is widely distributed in connective tissues and the CNS, and is a characteristic component of peripheral nerve [4]. As mentioned earlier, it is observed as a component of human melanoma [49] and is capable of evoking an immune response, although the response appears less reliable than against G_{M2} [85]. Along with other gangliosides, G_{D2} was observed to be highly elevated in human retinoblastoma [103]; moreover, it was found to be elevated in the sera of retinoblastoma patients, suggesting that it might be shed from the tumor, and that it might be a useful serum biomarker for the disease. It is also highly expressed in primary neuroblastoma, and has been suggested as both a marker and an immunotherapeutic target antigen for this tumor [104]. Notably, the levels of G_{D2} are elevated in serum and plasma samples from patients with neuroblastoma [19, 105, 106]. The phenomenon of ganglioside shedding by tumors of neuroectodermal origin, notably of G_{M2} and G_{D2} from medulloblastoma and neuroblastoma, respectively, of G_{M3} and G_{D3} from melanoma, and of G_{D3} from glioma, has been hypothesized to facilitate tumor survival and proliferation in the host by an immunosuppressive mechanism, possibly aided by angiogenic and platelet activating activities [18, 19, 21, 75, 105, 107–112].

8.4.2.2 Fucosyl-G_{M1}

Although detected in normal animal tissues and prominently, along with the α-galactosylated derivative Galα3(Fucα2)G_{M1} (B-G_{M1}, or IV^{3}Galα, IV^{2}Fucα, II^{3}NeuAcαGg$_4$Cer), in rat hepatoma and derived cell lines [11, 14, 113], fucosyl-G_{M1} (Fucα3G_{M1}, or IV^{2}Fucα, II^{3}NeuAcαGg$_4$Cer; **12**) has been regarded as an important specific marker of human small-cell lung carcinoma (SCLC), as originally reported [114–116]. Small amounts were detected immunohistochemically with an mAb, F12, by Nilsson et al. [115] in normal human pancreas, spleen, and brain, and later by Brezicka et al. [116] in thymus, spleen, lamina propria and intramural ganglionic cells of the small intestine, and islet cells of the pancreas. Zhang et al. [4] reported a much more restricted distribution of reactivity using mAb F12, confined to a few normal islet cells of pancreas and occasional dorsal root ganglion neurons. Among human cancer tissues in this study, only SCLC reacted.

Fucosyl-G_{M1} has been extensively studied as a potential immunodiagnostic serum marker and possible immunotherapeutic target for SCLC. Vangsted et al. [117] reported immunochemical detection of the antigen in sera of patients with SCLC. The protocol for these studies consisted of microscale extraction of total GSL from serum samples <1 ml, followed by detection via HPTLC overlay immunostaining with an antibody specific for fucosyl-G_{M1}. Using this assay, they were able to demonstrate shedding of fucosyl-G_{M1} from SCLC cell lines in vitro, from xenografts of SCLC cells in nude mice in vivo, and in sera of some human SCLC patients. A sensitive scintillation proximity assay was subsequently developed for this purpose [118, 119] and applied to a clinical diagnostic study showing good correlation between serum fucosyl-G_{M1} expression and SCLC pathology [120].

Interestingly, some reports have implicated autoantibodies to fucosyl-G_{M1} in sensory neuropathies [121]; another study found circulating antibodies to fucosyl-G_{M1} in a majority of neuropathy patients studied but no correlation with sensory impairment [122]. They found a correlation instead with autoantibodies recognizing disialosyl-gangliosides fucosyl-G_{D1b}, B-G_{D1b}, and G_{Q1b}, which also cross-reacted with G_{D1b}, G_{T1b}, and G_{D2}. However, this writer could find no studies describing detection of fucosyl-G_{M1} autoantibodies in SCLC patients. Nevertheless, fucosyl-G_{M1} has been proposed and studied as a key target of immunotherapy for SCLC [4, 86, 95, 119, 123].

8.4.3 Lacto-Series (Type 1 Chain; Lc_n) Antigens

A variety of peripheral core structures and epitopes of lacto-series antigens are shared with glycoprotein glycans and milk or other oligosaccharides.

8.4.3.1 Disialosyl-Lc_4Cer Identical in glycan structure to a previously described milk oligosaccharide [124], disialosyl-Lc_4Cer ($IV^3NeuAc\alpha,III^6NeuAc\alpha$-$Lc_4$Cer; **18**) was reported in the same year as an antigen in metastatic deposits of human colonic adenocarcinoma in liver, recognized by mAb FH9 [47], and of a human embryonal carcinoma cell line, PA1, recognized by mAb K4 [125].

8.4.3.2 Monosialosyl-Le^a Monosialosyl-Le^a (SLe^a; $IV^3NeuAc\alpha,III^6NeuAc\alpha$-$Lc_4$Cer; **19**) is well-known as the antigen recognized by mAb CA 19-9, produced by immunization of mice with the human colorectal carcinoma cell line SW 1116 [126, 127]. The epitope is shared by oligosaccharides, glycoprotein glycans, especially O-glycans carried on mucins, and higher molecular weight GSLs, all of which can react with CA 19-9 and many other antibodies that have been generated to this antigenic sequence, which a sialosylated derivative of the Lewis a (Le^a) structure. Expression on GSLs can, of course, be differentiated by extraction and characterization of antigens demonstrating reactivity in immuno-overlay staining of thin-layer chromatograms (TLC), as originally described by Magnani et al. [128], or other GSL-specific assays based on solid-phase immunoassay, for example, [129, 130]. The detection of the SLe^a ganglioside by CA 19-9 was the first major finding facilitated by application of TLC overlay staining [126]. The ganglioside was subsequently found to be expressed in lipid extracts of just over 50% of adenocarcinoma and pancreatic carcinoma samples, and from the majority of gastric adenocarcinomas, but was not detected in lipids of esophageal carcinomas or normal colon and gastric mucosa, pancreas, kidney, liver, or bone marrow [127].

Since that time, extensive studies of the expression, antigenicity, chemistry and biochemistry, genetics, possible functions, and the diagnostic and therapeutic value of SLe^a have produced a literature too extensive to cover in a short space (for a review, see, e.g., [131]). In the immunohistochemical study of Zhang et al. [5], which is mentioned here due to its explicit connection to a glycan-based tumor vaccine development program, CA 19-9 expression was detected in all colon tumor samples, about half of the breast, SCLC, pancreas, and gastric tumor samples, and in 1/5 prostate tumors, but not in cancers of neuroectodermal origin. It was observed in a wide variety of normal epithelial tissues but not in connective tissues, brain, kidney, testis, spleen, or muscle tissues. CA-19-9-defined SLe^a has been considered as a serum diagnostic marker for a variety of cancers, typically including pancreatic and gastrointestinal malignancies [132, 133]; however, recent indications are that its usefulness is considered limited to the former, and then only under a rather restricted set of circumstances [134]. By itself, it appears to be less useful as a target for immunotherapy, due to its extensive expression in normal epithelia, although it should be remembered that synthesis of Le^a-related antigens is abrogated in FUT3 ($\alpha 3/4$ fucosyltransferase) nonexpressors. Although it has been suggested as a target in colon cancer [5], it does not appear to have a high priority. On the other hand, along with sialosyl-Le^x (see later sections), it has been suggested to be a strong candidate model for conformation-based glycomimetic antiadhesion drugs [131].

8.4.3.3 Disialosyl-Le^a

A novel GSL antigen, disialosyl-Le^a (DSLe^a; III^{4}Fucα,IV^{3}NeuAcα,III^{6}NeuAcα-Lc$_4$Cer; 20), originally detected in liver deposits of metastatic colonic adenocarcinoma, was isolated, characterized, and then used to generate mAb FH7, strongly recognizing the immunogen, although exhibiting some cross-reactivity with monosialosyl-Le^a [135]. It was hoped that DSLe^a might be a cancer-associated marker and, by extension, FH7 might prove to be a useful immunodiagnostic reagent. However, a detailed immunohistochemical examination of both normal and cancerous tissues from colon and pancreas, comparing the expression of FH7 (DSLe^a), CA 19-9 (SLe^a), and nonsialylated Le^a epitopes, showed positive FH7 reactivity in a variety of normal cell types as well as tumor tissues [136]. The results indicated an intriguing complementarity of staining between FH7 and CA 19-9 in normal colon, pointing to expression of significant α2,6-sialyltransferase (α6-sialyl-T) activity in normal colonocytes that leads to high expression of FH7 and apparent deletion of CA 19-9 reactivity. In tumor tissues, co-expression of elevated CA 19-9 was observed along with FH7, which could result from transformation-related down-regulation of the α6-sialyl-T. In contrast, FH7 and CA 19-9 were observed to be co-expressed in both normal and cancerous pancreatic tissues. A related study of the expression of α6- and α3-sialosylated type 1 chain epitopes in serum mucin-type glycoproteins from normal subjects, cancer patients, and patients with nonmalignant disorders indicated that α6-sialosylated-Le^a, as detected by FH7, was expressed in many nonmalignant disorders [137]. It was also detected in all cases of benign prostate tumors examined by Young et al. [138]. All of these results tend to contraindicate the potential of DSLe^a or related FH7-reactive structures for cancer diagnostic or immunotherapeutic purposes.

8.4.4 Neolacto-Series (Type 2 Chain; nLc$_x$) Antigens

A variety of peripheral core structures and epitopes of neolacto-series antigens are shared with glycoprotein glycans and milk or other oligosaccharides.

8.4.4.1 Le^x, Le^x-Le^x, (Le^x)$_n$ Neutral GSLs

The Lewis x pentasaccharide (Le^x; III^{3}Fucα-nLc$_4$Cer; **21**) structure was first isolated and characterized as a GSL by Yang and Hakomori [139] from human adenocarcinoma tissue. The pentasaccharide itself was already known as a component of human milk characterized by Kobata and Ginsburg [140] a couple of years earlier, but the designation as a "Lewis" structure was not applied by either group. Rather, it was observed that, despite the obvious structural isomeric relationship to Le^a, neither the pentasaccharide nor the corresponding GSL exhibited Lewis or any other known serological activity. However, it was noted that the "ABH Lewis-inactive" GSL was detected and isolable in high yield from several cases of adenocarcinoma tissue [141], while being observed in low abundance at best in certain normal tissues. Significantly, it was thought "conceivable that this glycolipid may play a role in determining the immunopathological specificity of human tumors" [139]. The importance of the "X" or "Le^x" structure, as it was subsequently designated, as a developmentally regulated antigen was realized later, when it was observed that SSEA-1, defined by an mAb against an embryonal carcinoma cell line, F9 [142], reacted strongly with GSLs and oligosaccharides containing one or more Galβ4(Fucα3)GlcNAcβ3 trisaccharide determinants [143, 144]. The anti-SSEA-1 mAb recognized the corresponding antigen at the cell surface of the murine embryo beginning at the eight-cell stage, which was subsequently expressed only in a highly restricted manner [142, 145]; in addition, it reacted with murine teratocarcinoma and a variety of human cells and tumor-cell lines [146, 147].

With respect to isolated antigens, it had been observed that the reaction of the anti-SSEA-1 mAb with Le^x pentaglycosylceramide was relatively weak, unless arrayed in high density, while the reaction with other, higher molecular weight neutral GSL components in adenocarcinoma tissues was much stronger. This led to the isolation and characterization of a series of adenocarcinoma GSLs having two or more repeating Galβ4(Fucα3)GlcNAcβ3 units (**22, 24, 25, 27**) [143, 148, 149]. It had been proposed [143, 144] that the SSEA-1/Le^x determinants could be carried on either linear or branched poly-N-acetyllactosamine (poly-LacNAc) chains responsible for the i and I blood group activities, respectively, but all SSEA-1-reactive GSLs isolated to this point were carried on linear, i type, poly-LacNAc chains (also referred to as repeating type 2 chain or nLc$_n$Cer core-type antigens). During the same period, anti-SSEA-1-reactive Le^x carrying GSL antigens were isolated in low yield from human O erythrocytes and characterized as incompletely α3-fucosylated linear poly-LacNAc GSLs (repeating type 2 chain or nLc$_n$Cer core-type antigens) in which the free GlcNAc residues were those most proximal to the ceramide [150]. Subsequently, a variety of these antigens was isolated in higher yield from the neutral GSL fraction of normal human granulocytes (polymorphonuclear neutrophils, PMNs), and the question was raised whether the SSEA-1 active components thought to be in red blood cells were mainly impurities contributed by granulocytes commonly contaminating erythrocyte preparations [151, 152]. Similar components were observed in the neutral GSL fraction extracted from the HL-60 myeloid leukemia cell line [152]. SSEA-1 and other poly-Le^x/LAcNAc antigens have generated considerable interest due to their potential involvement in cancer pathobiology both as immunogens and as potential ligands for selectins (see below).

The discovery of di- and oligo-Le^x determinants in oligosaccharides and glycoconjugates stimulated an effort to generate mAbs more specific to the repeating Galβ4(Fucα3)GlcNAcβ3 motif. Fukushi et al. [153] succeeded in establishing an mAb, FH4 (IgG3), which reacted preferentially with GSLs bearing two or three Le^x trisaccharide repeats (e.g., **22, 24**) while not recognizing those with only a single Le^x determinant (e.g., **21, 23, 26**). This mAb, along with others directed to related type 2 chain determinants, such as sialosyl-Le^x, Le^y, and Le^y-Le^x (see below), has been used to map the expression of non-ABO poly-LacNAc antigenic structures during normal human development and differentiation and in malignant and nonmalignant tissues from various organs [154–158].

8.4.4.2 Le^x, Le^x-Le^x, (Le^x)$_n$ Gangliosides

Sialosyl-Le^x hexaglycosylceramide (SLe^x; III^{3}Fucα,IV^{3}NeuAcα-nLc$_4$Cer; **37**) was first described as a cell membrane constituent of human kidney [159]. Along with more complex analogs, and other glycoconjugates bearing analogous structures, it has been regarded as a model ligand for E-selectin [160, 161] and may contain a major part of the epitope recognized by P- and S-selectins (the literature on this subject is far too extensive to do it justice here; for a review of the chemistry, biology, and some potential therapeutic applications targeting SLe^x, see, e.g., [131]). It is noteworthy that although SLe^x and SLe^a are immunologically distinct (with few exceptions, antibodies to one do not cross-react with the other), E-selectin recognizes a conformationally determined epitope common to both [162, 163]. Livingston and Ragupathi [86] considered Le^a and sialosyl-Le^a more reasonable vaccine targets than Le^x and sialosyl-Le^x, based on the wider distribution of the latter types on normal or critical tissues. This should not exclude further exploration, particularly of more complex determinants whose expression may be more highly delimited.

Adenocarcinoma tisues are a source of both unbranched i and branched I core neolacto-series gangliosides, terminated by either NeuAcα3 and NeuAcα6 residues (**31–36**); these nLc$_n$Cer and iso-nLc$_n$Cer gangliosides are also found in GSLs of a variety of normal tissues, including human erythrocytes, granulocytes, and placenta. A major difference, however, in adenocarcinoma, and to some degree in granulocytes and other leukocytes, is the extensive elaboration of the linear nLc$_n$Cer cores with Fucα3 residues, generating NeuAcα3/6-terminated poly-LacNAc chains containing one or more Galβ4(Fucα3)GlcNAcβ3 units (**37–40, 42, 44–56**), analogous to those observed in the neutral GSL fractions. These can be grouped according to key structural differences, for example, NeuAcα3- or NeuAcα6-terminated, or arrangement of Fucα3 residues; in many cases, these correspond to epitopes that can be distinguished by mAbs, many of which have been generated and selected by immunization with adenocarcinoma membrane GSLs.

An SLe^x hexaglycosylceramide with a nonreducing terminal NeuAcα6 residue (i.e., III^{3}Fucα,IV^{6}NeuAcα-nLc$_4$Cer) has never been identified and is most likely biosynthetically impossible [164]. Until recently, the only NeuAcα6-terminated Fucα3-modified adenocarcinoma ganglioside was "6C" having the intriguing structure III^{3}Fucα,VI^{6}NeuAcα-nLc$_6$Cer (**42**), wherein the single Fucα3 residue is attached to the GlcNAc proximal to ceramide [9, 165]. This fucoganglioside was detected by

8.4 DISCUSSION OF SELECTED ANTIGENS

an mAb, 1B9, generally recognizing NeuAcα6-terminated linear nLc$_n$Cer antigens, which were observed to be accumulated in adenocarcinoma [10, 165]. The presence of the Fucα3 residue in 6C appeared neither essential nor inhibitory for the recognition by mAb 1B9, and the mAb also recognized NeuAcα6-terminated LacNAc chains of glycoprotein glycans. NeuAcα6-terminated linear nLc$_n$Cer antigens are also expressed on normal cells such as erythrocytes and granulocytes [151, 152]; however, the fucosylated derivative 6C has not been found in any other tissue except for human colonic adenocarcinoma. Despite the apparent uniqueness of its distribution and structure, which is firmly established [165], it has not been the object of much further interest. Interestingly, a recent glycosphingolipidomic high performance liquid chromatography/mass spectrometry (HPLC/MS) reexamination of colonic adenocarcinoma samples turned up two related NeuAcα6-terminated gangliosides, proposed to have the novel structures IV^{2}Fucα,IV^{6}NeuAcα-nLc$_4$Cer (**41**) and VI^{2}Fucα, III^{3}Fucα,VI^{6}NeuAcα-nLc$_6$Cer (**43**) [166]. These structures, which have not yet been completely and unambiguously confirmed, are the first reported instances of branched NeuAcα6(Fucα2)Galβ4GlcNAcβ3-terminated GSLs; the latter is particularly interesting as it is a further fucosylated derivative of 6C, but its existence does not contradict previous observations that the terminal NeuAcα6Galβ4 and penultimate Fucα3GlcNAcβ3 modifications are mutually exclusive on poly-LacNAc core structures of glycoprotein and GSL glycans [164, 165, 167].

The second class of poly-Le^x/LacNAc ganglioside antigens are those terminated by NeuAcα3; core gangliosides with two or more LacNAc units, and without Fucα3GlcNAcβ3 modifications, can be classified as either unbranched or branched, i or I, for example, **33** or **35/36**, respectively. Along with G$_{M3}$ ganglioside, human erythrocytes, placenta, and granulocytes, among other normal tissues, as well as colonic adenocarcinoma and HL-60 cells, express both types of NeuAcα3-terminated GSL antigens. In adenocarcinoma tissues, granulocytes, and HL-60 cells, these are accompanied by extensive arrays of GSL antigens generated by modification of the GlcNAcβ3 residues, chiefly in the linear chains, by addition of one or more Fucα3 residues (**37–40, 44–56**). Due to their potential involvement in cancer pathobiology both as immunogens and as potential ligands for selectins, which are involved in inflammatory/metastatic processes, this class of GSL antigens has been the subject of intense interest as targets for diagnostic and therapeutic strategies. The antigens may be further subdivided into four general subclasses according to structural features and differential recognition by certain mAbs.

The first subclass may be defined as those with at least one penultimate Fucα3GlcNAcβ3 modification adjacent to the terminal NeuAcα3Galβ4, recognized by mAbs with specificity for IV^{3}NeuAc-Le^x or VI^{3}NeuAc-Le^x-Le^x (as in, e.g., IV^{3}NeuAcα,III^{3}Fucα-nLc$_4$Cer or VI^{3}NeuAcα,III^{3}Fucα,V^{3}Fucα-nLc$_6$Cer; **37** or **40**, respectively). The latter was first isolated and characterized by Fukushi et al. [168] (see also [165]). Both react with an mAb defining "classical" SLe^x, such as CSLEX-1, while the latter exhibits stronger recognition by mAb FH6 [168]. Though the epitope of FH6 does not require the second III^{3}Fucα residue, it does require the VI^{3}NeuAcα and V^{3}Fucα residues on the *repeating* LacNAc tetrasaccharide core, that is, it is an SLe^x-i antibody.

These gangliosides are typically detected in gastrointestinal, colorectal, breast, and lung cancers. For example, in a study of human colonic tissues, comparing various types of tumors versus normal patterns of expression, mAb FH6 was found to be highly specific with respect to lack of staining of normal tissues, while it stained a very high proportion of cancerous tissues, with the notable exception of poorly differentiated cancers [156]. By comparison, in the same study, mAb FH4, directed against nonsialosylated di- or tri-Le^x, stained almost all types of cancers, including poorly differentiated ones, although it also stained a higher proportion of nonneoplastic diseased tissues.

The second subclass may be defined by having the GlcNAc residue nearest the terminal remain underivatized; a single Fucα3 residue is attached to the *second* GlcNAc from the nonreducing end (e.g., VIII^{3}NeuAcα,V^{3}Fucα-nLc$_8$Cer and X^{3}NeuAcα,VII^{3}Fucα-nLc$_{10}$Cer; **45** and **47**, respectively). These components have been detected and/or isolated from the leukocytes of patients with chronic myelogenous leukemia [169], human colonic adenocarcinoma tissues [170], and the human promyelocytic leukemia cell line HL-60 [169, 171–173]. The NeuAcα3Galβ4GlcNAc β3Galβ4(Fucα3)GlcNAcβ3 sequence is recognized by mAb VIM-2 [169]; another mAb, ACFH-18, also recognizes this sequence, but the NeuAcα3 residue is not an essential component of the epitope, and it will also react with neutral/desialosylated GSLs bearing a single Fucα3 on any GlcNac residue except the penultimate [170]. There appears to be no cross recognition between the VIM-2 and SLe^x epitopes.

The third subclass can be considered an extension of the second, with the additional feature that at least four unbranched repeating LacNAc units are required (**47, 49–56**) [174]. Typically, there is room for one or more *additional* Fucα3 units on GlcNAc residues further from the reducing end, although these are not considered structurally or functionally essential. A wide variety of these ganglioside components were isolated by Stroud et al. [171–173] from HL-60 cells as well as from normal human leukocytes, and called "myeloglycans," in consideration of their cell type origin. These cells react with a variety of anti-SLe^x mAbs, including CSLEX-1, FH6, SNH3, and SNH4, which were therefore long assumed to define the dominant functional surface epitope; however, although some fractions of myeloglycan contain cross-reacting SLe^x and SdiLe^x determinants on long-chain cores, the great majority of nonreducing end sequences are of the VIM2/ACFH-18 type. Overall, these myeloid cell gangliosides appeared to be particularly rich in unbranched longer chain core structures (nLc$_8$Cer, nLc$_{10}$Cer, nLc$_{12}$Cer, along with minor amounts of higher uncharacterized poly-LacNAc components) incorporating one or more Le^x-type Fucα3 residues, but rarely on the penultimate GlcNAc; shorter chain and branched core (nLc$_4$Cer, nLc$_6$Cer, and iso-nLc$_8$Cer) gangliosides were present, along with G$_{M3}$, but not fucosylated. The particular significance of the extended myeloglycan-defining sequence, namely, NeuAcα3Galβ4GlcNAcβ3Galβ4(Fucα3)GlcNAcβ3Galβ4-(±Fucα3)GlcNAcβ3Galβ4(±Fucα3)GlcNAcβ3, is that it has been hypothesized as required for true, in vivo binding to E-selectin (but not P-selectin) [171–174], mediating E-selectin-dependent tethering and rolling under physiological conditions of high shear stress [175]. Thus it may be a "metastatic/invasive" adhesion epitope for interaction of cancer cells with E-selectin.

The fourth subclass consists of minor components, characterized by a single Fucα3 modification two or more GlcNac residues away from the nonreducing terminal, presenting neither SLe^x nor VIM-2-reactive determinant, but expected to react with ACFH-18 (**46, 48**). Any special functional significance is unclear.

8.4.4.3 Le^y, Extended Le^y, and Le^y-Le^x

An mAb, AH-6, generated by immunization against membrane lipids of human gastric cancer cell line MKN74 was found to react preferentially with GSLs terminated by the Y or Le^y determinant [176]. The Y or Le^y tetrasaccharide Fucα2Galβ4(Fucα3)GlcNAcβ3 is a non-cross-reacting structural isomer of the Lewis b (Le^b) antigen. Like the isomeric Le^x and Le^a trisaccharides, both Le^y and Le^b are typically presented as nonreducing terminal structures carried on lactosyl, LacNAc, or poly-LacNAc chains of GSLs and glycoprotein glycans. It is generally accepted that the Le^y determinant is synthesized by Fucα3 addition to the type 2 chain H blood group trisaccharide Fucα2Galβ4GlcNAcβ3, and not by Fucα2 addition (catalyzed the O blood group gene encoded Fuc-T) to a preexisting terminal Le^x Galβ4(Fucα3)GlcNAcβ3 determinant [177]. The short-chain Le^y hexaglycosylceramide is a normal GSL antigen of dog intestine [178], and multiple Le^y-active components can be extracted from normal, particularly O blood group, erythrocytes [176], and granulocytes [152]. Significantly, AH-6 reacted strongly with multiple components of the immunizing cell line MKN74, as well as GSLs of human adenocarcinoma (the same colonic metastatic to liver as used in many previous studies) [176]; areas of normal colonic epithelia and benign polyps can be stained with AH-6, but this mAb strongly stained essentially all adenocarcinoma tissues (regardless of origin or blood type), and all colonic polyps exhibiting severe dysplasia, considered a predictor of metastatic potential [179].

Miyake et al. [180] examined the expression of AH-6/Le^y, FH-2 (anti-SSEA-1)/Le^x, and FH-6/SLe^x-i related staining of normal lung tissue development. In this study, lung adenocarcinoma generally expressed all three antigen types, but AH-6 stained most strongly and reliably all classes of adenocarcinoma. Some cases of squamous cell carcinoma weakly expressed one or two of the antigen types, with AH-6 being the most commonly expressed. Moderate expression of AH-6 was noted in 5/7 large-cell carcinoma samples, compared to occasional weak staining by FH-2 and FH-6, and small-cell carcinoma was occasionally stained only by AH-6.

Isolation and characterization of Le^y-active components of the human colonic adenocarcinoma (metastatic to liver), facilitated by their reaction with mAb AH-6, led to identification of the expected short-chain Le^y hexaglycosylceramide (**28**), identical in glycan structure to that found in dog intestine, as well as two novel components with lower Rf values in HPTLC [135]. The structures of comparatively minor and major extended Le^y components were elucidated as V^{3}Fucα,VI^{2}Fucα-nLc$_6$Cer (**29**) and III^{3}Fucα,V^{3}Fucα,VI^{2}Fucα-nLc$_6$Cer (**30**), respectively [135, 181]. Immunization of mice with III^{3}Fucα,V^{3}Fucα,VI^{2}Fucα-nLc$_6$Cer (Le^y-Le^x), and selection of hybridomas by positive reaction with this antigen, while excluding those reacting with Le^y hexaglycosyl- and octaglycosylceramide antigens, yielded an IgM mAb, KH-1, reacting with the trifucosylated Le^y-Le^x antigen [182]. KH-1 exhibited some cross-reactivity with di-Le^x (Le^x-Le^x), consistent with the shared inner sequence.

In a study of Le^x and Le^y antigen expression in human pancreas and pancreatic cancers, mAbs against Le^x, di-Le^x, and sialosyl-Le^x-i stained none of the normal pancreatic tissues, while mAbs against Le^y, extended Le^y, and Le^y-Le^x stained a significant proportion of normal pancreatic tissues [157]. In this study, Le^x- and Le^y-related antigens stained similar, higher proportions of pancreatic cancers, especially those that were well differentiated. Compared with normal tissues, the differential in expression between normal and cancerous tissues was clearly much higher with the Le^x-related antigens. On the other hand, the differential in expression of extended, complex Le^y-related antigens appeared to be more promising with human colonic cancers [183], comparable with the results obtained with mAbs to the extended, complex Le^x-related antigens noted by Itzkowitz et al. [156]. In particular, the mAb against Le^y-Le^x (KH-1), and others directed against extended-Le^y (CC1, CC2) preferentially stained malignant colonic versus normal tissues; the differential was much higher than with the general Le^y mAb AH-6 [183]. Livingston and Ragupathi [86] considered Le^y-terminated antigens to be reasonable vaccine targets.

8.5 OTHER ANTIGENS

The following extended lacto/neolacto-hybrid and repeating lacto-series antigens have not been extensively studied.

8.5.1 Le^a-Le^a and Le^b-Le^a

An mAb, NCC-ST-421 (IgG3), raised by immunization of mice with a xenograft of human gastric carcinoma was observed to have strong antitumor effects both in vitro and in an in vivo human-to-mouse tumor xenograft model system [184]. In HPTLC overlay immunostaining assays, NCC-ST-421 appeared to recognize strongly specific sets of Le^a cross-reactive neutral GSL components in a variety of tumor tissues [185]. Expression of these components in normal tissues was highly limited, confined to small intestine and pancreas. One set of bands migrated with approximately the same Rf as Le^a-pentaglycosylceramide, but two groups of reactive bands with successively lower Rf values were also observed. Reactive neutral GSL components with similar Rf values were also observable in excised xenograft tumors generated by inoculation of nude mice with Colo205 colonic adenocarcinoma cells. From extracts of these xenograft tumors, a component with Rf similar to an octaglycosylceramide with difucosylated nLc$_6$Cer cores was isolated and unambiguously characterized as a repeating Le^a-Le^a GSL, with the structure V^{4}Fucα,III^{4}Fucα-Lc$_6$Cer (**15**) [185, 186]. Small amounts of extended Le^x or Le^y, or repeating Le^x GSL were apparent, but only at a level of 5–10% visible in ^{1}H-NMR or mass spectra. The potential of mAb NCC-ST-421 as a passive immunotherapeutic has been examined experimentally [187, 188] but not, so far, immunization with the Le^a-Le^a antigen.

From the same source, Colo205 xenograft tumors, a second type of repeating Type 1 chain GSL was isolated and characterized as a trifucosylated Lc$_6$cer core with an Le^b-Le^a structure (VI^{2}Fucα,V^{4}Fucα,III^{4}Fucα-Lc$_6$Cer; **16**) [186, 189]. An mAb generated by immunization with this antigen, IMH2 (IgG3), recognized the immunogen, but also cross-reacted with Le^y-Le^x as well as Le^y or Le^b hexaglycosylceramides; more significantly, it displayed antitumor properties similar to those of NCC-ST-421 [190].

Previously, it had been proposed that type 1 chain GSLs could not be extended by repetitive core chain glycosylations beyond Lc$_4$Cer [191]. The repetitive Colo205 antigens are clear exceptions to this "rule," indicating a potential biosynthetic capability not significantly expressed in normal tissues, but activated under circumstances of oncological transformation. As with other "neo"-antigens, it may have more to do with deactivation of a normally competitive pathway, or with a breakdown in the topological organization of GSL biosynthetic pathways, which rearranges the normal substrate/transferase relationships, than with generation of a novel transferase activity.

8.5.2 Le^a-Le^x

A neutral difucosylated GSL with a hybrid para-lacto (type 1 on type 2 chain) core has not yet been observed in nature but was synthetically generated to answer a question about the epitope specificity of an mAb, 43-9F, recognizing a putative cancer-associated antigen [192, 193]. An mAb 43-9F (IgM), generated by immunization of mice with a human squamous lung carcinoma (SLC) cell line, was found to react differentially with the majority of human lung squamous and all adenocarcinoma tumor samples, while exhibiting no staining of SCLC samples (some other tumor and a highly delimited set of normal human cell types were also stained) [192]. The 43-9F antigen was also detectable on membrane-associated glycoproteins and soluble glycoconjugates secreted into media from cultured SLC cells. The mAb reacted with Le^a pentaglycosylceramide and Le^a pentasaccharide-BSA conjugate, and not with any Le^b, Le^x, or Le^y terminated structure, but additional evidence pointed to a more complex extended epitope, such as presented by the octasaccharide V^{4}Fucα,III^{3}Fucα-pLc$_6$, incorporating a nonreducing terminal Le^a-Le^x sequence, and a branched decasaccharide bearing the same Le^a-Le^x sequence, isolable from human milk [193] or feces of breast-milk-fed infants [194]. An octasaccharide from human milk was a major component retarded in liquid chromatography on a 43-9F affinity column, and characterization by fast-atom bombardment–mass spectrometry (FAB-MS) of the reduced permethylated oligosaccharide was consistent with the Le^a-Le^x sequence of V^{4}Fucα,III^{3}Fucα-pLc$_6$. A previously obtained sample of this octasaccharide was the most inhibitory (along with the branched decasaccharide) in a competitive binding assay with 43-9F. A biosynthetic GSL, V^{4}Fucα,III^{3}Fucα-pLc$_6$Cer (**57**), was generated, and this was used to provide additional confirmation of the specificity of 43-9F to the Le^a-Le^x structure, while ruling out cross-recognition of Le^a-Le^a, the isomeric antigen available from Colo205 cells [195]. Nevertheless, expression of the Le^a-Le^x determinant appears so far confined to glycoprotein carriers in vivo, and no native GSL with this structure has been isolated from any tissue. Interestingly, a potential monofucosylated hybrid core precursor, V^{4}Fucα-pLc$_6$Cer, was isolated in small quantities from human blood cell membranes [191].

ACKNOWLEDGMENT

This work has been supported by the Danish Medical Research Council for Technology and Innovation, the Stjerne Program of Excellence, and the Copenhagen Center for Glycomics.

REFERENCES

1. Hakomori S (1986) Tumor-associated glycolipid antigens, their metabolism and organization. *Chem. Phys. Lipids* 42:209–233.
2. Hakomori S (1996) Tumor malignancy defined by aberrant glycosylation and sphingo(glyco)lipid metabolism. *Cancer Res.* 56:5309–5318.
3. Hakomori S (1998) Cancer-associated glycosphingolipid antigens: Their structure, organization, and function. *Acta Anat.* 161:79–90.
4. Zhang S, Cordon-Cardo C, Zhang HS, Reuter VE, Adluri S, Hamilton WB, Lloyd KO, Livingston PO (1997) Selection of tumor antigens as targets for immune attack using immunohistochemistry: I. Focus on gangliosides. *Int. J. Cancer* 73:42–49.
5. Zhang S, Zhang HS, Cordon-Cardo C, Reuter VE, Singhal AK, Lloyd KO, Livingston PO (1997) Selection of tumor antigens as targets for immune attack using immunohistochemistry: II. Blood group-related antigens. *Int. J. Cancer* 73:50–56.
6. Hakomori S (2000) Traveling for the glycosphingolipid path. *Glycoconj. J.* 17:627–647.
7. Hakomori S (1999) Antigen structure and genetic basis of histo-blood groups A, B and O: Their changes associated with human cancer. *Biochim. Biophys. Acta* 1473:247–266.
8. Holmes EH, Hakomori S, Ostrander GK (1987) Synthesis of type 1 and 2 lacto series glycolipid antigens in human colonic adenocarcinoma and derived cell lines is due to activation of a normally unexpressed $\beta 1 \rightarrow 3N$-acetylglucosaminyltransferase. *J. Biol. Chem.* 262:15649–15658.
9. Hakomori S, Nudelman E, Levery SB, Patterson CM (1983) Human cancer-associated gangliosides defined by a monoclonal antibody (IB9) directed to sialosyl$\alpha 2 \rightarrow 6$galactosyl residue: A preliminary note. *Biochem. Biophys. Res. Commun.* 113:791–798.
10. Hakomori S, Patterson CM, Nudelman E, Sekiguchi K (1983) A monoclonal antibody directed to N-acetylneuraminosyl-$\alpha 2,6$-galactosyl residue in gangliosides and glycoproteins. *J. Biol. Chem.* 258:11819–11822.
11. Holmes EH, Hakomori S (1982) Isolation and characterization of a new fucoganglioside accumulated in precancerous rat liver and in rat hepatoma induced by N-2-acetylaminofluorene. *J. Biol. Chem.* 257:7698–7703.
12. Gessner P, Riedl S, Quentmaier A, Kemner W (1993) Enhanced activity of CMP-NeuAc: Galβ1-4GlcNAc: α2-6-sialyltransferase in metastasizing human colorectal tumor tissue and serum of tumor patients. *Cancer Lett.* 75:143–149.
13. Dall'Olio F, Malagolini N, DiStefano G, Minni F, Marrano D, Serafini-Cessi F (1989) Increased CMP-NeuAc:Galβ1,4GlcNAc-R α2-6-sialyltransferase activity in human colorectal cancer tissues. *Int. J. Cancer* 44:434–439.
14. Holmes EH, Hakomori S (1983) Enzymatic basis for changes in fucoganglioside during chemical carcinogenesis. Induction of a specific α-fucosyltransferase and status of an α-galactosyltransferase in precancerous rat liver and hepatoma. *J. Biol. Chem.* 258:3706–3713.

15. Holmes EH, Hakomori S (1987) The chemical carcinogen-induced enzyme, GDP-fucose: GM1 α1-2 fucosyltransferase in rat liver and hepatoma: Modulation by and association with phospholipids. *J. Biochem. (Tokyo)* 101:1095–1105.
16. Black PH (1980) Shedding from normal and cancer-cell surfaces [editorial]. *N. Engl. J. Med.* 303:1415–1416.
17. Black PH (1980) Shedding from the cell surface of normal and cancer cells. *Adv. Cancer Res.* 32:75–199.
18. Ladisch S, Gillard B, Wong C, Ulsh L (1983) Shedding and immunoregulatory activity of YAC-1 lymphoma cell gangliosides. *Cancer Res.* 43:3808–3813.
19. Ladisch S, Wu ZL, Feig S, Ulsh L, Schwartz E, Floutsis G, Wiley F, Lenarsky C, Seeger R (1987) Shedding of GD2 ganglioside by human neuroblastoma. *Int. J. Cancer* 39:73–76.
20. Li RX, Ladisch S (1991) Shedding of human neuroblastoma gangliosides. *Biochim. Biophys. Acta* 1083:57–64.
21. Chang F, Li R, Ladisch S (1997) Shedding of gangliosides by human medulloblastoma cells. *Exp. Cell Res.* 234:341–346.
22. Ladisch S, Ulsh L, Gillard B, Wong C (1984) Modulation of the immune response by gangliosides. Inhibition of adherent monocyte accessory function in vitro. *J. Clin. Invest.* 74:2074–2081.
23. Floutsis G, Ulsh L, Ladisch S (1989) Immunosuppressive activity of human neuroblastoma tumor gangliosides. *Int. J. Cancer* 43:6–9.
24. Grayson G, Ladisch S (1992) Immunosuppression by human gangliosides. II. Carbohydrate structure and inhibition of human NK activity. *Cell Immunol.* 139:18–29.
25. Li R, Villacreses N, Ladisch S (1995) Human tumor gangliosides inhibit murine immune responses in vivo. *Cancer Res.* 55:211–214.
26. Karlsson KA, Larson G (1981) Potential use of glycosphingolipids of human meconium for blood group chemotyping of single individuals. *FEBS Lett.* 128:71–74.
27. Andrews PW, Goodfellow PN, Shevinsky LH, Bronson DL, Knowles BB (1982) Cell-surface antigens of a clonal human embryonal carcinoma cell line: Morphological and antigenic differentiation in culture. *Int. J. Cancer* 29:523–531.
28. Shevinsky LH, Knowles BB, Damjanov I, Solter D (1982) Monoclonal antibody to murine embryos defines a stage-specific embryonic antigen expressed on mouse embryos and human teratocarcinoma cells. *Cell* 30:697–705.
29. Kannagi R, Levery SB, Ishigami F, Hakomori S, Shevinsky LH, Knowles BB, Solter D (1983) New globoseries glycosphingolipids in human teratocarcinoma reactive with the monoclonal antibody directed to a developmentally regulated antigen, stage-specific embryonic antigen 3. *J. Biol. Chem.* 258:8934–8942.
30. Bremer EG, Levery SB, Sonnino S, Ghidoni R, Canevari S, Kannagi R, Hakomori S (1984) Characterization of a glycosphingolipid antigen defined by the monoclonal antibody MBr1 expressed in normal and neoplastic epithelial cells of human mammary gland. *J. Biol. Chem.* 259:14773–14777.
31. Menard S, Tagliabue E, Canevari S, Fossati G, Colnaghi MI (1983) Generation of monoclonal antibodies reacting with normal and cancer cells of human breast. *Cancer Res.* 43:1295–1300.

32. Canevari S, Fossati G, Balsari A, Sonnino S, Colnaghi MI (1983) Immunochemical analysis of the determinant recognized by a monoclonal antibody (MBr1) which specifically binds to human mammary epithelial cells. *Cancer Res.* 43:1301–1305.

33. Holgersson J, Jovall PA, Samuelsson BE, Breimer ME (1991) Blood group type glycosphingolipids of human kidneys. Structural characterization of extended globo-series compounds. *Glycoconj. J.* 8:424–433.

34. Blomberg J, Breimer ME, Karlsson KA (1982) Glycosphingolipids of a green monkey kidney cell line (GMK AH-1). Evidence for a novel pentaglycosylceramide based on globotetraosylceramide. *Biochim. Biophys. Acta* 711:466–477.

35. Siddiqui B, Hakomori S (1971) A revised structure for the Forssman glycolipid hapten. *J. Biol. Chem.* 246:5766–5769.

36. Kannagi R, Cochran NA, Ishigami F, Hakomori S, Andrews PW, Knowles BB, Solter D (1983) Stage-specific embryonic antigens (SSEA-3 and -4) are epitopes of a unique globo-series ganglioside isolated from human teratocarcinoma cells. *EMBO J.* 2:2355–2361.

37. Chien JL, Hogan EL (1983) Novel pentahexosyl ganglioside of the globo series purified from chicken muscle. *J. Biol. Chem.* 258:10727–10730.

38. Saito S, Levery SB, Salyan ME, Goldberg RI, Hakomori S (1994) Common tetrasaccharide epitope NeuAcα2 → 3Galβ1 → 3(NeuAcα2 → 6)GalNAc, presented by different carrier glycosylceramides or *O*-linked peptides, is recognized by different antibodies and ligands having distinct specificities. *J. Biol. Chem.* 269:5644–5652.

39. Stroud MR, Stapleton AE, Levery SB (1998) The P histo-blood group-related glycosphingolipid sialosyl galactosyl globoside as a preferred binding receptor for uropathogenic *Escherichia coli*: Isolation and structural characterization from human kidney. *Biochemistry* 37:17420–17428.

40. Satoh M, Handa K, Saito S, Tokuyama S, Ito A, Miyao N, Orikasa S, Hakomori S (1996) Disialosyl galactosylgloboside as an adhesion molecule expressed on renal cell carcinoma and its relationship to metastatic potential. *Cancer Res.* 56:1932–1938.

41. Ito A, Saito S, Masuko T, Oh-Eda M, Matsuura T, Satoh M, Nejad FM, Enomoto T, Orikasa S, Hakomori SI (2001) Monoclonal antibody (5F3) defining renal cell carcinoma-associated antigen disialosyl globopentaosylceramide (V3NeuAcIV6NeuAcGb5), and distribution pattern of the antigen in tumor and normal tissues. *Glycoconj. J.* 18:475–485.

42. Kundu SK, Samuelsson BE, Pascher I, Marcus DM (1983) New gangliosides from human erythrocytes. *J. Biol. Chem.* 258:13857–13866.

43. Dasgupta S, Chien JL, Hogan EL, van Halbeek H (1991) A disialoganglioside of the globo-series from chicken skeletal muscle. *J. Lipid Res.* 32:499–506.

44. Levery SB, Salyan ME, Steele SJ, Kannagi R, Dasgupta S, Chien J-L, Hogan EL, van Halbeek H, Hakomori S (1994) A revised structure for the disialosyl globo-series gangliosides of human erythrocytes and chicken skeletal muscle. *Arch. Biochem. Biophys.* 312:125–134.

45. Hakomori S, Zhang Y (1997) Glycosphingolipid antigens and cancer therapy. *Chem. Biol.* 4:97–104.

46. Ito A, Levery SB, Saito S, Satoh M, Hakomori S (2001) A novel ganglioside isolated from renal cell carcinoma. *J. Biol. Chem.* 276:16695–16703.

47. Fukushi Y, Nudelman E, Levery SB, Higuchi T, Hakomori S (1986) A novel disialoganglioside (IV^{3}NeuAcIII^{6}NeuAcLc$_4$) of human adenocarcinoma and the monoclonal antibody (FH9) defining this disialosyl structure. *Biochemistry* 25:2859–2866.
48. Fredman P, Mansson JE, Wikstrand CJ, Vrionis FD, Rynmark BM, Bigner DD, Svennerholm L (1989) A new ganglioside of the lactotetraose series, GalNAc-3'-isoLM1, detected in human meconium. *J. Biol. Chem.* 264:12122–12125.
49. Portoukalian J, Zwingelstein G, Dore JF (1979) Lipid composition of human malignant melanoma tumors at various levels of malignant growth. *Eur. J. Biochem.* 94:19–23.
50. Hamilton WB, Helling F, Lloyd KO, Livingston PO (1993) Ganglioside expression on human malignant melanoma assessed by quantitative immune thin-layer chromatography. *Int. J. Cancer* 53:566–573.
51. Kawamura K, Ohyama C, Watanabe R, Satoh M, Saito S, Hoshi S, Gasa S, Orikasa S (2001) Glycolipid composition in bladder tumor: A crucial role of G$_{M3}$ ganglioside in tumor invasion. *Int. J. Cancer* 94:343–347.
52. Nores GA, Dohi T, Taniguchi M, Hakomori S (1987) Density-dependent recognition of cell surface GM3 by a certain anti-melanoma antibody, and GM3 lactone as a possible immunogen: Requirements for tumor-associated antigen and immunogen. *J. Immunol.* 139:3171–3176.
53. Otsuji E, Park YS, Tashiro K, Kojima N, Toyokuni T, Hakomori S (1995) Inhibition of B16 melanoma metastasis by administration of GM3- or Gg3-liposomes: Blocking adhesion of melanoma cells to endothelial cells (anti-adhesion theraoy) via inhibition of GM3-Gg3Cer or GM3-LacCer interaction. *Int. J. Oncol.* 6:319–327.
54. Estevez F, Carr A, Valiente O, Mesa C, Barroso O, Sierra GV, Fernandez LE (1999) Enhancement of the immune response to poorly immunogenic gangliosides after incorporation onto very small size proteoliposomes (VSSP). *Vaccine* 18:190–197.
55. Gabri MR, Ripoll GV, Alonso DF, Gomez DE (2002) Role of cel surface GM3 ganglioside and sialic acid in the anti-tumor activity of a GM3-based vaccine in the murine B16 melanoma model. *J. Cancer Res. Clin. Oncol.* 128:669–677.
56. Saito S, Nojiri H, Satoh M, Ito A, Ohyama C, Orikasa S (2000) Inverse relationship of expression between GM3 and globo-series ganglioside in human renal cell carcinoma. *Tohoku J. Exp. Med.* 190:271–278.
57. Watanabe R, Ohyama C, Aoki H, Takahashi T, Satoh M, Saito S, Hoshi S, Ishii A, Saito M, Arai Y (2002) Ganglioside GM3 overexpression induces apoptosis and reduces malignant potential in murine bladder cancer. *Cancer Res.* 62:3850–3854.
58. Irie A, Koyama S, Kozutsumi Y, Kawasaki T, Suzuki A (1998) The molecular basis for the absence of N-glycolylneuraminic acid in humans. *J. Biol. Chem.* 273:15866–15871.
59. Chou HH, Takematsu H, Diaz S, Iber J, Nickerson E, Wright KL, Muchmore EA, Nelson DL, Warren ST, Varki A (1998) A mutation in human CMP-sialic acid hydroxylase occurred after the Homo-Pan divergence. *Proc. Natl. Acad. Sci. U.S.A.* 95:11751–11756.
60. Higashi H, Nishi Y, Fukui Y, Ikuta K, Ueda S, Kato S, Fujita M, Nakano Y, Taguchi T, Sakai S (1984) Tumor-associated expression of glycosphingolipid Hanganutziu-Deicher antigen in human cancers. *Gann.* 75:1025–1029.
61. Hirabayashi Y, Higashi H, Kato S, Taniguchi M, Matsumoto M (1987) Occurrence of tumor-associated ganglioside antigens with Hanganutziu-Deicher antigenic activity on human melanomas. *Jpn. J. Cancer Res.* 78:614–620.

62. Hirabayashi Y, Kasakura H, Matsumoto M, Higashi H, Kato S, Kasai N, Naiki M (1987) Specific expression of unusual GM2 ganglioside with Hanganutziu-Deicher antigen activity on human colon cancers. *Jpn. J. Cancer Res.* 78:251–260.
63. Fukui Y, Maru M, Ohkawara K, Miyake T, Osada Y, Wang DQ, Ito T, Higashi H, Naiki M, Wakamiya N (1989) Detection of glycoproteins as tumor-associated Hanganutziu-Deicher antigen in human gastric cancer cell line, NUGC4. *Biochem. Biophys. Res. Commun.* 160:1149–1154.
64. Miyake M, Hashimoto K, Ito M, Ogawa O, Arai E, Hitomi S, Kannagi R (1990) The abnormal occurrence and the differentiation-dependent distribution of N-acetyl and N-glycolyl species of the ganglioside GM2 in human germ cell tumors. A study with specific monoclonal antibodies. *Cancer* 65:499–505.
65. Kawai T, Kato A, Higashi H, Kato S, Naiki M (1991) Quantitative determination of N-glycolylneuraminic acid expression in human cancerous tissues and avian lymphoma cell lines as a tumor-associated sialic acid by gas chromatography-mass spectrometry. *Cancer Res.* 51:1242–1246.
66. Bardor M, Nguyen DH, Diaz S, Varki A (2005) Mechanism of uptake and incorporation of the non-human sialic acid N-glycolylneuraminic acid into human cells. *J. Biol. Chem.* 280:4228–4237.
67. Marquina G, Waki H, Fernandez LE, Kon K, Carr A, Valiente O, Perez R, Ando S (1996) Gangliosides expressed in human breast cancer. *Cancer Res.* 56:5165–5171.
68. Carr A, Mullet A, Mazorra Z, Vazquez AM, Alfonso M, Mesa C, Rengifo E, Perez R, Fernandez LE (2000) A mouse IgG1 monoclonal antibody specific for N-glycolyl GM3 ganglioside recognized breast and melanoma tumors. *Hybridoma* 19:241–247.
69. Higashi H, Sasabe T, Fukui Y, Maru M, Kato S (1988) Detection of gangliosides as N-glycolylneuraminic acid-specific tumor-associated Hanganutziu-Deicher antigen in human retinoblastoma cells. *Jpn. J. Cancer Res.* 79:952–956.
70. de Leon J, Fernandez A, Mesa C, Clavel M, Fernandez LE (2006) Role of tumour-associated N-glycolylated variant of GM3 ganglioside in cancer progression: Effect over CD4 expression on T cells. *Cancer Immunol. Immunother.* 55:443–450.
71. Oliva JP, Valdes Z, Casaco A, Pimentel G, Gonzalez J, Alvarez I, Osorio M, Velazco M, Figueroa M, Ortiz R, Escobar X, Orozco M, Cruz J, Franco S, Diaz M, Roque L, Carr A, Vazquez AM, Mateos C, Rubio MC, Perez R, Fernandez LE (2006) Clinical evidences of GM3 (NeuGc) ganglioside expression in human breast cancer using the 14F7 monoclonal antibody labelled with (99 m)Tc. *Breast Cancer Res. Treat.* 96:115–121.
72. Portoukalian J, Zwingelstein G, Dore JF, Bourgoin JJ (1976) Studies of a ganglioside fraction extracted from human malignant melanoma. *Biochimie* 58:1285–1287.
73. Dippold WG, Knuth A, Meyer zum Buschenfelde KH (1984) Inhibition of human melanoma cell growth in vitro by monoclonal anti-GD3-ganglioside antibody. *Cancer Res.* 44:806–810.
74. Portoukalian J, Zwingelstein G, Abdul-Malak N, Dore JF (1978) Alteration of gangliosides in plasma and red cells of humans bearing melanoma tumors. *Biochem. Biophys. Res. Commun.* 85:916–920.
75. Bernhard H, Meyer zum Buschenfelde KH, Dippold WG (1989) Ganglioside GD3 shedding by human malignant melanoma cells. *Int. J. Cancer* 44:155–160.

76. Cheresh DA, Varki AP, Varki NM, Stallcup WB, Levine J, Reisfeld RA (1984) A monoclonal antibody recognizes an O-acylated sialic acid in a human melanoma-associated ganglioside. *J. Biol. Chem.* 259:7453–7459.
77. Cheresh DA, Reisfeld RA, Varki AP (1984) O-acetylation of disialoganglioside GD3 by human melanoma cells creates a unique antigenic determinant. *Science* 225:844–846.
78. Tsuchida T, Saxton RE, Morton DL, Irie RF (1987) Gangliosides of human melanoma. *J. Natl. Cancer Inst.* 78:45–54.
79. Sjoberg ER, Manzi AE, Khoo KH, Dell A, Varki A (1992) Structural and immunological characterization of O-acetylated GD2. Evidence that GD2 is an acceptor for ganglioside O-acetyltransferase in human melanoma cells. *J. Biol. Chem.* 267:16200–16211.
80. Ye JN, Cheung NK (1992) A novel O-acetylated ganglioside detected by anti-GD2 monoclonal antibodies. *Int. J. Cancer* 50:197–201.
81. Ravindranath MH, Morton DL, Irie RF (1989) An epitope common to gangliosides O-acetyl-GD3 and GD3 recognized by antibodies in melanoma patients after active specific immunotherapy. *Cancer Res.* 49:3891–3897.
82. Ritter G, Boosfeld E, Markstein E, Yu RK, Ren SL, Stallcup WB, Oettgen HF, Old LJ, Livingston PO (1990) Biochemical and serological characteristics of natural 9-O-acetyl GD3 from human melanoma and bovine buttermilk and chemically O-acetylated GD3. *Cancer Res.* 50:1403–1410.
83. Ritter G, Ritter-Boosfeld E, Adluri R, Calves M, Ren S, Yu RK, Oettgen HF, Old LJ, Livingston PO (1995) Analysis of the antibody response to immunization with purified O-acetyl G_{D3} gangliosides in patients with malignant melanoma. *Int. J. Cancer* 62:668–672.
84. Livingston PO (1988) Immunization with purified gangliosides. *J. Cell Biochem.* 12E:S-123.
85. Tai T, Cahan LD, Tsuchida T, Saxton RE, Irie RF, Morton DL (1985) Immunogenicity of melanoma-associated gangliosides in cancer patients. *Int. J. Cancer* 35:607–612.
86. Livingston PO, Ragupathi G (1997) Carbohydrate vaccines that induce antibodies against cancer. 2. Previous experience and future plans. *Cancer Immunol. Immunother.* 45:10–19.
87. Ritter G, Boosfeld E, Calves MJ, Oettgen HF, Old LJ, Livingston PO (1990) Antibody response to immunization with purified GD3 ganglioside and GD3 derivatives (lactones, amide and gangliosidol) in the mouse. *Immunobiology* 182:32–43.
88. Chapman PB, Wu D, Ragupathi G, Lu S, Williams L, Hwu WJ, Johnson D, Livingston PO (2004) Sequential immunization of melanoma patients with GD3 ganglioside vaccine and anti-idiotypic monoclonal antibody that mimics GD3 ganglioside. *Clin. Cancer Res.* 10:4717–4723.
89. Livingston PO, Natoli EJ, Calves MJ, Stockert E, Oettgen HF, Old LJ (1987) Vaccines containing purified GM2 ganglioside elicit GM2 antibodies in melanoma patients. *Proc. Natl. Acad. Sci. U.S.A.* 84:2911–2915.
90. Livingston PO, Calves MJ, Natoli EJJ (1987) Approaches to augmenting the immunogenicity of the ganglioside GM2 in mice: Purified GM2 is superior to whole cells. *J. Immunol.* 138:1524–1529.
91. Livingston PO, Ritter G, Srivastava P, Padavan M, Calves MJ, Oettgen HF, Old LJ (1989) Characterization of IgG and IgM antibodies induced in melanoma patients by immunization with purified GM2 ganglioside. *Cancer Res.* 49:7045–7050.

92. Livingston PO, Wong GY, Adluri S, Tao Y, Padavan M, Parente R, Hanlon C, Calves MJ, Helling F, Ritter G (1994) Improved survival in stage III melanoma patients with GM2 antibodies: A randomized trial of adjuvant vaccination with GM2 ganglioside. *J. Clin. Oncol.* 12:1036–1044.
93. Livingston PO, Adluri S, Helling F, Yao TJ, Kensil CR, Newman MJ, Marciani D (1994) Phase 1 trial of immunological adjuvant QS-21 with a GM2 ganglioside-keyhole limpet haemocyanin conjugate vaccine in patients with malignant melanoma. *Vaccine* 12:1275–1280.
94. Kitamura K, Livingston PO, Fortunato SR, Stockert E, Helling F, Ritter G, Oettgen HF, Old LJ (1995) Serological response patterns of melanoma patients immunized with a GM2 ganglioside conjugate vaccine. *Proc. Natl. Acad. Sci. U.S.A.* 92:2805–2809.
95. Livingston PO, Zhang S, Lloyd KO (1997) Carbohydrate vaccines that induce antibodies against cancer. 1. Rationale. *Cancer Immunol. Immunother.* 45:1–9.
96. Livingston P (1998) Ganglioside vaccines with emphasis on GM2. *Semin. Oncol.* 25:636–645.
97. Kundu SK, Diego I, Osovitz S, Marcus DM (1985) Glycosphingolipids of human plasma. *Arch. Biochem. Biophys.* 238:388–400.
98. Lin B, Kubushiro K, Akiba Y, Cui Y, Tsukazaki K, Nozawa S, Iwamori M (1997) Alteration of acidic lipids in human sera during the course of pregnancy: Characteristic increase in the concentration of cholesterol sulfate. *J. Chromatogr. B Biomed. Sci. Appl.* 704:99–104.
99. Sela BA, Iliopoulos D, Guerry D, Herlyn D, Koprowski H (1989) Levels of disialogangliosides in sera of melanoma patients monitored by sensitive thin-layer chromatography and immunostaining. *J. Natl. Cancer Inst.* 81:1489–1492.
100. Cahan LD, Irie RF, Singh R, Cassidenti A, Paulson JC (1982) Identification of a human neuroectodermal tumor antigen (OFA-I-2) as ganglioside GD2. *Proc. Natl. Acad. Sci. U.S.A.* 79:7629–7633.
101. Higashi H, Hirabayashi Y, Hirota M, Matsumoto M, Kato S (1987) Detection of ganglioside GM2 in sera and tumor tissues of hepatoma patients. *Jpn. J. Cancer Res.* 78:1309–1313.
102. Chang F, Li R, Noon K, Gage D, Ladisch S (1997) Human medulloblastoma gangliosides. *Glycobiology* 7:523–530.
103. Portoukalian J, David MJ, Gain P, Richard M (1993) Shedding of GD2 ganglioside in patients with retinoblastoma. *Int. J. Cancer* 53:948–951.
104. Wu ZL, Schwartz E, Seeger R, Ladisch S (1986) Expression of GD2 ganglioside by untreated primary human neuroblastomas. *Cancer Res.* 46:440–443.
105. Valentino L, Moss T, Olson E, Wang HJ, Elashoff R, Ladisch S (1990) Shed tumor gangliosides and progression of human neuroblastoma. *Blood* 75:1564–1567.
106. Valentino LA, Ladisch S (1992) Localization of shed human tumor gangliosides: Association with serum lipoproteins. *Cancer Res.* 52:810–814.
107. Alessandri G, Filippeschi S, Sinibaldi P, Mornet F, Passera P, Spreafico F, Cappa PM, Gullino PM (1987) Influence of gangliosides on primary and metastatic neoplastic growth in human and murine cells. *Cancer Res.* 47:4243–4247.
108. Valentino LA, Ladisch S (1994) Circulating tumor gangliosides enhance platelet activation. *Blood* 83:2872–2877.

109. Ladisch S, Li R, Olson E (1994) Ceramide structure predicts tumor ganglioside immunosuppressive activity. *Proc. Natl. Acad. Sci. U.S.A.* 91:1974–1978.
110. Valentino LA, Ladisch S (1996) Tumor gangliosides enhance alpha2 beta1 integrin-dependent platelet activation. *Biochim. Biophys. Acta* 1316:19–28.
111. Koochekpour S, Pilkington GJ (1996) Vascular and perivascular GD3 expression in human glioma. *Cancer Lett.* 104:97–102.
112. Koochekpour S, Merzak A, Pilkington GJ (1996) Vascular endothelial growth factor production is stimulated by gangliosides and TGF-beta isoforms in human glioma cells in vitro. *Cancer Lett.* 102:209–215.
113. Baumann H, Nudelman E, Watanabe K, Hakomori S (1979) Neutral fucolipids and fucogangliosides of rat hepatoma HTC and H35 cells, rat liver, and hepatocytes. *Cancer Res.* 39:2637–2643.
114. Nilsson O, Månsson JE, Brezicka FT, Holmgren J, Lindholm L, Sörensen S, Yngvason F, Svennerholm L (1984) Fucosyl-GM 1: A ganglioside associated with small cell lung carcinomas. *Glycoconj. J.* 1:43–49.
115. Nilsson O, Brezicka FT, Holmgren J, Sorenson S, Svennerholm L, Yngvason F, Lindholm L (1986) Detection of a ganglioside antigen associated with small cell lung carcinomas using monoclonal antibodies directed against fucosyl-G_{M1}. *Cancer Res.* 46:1403–1407.
116. Brezicka FT, Olling S, Nilsson O, Bergh J, Holmgren J, Sorenson S, Yngvason F, Lindholm L (1989) Immunohistological detection of fucosyl-G_{M1} ganglioside in human lung cancer and normal tissues with monoclonal antibodies. *Cancer Res.* 49:1300–1305.
117. Vangsted AJ, Clausen H, Kjeldsen TB, White T, Sweeney B, Hakomori S, Drivsholm L, Zeuthen J (1991) Immunochemical detection of a small cell lung cancer-associated ganglioside (FucG_{M1}) antigen in serum. *Cancer Res.* 51:2879–2884.
118. Pallesen T, Vangsted A, Drivsholm L, Clausen H, Zeuthen J, Wallin H (1992) Serum immunoassay of a small cell lung cancer associated ganglioside: Development of a sensitive scintillation proximity assay. *Glycoconj. J.* 9:331–335.
119. Vangsted AJ, Zeuthen J (1993) Monoclonal antibodies for diagnosis and potential therapy of small cell lung cancer—the ganglioside antigen fucosyl-G_{M1}. *Acta Oncol.* 32:845–851.
120. Vangsted A, Drivsholm L, Andersen E, Pallesen T, Zeuthen J, Wallin H (1994) New serum markers for small-cell lung cancer. I. The ganglioside fucosyl-G_{M1}. *Cancer Detect. Prev.* 18:221–229.
121. Yoshino H, Ariga T, Latov N, Miyatake T, Kushi Y, Kasama T, Handa S, Yu RK (1993) Fucosyl-G_{M1} in human sensory nervous tissue is a target antigen in patients with autoimmune neuropathies. *J. Neurochem.* 61:658–663.
122. Yuki N, Ariga T (1997) Antibodies to fucogangliosides in neurological diseases. *J. Neurol. Sci.* 150:81–84.
123. Livingston PO, Hood C, Krug LM, Warren N, Kris MG, Brezicka T, Ragupathi G (2005) Selection of GM2, fucosyl GM1, globo H and polysialic acid as targets on small cell lung cancers for antibody mediated immunotherapy. *Cancer Immunol. Immunother.* 54:1018–1025.
124. Smith DF, Zopf DA, Ginsburg V (1978) Fractionation of sialyl oligosaccharides of human milk by ion-exchange chromatography. *Anal. Biochem.* 85:602–608.
125. Fukuda MN, Bothner B, Lloyd KO, Rettig WJ, Tiller PR, Dell A (1986) Structure of glycosphingolipids isolated from human embryonal carcinoma cells. The presence

of mono- and disialosyl glycolipids with blood group type 1 sequence. *J. Biol. Chem.* 261:5145–5153.

126. Magnani JL, Brockhaus M, Smith DF, Ginsburg V, Blaszczyk M, Mitchell KF, Steplewski Z, Koprowski H (1981) A monosialoganglioside is a monoclonal antibody-defined antigen of colon carcinoma. *Science* 212:55–56.

127. Magnani JL, Nilsson B, Brockhaus M, Zopf D, Steplewski Z, Koprowski H, Ginsburg V (1982) A monoclonal antibody-defined antigen associated with gastrointestinal cancer is a ganglioside containing sialylated lacto-N-fucopentaose II. *J. Biol. Chem.* 257:14365–14369.

128. Magnani JL, Smith DF, Ginsburg V (1980) Detection of gangliosides that bind toxin: Direct binding of 125 I-labeled toxin to thin-layer chromatography. *Anal. Biochem.* 109:399–402.

129. Young WW, Jr, MacDonald EM, Nowinski RC, Hakomori SI (1979) Production of monoclonal antibodies specific for two distinct steric portions of the glycolipid ganglio-N-triosylceramide (asialo GM2). *J. Exp. Med.* 150:1008–1019.

130. Brockhaus M, Magnani JL, Blaszczyk M, Steplewski Z, Koprowski H, Karlsson KA, Larson G, Ginsburg V (1981) Monoclonal antibodies directed against the human Leb blood group antigen. *J. Biol. Chem.* 256:13223–13225.

131. Magnani JL (2004) The discovery, biology, and drug development of sialyl Lea and sialyl Lex. *Arch. Biochem. Biophys.* 426:122–131.

132. Steinberg W (1990) The clinical utility of the CA 19-9 tumor-associated antigen. *Am. J. Gastroenterol.* 85:350–355.

133. Perkins GL, Slater ED, Sanders GK, Prichard JG (2003) Serum tumor markers. *Am. Fam. Physician.* 68:1075–1082.

134. Locker GY, Hamilton S, Harris J, Jessup JM, Kemeny N, Macdonald JS, Somerfield MR, Hayes DF, Bast RC, Jr. (2006) ASCO 2006 update of recommendations for the use of tumor markers in gastrointestinal cancer. *J. Clin. Oncol.* 24:5313–5327.

135. Nudelman E, Fukushi Y, Levery SB, Higuchi T, Hakomori S (1986) Novel fucolipids of human adenocarcinoma: IV. Disialosyl Le^a antigen (III^{4}FucIII^{6}NeuAcIV^{3}NeuAcLc$_4$) of human colonic adenocarcinoma and the monoclonal antibody (FH7) defining this structure. *J. Biol. Chem.* 261:5487–5495.

136. Itzkowitz SH, Yuan M, Fukushi Y, Lee H, Shi ZR, Zurawski V, Jr, Hakomori S, Kim YS (1988) Immunohistochemical comparison of Le^a, monosialosyl Le^a (CA 19-9), and disialosyl Le^a antigens in human colorectal and pancreatic tissues. *Cancer Res.* 48:3834–3842.

137. Kannagi R, Kitahara A, Itai S, Zenita K, Shigeta K, Tachikawa T, Noda A, Hirano H, Abe M, Shin S, Fukushi Y, Hakomori S, Imura H (1988) Quantitative and qualitative characterization of human cancer-associated serum glycoprotein antigens expressing epitopes consisting of sialyl or sialyl-fucosyl type 1 chain. *Cancer Res.* 48:3856–3863.

138. Young WW, Jr., Mills SE, Lippert MC, Ahmed P, Lau SK (1988) Deletion of antigens of the Lewis a/b blood group family in human prostatic carcinoma. *Am. J. Pathol.* 131:578–586.

139. Yang H-J, Hakomori S (1971) A sphingolipid having a novel type of ceramide and lacto-*N*-fucopentaose III. *J. Biol. Chem.* 246:1192–1200.

140. Kobata A, Ginsburg V (1969) Oligosaccharides of human milk. II. Isolation and characterization of a new pentasaccharide, lacto-N-fucopentaose 3. *J. Biol. Chem.* 244:5496–5502.

141. Hakomori S, Andrews HD (1970) Sphingoglycolipids with Leb activity, and the co-presence of Lea-, Leb-glycolipids in human tumor tissue. *Biochim. Biophys. Acta* 202:225–228.
142. Solter D, Knowles BB (1978) Monoclonal antibody defining a stage-specific mouse embryonic antigen (SSEA-1). *Proc. Natl. Acad. Sci. U.S.A.* 75:5565–5569.
143. Hakomori S, Nudelman E, Levery SB, Solter D, Knowles BB (1981) The hapten structure of a developmentally regulated glycolipid antigen (SSEA-1) isolated from human erythrocytes and adenocarcinoma: A preliminary note. *Biochem. Biophys. Res. Commun.* 100:1578–1586.
144. Gooi HC, Feizi T, Kapadia A, Knowles BB, Solter D, Evans MJ (1981) Stage-specific embryonic antigen involves alpha 1 goes to 3 fucosylated type 2 blood group chains. *Nature* 292:156–158.
145. Solter D, Knowles BB (1979) Developmental stage-specific antigens during mouse embryogenesis. *Curr. Top. Dev. Biol.* 13(Pt 1):139–165.
146. Knowles BB, Aden DP, Solter D (1978) Monoclonal antibody detecting a stage-specific embryonic antigen (SSEA-1) on preimplantation mouse embryos and teratocarcinoma cells. *Curr. Top. Microbiol. Immunol.* 81:51–53.
147. Knowles BB, Rappaport J, Solter D (1982) Murine embryonic antigen (SSEA-1) is expressed on human cells and structurally related human blood group antigen I is expressed on mouse embryos. *Dev. Biol.* 93:54–58.
148. Hakomori S, Nudelman E, Kannagi R, Levery SB (1982) The common structure in fucosyllactosaminolipids accumulating in human adenocarcinomas, and its possible absence in normal tissue. *Biochem. Biophys. Res. Commun.* 109:36–44.
149. Hakomori S, Nudelman E, Levery SB, Kannagi R (1984) Novel fucolipids accumulating in human adenocarcinoma. I. Glycolipids with di- or trifucosylated type 2 chain. *J. Biol. Chem.* 259:4672–4680.
150. Kannagi R, Nudelman E, Levery SB, Hakomori S (1982) A series of human erythrocyte glycosphingolipids reacting to the monoclonal antibody directed to a developmentally regulated antigen SSEA-1. *J. Biol. Chem.* 257:14865–14874.
151. Fukuda MN, Dell A, Oates JE, Wu P, Klock JC, Fukuda M (1985) Structures of glycosphingolipids isolated from human granulocytes. The presence of a series of linear poly-N-acetyllactosaminylceramide and its significance in glycolipids of whole blood cells. *J. Biol. Chem.* 260:1067–1082.
152. Symington FW, Hedges DL, Hakomori S (1985) Glycolipid antigens of human polymorphonuclear neutrophils and the inducible HL-60 myeloid leukemia line. *J. Immunol.* 134:2498–2506.
153. Fukushi Y, Hakomori S, Nudelman E, Cochran N (1984) Novel fucolipids accumulating in human adenocarcinoma. II. Selective isolation of hybridoma antibodies that differentially recognize mono-, di-, and trifucosylated type 2 chain. *J. Biol. Chem.* 259:4681–4685.
154. Fukushi Y, Hakomori S, Shepard T (1984) Localization and alteration of mono-, di-, and trifucosyl $\alpha 1 - 3$ type 2 chain structures during human embryogenesis and in human cancer. *J. Exp. Med.* 160:506–520.
155. Fukushi Y, Orikasa S, Shepard T, Hakomori S (1986) Changes of Lex and dimeric Lex haptens and their sialylated antigens during development of human kidney and kidney tumors. *J. Urol.* 135:1048–1056.

156. Itzkowitz SH, Yuan M, Fukushi Y, Palekar A, Phelps PC, Shamsuddin AM, Trump BF, Hakomori S, Kim YS (1986) Lewisx- and sialylated Lewisx-related antigen expression in human malignant and nonmalignant colonic tissues. *Cancer Res.* 46:2627–2632.
157. Kim YS, Itzkowitz SH, Yuan M, Chung Y, Satake K, Umeyama K, Hakomori S (1988) Lex and Ley antigen expression in human pancreatic cancer. *Cancer Res.* 48:475–482.
158. Nakasaki H, Mitomi T, Noto T, Ogoshi K, Hanaue H, Tanaka Y, Makuuchi H, Clausen H, Hakomori S (1989) Mosaicism in the expression of tumor-associated carbohydrate antigens in human colonic and gastric cancers. *Cancer Res.* 49:3662–3669.
159. Rauvala H (1976) Gangliosides of human kidney. *J. Biol. Chem.* 251:7517–7520.
160. Phillips ML, Nudelman E, Gaeta FC, Perez M, Singhal AK, Hakomori S, Paulson JC (1990) ELAM-1 mediates cell adhesion by recognition of a carbohydrate ligand, sialyl-Lex. *Science* 250:1130–1132.
161. Tiemeyer M, Swiedler SJ, Ishihara M, Moreland M, Schweingruber H, Hirtzer P, Brandley BK (1991) Carbohydrate ligands for endothelial-leukocyte adhesion molecule 1. *Proc. Natl. Acad. Sci. U.S.A.* 88:1138–1142.
162. Berg EL, Robinson MK, Mansson O, Butcher EC, Magnani JL (1991) A carbohydrate domain common to both sialyl Le(a) and sialyl Le(X) is recognized by the endothelial cell leukocyte adhesion molecule ELAM-1. *J. Biol. Chem.* 266:14869–14872.
163. Takada A, Ohmori K, Takahashi N, Tsuyuoka K, Yago A, Zenita K, Hasegawa A, Kannagi R (1991) Adhesion of human cancer cells to vascular endothelium mediated by a carbohydrate antigen, sialyl Lewis A. *Biochem. Biophys. Res. Commun.* 179:713–719.
164. Paulson JC, Prieels JP, Glasgow LR, Hill RL (1978) Sialyl- and fucosyltransferases in the biosynthesis of asparaginyl-linked oligosaccharides in glycoproteins. Mutually exclusive glycosylation by beta-galactoside alpha2 goes to 6 sialyltransferase and N-acetylglucosaminide alpha1 goes to 3 fucosyltransferase. *J. Biol. Chem.* 253:5617–5624.
165. Levery SB, Nudelman E, Kannagi R, Symington FW, Andersen NH, Clausen H, Baldwin M, Hakomori S (1988) ^{1}H-NMR analysis of type-2 chain lacto-gangliosides. Confirmation of structure of a novel cancer-associated fucoganglioside, α-NeuAc-(2 → 6)-β-D-Gal$p$-(1 → 4)-β-D-Glc$p$NAc-(1 → 3)-β-D-Gal$p$-(1 → 4)-[α-L-Fuc$p$-(1 → 3)]-β-D-Glc$p$NAc-(1 → 3)-β-D-Gal$p$-(1 → 4)-β-D-Glc$p$-(1 → 1)-Cer (VI^{6}NeuAcIII^{3}FucnLc$_6$ Cer). *Carbohydr. Res.* 178:121–144.
166. Korekane H, Tsuji S, Noura S, Ohue M, Sasaki Y, Imaoka S, Miyamoto Y (2007) Novel fucogangliosides found in human colon adenocarcinoma tissues by means of glycomic analysis. *Anal. Biochem.* 364:37–50.
167. Holmes EH, Ostrander GK, Hakomori S (1985) Enzymatic basis for the accumulation of glycolipids with X and dimeric X determinants in human lung cancer cells (NCI-H69). *J. Biol. Chem.* 260:7619–7627.
168. Fukushi Y, Nudelman E, Levery SB, Hakomori S, Rauvala H (1984) Novel fucolipids accumulating in human adenocarcinoma. III. A hybridoma antibody (FH6) defining a human cancer-associated difucoganglioside (VI^{3}NeuAcV^{3}III^{3}Fuc$_2$nLc$_6$). *J. Biol. Chem.* 259:10511–10517.
169. Macher BA, Buehler J, Scudder P, Knapp W, Feizi T (1988) A novel carbohydrate, differentiation antigen on fucogangliosides of human myeloid cells recognized by monoclonal antibody VIM-2. *J. Biol. Chem.* 263:10186–10191.

170. Nudelman ED, Levery SB, Stroud MR, Salyan ME, Abe K, Hakomori S (1988) A novel tumor-associated, developmentally regulated glycolipid antigen defined by monoclonal antibody ACFH-18. *J. Biol. Chem.* 263:13942–13951.
171. Stroud MR, Handa K, Ito K, Salyan ME, Fang H, Levery SB, Hakomori S, Reinhold BB, Reinhold VN (1995) Myeloglycan, a series of E-selectin-binding polylactosaminolipids found in normal human leukocytes and myelocytic leukemia HL60 cells. *Biochem. Biophys. Res. Commun.* 209:777–787.
172. Stroud MR, Handa K, Salyan ME, Ito K, Levery SB, Hakomori S, Reinhold BB, Reinhold WN (1996) Monosialogangliosides of human myelogenous leukemia HL60 cells and normal human leukocytes. 1. Separation of E-selectin binding from nonbinding gangliosides, and absence of sialosyl-Le^x having tetraosyl to octaosyl core. *Biochemistry* 35:758–769.
173. Stroud MR, Handa K, Salyan ME, Ito K, Levery SB, Hakomori S, Reinhold BB, Reinhold VN (1996) Monosialogangliosides of human myelogenous leukemia HL60 cells and normal human leukocytes. 2. Characterization of E-selectin binding fractions, and structural requirements for physiological binding to E-selectin. *Biochemistry* 35:770–778.
174. Hakomori S (2003) Structure, organization, and function of glycosphingolipids in membrane. *Curr. Opin. Hematol.* 10:16–24.
175. Handa K, Stroud MR, Hakomori S (1997) Sialosyl-fucosyl Poly-LacNAc without the sialosyl-Le^x epitope as the physiological myeloid cell ligand in E-selectin-dependent adhesion: Studies under static and dynamic flow conditions. *Biochemistry* 36:12412–12420.
176. Abe K, McKibbin JM, Hakomori S (1983) The monoclonal antibody directed to difucosylated type 2 chain (Fuc1α → 2Galβ1→ 4[Fucα1 → 3]GlcNAc; Y Determinant). *J. Biol. Chem.* 258:11793–11797.
177. Clausen H, Hakomori S (1989) ABH and related histo-blood group antigens: Immunochemical differences in carrier isotypes and their distribution. *Vox Sang.* 56:1–20.
178. McKibbin JM, Spencer WA, Smith EL, Mansson JE, Karlsson KA, Samuelsson BE, Li YT, Li SC (1982) Lewis blood group fucolipids and their isomers from human and canine intestine. *J. Biol. Chem.* 257:755–760.
179. Abe K, Hakomori S, Ohshiba S (1986) Differential expression of difucosyl type 2 chain (LeY) defined by monoclonal antibody AH6 in different locations of colonic epithelia, various histological types of colonic polyps, and adenocarcinomas. *Cancer Res.* 46:2639–2644.
180. Miyake M, Zenita K, Tanaka O, Okada Y, Kannagi R (1988) Stage-specific expression of SSEA-1-related antigens in the developing lung of human embryos and its relation to the distribution of these antigens in lung cancers. *Cancer Res.* 48:7150–7158.
181. Levery SB, Nudelman ED, Andersen NH, Hakomori S (1986) ^{1}H-NMR analysis of glycolipids possessing mono- and multimeric X and Y haptens: Characterization of two novel extended Y structures from human adenocarcinoma. *Carbohydr. Res.* 151:311–328.
182. Kaizu T, Levery SB, Nudelman E, Stenkamp RE, Hakomori S (1986) Novel fucolipids of human adenocarcinoma: VI. Monoclonal antibody specific for trifucosyl Le^y (III^{3}FucV^{3}FucVI^{2}FucnLc$_6$) and a possible three-dimensional epitope structure. *J. Biol. Chem.* 261:11254–11258.

183. Kim YS, Yuan M, Itzkowitz SH, Sun QB, Kaizu T, Palekar A, Trump BF, Hakomori S (1986) Expression of LeY and extended LeY blood group-related antigens in human malignant, premalignant, and nonmalignant colonic tissues. *Cancer Res.* 46:5985–5992.
184. Watanabe M, Ohishi T, Kuzuoka M, Nudelman ED, Stroud MR, Kubota T, Kodairo S, Abe O, Hirohashi S, Shimosato Y, Hakomori S (1991) In vitro and in vivo antitumor effects of murine monoclonal antibody NCC-ST-421 reacting with dimeric Le(a) (Le(a)/Le(a)) epitope. *Cancer Res.* 51:2199–2204.
185. Stroud MR, Levery SB, Nudelman ED, Salyan ME, Towell JA, Roberts CE, Watanabe M, Hakomori S (1991) Extended type 1 chain glycosphingolipids: dimeric Le^a (III^{4}V^{4}Fuc$_2$Lc$_6$) as human tumor-associated antigen. *J. Biol. Chem.* 266:8439–8446.
186. Stroud MR, Levery SB, Hakomori S (1993) Extended type 1 chain glycosphingolipids: Le^a-Le^a (dimeric Le^a) and Le^b-Le^a as human tumor associated antigens. In Garegg PJ, Lindberg AA, Eds. *Carbohydrate Antigens*. American Chemical Society Monographs, Vol. 519. American Chemical Society, Washington, DC, pp. 159–175.
187. Ishizuka H, Watanabe M, Kubota T, Matsuzaki SW, Otani Y, Kitajima M (1998) Antitumor activity of murine monoclonal antibody NCC-ST-421 on human cancer cells by inducing apoptosis. *Anticancer Res.* 18:2513–2518.
188. Kawano Y, Watanabe M, Kubota T, Nishibori H, Kurihara N, Teramoto T, Kitajima M (1997) A streptococcal preparation (OK-432) enhances monoclonal antibody NCC-ST-421 cytotoxicity against human colon cancer. *Anticancer Res.* 17:2449–2453.
189. Stroud MR, Levery SB, Salyan ME, Roberts CE, Hakomori S (1992) Extended type-1 chain glycosphingolipid antigens. Isolation and characterization of trifucosyl-Le^b antigen (III^{4}V^{4}VI^{2}Fuc$_3$Lc$_6$). *Eur. J. Biochem.* 203:577–586.
190. Ito H, Tashiro K, Stroud MR, Orntoft TF, Meldgaard P, Singhal AK, Hakomori S (1992) Specificity and immunobiological properties of monoclonal antibody IMH2, established after immunization with Le(b)/Le(a) glycosphingolipid, a novel extended type 1 chain antigen. *Cancer Res.* 52:3739–3745.
191. Kannagi R, Levery SB, Hakomori S (1985) Le^a-active heptaglycosylceramide, a hybrid of type 1 and type 2 chain, and the pattern of glycolipids with Le^a, Le^b, X (Le^x), and Y (Le^y) determinants in human blood cell membranes (ghosts). Evidence that type 2 chain can elongate repetitively but type 1 chain cannot. *J. Biol. Chem.* 260:6410–6415.
192. Pettijohn DE, Stranahan PL, Due C, Ronne E, Sorensen HR, Olsson L (1987) Glycoproteins distinguishing non-small cell from small cell human lung carcinoma recognized by monoclonal antibody 43-9F. *Cancer Res.* 47:1161–1169.
193. Martensson S, Due C, Pahlsson P, Nilsson B, Eriksson H, Zopf D, Olsson L, Lundblad A (1988) A carbohydrate epitope associated with human squamous lung cancer. *Cancer Res.* 48:2125–2131.
194. Sabharwal H, Nilsson B, Gronberg G, Chester MA, Dakour J, Sjoblad S, Lundblad A (1988) Oligosaccharides from feces of preterm infants fed on breast milk. *Arch. Biochem. Biophys.* 265:390–406.
195. Stroud MR, Levery SB, Martensson S, Salyan ME, Clausen H, Hakomori S (1994) Human tumor-associated Le^a-Le^x hybrid carbohydrate antigen IV^3(Galβ1 → 3[Fucα1 → 4]GlcNAc)III^{3}FucnLc$_4$ defined by monoclonal antibody 43-9F: Enzymatic synthesis, structural characterization, and comparative reactivity with various antibodies. *Biochemistry* 33:10672–10680.

9

SEMISYNTHETIC AND FULLY SYNTHETIC CARBOHYDRATE-BASED CANCER VACCINES

Therese Buskas, Pamela Thompson, and Geert-Jan Boons
Complex Carbohydrate Research Center, University of Georgia, Athens, Georgia 30602

9.1 INTRODUCTION TO CANCER VACCINES

Traditionally, the treatment of cancer has primarily relied on surgery, chemotherapy, and radiation. As research has unraveled more details of the intrinsic underlying biological and immunological mechanisms of cancer, new approaches such as angiogenesis inhibitor therapy, gene therapy, and biological or immunotherapy [1–7] have emerged as possible treatments. One example of immunotherapy is the passive administration of antibodies, which has been shown to mediate tumor regression in certain patients [8–10]. Currently, Rituxan (Genentec Biogen Idec) and Herceptin (Genentec Oncology) are two monoclonal antibody (mAbs) drugs that are used for the treatment of non-Hodgkin's lymphoma and breast cancer, respectively [9].

Cancer vaccines are also emerging as a possible treatment for cancer [1, 2, 7, 11–15]. Today, two types of prophylactic cancer vaccines have been approved for human use. These vaccines protect individuals against contracting viral infections that have been associated with an increased chance of developing certain types of cancers. These vaccines are aimed at the hepatitis B virus (HBV), associated with certain liver cancers (Engerix-B and Twinrix, GlaxoSmithKline Biologicals), and the

Carbohydrate-Based Vaccines and Immunotherapies. Edited by Zhongwu Guo and Geert-Jan Boons
Copyright © 2009 John Wiley & Sons, Inc.

human papilloma virus (HPV), which is closely linked to cervical carcinomas (Gardasil, Merck and Company; Cervarix, GlaxoSmithKline Biologicals). The HPV vaccine protects against two types of virus, HPV 16 and 18, which are the cause of 70% of cervical cancer cases [16].

Most of the experimental cancer vaccines are therapeutic vaccines, that is, vaccines that are administered to an already diseased person [11, 17]. The concept of these vaccines is that upon administration, the vaccine will trigger an immune response that can reduce the tumor size or target metastasized tumor cells that may have evaded surgery or other primary therapies. A prerequisite of a cancer vaccine is the ability to distinguish tumor cells from normal cells. Therefore, an ideal target antigen should be expressed exclusively, or in abundance, by malignant cells and be accessible to the immune system on the cell surface. Several therapeutic cancer vaccines, which are aimed at, for example, breast, lung, colon, and prostate cancer, have reached evaluation in clinical trials.

Cancer vaccines derived from whole tumor cells, irradiated to render them unable to replicate, were among the first therapeutic cancer vaccines tested [18]. An advantage of using tumor cells is that they express multiple antigens, and thus a specific antigen does not need to be identified. The primary tumor and metastases may display different antigens, and therefore cell-based vaccines may reduce the possibility of antigenic escape. However, tumor cells can possess properties that allow for immune evasion and as a result may have limited immunogenicity. This has prompted the use of gene transfer strategies to tumor cell lines to enhance the immune response by providing immunostimulatory molecules [19–23]. One such vaccine based on genetically modified prostate cancer cell lines, GVAX immunotheraphy for prostate cancer (Cell Genesys), has reached phase III clinical trials [24].

The identification of tumor-associated antigens makes it possible to develop subunit vaccines. Tumor antigens may be tumor restricted or more widely expressed, they may be mutated, over- or aberrantly expressed [25]. Although many tumor-specific mutations have been identified in intracellular proteins, a significant number have been observed in the extra-cellular domain of membrane proteins. Consequently, several protein- and peptide-based cancer vaccines have been designed that are aimed at exploiting cellular immune responses toward this type of antigens [26, 27].

It is also well established that tumor cells may display aberrant glycosylation patterns due to up- and/or down-regulation of glycosyl transferases, which may lead to both overexpression and truncation of oligosaccharides of cell surface glycoproteins and glycolipids [28–30]. Truncated glycosylation also renders parts of the peptide backbone, which normally is shielded by the glycan, more "visible" to the immune system. Thus, tumor-associated carbohydrate and glycopeptide antigens provide a viable target for the development of tumor-selective or tumor-specific carbohydrate-based cancer vaccines [31–38].

9.2 TUMOR-ASSOCIATED CARBOHYDRATE ANTIGENS (TACAs)

It is well established that protein- and lipid-bound oligosaccharides found on the surface of cells are involved in many essential processes impacting eukaryotic biology

9.2 TUMOR-ASSOCIATED CARBOHYDRATE ANTIGENS (TACAs)

and disease [39–42]. Examples of such processes include fertilization, embryogenesis, neuronal development, hormone activities, the proliferation of cells, and their organization into specific tissues. Furthermore, the carbohydrates of host cells are employed by pathogens for cell entry or immunological evasion. Carbohydrates are also capable of inducing a protective antibody response, which is crucial to the survival of the organism during infection [43–45].

It is not surprising that malignant cells, which display differences in cell adhesion and cell motility, display altered cell surface glycosylation. Indeed, remarkable changes in cell surface carbohydrates occur with tumor progression, which appears to be intimately associated with the dreaded state of metastasis [46]. This abnormal glycosylation is an important criterion for the stage, direction, and fate of tumor progression. Numerous studies have shown that abnormal glycosylation in primary tumors is strongly correlated with the poor survival rates of patients.

Several tumor-associated glycosphingolipids have been identified as adhesion molecules and, not surprisingly, these compounds promote tumor cell invasion and metastasis [47]. For example, the Lewis antigens sialyl Le^a (SLe^a), SLe^x, SLe^x-Le^x, and Le^y are identified as human tumor-associated antigens (Fig. 9.1) [48, 49]. The SLe^a saccharide is found overexpressed on a range of tumor cells, and it has been found that patients with pancreatic, prostate, and colon cancer have elevated levels of this antigen in their serum [50]. SLe^a was found as the epitope for the CA19-9 monoclonal antibody (mAb) [50], which consequently has been developed as a serum-based diagnostic assay particularly for pancreatic tumors [51]. The Le^y tetrasaccharide is overexpressed on a range of carcinomas including ovary, breast, colon, prostate, and non-small-cell lung cancers. However, the Lewis antigens have also been

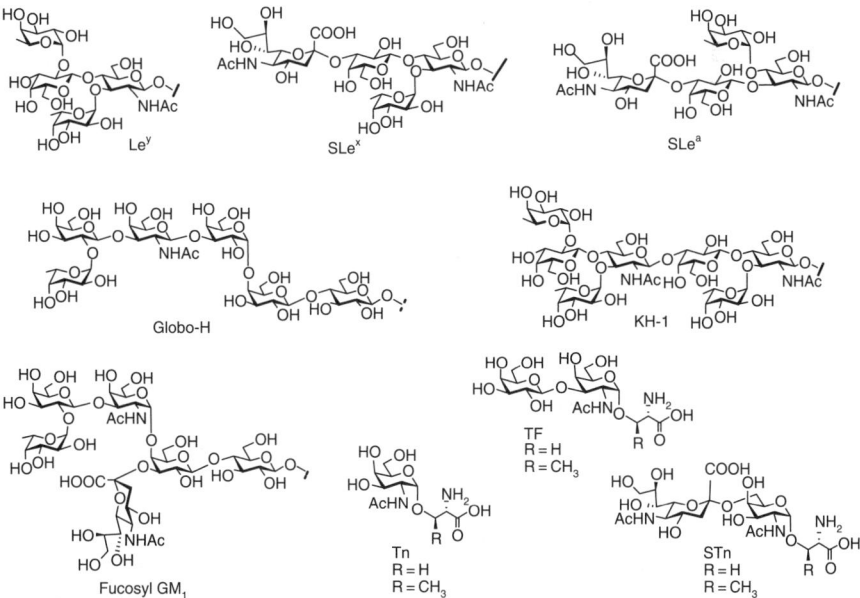

Figure 9.1 Human tumor-associated antigens.

identified as ligands for the endothelial cell surface receptors E- and P-selectin that mediate the adhesion of circulating leukocytes [46, 52]. The binding of cancer cells to the endothelium and their transport through endothelial cells lining the capillary wall are critical for the escape of metastatic cancer cells from circulation and their spread to other tissue. Tumor cells that express the Lewis antigens have the ability to utilize the adhesion cascade available to normal circulating cells, and their overexpression is consequently intimately linked with metastasis [47, 49].

Other examples of well-documented tumor-associated carbohydrate antigens include the KH-1 antigen and Globo-H (Fig. 9.1). The KH-1 antigen, which displays the heterodimeric Le^y-Le^x heptasaccharide, was isolated from human colonic adenocarcinoma cells [53]. The antigen has only been found on the surface of these cells and has never been isolated from normal colonic tissue, thus providing a highly specific marker for colon malignancies [54, 55]. Globo-H [56], also known as the MBr-1 antigen, was isolated from human breast cancer cells using a monoclonal antibody, MBr-1 [57, 58]. This oligosaccharide has also been identified as a tumor-associated antigen for ovary, colon, prostate, lung, and small-cell lung cancers [59].

A large number of epithelial cancers such as breast, ovarian, colorectal, pancreatic, and prostate cancers exhibit striking alterations in the level of expression and glycosylation profile of mucins [60, 61]. Mucins are high-molecular-weight glycoproteins containing numerous O-linked carbohydrate side chains and are found at the apical surface of epithelial cells or as extracellular secreted glycoproteins. Twenty different human mucins are known, which posses a similar overall architecture with an N-terminal region followed by a region containing a variable number of tandem repeat (VNTR) units. The tandem repeats of the different mucins, which are rich in threonine, serine, and proline residues, contain 8–23 amino acids. The cell membrane tethered mucins are involved in diverse functions ranging from shielding the airway epithelium against pathogenic infection to regulating cellular signaling and transcription [61]. Many of the mucins have been implicated in disease such as pulmonary diseases and cancer. For example, MUC-1 (polymorphic epithelial mucin, PEM) is found overexpressed in more than 90% of breast carcinomas [62, 63] and is also found overexpressed in patient sera and have, as a consequence, found clinical use as a marker (CA15-3, Truquant, CASA) [64] for breast cancer. MUC-1 has also been associated with other carcinomas such as ovarian, lung, colon, and pancreatic carcinomas. MUC-1 is a transmembrane protein with a large and highly glycosylated extra-cellular domain consisting of multiple 20-amino-acid repeating units (HGVTSAPDTRP-APGSTAPPA) of which each repeat has 5 potential sites for O-glycosylation [65]. In cancer cells, MUC-1 is overexpressed and deficiently glycosylated due to a down-regulation of glucosylaminyl transferase 1 (GnT-1) [66–68]. As a result, tumor-associated MUC-1 carries the antigens Tn (αGalNAc-Thr), STn (αNeu5Ac-(2,6)-αGalNAc-Thr), and the Thomsen–Friedenreich (TF or T) antigen [αGal(1,3)-αGalNAc-Thr (Fig. 9.1)] [65, 69–71].

Humoral responses to MUC-1 have been observed in benign diseases and carcinoma patients, and it has been found that the presence of circulating antibodies against MUC-1 at the time of cancer diagnosis correlates with a favorable disease outcome in breast cancer patients [60]. Antibodies induced by MUC-1 isolated from tumor tissues have identified the PDTRP (Pro-Asp-Thr-Arg-Pro) peptide motif as the immunodominant domain of the MUC-1 tandem repeat [72, 73]. The specificity of these anti-MUC-1 antibodies has been verified employing synthetic Tn- and T antigens [74–77]. Furthermore, conformational studies by nuclear magnetic resonance (NMR) complemented by light-scattering measurements have indicated that de-glycosylation of MUC-1 results in a less extended and more globular structure [78]. Similar studies using MUC-1-related O-glycopeptides have shown that the carbohydrate moieties exert conformational effects, which may rationalize differences observed in antibody binding of MUC-1-related glycopeptides and peptides [78].

9.3 CARBOHYDRATE-BASED CANCER VACCINES

The development of carbohydrate-based cancer vaccines is by no means a trivial task and several hurdles need to be overcome. The heterogeneity of cell surface glycosylation makes the isolation of tumor-associated carbohydrate antigens in well-defined forms and reasonable amounts an almost impossible task. Fortunately, this obstacle can be addressed by synthetic organic chemistry, which can provide homogeneous oligosaccharide antigens of high purity, indisputable structural integrity, and in relatively large amounts. Recent advances in organic synthesis of oligosaccharides has equipped chemists with more sophisticated tools, yet the preparation of many of these large complex antigens still represents a considerable undertaking.

The low antigenicity of tumor-associated carbohydrate antigens signifies an additional hurdle. This observation is not surprising as tumor-associated saccharides are "self-antigens," that is, they are present in low concentration on the surface of certain normal cells, and consequently receive tolerance by the immune system. This immunotolerance is further reinforced as the growing tumor is shedding these antigens into the bloodstream. As a consequence, induction of high-affinity IgG antibodies against tumor-associated carbohydrate antigens has proven more challenging than the induction of similar antibodies against viral and bacterial carbohydrate antigens. Indeed, high titers of IgG antibodies have been referred to as the "holy grail" for carbohydrate-based tumor vaccines [79]. The question thus posed is how to trick the immune system to break tolerance and induce a response to these tumor-associated antigens. The inherently T-cell-independent nature of oligosaccharides further complicates carbohydrate-based cancer vaccine development. The production of high-affinity IgG antibodies requires antigen recognition by B and T lymphocytes and dendritic cells. Consequently, the inability of carbohydrates to activate T lymphocytes results in formation of exclusively low-affinity IgM antibodies and lack of immunological memory. Below the immune response to carbohydrates and the interaction between the innate and adaptive immune responses will be outlined.

9.4 HUMORAL IMMUNE RESPONSE TO CARBOHYDRATES

Antibodies that target tumor-related carbohydrate and glycopeptide antigens have been shown in preclinical and clinical settings to have the ability to eliminate circulating tumor cells [13, 80–82]. The antibodies can be acquired naturally, by passive immunization, or induced by active specific immunization with a vaccine containing a carbohydrate epitope. The antibodies mediate elimination of tumor cells by complement-dependent cytotoxicity (CDC) and by antibody-dependant cellular cytotoxicity (ADCC) conferred by natural killer cells and macrophages. Other anticancer properties that have been ascribed to these antibodies include interference with receptor-mediated signaling, adhesion, and metastasis.

Antibodies are produced by B cells that have been activated with their cognate antigen. The B lymphocytes carry membrane-bound Ig proteins that can recognize a wide variety of compounds. Carbohydrates, for example, can bind to receptors of B lymphocytes, induce crosslinking of the Ig proteins, which will lead to activation of the B cell and production of low-affinity IgM antibodies [83]. To achieve a class switch to high-affinity IgG antibodies, the B cells need to interact with helper T cells (Fig. 9.2) [84, 85]. Activation of helper T cells requires, in turn, the involvement of antigen-presenting cells (APCs). The most highly specialized APCs are dendritic cells, which are capable of capturing protein antigens that, after internalization and

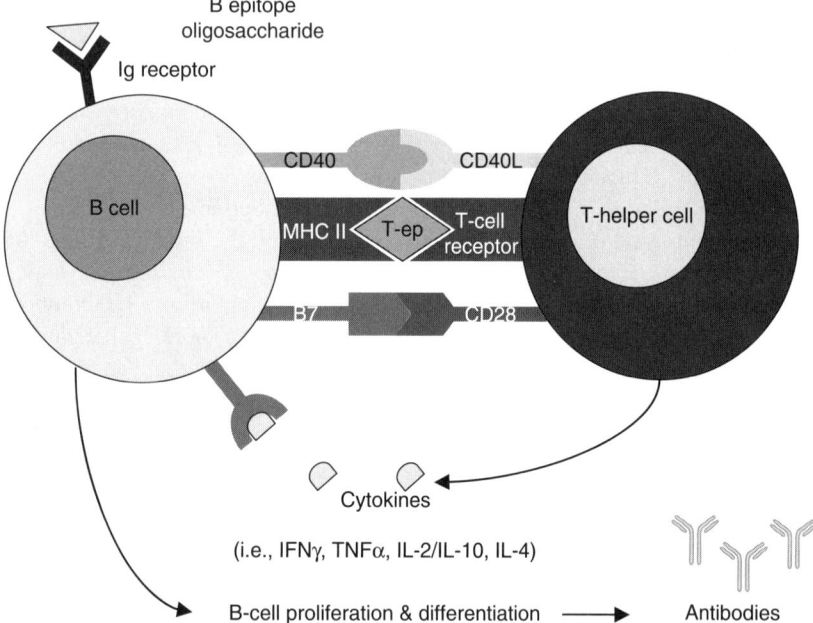

Figure 9.2 Schematic presentation of the interaction between B cells and helper T cells. Activation of B cells is initiated by specific recognition of antigens by the Ig receptors, which will lead to production of low-affinity IgM antibodies. To achieve a class-switch to specific IgG antibodies, the B cell needs activation by an activated helper T cell. In this context, B cells display processed protein antigens on their cell surface in the form of a peptide–MHC II complex and express the co-stimulator B7. Helper T cells that have been activated with the same peptide epitope by an APC, recognize the peptide antigen and B7 presented by the B cell. The T-helper cell is stimulated to express CD40 ligand (CD40L) and to secret cytokines. CD40L then binds to CD40 on the B cells and initiates B-cell proliferation and differentiation into an antibody secreting cell. T-ep = T-helper epitope. (See color insert.)

proteolytic cleavage into peptides, are presented on the surface of the APC as a complex with class II MHC molecules. Subsequently, the APCs will migrate to lymphoid organs where the peptide complexed to class II major histocompatibility complex (MHC) peptide will interact with the T-cell receptors of naïve T lymphocytes, resulting in their activation [86, 87]. A similar type of interaction via MHC class II exists between B cells and T cells. Naïve B and helper T cells reside in different compartments of the lymphatic system and are induced to migrate toward one another only after activation by an antigen, ensuring that the cells come together only when needed. Thus, activation of naïve T cells induce migration to the T-cell zone where the T-helper cell will interact with B cells [88]. The class II MHC–peptide complex presented by a B cell will mediate an interaction with the helper T cells, which will lead to expression of co-stimulatory proteins, further augmenting the interaction between the two cell types. Activated helper T cells express CD40L, which will bind with CD40 on the B cell resulting in cytokine production by the T cell [89]. A combination of binding to CD40 and cytokine signaling will stimulate the B cell to proliferate and differentiate into antibody-secreting cells. In addition, memory B cells will be formed that live for a long time and respond rapidly to subsequent exposures of antigen by differentiating into high-affinity (IgG) antibody secretors [90].

The MHC molecules show a relatively broad specificity for peptide binding, and the fine specificity of antigen recognition resides largely in the antigen receptor of the T lymphocyte. However, peptides that can bind to MHC share many structural features that promote binding interactions. In general, class II MHC requires peptides of 12–20 amino acids for optimum binding. MHC genes are polymorphic with more than 250 alleles for some of these genes in the population. As a result different persons recognize different peptides as T-helper epitopes. Structural studies have, however, identified peptide sequences that are recognized by many individuals. These peptides are named universal or promiscuous peptide T-helper epitope and have garnered attention for vaccine development.

9.5 MHC-MEDIATED IMMUNE RESPONSE TO GLYCOPEPTIDES

$CD8^+$ and $CD4^+$ T cells recognize protein antigens presented as peptides bound to MHC class I and II molecules, respectively. Extra-cellular protein antigens are recognized by APCs, which generate class-II-restricted peptide epitopes through a multistep process. The process culminates in transport of an MHC II–peptide complex to the cell surface and presentation of the peptide antigen by MHC II to $CD4^+$ T helper cells. MHC class I molecules, on the other hand, form a complex with peptides derived from intracellular microbial proteins. When this complex is presented on the cell surface, the peptide antigen can be recognized by $CD8^+$ cytotoxic T cells, which leads to their activation and proliferation, resulting in killing of infected cells.

Contrary to previous understanding, it is now apparent that glycopeptides can mediate classical MHC-mediated immune responses [91–94]. In addition to the peptide backbone that provides the binding motif for the MHC molecule, the glycan moiety can facilitate the recognition of T cells and stimulate immune responses specific for a carbohydrate antigen. For example, this feature has been shown in studies employing glycopeptides derived from type II collagen [95] and the human immunodeficiency virus (HIV) envelope glycoprotein [96]. These studies revealed that the peptide backbone binds to the MHC class II groove and that the sugar moiety was recognized by T-cell receptor on T cells. Studying endosomal processing of MUC-1 by APCs showed that the carbohydrates survive the cellular processing of the glycoprotein by dendritic cells (DCs) for presentation of the generated glycopeptide antigens to MHC II [97].

Dendritic cells have long been known to activate helper T cells through the MHC II pathway. On the other hand, it was believed that MHC class I presentation of endogenous peptide antigens is performed by nucleated somatic cells. However, it is now apparent that in addition to class-II-restricted epitopes, DCs can acquire exogenous antigens and generate MHC class-I-restricted peptides and present these to $CD8^+$ cytotoxic T cells in a process termed *cross-presentation* [98–100]. Adenocarcinomas express only low levels of MHC class II molecules and, in addition, $CD4^+$ helper T cells are not expected to react with tumor cells. Cytotoxic T lymphocytes (CTL), on the other hand, are expected to have a direct effect on tumor cells and represent an opportunity for the development of glycopeptide-based cancer vaccines [101]. Native MUC-1 glycopeptides have been shown to bind to MHC class I molecules both in vitro and in vivo [102]. The binding affinity was higher for the glycopeptide compared to peptide alone. MUC-1 glycopeptides carrying the TF or Tn antigen designed to have a high affinity for MHC class I molecules were used to induce a carbohydrate-specific cytotoxic T-cell response in mice [103]. It was found that the CTL lines generated with TF and Tn antigen, cross-reacted with Tn or TF, respectively, which led to the conclusion that the GalNAc residue is highly immunogenic. However, helper T cells are required to sustain $CD8^+$ cells and to ensure the development of memory $CD8^+$ cells [88, 104]. Therefore, Gendler and co-workers designed and immunologically elucidated a di-epitope vaccine candidate that incorporates an MUC-1 CTL epitope and a universal helper T-cell epitope derived from the hepatitis B core antigen sequence [105]. The vaccine was administered in combination with granulocyte macrophage colony-stimulating factor (GM-CSF) and phosphate-guanine-containing oligodeoxynucleotides (CpG-ODN) as adjuvants. The vaccine was tested in a therapeutic and prophylactic setting in a mouse tumor model, using MC38 colon cancer cells, and lead to a reduction of tumor burden and complete tumor rejection, respectively.

9.6 TOLL-LIKE RECEPTORS AND THE LINK BETWEEN INNATE AND ADAPTIVE IMMUNITY

In addition to activation of B and T lymphocytes, adaptive immune responses require danger signals that are provided by the innate immune system. This mode of activation is called the two-signal hypothesis for lymphocyte activation and ensures that immune responses are not induced against harmless substances or self-antigens. In a vaccine setting, an adjuvant is included to provide the necessary danger signals.

The innate immune system is an evolutionarily ancient system designed to detect the presence of microbial invaders and activate protective responses [106]. The innate immune system responds rapidly to families of highly conserved compounds, which are integral parts of pathogens and perceived danger signals by the host. Recognition of these pathogen-associated molecular patterns (PAMPs) is mediated by sets of highly conserved receptors [107], whose activation results in acute inflammatory responses. These transmembrane receptor proteins are referred to as Toll-like receptors (TLRs). The responses mediated by TLRs include direct local attack against the invading pathogen and the production of a diverse set of cytokines and chemokines [108–110]. Apart from antimicrobial properties, the cytokines also activate and regulate the adaptive component of the immune system [111, 112]. Thus, while the innate and adaptive components of the immune system are often depicted as being distinct entities, in effect they complement and compensate for each other.

The discovery of the TLRs less than a decade ago has advanced our understanding of early events in microbial recognition and response and the subsequent development of an adaptive immune response [113–119]. To date, 11 members of the mammalian TLR family have been identified, each potentially recognizing a discrete class of PAMP [120]. For example, lipopeptides such as Pam_3Cys are recognized by TLR2/6, Pam_3Cys derivatives by TLR1/2, lipopolysaccharide (LPS) by TLR4/MD2, bacterial flagellin by TLR5, double-stranded RNA (ribonucleic acid) by TLR3 [121], and bacterial DNA (deoxyribonucleic acid) by TLR9.

There is emerging evidence that cytokines, produced by activation of TLRs, play crucial roles in the initiation and control of adaptive immune responses. For example, antigen-presenting cells and naïve T cells need to be stimulated to produce a number of co-stimulatory proteins for optimum interaction between T-helper cells and B and antigen-presenting cells [122, 123]. Other cytokines are important for directing the effector T-cell response toward a T-helper-1 (Th-1) or T-helper-2 (Th-2) phenotype [124]. During the last few years, it has become clear that many adjuvants used for immunization contain ligands for TLRs. For example, the active component of complete Freund's adjuvant (CFA) is heat-killed mycobacterium tuberculosis that has several ligands for TLRs. Recently, TLR9 agonists have demonstrated potential for treatment of cancer both as monotherapy and in combination with other immunotherapies such as therapeutic vaccines [125]. The agonists induce activation and maturation of dendritic cells, which initiate activation of natural killer cells and the expansion of Th1 cells and CTLs. TLR9 agonists also enhance the differentiation of B cells into antibody-secreting cells. A range of synthetic CpG-ODNs are currently in clinical trials. Their use as vaccine adjuvant have shown to enhance antibody titers and antigen-specific $CD8^+$ T cells. As mentioned above, an MUC-1 glycopeptide vaccine co-administered with CpG-ODN was shown to reduce the tumor burden in mice [105].

9.7 CHEMICAL SYNTHESIS OF TUMOR-ASSOCIATED CARBOHYDRATES AND GLYCOPEPTIDES

Efficient methods for the preparation of tumor-associated carbohydrates are critical for the development of carbohydrate-based cancer vaccines. Considerable improvements have been made in this field [126–137], yet the construction of oligosaccharides and glycopeptides remains a challenging task due to the combined demands of elaborate procedures for glycosyl donor and acceptor preparation and the requirements of regio- and stereo-selectivety in glycoside bond formation. Many new leaving groups for the anomeric center have been developed, which can be introduced under mild reaction conditions and are sufficiently stable for purification and storage for a considerable period of time. The most commonly employed glycosyl donors include anomeric fluorides [138], trichloroacetimidates [139], and thioglycosides [140]. These approaches, under the appropriate reaction conditions, can give high yields and good anomeric ratios. The glycal assembly strategy [141], the use of anomeric sulfoxides [142], and dehydrative glycosylation protocols [143–148] are also emerging as attractive tools for the assembly of complex oligosaccharides. Furthermore, these leaving groups can be activated under mild reaction conditions and guarantee high yields and good anomeric ratios when performed under the appropriate reaction conditions. Convergent synthetic strategies that allow the convenient assembly of complex oligosaccharides from properly protected building units involving a minimum number of synthetic steps have become available. In particular, one-pot multistep approaches for selective monosaccharide protection [149, 150] and oligosaccharide assembly are being pursued, which do not require intermediate work-up and purification steps and hence speed up the process of chemical synthesis considerably. Several research groups have demonstrated that chemoselective, orthogonal, and iterative glycosylation strategies, which exploit differential reactivities of anomeric leaving groups, allow several selected glycosyl donors to react in a specific order, resulting in a single oligosaccharide product [151–156]. Methods for solid-phase oligosaccharide synthesis have been reported, and these procedures shorten oligosaccharide synthesis by removing the need to purify intermediate derivatives [157, 158]. Recently, automated solid-phase synthesis was used for the preparation of a branched dodecasaccharide [159].

A crucial step in the chemical synthesis of glycopeptide vaccine candidates is the merger of carbohydrate and peptide chemistry [160]. Different synthetic approaches can be envisaged for the preparation of glycopeptides. For example, a protected (or unprotected) oligosaccharide can be linked to the side chain of an amino acid and then be incorporated by solid-phase glycopeptide synthesis. Alternatively, an unprotected oligosaccharide equipped with a proper functional group can be conjugated to a peptide using well-established conjugation chemistry.

The combination of carbohydrate and peptide chemistry is also not a trivial task, due to the additional complexity and lability conferred by the carbohydrate group. The O-glycosidic bond, in particular the α-fucosidic linkage, is acid labile [161–163] and an O-linked glycopeptide can undergo β elimination [164] upon treatment with strong bases. Strong bases can also racemize the stereogenic centers of peptides. Thus, many methods and reaction conditions that are commonly employed in solid-phase peptide synthesis (SPPS) are not suitable for the preparation of glycopeptides [160, 165, 166]. Therefore, special care must be taken in choosing protecting groups for the glycosylated amino acids. Presently, the use of acetyl esters for hydroxyl protection of the oligosaccharide moiety and N^{α}-Fmoc protected amino acids [167], combined with coupling reagents such as DIC/HOBt [168, 169],

PyBOP/HOBt [170], HBTU/HOBt [171], and HATU/HOAt [172, 173], have emerged as a standard technique for solid-phase glycopeptide synthesis.

In the Fmoc protocol, the iterative removal of the N^α-Fmoc group needed for peptide elongation can be achieved using mild bases such as piperidine and morpholine, without affecting the O-acetyl groups of the saccharide or inducing β elimination of the glycan [174, 175]. However, in the Boc/benzyl protocol the recurring cleavage of the N^α-Boc group is accomplished by trifluoroacetic acid treatment, and such strongly acidic conditions are incompatible with the presence of acid-labile glycosidic linkages. The use of O-acylated glycosyl amino acids in glycopeptide synthesis is, however, not free of complications. One such complication is that the basic conditions required for deacetylation may affect the side chains of amino acids such as cysteine [175] and asparagine [163]. In light of these problems, methods to use of amino acids carrying unprotected glycans have been developed [176].

The α-GalNAc-Ser/Thr glycosyl amino acid known as the Tn antigen and the βGal(1-3)GalNAc-Ser/Thr known as the TF antigen are key structures for the mucin-type O-linked glycopeptides. The O-glycosidic bond of the Tn antigen is commonly installed by glycosylation of the hydroxyl group of a properly protected Ser or Thr using a 2-azido-2-deoxygalactose donor, such as a halo [177–183], trichloroacetimidate [182, 183], or a thioglycoside [184–186] donor (Scheme 9.1). The glycosylation of an amino acid alcohol with a more complex oligosaccharide donor to prepare a glycosyl amino acid building block can, however, be unpredictable in regard to both yields and stereoselectivity. To this end, Meldal and co-workers [187] introduced what was later termed "the cassette method" [182], an approach that utilizes a common $GalN_3$-Thr building block that can be extended by a more complex oligosaccharide by successive glycosylations and deprotections (Scheme 9.2). The use of enzymatic transfer of individual monosaccharides (or large oligosaccharides) to glycopeptides containing a simple glycan also offers an attractive alternative to the tedious synthesis of complex oligosaccharides [188].

Apart from natural mucin-type linkages, large oligosaccharides can also be linked to the side chain of nonnatural amino acids by, for example, olefin cross-metathesis or a Horner–Emmons reaction and then incorporated into the backbone of a growing peptide chain [189].

An oligosaccharide (or glycopeptide) and a peptide epitope can also be stitched together using a variety of linkers that are commonly employed in traditional conjugation chemistry [190, 191] such as disulfide and thioether formation and oxime chemistry (Table 9.1).

One advantage of synthetic oligosaccharides is that a linker carrying a specific functional group can be introduced during synthesis. This linker functionality can facilitate the chemoselective ligation of an unprotected oligosaccharide to a peptide epitope. Oxime chemistry in which an aldehyde or ketone reacts chemoselectively with an aminooxy functionality at near neutral conditions offers an appealing route to large synthetic multiepitope vaccines [192, 193]. An oxo-aldehyde is classically obtained by periodate oxidation of an N-terminal serine [194]. Masked aldehydes can be incorporated into a growing peptide chain by using, for example, specialty linkers and amino acid building blocks [195, 196]. It has also been shown that it is

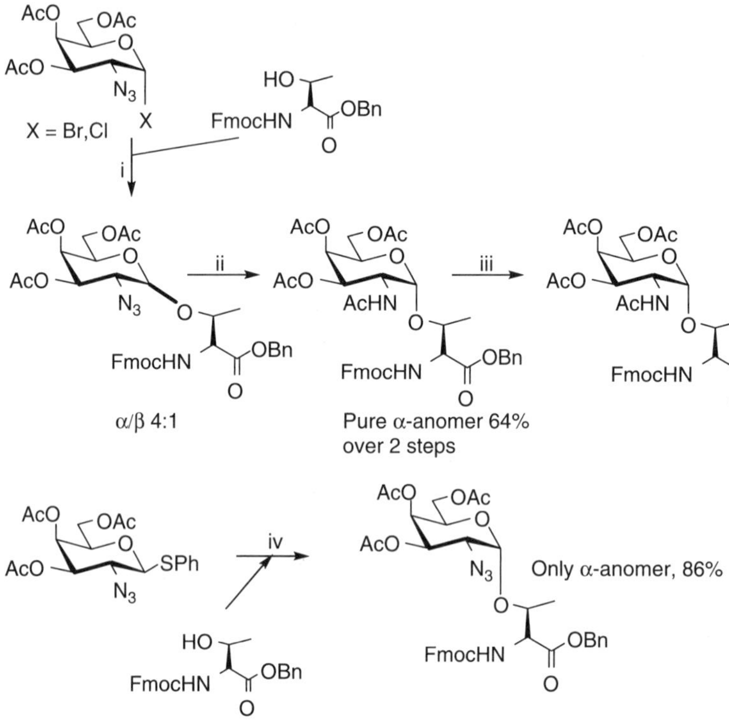

Reaction conditions: (i) AgClO$_4$/Ag$_2$CO$_3$; (ii) Zn-dust, HOAc/Ac$_2$O/THF; (iii) H$_2$, Pd/C, pyridine/iPrOH1:1; (iv) Ph$_2$SO, Tf$_2$O, DCM, −60°C.

Scheme 9.1

possible to synthesize a peptide that incorporates an amino acid carrying an unprotected ketone functionality in the side chain [197]. Recently, native chemical ligation [198, 199] and "click chemistry" [200, 201] (Cu(I)-mediated Huisgen cyclo-addition [200, 202, 203]) have emerged as powerful tools for chemoselective ligations. In the Huisgen cyclo-addition, an azide and an alkyne group react, typically in the

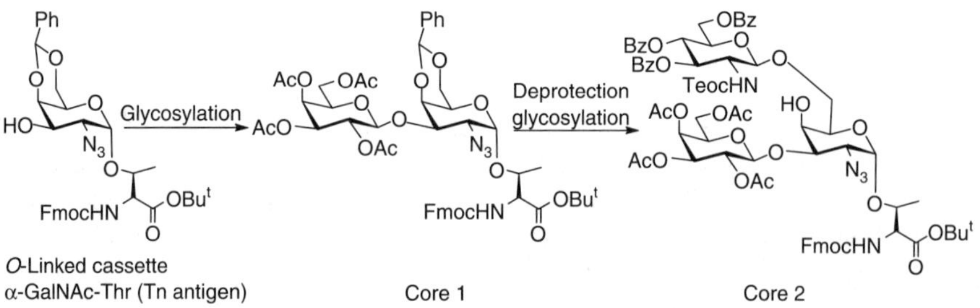

Scheme 9.2

9.7 SYNTHESIS OF TUMOR-ASSOCIATED CARBOHYDRATES AND GLYCOPEPTIDES

Table 9.1 Conjugation Chemistry for Ligating a Peptide Epitope and a Oligosaccharide (or Glycopeptide)

Reaction	Functional Group 1	Functional Group 2	Product
Thioalkylation	R–SH	X–CH$_2$–C(O)–R	R–S–CH$_2$–C(O)–R
Thiol addition	R–SH	maleimide-R	R-S-succinimide-R
Disulfide formation	R–SH	R′–SH	R–S–S–R′
Oxime formation	R–CHO	H$_2$N–O–R	R–CH=N–O–R
Hydrazone formation	R–CHO	H$_2$N–NH–R	R–CH=N–NH–R
Huisgen cyclo-addition Triazole formation	R–N$_3$	≡–R	triazole(R, R)

presence of Cu(I), to form a triazole moiety (Table 9.1). Although attractive, it should be noted that the click reaction introduces a rigid triazole moiety, which may be immunogenic and thus further suppress the low immunogenicity of tumor-associated carbohydrate antigens. Native chemical ligation (NCL), on the other hand, is a chemoselective reaction that results in the formation of an amide bond (Scheme 9.3). The reaction occurs at physiological pH between an N-terminal cysteine residue and (glyco)peptide thioesters [204]. The first step is a trans-thioesterification between the thioester and the sulfhydryl group of the N-terminal cysteine. The ligated peptide thioester then undergoes a rapid, irreversible, and spontaneous intramolecular S → N shift, which generates the thermodynamically favored native amide bond at the ligation junction. NCL has been shown to occur uniquely at the N-terminal cysteine residue regardless of any additional internal cysteine residues. Easy access to C-terminal glycopeptide thioesters is crucial to NCL. Initially, preparation of the peptide thioesters relied on the Boc strategy due to the inherent base sensitivity of the thioesters. However, as it is preferable to use the Fmoc strategy for the acid-sensitive glycopeptides, several new methods have been developed for their preparation [205–213]. Tandem NCL has been introduced to synthesize more complex structures by using a (glyco)peptide that carries a C-terminal thioester and an N-terminal cysteine with an orthogonal thiol-protecting group. These features make it possible to perform serial ligation reactions [214, 215].

Several of the approaches mentioned in this section have been used to obtain semisynthetic (glyco-protein conjugates) and fully synthetic carbohydrate-based cancer vaccines and a few examples will be highlighted in the following sections.

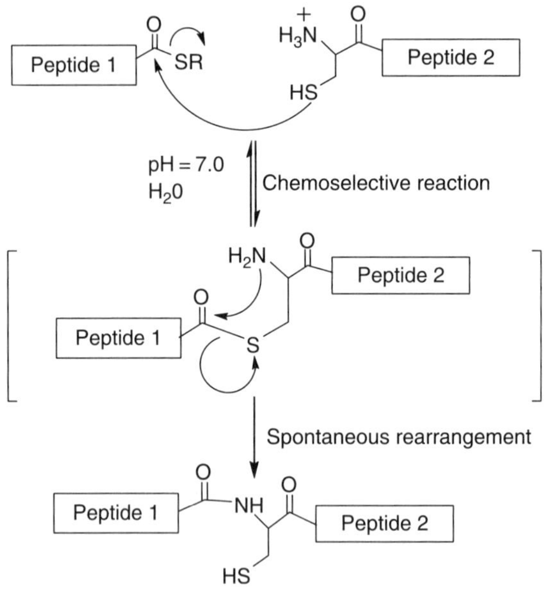

Scheme 9.3

9.8 SEMISYNTHETIC CARBOHYDRATE-BASED CANCER VACCINES

More than 70 years ago, it was shown that low-immunogenicity and T-cell-independent saccharides can be overcome by conjugation to an immunogenic carrier protein [216]. In this approach, the carrier protein enhances the presentation of the carbohydrate to the immune system and provides helper T epitopes (peptide fragments of 12–15 amino acids) that can activate T-helper cells. The protein also can possess mitogenic and adjuvant-like properties that stimulate the innate immune response. As a result, the class switch from low-affinity IgM to high-affinity IgG antibodies is accomplished. This strategy has been succesfully employed for vaccine development for target bacterial (poly)oligosaccharides. Not surprisingly, the use of foreign protein carriers [e.g., keyhole limpet hemocyanin (KLH), bovine serum albumin (BSA), detoxified tetanus toxoid] was implemented early on in the efforts to create cancer vaccines that target tumor-associated oligosaccharides [2, 43, 80, 217–219].

Helling and co-workers established the immunological effect of conjugating a tumor-associated saccharide moiety to a carrier protein [220]. They obtained a saccharide containing an aldehyde moiety by ozonolysis of the double bond of the ceramide backbone of GD3. The saccharide derivative was coupled by reductive amination to amino functionalities present on a carrier protein. The immunogenicity of this conjugate was examined by the immunization of mice in the presence of the immunoadjuvant QS-21. In most mice, high titers of IgM and low titers of IgG against GD3 were observed. Importantly, these antibodies could induce complement-mediated lysis of human melanoma cells expressing GD3.

9.8 SEMISYNTHETIC CARBOHYDRATE-BASED CANCER VACCINES

When relatively short carbohydrate fragments are isolated from natural sources and linked to a carrier protein, it is likely that vital recognition elements may be destroyed during this process, which will lead to a decrease or complete loss of immunogenicity. Furthermore isolation and purification of TACAs from nature with structural integrity and in reasonable amount is an almost futile endeavor. Organic synthesis allows the incorporation of a functional group with unique reactivity that allows selective conjugation to a carrier protein. Several of the more complex tumor-associated carbohydrate antigens have been prepared by organic synthesis. Most notable is the collaboration of the Livingston and Danishefsky research teams at Memorial Sloan-Kettering Cancer Center, which have resulted in the synthesis of Globo-H [221], GM1 [222], Lewis^y [223–225], Lewis^b [226, 227], KH-1 (Le^y-Le^x) [228, 229], MUC-1 [230, 231], and TN and TF antigens [183] (see Scheme 9.4 for the synthesis of Globo-H). These

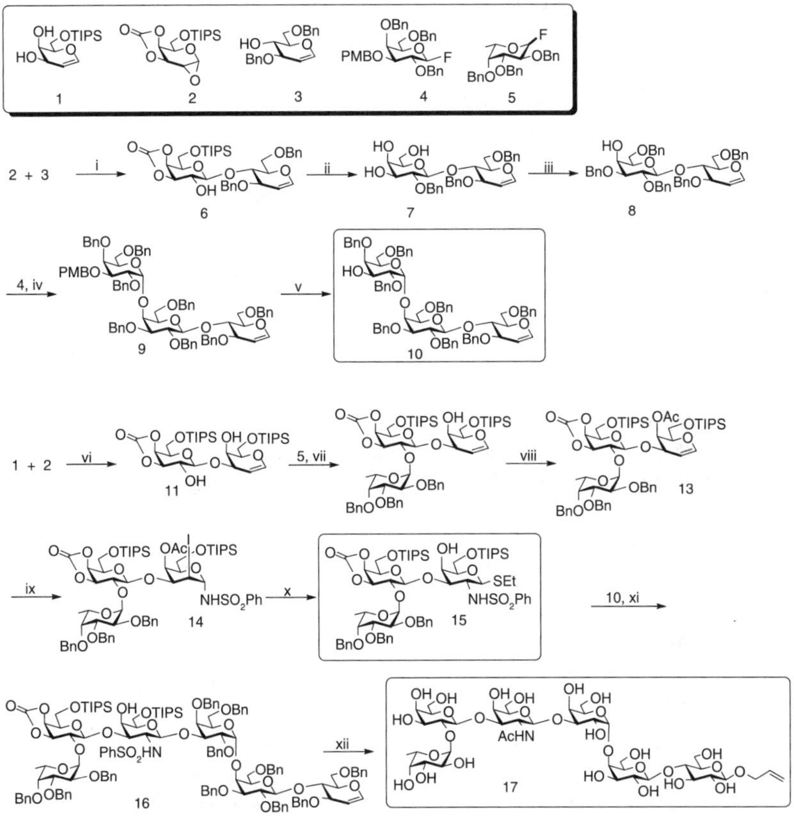

Reaction conditions: (i) ZnCl$_2$, THF; (ii) 1. BnBr, NaH, DMF; 2. TBAF, THF, then MeOH, MeONa; (iii) (n-Bu$_3$Sn)$_2$O, Bu$_2$SnO, PhH, then TBABr, BnBr; (iv) AgClO$_4$, SnCl$_2$, di-tert-butylpyridine, 4ÅMS, Et$_2$O; (v) DDQ, DCM, H$_2$O; (vi) ZnCl$_2$, THF; (vii) AgClO$_4$, SnCl$_2$, di-tert-butylpyridine; (viii) Ac$_2$O, pyridine, cat. DMAP; (ix) I(coll)$_2$ClO$_4$, PhSO$_2$NH$_2$, 4ÅMS, THF; (x) EtSH, LHMDS, DMF; (xi) MeOTf, 4ÅMS, Et$_2$O-DCM (2:1); (xii) 1. TBAF, THF; 2. Na, NH$_3$, THF; 3. Ac$_2$O, TEA, DMAP, THF, DMF; 4. 3,3'-dimethyldioxirane, DCM, 5. allyl alcohol; 6. NaOMe, MeOH.

Scheme 9.4

compounds were equipped with a linker that allowed selective conjugation to a carrier protein using their olefin-ozonolysis-reductive amination [221–225] or maleimide-thiol [230, 232] conjugation protocols. The conjugates raised in mice high titers of IgM and low titers of IgG antibodies, which could induce a complement-mediated lysis of cancer cells.

Several other research groups have also reported elegant syntheses and immunological evaluations of tumor-associated saccharide antigens [233–250].

A number of carbohydrate–protein conjugates have been examined in preclinical and phases I, II, and III clinical trials [217, 251–271]. The results reported to date indicate that carbohydrate conjugate vaccines are most promising when used in combination with a potent adjuvant such as the saponin QS-21, the immunomodulator cyclophosphamide (CY) [272], and stem cell rescue [273]. Low toxicity and correlations between vaccine-induced antibody responses and clinical course after immunization have been found. However, even when an optimized immunization protocol was used, it was difficult to induce high titers of high-affinity IgG antibodies in most patients.

The results of the preclinical and clinical studies have indicated that many factors that can influence the antigenicity of tumor-associated antigens conjugated to carrier proteins. For example, the choice of carrier protein and immunoadjuvant can greatly influence the magnitude and specificity of the elicited immune response.

A major drawback of the use of carrier proteins is they are highly immunogenic and will inevitably elicit strong B-cell responses. This feature can lead to carrier-induced epitope suppression, which in particular is a problem when self-antigens such as tumor-associated carbohydrates are employed. To address this issue, a recombinant supercarrier protein, consisting of a string of well-documented promiscuous helper T epitopes, has been developed and was used to provide help to an oligosaccharide from *Haemophilus influenzae* type b [274]. An appealing aspect of this strategy is that the multiple $CD4^+$ peptides bind to a broad range of human MHC class II molecules, thus rendering the vaccine effective in a broad spectrum of the human population.

The conjugation of carbohydrates to a carrier protein also introduces several problems, which are particularly pronounced for tumor-associated carbohydrate antigens. In general, carbohydrate–protein conjugation chemistry is difficult to control, which results in conjugates with ambiguities in composition and structure and batchwise variations of prepared glycoconjugates. In addition, the linkers that are employed for the conjugation of carbohydrates to proteins can be immunogenic, leading to epitope suppression [247, 275]. For example, it has been shown that a Le^y–KLH conjugate with the rigid cyclohexyl-maleimide linker elicited an antibody response largely toward the linker [247]. However, if the conjugate was prepared using a more flexible propionate linker, the immune response toward the linker was negligible and instead directed against the Le^y tetrasaccharide. Still, as the carrier protein contains multiple B epitopes, antibodies were mainly elicited against the protein.

It is clear that the successful development of carbohydrate-based cancer vaccines requires novel strategies for more efficient presentation of tumor-associated carbohydrate epitopes resulting in the more efficient class switch to IgG antibodies.

9.9 FULLY SYNTHETIC CARBOHYDRATE-BASED CANCER VACCINES

A remedy for highly immunogenic protein carriers is the use of fully synthetic vaccines. This strategy has been pursued in the field of peptide-based vaccines, perhaps reflecting the relative ease of synthesizing peptide haptens [27, 276–278]. Major advantages of fully synthetic vaccine candidates are that they can be designed to incorporate only those elements required for a desired immune response, and they can be produced, in a reproducible fashion to give chemically well-defined compounds. Synthetic peptide B epitopes and peptide helper T epitopes have been combined on immunosilent carriers such as poly-lysine scaffolds [276] and lipids [277]. Vaccine constructs carrying lipids have subsequently been incorporated into liposomes [279], which then display the antigens on their surface, providing a multivalent presentation similar to that of a protein conjugate. Such a multivalent presentation is important for B-cell clustering and their subsequent activation. The lipid anchor can be chosen to consist of an immunoadjuvant such as a TLR ligand, giving access to a construct that carries a built-in immunoadjuvant. The use of synthetic immunosilent carriers for tumor-related carbohydrate antigens of low immunogenicity presents an appealing opportunity to achieve a more focused immune response toward the carbohydrate antigen. Attempts at realizing these strategies have been pursued and will be discussed in further detail below.

9.10 B-EPITOPE AND RECEPTOR LIGAND DI-EPITOPE CONSTRUCTS

A number of fully synthetic scaffolds for tumor-associated carbohydrate epitopes have been examined with the goal to improve the presentation to relevant immune cells and thereby inducing a more effective class switch to IgG antibodies. One approach is to link the tumor-associated epitope to a receptor ligand that targets or activates appropriate immune cells. For example, it has been shown that mannosylation of antigens may result in selective targeting of antigen presentation cells (APC), which carry mannose receptors [280]. In one clinical trial, relatively high titers of IgG1 antibodies were elicited by immunization of patients with a mannan-MUC-1 fusion protein [281]. The antibodies were, however, not directed against the important glycosylated PDTRP sequence but rather to adjacent STAPPAHG and PAPGSTAP sequences. Unfortunately, only 4 of 15 patients exhibited MUC-1-specific T-cell proliferation.

Pam$_3$Cys, which is a ligand for Toll-like receptor 2 (TLR2) and a potent B-cell activator has also been utilized as a "carrier" for targeting of APCs [282–285]. TLRs are expressed largely on dendritic cells and macrophages, but also on B cells albeit in lower levels. The cytokines produced after activation of TLRs are crucial in the initiation and control of the adaptive immune responses. Toyukuni et al. linked a dimeric Tn antigen cluster to Pam$_3$Cys and the vaccine construct elicited mainly IgM but also IgG antibodies in mice (Fig. 9.3) [286, 287]. This study showed, perhaps for the first time, that a small synthetic carbohydrate antigen could generate an immune

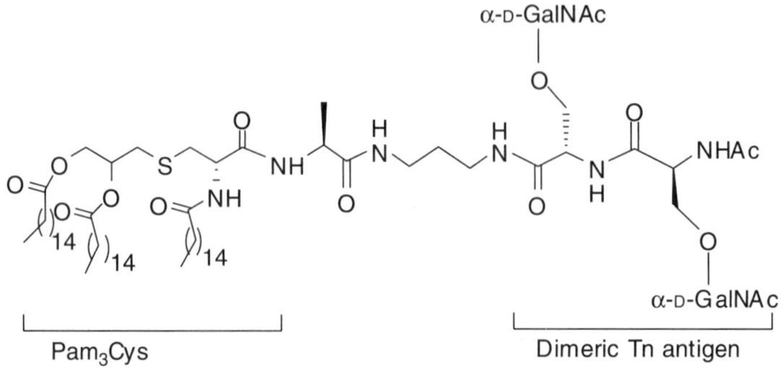

Figure 9.3 Cancer vaccine candidate comprising a dimeric Tn antigen and the TLR agonist Pam$_3$Cys.

response against the carbohydrate antigen without a macromolecular carrier or additional adjuvant.

Using a similar approach, the Danishefsky research team synthesized a monomeric and a trimeric Le^y cluster, which was coupled to the synthetic immunoadjuvant Pam$_3$Cys (Fig. 9.4) [225, 288, 289]. Mice immunized with the vaccine constructs elicited antibodies that reacted with a Le^y expressing tumor cell line. However, only IgM antibodies were elicited due to the lack of a helper T epitope required to induce a class switch to IgG antibodies and affinity maturation of B cells. Administering the Pam$_3$Cys-linked construct together with the external adjuvant QS-21 also resulted

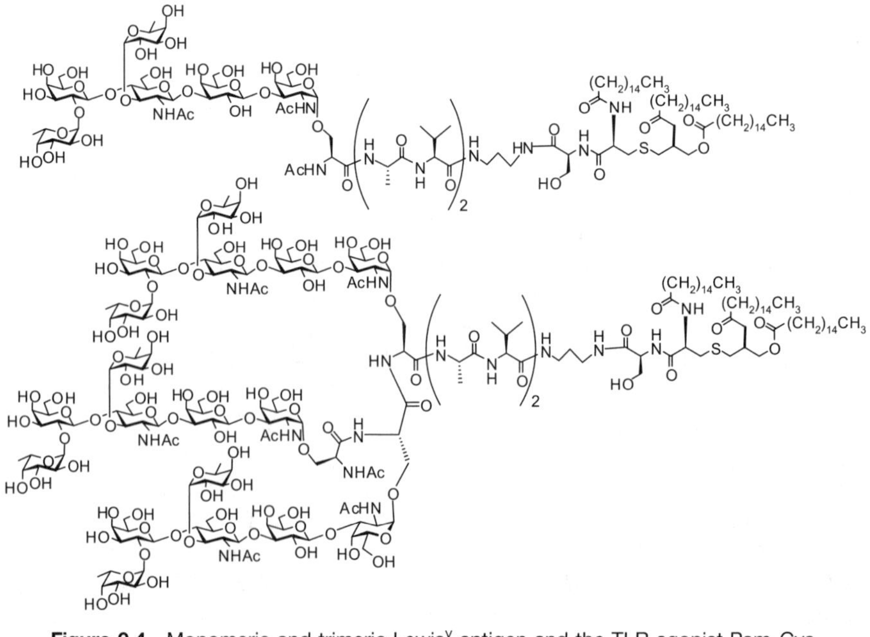

Figure 9.4 Monomeric and trimeric Lewis^y antigen and the TLR agonist Pam$_3$Cys.

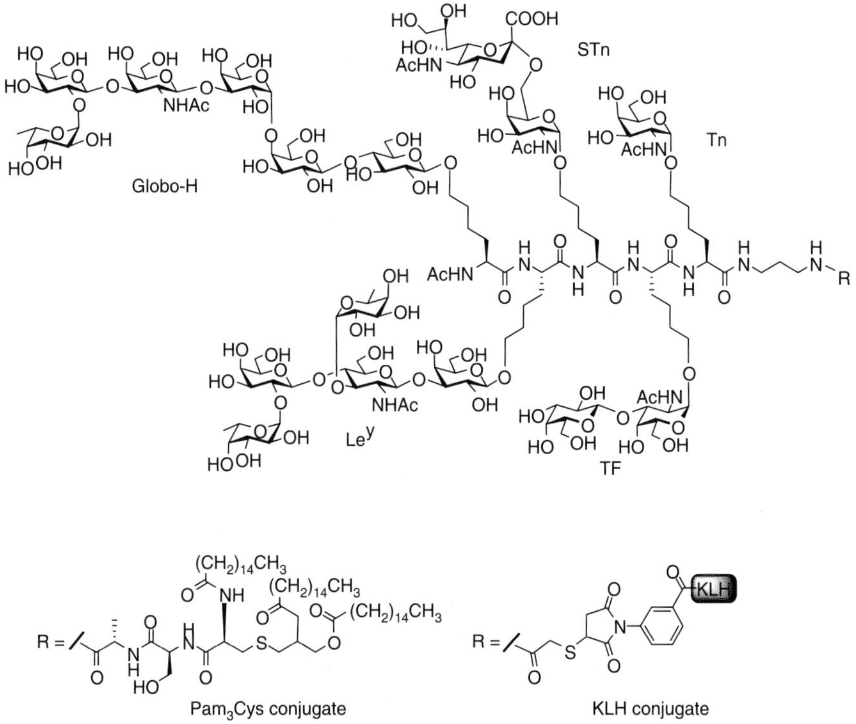

9.10 B-EPITOPE AND RECEPTOR LIGAND DI-EPITOPE CONSTRUCTS

Figure 9.5 Unimolecular pentavalent antigenic vaccine construct.

in mainly IgM antibodies. On the other hand, by coupling the monomeric and trimeric Le^y antigens to KLH both IgM and IgG antibodies could be produced in mice.

Antigenic responses against the Tn antigen could also be achieved by epitope clustering [231]. In addition to higher titers of antibodies, inoculations with a trimeric Tn-antigen conjugate vaccine resulted in antisera that recognized better Tn-positive

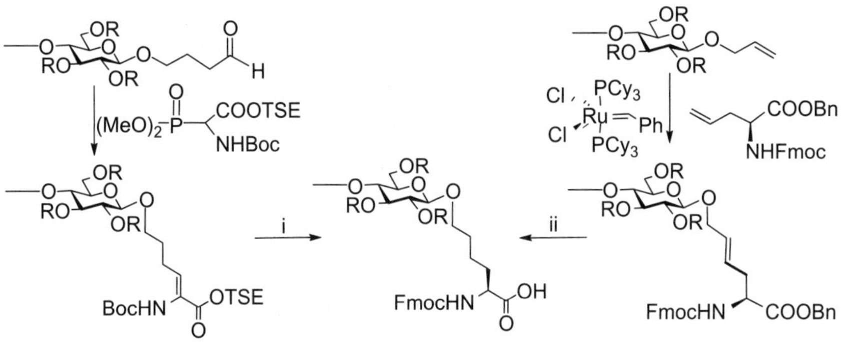

Reaction conditions: (i) 1. (S, S)-Et-DuPHOS-Rh$^+$, H$_2$, 50 psi, 2. TBAF, 3. TFA, 4. Fmoc-Su, NaHCO$_3$; (ii) H$_2$, Pt-C.

Scheme 9.5

Table 9.2 Reciprocal Median Peak ELISA^a Titers After Immunization of Groups of Five Mice with a Pentavalent Vaccine Candidate Together with QS-21 (10 μg)

Vaccine Candidate	Amount Injected (μg)	Globo-H Ceramide		Le^y Ceramide		STn (OSM)		TF (dPSM)		Tn (dOSM)	
		IgM	IgG	IgM	IgG	IgM	IgG	IgM	IgG	IgM	IgG
Pentavalent	10	100	0	0	0	0	0	0	0	0	0
Pentavalent − KLH conjugate	10	6400	400	0	0	0	800	200	400	200	0
Pentavalent + KLH	10	400	0	100	0	0	0	0	0	400	0
Pentavalent-Pam3Cys conjugate	30	800	0	800	0	0	0	0	100	800	0

^{a}ELISA, enzyme-linked immunosorbent assay.

cancer cell lines compared to antisera produced against a monomeric Tn vaccine. Furthermore, promising results were achieved when a multiantigenic glycopeptide containing several Tn antigens was used [183].

A commendable chemical synthesis was undertaken to obtain a unimolecular multiantigenic construct comprising the Globo-H, Lewis^y, STn, TF, and Tn antigens all attached to a single peptide backbone (Fig. 9.5) [189, 290, 291]. The rational of a polyantigenic construct [291–293] is that it combines carbohydrate antigens that are all closely associated with a particular type of cancer, in this case prostate cancer. The oligosaccharides were synthesized using the glycal assembly method and equipped with pentenyl or allyl spacers, which were then utilized to produce, via olefin cross-metathesis [294, 295] or Wittig-like reactions followed by asymmetric hydrogenation, side-chain glycosylated norleucine amino acid building blocks (Scheme 9.5) [296]. The glycosylated amino acid building blocks were then coupled using traditional peptide chemistry. The fully assembled unimolecular pentavalent construct was linked to Pam$_3$Cys through an amide bond and its immunogenicity evaluated in a murine host. Antibodies were elicited against all the antigens, however, mainly IgM antibodies were detected despite the co-administration with the additional adjuvant, QS-21 (Table 9.2). The multiantigenic construct was also linked to KLH using a maleimide-based linker, and the resulting conjugate elicited both IgM and IgG antibodies in the murine host when co-administered with QS-21. The elicited antibodies reacted with three different tumor cell lines, which all express two or more of the five antigens on their cell surface.

Calixarenes, which represent nonpeptidic scaffolds with a low toxicity, have been utilized to design a vaccine candidate based on the Tn antigen [297]. A Tn antigen mimetic, where the anomeric oxygen was replaced with sulfur (S-Tn) was coupled to the wide rim of a calix[4]arene derivative through a glycine residue using amide bond chemistry (Fig. 9.6). A single Pam$_3$Cys moiety was then introduced at the narrow rim of the calix[4]arene scaffold. The antigenicity of the tetravalent calixarene vaccine candidate was compared in a murine host to a monovalent construct where the Tn antigen mimetic was linked directly to Pam$_3$Cys. A four-fold increase in IgG antibody titers indicated that the multivalency introduced by the calixarene scaffold somewhat enhanced the antigenicity.

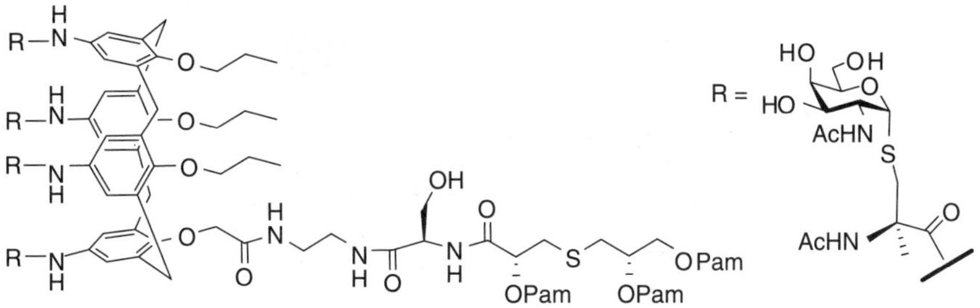

Figure 9.6 Anticancer vaccine candidate based on the Calix[4]arene scaffold, which carry the tumor-related S-Tn glycomimetic antigen and the immunoadjuvant Pam$_3$CysSer.

9.11 B- AND T-CELL DI-EPITOPE CONSTRUCTS

In order to stimulate a T-cell-dependent immune response, B cells need to be stimulated by activated $CD4^+$ helper T cells to proliferate and differentiate into antibody-secreting cells. As described above, a helper T cell is in turn activated when its TcR binds to a helper T peptide presented by MHC II molecules of an APCs. MHC II genes are polymorphic, and as a result different individuals recognize different peptides as helper-T-cell epitopes. These peptides share common features for binding and several peptides, which are recognized by a broad spectrum of MHC class II molecules, have been identified. These peptides have been coined as universal or "promiscuous" helper-T-cell peptides.

The approach of combining a tumor-associated carbohydrate or glycopeptide B epitope with a universal T-helper epitope peptide has been explored by several groups. In one attempt, an MUC-1 peptide carrying a single STn moiety was linked to a Th epitope derived from ovalbumin through a polar, nonimmunogenic linker (Fig. 9.7) [298]. The vaccine candidate was assembled by traditional peptide synthesis and fragment condensation and administered, together with complete Freund's adjuvant, to transgenic mice expressing T-cell receptors specific for the ovalbumin T epitope. An IgG antibody response was observed and the concentration of serum antibodies were increased between the boosts. It was also shown by competitive studies with unglycosylated peptides and a nonrelated glycopeptide derived from MUC-4 that the elicited antibodies were highly specific for the glycosylated MUC-1 peptide. To target the differently glycosylated tumor-associated epitopes related to MUC-1, a tri-antigenic glycopeptide containing unglycosylated, Tn and TF MUC-1 was constructed (Scheme 9.6) [299]. A peptide backbone incorporating one copy of the universal pan DR epitope (PADRE) peptide and two masked aldehyde groups was synthesized on a polyethylene glycol polyamide (PEGA) resin employing a traditional protocol [196]. The vaccine candidate was then elegantly assembled using unprotected Tn- and TF-MUC-1 glycopeptides and tandem oxime ligations. Evaluation in mice in the presence of the mild adjuvant alumn revealed that the vaccine construct was highly immunogenic and IgG antibodies were raised toward all three forms of the MUC-1 fragment. The antisera were shown to recognize native tumor-associated epitopes expressed by human mammary adenocarcinoma cells.

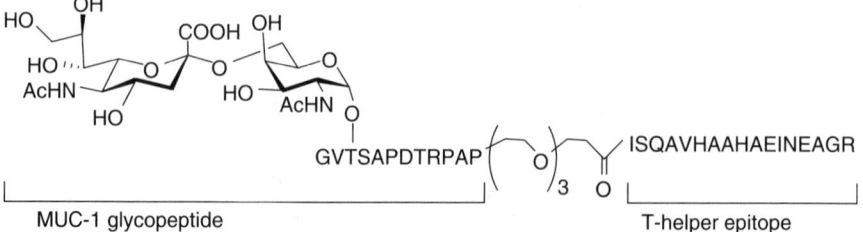

Figure 9.7 Cancer vaccine candidate consisting of a T-helper cell epitope derived from ovalbumin and the tumor-related glycopeptide antigen from MUC-1.

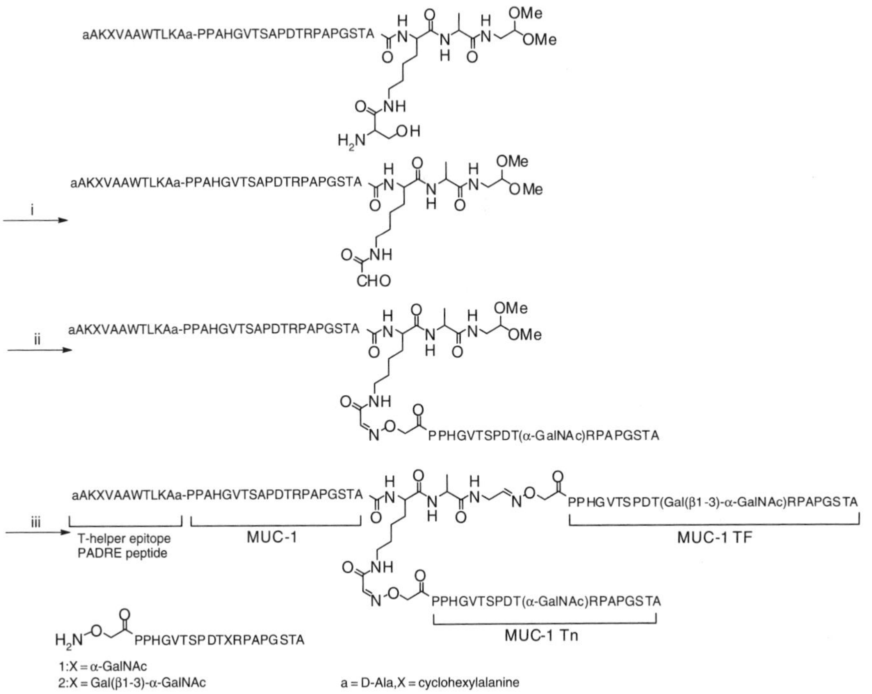

9.11 B- AND T-CELL DI-EPITOPE CONSTRUCTS

Scheme 9.6

Reaction conditions: (i) NaIO$_4$, HPLC purification; (ii) **1**, NaOAc (0.1 M, pH 4.6), HPLC purification; (iii) 10% TFA, **2**, HPLC purification.

Multiantigenic peptide (MAP) scaffolds based on a polylysine core was developed by Tam and co-workers for use as peptide vaccines [276]. Multiple copies, typically four or eight, of B and helper T epitopes were synthesized on the arms of a lysine scaffold. When glycopeptide B epitopes were incorporated, the corresponding scaffolds were coined multiantigenic glycopeptides (MAGs) [300]. The use of MAGs has successfully been pursued for the tumor-associated Tn antigen. Thus, an oligolysine core with four arms was extended with a CD4$^+$ peptide epitope derived from polio-virus (PV) and the tumor-related B-epitope glycopeptide (Fig. 9.8). Initial studies with a MAG carrying a monomeric Tn antigen raised an immune response, which increased the survival rate of tumor-bearing mice [301]. However, the monomeric Tn-MAG was not able to inhibit the binding of the Tn-antigen-specific MLS128 antibody to native Tn antigens expressed on human Jurkat cells. Therefore, to improve the vaccine candidate and to mimic the natural clustered presentation of Tn antigen on tumor cells, a cluster of three Tn antigens was introduced at the N-terminal end of the CD4$^+$ epitope [302]. The vaccine candidate was extensively tested in mice. Immunizations together with alumn gave Tn-specific antibodies of both IgM and IgG class with titers that were persistent for several months. Isotyping of the IgG antibodies pointed to an anticarbohydrate response, as mainly IgG1 but also IgG3 antibodies were observed. The IgM and IgG antibodies recognized tumor cells

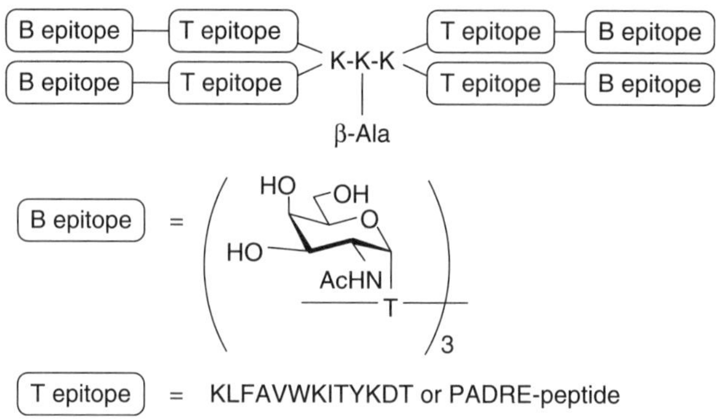

Figure 9.8 Schematic representation of a multiple antigen glycopeptide cancer vaccine candidate containing a tumor-associated Tn-antigen cluster and a universal T-cell epitope.

expressing the Tn antigen. Depleting CD8 cells did not affect the antibody response, whereas depleting the CD4 cells abolished the response, clearly demonstrating a helper T-cell-dependent immune response. The protective ability of the clustered Tn-MAG was investigated by tumor challenges in mice. Increased survival rates were observed in both a prophylactic and therapeutic setting. In the prophylactic setting, the survival rate was increased in a dose-dependent manner by as much as 90%. However, the concept of using immunotherapy to ease a patient's tumor burden or eradicate metastases relies on using a vaccine candidate in a therapeutic setting. Mice were injected with murine adenocarcinoma TA3/Ha cells and then received the vaccine in the presence of Freund's adjuvant or alumn. In this setting, a 40% survival rate was observed regardless of adjuvant. Administration of CY, which is reported to increase antitumor response, with the Tn-clustered MAG increased the survival rate to 80% indicating that CY can be used to improve the efficacy of the MAG vaccine. The therapeutic protection conferred by the clustered Tn-MAG vaccine was twice as effective compared to the protection offered by the corresponding monomeric Tn-MAG. Furthermore, the clustered MAG construct induced superior titers of anti-Tn IgG antibodies when compared to a KLH conjugate carrying the trimeric Tn cluster [231]. The MAG construct elicited good titers of IgG antibodies using the mild alumn adjuvant, whereas the clustered Tn–KLH conjugate required co-administration of the much more potent adjuvant QS-21 to achieve IgG anti-Tn antibodies. Furthermore, upon booster injections, a memory response was elicited in the mice that received initial immunizations with the MAG vaccine. No memory response was seen in the mice that initially received the KLH conjugate. The MAG was further evaluated in nonhuman primates [303]. For this purpose, the murine helper T epitope from polio virus was exchanged by either the PADRE peptide or a peptide-derived from tetanus toxin (TT), which are capable of binding to human MHC class II molecules. Both "humanized" MAG vaccine candidates induced an anti-Tn IgG antibody response, which recongnized the Tn antigen in its natural form on human tumor cells. Importantly, it was shown that the elicited antibodies mediated antibody-dependant cell cytotoxicity in the presence of natural killer (NK) cells.

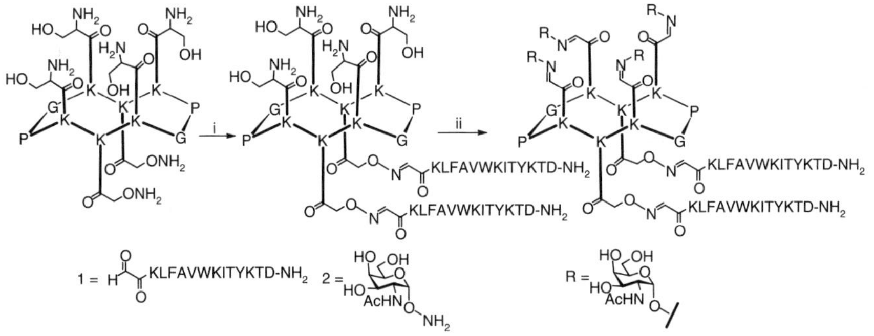

9.11 B- AND T-CELL DI-EPITOPE CONSTRUCTS

Reaction conditions: (i) 1, NaOAc buffer 0.1 M pH 4; (ii) 1. NaIO$_4$, phosphate buffer 0.01 M, pH 7.3; 2. 2, NaOAc buffer 0.1 M pH 4.

Scheme 9.7

The regioselectively addressable functionalized template (RAFT) is a recently introduced multimeric peptide-based scaffold [304]. The RAFT is a cyclic decapeptide consisting of proline, glycine, and lysine residues (Scheme 9.7). The Pro-Gly moiety induces β turns, which facilitate cyclization and provide stability in solution. The side chains of the lysine residues provide opportunities for selective incorporation of different antigens on opposite faces of the RAFT via classical ligation chemistry. A cancer vaccine candidate based on the RAFT scaffold, which contained four copies of a Tn antigen analog and two copies of a CD4$^+$ helper T-cell epitope derived from PV, was synthesized using orthogonally protected lysine side chains and tandem oxime chemistry [305]. The vaccine candidate stimulated PV-specific T cells in vitro and raised in a murine model, Tn-specific IgG antibodies. The antibodies recognized native Tn antigens on human Jurkat cells and thus, the nonnatural oxime bond did not interfere with the immune response. Importantly, it was shown that the elicited antibodies had low reactivity toward the RAFT scaffold itself, proving that it is nonimmunogenic.

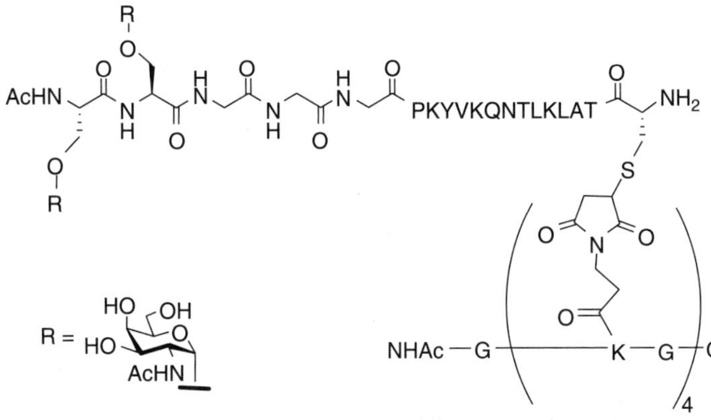

Figure 9.9 Structure of comblike glycodendrimer.

288 SEMISYNTHETIC AND FULLY SYNTHETIC CARBOHYDRATE-BASED CANCER VACCINES

A linear peptide backbone consisting of Gly-Lys repeating units was investigated as a potential carrier (Fig. 9.9) [306]. Comblike dendrimers bearing dimeric Tn antigen and a influenza virus hemagglutinin (IVH) helper T-cell epitope were assembled using cysteine-modified glycopeptides and maleimidopropionate-modified linear synthetic carrier. The conjugate elicited Tn-specific IgM, IgG, and IgA antibodies in mice. Presensitizing the mice with an IVH vaccine increased the levels of anti-Tn IgM and IgA antibodies, whereas the IgG levels remained the same.

9.12 TRICOMPONENT VACCINES

A tricomponent vaccine that contains a carbohydrate B-cell epitope, a helper T-cell epitope, and a potent immune activator/modulator such as a TLR ligand or a cytokine would incorporate the minimal subunits necessary to evoke an immune response against a carbohydrate [307–309].

In a first report, a fully synthetic three-component anticancer vaccine composed of the Tn antigen, a helper T-epitope derived from *Neisseria meningitis*, and the TLR

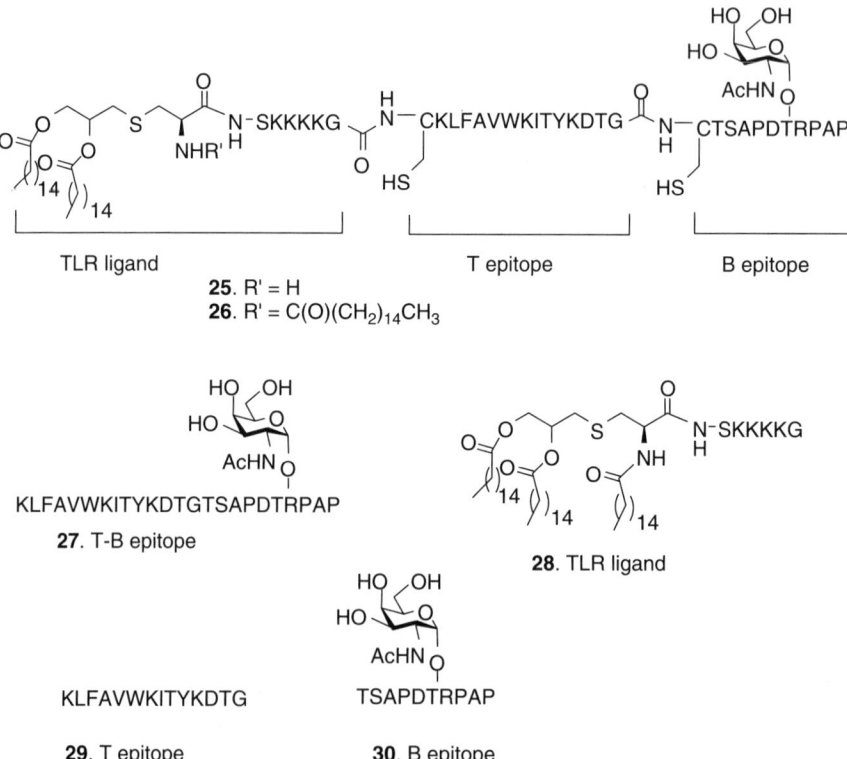

Figure 9.10 Structures of three component vaccine candidates and part structures used to elucidate the influence of covalent attachment on antigenic response.

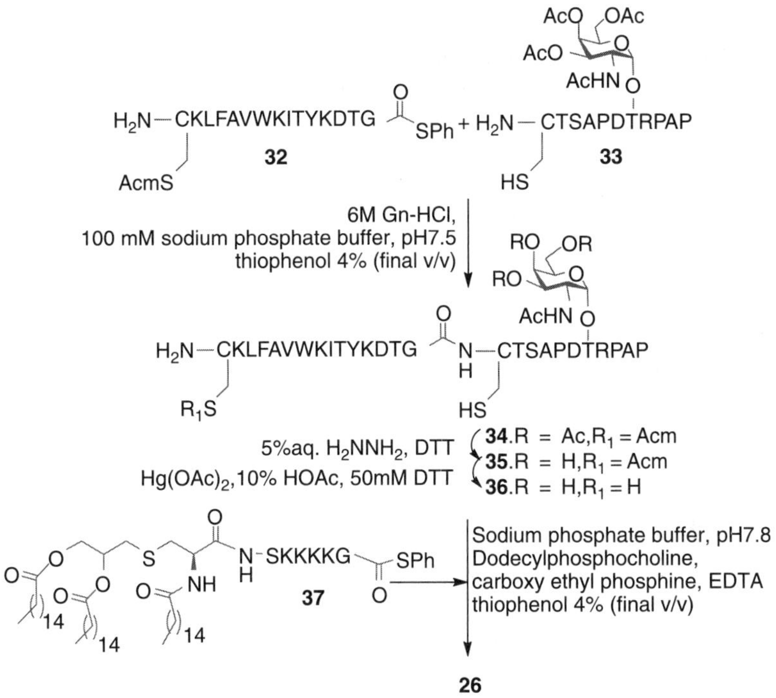

9.12 TRICOMPONENT VACCINES

[Scheme 9.8 depicting synthesis of compounds 32, 33, 34, 35, 36, 37, and 26 with reagents including 6M Gn-HCl, 100 mM sodium phosphate buffer pH 7.5, thiophenol 4% (final v/v); 5% aq. H$_2$NNH$_2$, DTT; Hg(OAc)$_2$, 10% HOAc, 50 mM DTT; Sodium phosphate buffer pH 7.8, Dodecylphosphocholine, carboxy ethyl phosphine, EDTA, thiophenol 4% (final v/v)]

34. R = Ac, R$_1$ = Acm
35. R = H, R$_1$ = Acm
36. R = H, R$_1$ = H

Scheme 9.8

ligand Pam$_3$Cys was designed and synthesized, using a block synthetic approach [308]. The vaccine candidate was included in phospho-lipid-based liposomes and then evaluated for its immunogenicity in mice, in the presence or absence of the external adjuvant QS-21. Although only low and moderate titers of IgG antibodies were raised against the Tn antigen, the results indicated promising possibilities of further

Table 9.3 ELISA anti-MUC-1 Antibody Titers[a] After Four Immunizations

Immunization	Total IgG[b]	IgG1	IgG2a	IgG2b	IgG3	IgM
Pam$_2$CysSK$_4$ vaccine 25	20,900	66,900	700	900	7,300	1,400
Pam$_2$CysSK$_4$ vaccine 25 + QS-21	30,200	113,100	23,000	6,600	17,800	1,100
Pam$_3$CysSK$_4$ vaccine 26	169,600	389,300	56,500	42,700	116,800	7,200
Pam$_3$CysSK$_4$ vaccine 26 + QS-21	322,800	371,300	378,900	56,800	263,500	5,000

[a]Anti-MUC-1 antibody titers are presented as the median for groups of five mice. ELISA plates were coated with BSA–BrAc–MUC-1 conjugate and titers were determined by linear regression analysis, plotting dilution vs. absorbance. Titers are defined as the highest dilution yielding an optical density of 0.1 or greater over that of normal control mouse sera.
[b]A statistical significant difference ($p < 0.05$) was observed between **25** and **26**.

developing this strategy. In a subsequent study, two additional tricomponent vaccine candidates composed of the tumor-related MUC-1 glycopeptide, a well-documented helper T-cell epitope from PV, and either Pam_2CysSK_4 (**25**) or Pam_3CysSK_4 (**26**) as built-in immunoadjuvants, were designed (Fig. 9.10) [310, 311]. Pam_2CysSK_4 is a potent activator of TLR2 and TLR6, whereas Pam_3CysSK_4 induces cellular activation through TLR1 and TLR2 [285]. Compound **25** was assembled through traditional solid-phase peptide synthesis. A similar linear synthesis of **26** failed to

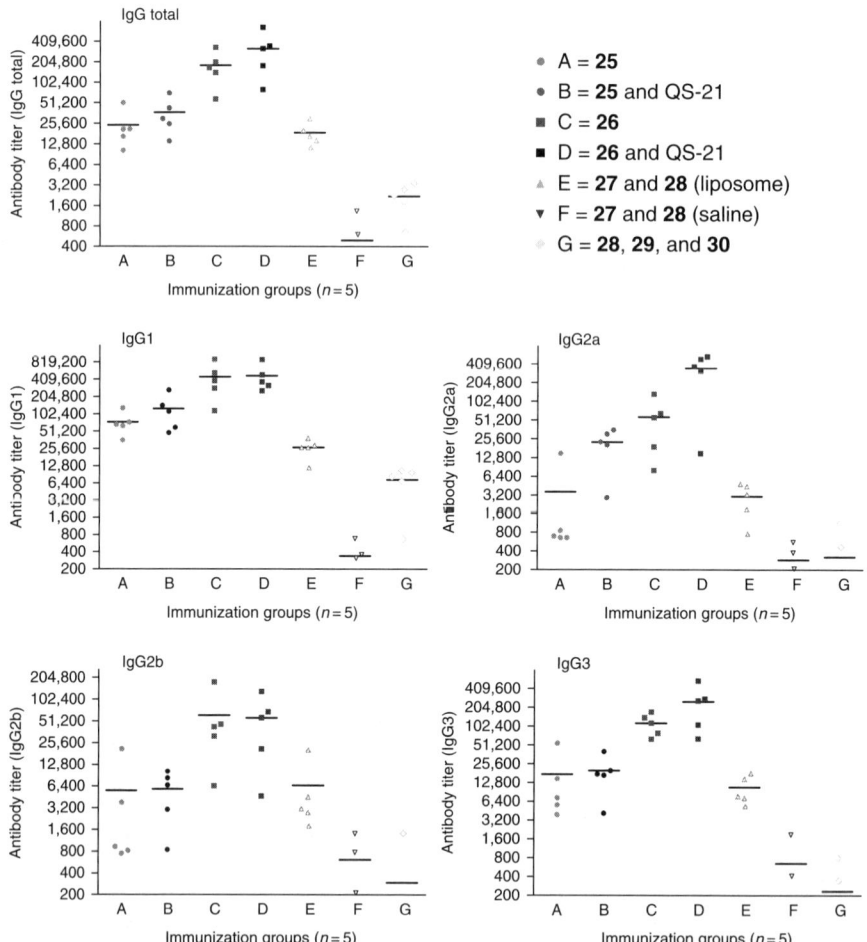

Figure 9.11 ELISA anti-MUC-1 IgG antibody titers after four immunizations with **25**, **25**/QS-21, **26**, **26**/QS-21, **27/28** (liposome), **27/28** (saline), and **28/29/30**. ELISA plates were coated with BSA–BrAc–MUC-1 conjugate, and titers were determined by linear regression analysis, plotting dilution *vs.* absorbance. Titers were defined as the highest dilution yielding an optical density of 0.1 or greater over that of normal control mouse sera. Each data point represents the titer for an individual mouse after four immunizations and the horizontal lines indicate the mean for the group of five mice.

produce a product of reasonable purity, and thus this compound was assembled through tandem native chemical ligation of building blocks **32**, **33**, and **37** (Scheme 9.8) [311]. The vaccine candidates were incorporated into liposomes consisting of phosphatidylcholine, phophatidylglycerol, and cholesterol and their antigenicity studied in murine hosts. Compound **26** induced exceptionally high IgG antibody titers (Table 9.3). Further subtyping of the antibodies revealed high titers of IgG3 antibodies, which are typical for an anticarbohydrate response, and a bias toward a Th2 response, as the levels of IgG1 antibodies were high. Co-administration with the external saponin immunoadjuvant QS-21 did not alter the titers of IgG antibodies, however, a shift toward a mixed Th1/Th2 response was induced. Interestingly, it was found that vaccine candidate **25**, which incorporates the TLR2 and TL6 ligand Pam_2CysSK_4 raised lower titers of anti-MUC-1 IgG antibodies. The elicited antibodies were shown to bind to human adenocarcinoma cell line (MCF7) tumor cells, which express the MUC-1 antigen.

The influence of covalent attachment of the various components of the vaccine candidate on antigenic responses and the importance of the liposomal presentation of the vaccine were further investigated in mice (Fig. 9.10). Uptake and proteolytic processing of antigen for subsequent presentation of a peptide–MHC class II complex on the cell surface of APCs is critical for eliciting IgG antibodies. It could be argued that by incorporating the three components into a liposome, proteolytic processing would be rendered unnecessary and thus a more robust immune response would be seen. However, it was shown that both the covalent attachment of the three components and the liposomal presentation were critical for achieving good antibody titers (Figs. 9.11 and 9.12). The lipid adjuvant moiety of the vaccine candidate facilitates the retention in the liposomes and aids in presenting the tumor-related antigen in a multivalent fashion to B-cell Ig receptors, which is required to be clustered to

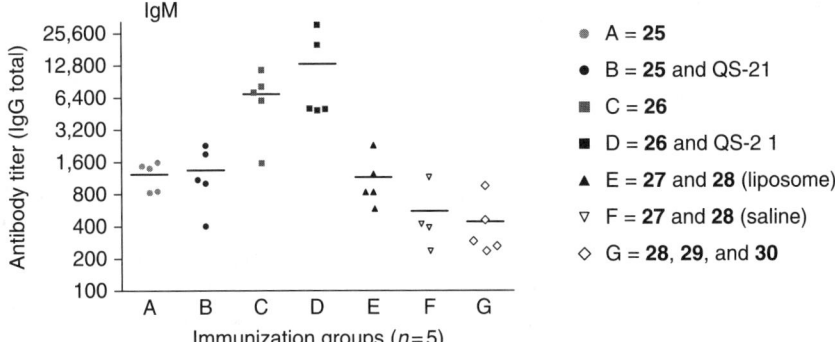

Figure 9.12 ELISA anti-MUC-1 IgM antibody titers after four immunizations with **25**, **25**/QS-21, **26**, **26**/QS-21, **27**/**28** (liposome), **27**/**28** (saline), and **28**/**29**/**30**. ELISA plates were coated with BSA–BrAc–MUC-1 conjugate, and titers were determined by linear regression analysis, plotting dilution vs. absorbance. Titers were defined as the highest dilution yielding an optical density of 0.1 or greater over that of normal control mouse sera. Each data point represents the titer for an individual mouse after four immunizations and the horizontal lines indicate the mean for the group of five mice.

induce activation of B cells. It was also shown that the TLR2 ligand Pam_3CysSK_4 induces cytokines, such as tumor necrosis factor-α (TNF-α), in a TLR2-dependent manner and facilitated uptake and internalization of the vaccine candidate by cells expressing TLR2. The covalent attachment of the lipid adjuvant thus also ensures that the cytokines are produced locally at the site where the vaccine interacts with relevant immune cells and facilitates uptake by APCs that express TLR2.

Although not a cancer vaccine, Bundle and co-workers reported recently the synthesis and immunological evaluation of a tricomponent composed of a $(1\rightarrow 2)\beta$-mannan trisaccharide derived from the yeast *Candida albicans*, a helper T epitope from the murine self 60-kDa heat-shock protein (hsp60) and an immunostimulatory nonapeptide from human interleukin-1β (IL-1β) [312]. The vaccine was formulated in Freund's incomplete adjuvant and tested for immunogenicity in mice. Moderate titers of antisaccharide IgG antibodies were elicted, but importantly it was shown that the antibodies recognized a crude cell wall extract of *C. albicans*.

Recently, a multiepitope vaccine consisting of a cluster of the Tn antigen as a B epitope, a $CD4^+$ T-cell epitope, a $CD8^+$ T-cell epitope, and a palmitic acid, serving as a built-in adjuvant, was reported [313]. The vaccine was based on the RAFT scaffold and assembled using oxime chemistry. The vaccine was delivered in an adjuvant-free setting and showed no adverse effects in the murine host. The elicited antibodies were shown to recognize human breast tumor cells MCF7 expressing the Tn antigen. The vaccine also induced strong specific $CD4^+$ T-cell and $CD8^+$ T-cell responses. In prophylactic tumor studies with MO5 tumor cells, none of 20 mice developed a tumor in the monitoring period of 90 days. In contrast, the survival rate for mice immunized with a vaccine candidate lacking the palmitic acid adjuvant and CpG as an external adjuvant was determined to be 80%.

REFERENCES

1. Kruger C, Greten TF, Korangy F (2007) Immune based therapies in cancer. *Histol. Histopathol.* 22:687–696.
2. Morse MA, Lyerly HK, Clay TM, Abdel-Wahab O, Chui SY, Garst J, Gollob J, Grossi PM, Kalady M, Mosca PJ, Onaitis M, Sampson JH, Seigler HF, Toloza EM, Tyler D, Vieweg J, Yang YP (2004) How does the immune system attack cancer? *Curr. Probl. Surg.* 41:9–132.
3. Finn OJ (2008) Tumor immunology top 10 list. *Immunol. Rev.* 222:5–8.
4. Ferguson AR, Nichols LA, Zarling AL, Thompson ED, Brinkman CC, Hargadon KM, Bullock TN, Engelhard VH (2008) Strategies and challenges in eliciting immunity to melanoma. *Immunol. Rev.* 222:28–42.
5. Hung CF, Wu TC, Monie A, Roden R (2008) Antigen-specific immunotherapy of cervical and ovarian cancer. *Immunol. Rev.* 222:43–69.
6. Mitchell DA, Fecci PE, Sampson JH (2008) Immunotherapy of malignant brain tumors. *Immunol. Rev.* 222:70–100.
7. Guinn BA, Kasahara N, Farzaneh F, Habib NA, Norris JS, Deisseroth AB (2007) Recent advances and current challenges in tumor immunology and immunotherapy. *Mol. Ther.* 15:1065–1071.

8. von Mehren M, Adams GP, Weiner LM (2003) Monoclonal antibody therapy for cancer. *Annu. Rev. Med.* 54:343–369.
9. Harding TA, Gallati C, Horlacher M, Becker A, Mousa SA (2008) Monoclonal antibodies in oncological malignancies: Current status and future directions. *Drug Future* 33:361–369.
10. Nicodemus CF, Smith LM, Schultes BC (2007) Role of monoclonal antibodies in tumor-specific immunity. *Exp. Opin. Biol. Ther.* 7:331–343.
11. Chamberlain RS (1999) Prospects for the therapeutic use of anticancer vaccines. *Drugs* 57:309–325.
12. Sinkovics JG, Horvath JC (2000) Vaccination against human cancers (review). *Int. J. Oncol.* 16:81–96.
13. Finn OJ (2003) Cancer vaccines: Between the idea and the reality. *Nat. Rev. Immunol.* 3:630–641.
14. Pazdur MP, Jones JL (2007) Vaccines: An innovative approach to treating cancer. *J. Infus. Nurs.* 30:173–178.
15. Giarelli E (2007) Cancer vaccines: A new frontier in prevention and treatment. *Oncology (Williston Park)* 21:11–17; discussion 18.
16. Hung CF, Ma B, Monie A, Tsen SW, Wu TC (2008) Therapeutic human papillomavirus vaccines: Current clinical trials and future directions. *Expert Opin. Biol. Ther.* 8:421–439.
17. Morse MA, Chui S, Hobeika A, Lyerly HK, Clay T (2005) Recent developments in therapeutic cancer vaccines. *Nat. Clin. Pract. Oncol.* 2:108–113.
18. Ward S, Casey D, Labarthe MC, Whelan M, Dalgleish A, Pandha H, Todryk S (2002) Immunotherapeutic potential of whole tumour cells. *Cancer Immunol. Immunother.* 51:351–357.
19. Pardoll DM (1995) Paracrine cytokine adjuvants in cancer immunotherapy. *Annu. Rev. Immunol.* 13:399–415.
20. Glick RP, Lichtor T, Mogharbel A, Taylor CA, Cohen EP (1997) Intracerebral versus subcutaneous immunization with allogeneic fibroblasts genetically engineered to secrete interleukin-2 in the treatment of central nervous system glioma and melanoma. *Neurosurgery* 41:898–906; discussion 906–907.
21. Coughlin CM, Salhany KE, Wysocka M, Aruga E, Kurzawa H, Chang AE, Hunter CA, Fox JC, Trinchieri G, Lee WM (1998) Interleukin-12 and interleukin-18 synergistically induce murine tumor regression which involves inhibition of angiogenesis. *J. Clin. Invest.* 101:1441–1452.
22. Dranoff G, Jaffee E, Lazenby A, Golumbek P, Levitsky H, Brose K, Jackson V, Hamada H, Pardoll D, Mulligan RC (1993) Vaccination with irradiated tumor cells engineered to secrete murine granulocyte-macrophage colony-stimulating factor stimulates potent, specific, and long-lasting anti-tumor immunity. *Proc. Natl. Acad. Sci. U.S.A.* 90:3539–3543.
23. Salgia R, Lynch T, Skarin A, Lucca J, Lynch C, Jung K, Hodi FS, Jaklitsch M, Mentzer S, Swanson S, Lukanich J, Bueno R, Wain J, Mathisen D, Wright C, Fidias P, Donahue D, Clift S, Hardy S, Neuberg D, Mulligan R, Webb I, Sugarbaker D, Mihm M, Dranoff G (2003) Vaccination with irradiated autologous tumor cells engineered to secrete granulocyte-macrophage colony-stimulating factor augments antitumor immunity in some patients with metastatic non-small-cell lung carcinoma. *J. Clin. Oncol.* 21:624–630.

24. Ward JE, McNeel DG (2007) GVAX: An allogeneic, whole-cell, GM-CSF-secreting cellular immunotherapy for the treatment of prostate cancer. *Expert Opin. Biol. Ther.* 7:1893–1902.
25. Stevanovic S (2002) Identification of tumour-associated T-cell epitopes for vaccine development. *Nat. Rev. Cancer* 2:514–520.
26. Rosenberg SA, Yang JC, Restifo NP (2004) Cancer immunotherapy: Moving beyond current vaccines. *Nat. Med.* 10:909–915.
27. Bijker MS, Melief CJ, Offringa R, van der Burg SH (2007) Design and development of synthetic peptide vaccines: Past, present and future. *Expert Rev. Vaccines* 6:591–603.
28. Springer GF (1997) Immunoreactive T and Tn epitopes in cancer diagnosis, prognosis, and immunotherapy. *J. Mol. Med.* 75:594–602.
29. Hakomori S (1998) Cancer-associated glycosphingolipid antigens: Their structure, organization, and function. *Acta Anat.* 161:79–90.
30. Brooks SA, Carter TM, Royle L, Harvey DJ, Fry SA, Kinch C, Dwek RA, Rudd PM (2008) Altered glycosylation of proteins in cancer: What is the potential for new antitumour strategies. *Anticancer Agents Med. Chem.* 8:2–21.
31. Livingston PO, Zhang S, Lloyd KO (1997) Carbohydrate vaccines that induce antibodies against cancer. 1. Rationale. *Cancer Immunol. Immunother.* 45:1–9.
32. Dube DH, Bertozzi CR (2005) Glycans in cancer and inflammation. Potential for therapeutics and diagnostics. *Nat. Rev. Drug Discov.* 4:477–488.
33. Danishefsky SJ, Allen JR (2000) From the laboratory to the clinic: A retrospective on fully synthetic carbohydrate-based anticancer vaccines. *Angew. Chem., Int. Ed.* 39:836–863.
34. Slovin SF, Keding SJ, Ragupathi G (2005) Carbohydrate vaccines as immunotherapy for cancer. *Immunol. Cell Biol.* 83:418–428.
35. Kobata A, Amano J (2005) Altered glycosylation of proteins produced by malignant cells, and application for the diagnosis and immunotherapy of tumours. *Immunol. Cell Biol.* 83:429–439.
36. Xu Y, Sette A, Sidney J, Gendler SJ, Franco A (2005) Tumor-associated carbohydrate antigens: A possible avenue for cancer prevention. *Immunol. Cell Biol.* 83:440–448.
37. Freire T, Bay S, Vichier-Guerre S, Lo-Man R, Leclerc C (2006) Carbohydrate antigens: Synthesis aspects and immunological applications in cancer. *Mini Rev. Med. Chem.* 6:1357–1373.
38. Cipolla L, Peri F, Airoldi C (2008) Glycoconjugates in cancer therapy. *Anticancer Agents Med. Chem.* 8:92–121.
39. Rudd PM, Elliott T, Cresswell P, Wilson IA, Dwek RA (2001) Glycosylation and the immune system. *Science* 291:2370–2376.
40. Varki A (1993) Biological roles of oligosaccharides—all of the theories are correct. *Glycobiology* 3:97–130.
41. Dwek RA (1996) Glycobiology: Toward understanding the function of sugars. *Chem. Rev.* 96:683–720.
42. Ohtsubo K, Marth JD (2006) Glycosylation in cellular mechanisms of health and disease. *Cell* 126:855–867.

43. Kuberan B, Linhardt RJ (2000) Carbohydrate based vaccines. *Curr. Org. Chem.* 4:653–677.
44. Vliegenthart JF (2006) Carbohydrate based vaccines. *FEBS Lett.* 580:2945–2950.
45. Lucas AH, Apicella MA, Taylor CE (2005) Carbohydrate moieties as vaccine candidates. *Clin. Infect. Dis.* 41:705–712.
46. Sanders DSA, Kerr MA (1999) Lewis blood group and CEA related antigens; coexpressed cell-cell adhesion molecules with roles in the biological progression and dissemination of tumours. *J. Clin. Pathol. Mol. Pathol.* 52:174–178.
47. Kobayashi H, Boelte KC, Lin PC (2007) Endothelial cell adhesion molecules and cancer progression. *Curr. Med. Chem.* 14:377–386.
48. Hakomori S (1999) Antigen structure and genetic basis of histo-blood groups A, B and O: Their changes associated with human cancer. *Biochim. Biophys. Acta* 1473:247–266.
49. Glinsky GV, Ivanova AB, Welsh J, McClelland M (2000) The role of blood group antigens in malignant progression, apoptosis resistance, and metastatic behavior. *Transfus. Med. Rev.* 14:326–350.
50. Magnani JL, Steplewski Z, Koprowski H, Ginsburg V (1983) Identification of the gastrointestinal and pancreatic cancer-associated antigen detected by monoclonal antibody 19-9 in the sera of patients as a mucin. *Cancer Res.* 43:5489–5492.
51. Koprowski H, Herlyn M, Steplewski Z, Sears HF (1981) Specific antigen in serum of patients with colon carcinoma. *Science* 212:53–55.
52. Bertozzi CR (1995) Cracking the carbohydrate code for selectin recognition. *Chem. Biol.* 2:703–708.
53. Nudelman E, Levery SB, Kaizu T, Hakomori S (1986) Novel fucolipids of human adenocarcinoma: Characterization of the major Ley antigen of human adenocarcinoma as trifucosylnonaosyl Ley glycolipid (III3FucV3FucVI2FucnLc6). *J. Biol. Chem.* 261:11247–11253.
54. Kaizu T, Levery SB, Nudelman E, Stenkamp RE, Hakomori S (1986) Novel fucolipids of human adenocarcinoma: Monoclonal antibody specific for trifucosyl Ley (III3FucV3FucVI2FucnLc6) and a possible three-dimensional epitope structure. *J. Biol. Chem.* 261:11254–11258.
55. Kim YS, Yuan M, Itzkowitz SH, Sun Q, Kaizu T, Palekar A, Trump BF, Hakomori S (1986) Expression of Ley and extended Ley blood group-related antigens in human-malignant, premalignant, and nonmalignant colonic tissues. *Cancer Res.* 46:5985–5992.
56. Kannagi R, Levery SB, Ishigami F, Hakomori SI, Shevinsky LH, Knowles BB, Solter D (1983) New Globoseries glycosphingolipids in human teratocarcinoma reactive with the monoclonal-antibody directed to a developmentally regulated antigen, stage-specific embryonic antigen-3. *J. Biol. Chem.* 258:8934–8942.
57. Menard S, Tagliabue E, Canevari S, Fossati G, Colnaghi MI (1983) Generation of monoclonal-antibodies reacting with normal and cancer-cells of human-breast. *Cancer Res.* 43:1295–1300.
58. Bremer EG, Levery SB, Sonnino S, Ghidoni R, Canevari S, Kannagi R, Hakomori SI (1984) Characterization of a glycosphingolipid antigen defined by the monoclonal-antibody Mbr1 expressed in normal and neoplastic epithelial-cells of human mammary-gland. *J. Biol. Chem.* 259:4773–4777.

59. Zhang S, Cordon-Cardo C, Zhang HS, Reuter VE, Adluri S, Hamilton WB, Lloyd KO, Livingston PO (1997) Selection of tumor antigens as targets for immune attack using immunohistochemistry: I. Focus on gangliosides. *Int. J. Cancer* 73:42–49.
60. Baldus SE, Engelmann K, Hanisch FG (2004) MUC1 and the MUCs: A family of human mucins with impact in cancer biology. *Crit. Rev. Clin. Lab. Sci.* 41:189–231.
61. Hattrup CL, Gendler SJ (2008) Structure and function of the cell surface (tethered) mucins. *Annu. Rev. Physiol.* 70:431–457.
62. Gendler SJ (2001) MUC1, the renaissance molecule. *J. Mammary Gland Biol. Neoplasia* 6:339–353.
63. Apostolopoulos V, Pietersz GA, McKenzie IF (1999) MUC1 and breast cancer. *Curr. Opin. Mol. Ther.* 1:98–103.
64. Bast RC, Jr, Badgwell D, Lu Z, Marquez R, Rosen D, Liu J, Baggerly KA, Atkinson EN, Skates S, Zhang Z, Lokshin A, Menon U, Jacobs I, Lu K (2005) New tumor markers: CA125 and beyond. *Int. J. Gynecol. Cancer* 15(Suppl 3):274–281.
65. Swallow DM, Gendler S, Griffiths B, Corney G, Taylor-Papadimitriou J, Bramwell ME (1987) The human tumour-associated epithelial mucins are coded by an expressed hypervariable gene locus PUM. *Nature* 328:82–84.
66. Brockhausen I (1999) Pathways of O-glycan biosynthesis in cancer cells. *Biochim. Biophys. Acta* 1473:67–95.
67. Lloyd KO, Burchell J, Kudryashov V, Yin BW, Taylor-Papadimitriou J (1996) Comparison of O-linked carbohydrate chains in MUC-1 mucin from normal breast epithelial cell lines and breast carcinoma cell lines. Demonstration of simpler and fewer glycan chains in tumor cells. *J. Biol. Chem.* 271:33325–33334.
68. Burchell JM, Mungul A, Taylor-Papadimitriou J (2001) O-linked glycosylation in the mammary gland: Changes that occur during malignancy. *J. Mammary Gland Biol. Neoplasia* 6:355–364.
69. Springer GF (1984) T and Tn, general carcinoma autoantigens. *Science* 224:1198–1206.
70. Gendler SJ, Lancaster CA, Taylor-Papadimitriou J, Duhig T, Peat N, Burchell J, Pemberton L, Lalani EN, Wilson D (1990) Molecular cloning and expression of human tumor-associated polymorphic epithelial mucin. *J. Biol. Chem.* 265:15286–15293.
71. Taylor-Papadimitriou J, Burchell J, Miles DW, Dalziel M (1999) MUC-1 and cancer. *Biochim. Biophys. Acta* 1455:301–313.
72. Burchell J, Taylor-Papadimitriou J, Boshell M, Gendler S, Duhig T (1989) A short sequence, within the amino acid tandem repeat of a cancer-associated mucin, contains immunodominant epitopes. *Int. J. Cancer* 44:691–696.
73. Gendler S, Taylor-Papadimitriou J, Duhig T, Rothbard J, Burchell J (1988) A highly immunogenic region of a human polymorphic epithelial mucin expressed by carcinomas is made up of tandem repeats. *J. Biol. Chem.* 263:12820–12823.
74. von Mensdorff-Pouilly S, Snijdewint FG, Verstraeten AA, Verheijen RH, Kenemans P (2000) Human MUC-1 mucin: A multifaceted glycoprotein. *Int. J. Biol. Markers* 15:343–356.
75. Hanisch FG, Muller S (2000) MUC-1: The polymorphic appearance of a human mucin. *Glycobiology* 10:439–449.
76. Muller S, Hanisch FG (2002) Recombinant MUC-1 probe authentically reflects cell-specific O-glycosylation profiles of endogenous breast cancer mucin. High density and prevalent core 2-based glycosylation. *J. Biol. Chem.* 277:26103–26112.

77. Karsten U, Serttas N, Paulsen H, Danielczyk A, Goletz S (2004) Binding patterns of DTR-specific antibodies reveal a glycosylation-conditioned tumor-specific epitope of the epithelial mucin (MUC-1). *Glycobiology* 14:681–692.
78. Braun P, Davies GM, Price MR, Williams PM, Tendler SJ, Kunz H (1998) Effects of glycosylation on fragments of tumour associated human epithelial mucin MUC-1. *Bioorg. Med. Chem.* 6:1531–1545.
79. Bundle DR (2007) A carbohydrate vaccine exceeds the sum of its parts. *Nat. Chem. Biol.* 3:604–606.
80. Livingston PO, Ragupathi G (1997) Carbohydrate vaccines that induce antibodies against cancer. 2. Previous experience and future plans. *Cancer Immunol. Immunother.* 45:10–19.
81. Ragupathi G (1996) Carbohydrate antigens as targets for active specific immunotherapy. *Cancer Immunol.* 43:152–157.
82. von Mensdorff-Pouilly S, Petrakou E, Kenemans P, van Uffelen K, Verstraeten AA, Snijdewint FG, van Kamp GJ, Schol DJ, Reis CA, Price MR, Livingston PO, Hilgers J (2000) Reactivity of natural and induced human antibodies to MUC1 mucin with MUC1 peptides and n-acetylgalactosamine (GalNAc) peptides. *Int. J. Cancer* 86:702–712.
83. DeFranco AL (2000) B-cell activation 2000. *Immunol. Rev.* 176:5–9.
84. Stavnezer J (1996) Antibody class switching. *Adv. Immunol.* 61:79–146.
85. Honjo T, Kinoshita K, Muramatsu M (2002) Molecular mechanism of class switch recombination: Linkage with somatic hypermutation. *Annu. Rev. Immunol.* 20:165–196.
86. Jelley-Gibbs DM, Strutt TM, McKinstry KK, Swain SL (2008) Influencing the fates of CD4 T cells on the path to memory: Lessons from influenza. *Immunol. Cell Biol.* 86:343–352.
87. Belz GT (2008) Getting together: Dendritic cells, T cells, collaboration and fates. *Immunol. Cell Biol.* 86:310–311.
88. Kennedy R, Celis E (2008) Multiple roles for CD4+ T cells in anti-tumor immune responses. *Immunol. Rev.* 222:129–144.
89. Foy TM, Aruffo A, Bajorath J, Buhlmann JE, Noelle RJ (1996) Immune regulation by CD40 and its ligand GP39. *Annu. Rev. Immunol.* 14:591–617.
90. Campos M, Godson DL (2003) The effectiveness and limitations of immune memory: Understanding protective immune responses. *Int. J. Parasitol.* 33:655–661.
91. Kihlberg JO, Elofsson M (1997) Solid-phase synthesis of glycopeptides: Immunological studies with T cell stimulating glycopeptides. *Curr. Med. Chem.* 4:85–116.
92. Dengjel J, Stevanovic S (2006) Naturally, presented MHC ligands carrying glycans. *Trans. Med. Hemother.* 33:38–44.
93. Werdelin O, Meldal M, Jensen T (2002) Processing of glycans on glycoprotein and glycopeptide antigens in antigen-presenting cells. *Proc. Natl. Acad. Sci. U.S.A.* 99:9611–9613.
94. Hanisch FG, Ninkovic T (2006) Immunology of O-glycosylated proteins: Approaches to the design of a MUC1 glycopeptide-based tumor vaccine. *Curr. Protein Pept. Sci.* 7:307–315.
95. Backlund J, Treschow A, Bockermann R, Holm B, Holm L, Issazadeh-Navikas S, Kihlberg J, Holmdahl R (2002) Glycosylation of type II collagen is of major importance for T cell tolerance and pathology in collagen-induced arthritis. *Eur. J. Immunol.* 32:3776–3784.

96. Surman S, Lockey TD, Slobod KS, Jones B, Riberdy JM, White SW, Doherty PC, Hurwitz JL (2001) Localization of CD4+ T cell epitope hotspots to exposed strands of HIV envelope glycoprotein suggests structural influences on antigen processing. *Proc. Natl. Acad. Sci. U.S.A.* 98:4587–4592.
97. Vlad AM, Muller S, Cudic M, Paulsen H, Otvos L, Jr, Hanisch FG, Finn OJ (2002) Complex carbohydrates are not removed during processing of glycoproteins by dendritic cells: Processing of tumor antigen MUC1 glycopeptides for presentation to major histocompatibility complex class II-restricted T cells. *J. Exp. Med.* 196:1435–1446.
98. Brode S, Macary PA (2004) Cross-presentation: Dendritic cells and macrophages bite off more than they can chew! *Immunology* 112:345–351.
99. Belz GT, Carbone FR, Heath WR (2002) Cross-presentation of antigens by dendritic cells. *Crit. Rev. Immunol.* 22:439–448.
100. Gromme M, Neefjes J (2002) Antigen degradation or presentation by MHC class I molecules via classical and non-classical pathways. *Mol. Immunol.* 39:181–202.
101. Franco A (2005) CTL-based cancer preventive/therapeutic vaccines for carcinomas: Role of tumour-associated carbohydrate antigens. *Scand. J. Immunol.* 61:391–397.
102. Apostolopoulos V, Yuriev E, Ramsland PA, Halton J, Osinski C, Li W, Plebanski M, Paulsen H, McKenzie IF (2003) A glycopeptide in complex with MHC class I uses the GalNAc residue as an anchor. *Proc. Natl. Acad. Sci. U.S.A.* 100:15029–15034.
103. Xu Y, Gendler SJ, Franco A (2004) Designer glycopeptides for cytotoxic T cell-based elimination of carcinomas. *J. Exp. Med.* 199:707–716.
104. Knutson KL, Disis ML (2005) Tumor antigen-specific T helper cells in cancer immunity and immunotherapy. *Cancer Immunol. Immunother.* 54:721–728.
105. Mukherjee P, Pathangey LB, Bradley JB, Tinder TL, Basu GD, Akporiaye ET, Gendler SJ (2007) MUC1-specific immune therapy generates a strong anti-tumor response in a MUC1-tolerant colon cancer model. *Vaccine* 25:1607–1618.
106. Beutler B (2004) Innate immunity: An overview. *Mol. Immunol.* 40:845–859.
107. van Amersfoort ES, van Berkel TJC, Kuiper J (2003) Receptors, mediators, and mechanisms involved in bacterial sepsis and septic shock. *Clin. Microbiol. Rev.* 16:379–414.
108. Akira S, Takeda K (2004) Toll-like receptor signalling. *Nat. Rev. Immunol.* 4:499–511.
109. Beutler B (2004) Inferences, questions and possibilities in Toll-like receptor signalling. *Nature* 430:257–263.
110. O'Neill LA (2006) How Toll-like receptors signal: What we know and what we don't know. *Curr. Opin. Immunol.* 18:3–9.
111. Lee HK, Iwasaki A (2007) Innate control of adaptive immunity: Dendritic cells and beyond. *Semin. Immunol.* 19:48–55.
112. Akira S, Takeda K, Kaisho T (2001) Toll-like receptors: Critical proteins linking innate and acquired immunity. *Nat. Immunol.* 2:675–680.
113. Beutler B, Hoebe K, Du X, Ulevitch RJ (2003) How we detect microbes and respond to them: The Toll-like receptors and their transducers. *J. Leukocyte Biol.* 74:479–485.
114. Lien E, Ingalls RR (2002) Toll-like receptors. *Criti. Care Med.* 30:S1–S11.
115. O'Neill LAJ (2004) After the Toll rush. *Science* 303:1481–1482.
116. O'Neill LAJ (2004) TLRs: Professor Mechnikov, sit on your hat. *Trends Immunol.* 25:687–693.

117. Pasare C, Medzhitov R (2003) Toll-like receptors: Balancing host resistance with immune tolerance. *Curr. Opin. Immunol.* 15:677–682.
118. Check W (2004) Innate immunity depends on Toll-like receptors. *Am. Soc. Microbiol. News* 70:317–322.
119. Schmitz F, Mages J, Heit A, Lang R, Wagner H (2004) Transcriptional activation induced in macrophages by Toll-like receptor (TLR) ligands: From expression profiling to a model of TLR signaling. *Eur. J. Immunol.* 34:2863–2873.
120. Tsan MF, Gao B (2004) Endogenous ligands of Toll-like receptors. *J. Leukoc. Biol.* 76:514–519.
121. Sarkar SN, Peters KL, Elco CP, Sakamoto S, Pal S, Sen GC (2004) Novel roles of TLR3 tyrosine phosphorylation and PI3 kinase in double-stranded RNA signaling. *Nat. Struct. Mol. Biol.* 11:1060–1067.
122. Pasare C, Medzhitov R (2004) Toll-like receptors and acquired immunity. *Semin. Immunol.* 16:23–26.
123. Pasare C, Medzhitov R (2005) Control of B-cell responses by Toll-like receptors. *Nature* 438:364–368.
124. Dabbagh K, Lewis DB (2003) Toll-like receptors and T-helper-1/T-helper-2 responses. *Curr. Opin. Infect. Dis.* 16:199–204.
125. Krieg AM (2008) Toll-like receptor 9 (TLR9) agonists in the treatment of cancer. *Oncogene* 27:161–167.
126. Schmidt RR (1986) New methods for the synthesis of glycosides and oligosaccharides—are there alternatives to the Koenigs-Knorr method. *Angew. Chem. Int. Ed.* 25:212–235.
127. Fraser-Reid B, Udodong UE, Wu Z, Ottoson H, Merritt JR, Rao S, Roberts C, Madsen R (1992) n-Pentenyl glycosides in organic chemistry: A contemporary example of serendipity. *Synlett.* 927–942.
128. Toshima K, Tatsuta K. (1993) Recent progress in O-glycosylation methods and its application to natural-products synthesis. *Chem. Rev.* 93:1503–1531.
129. Boons GJ (1996) Strategies in oligosaccharide synthesis. *Tetrahedron* 52:1095–1121.
130. Bartolozzi A, Seeberger PH (2001) New approaches to the chemical synthesis of bioactive oligosaccharides. *Curr. Opin. Struct. Biol.* 11:587–592.
131. Karst NA, Linhardt RJ (2003) Recent chemical and enzymatic approaches to the synthesis of glycosaminoglycan oligosaccharides. *Curr. Med. Chem.* 10:1993–2031.
132. Demchenko AV (2003) Stereoselective chemical 1,2-cis O-glycosylation: From "sugar ray" to modern techniques of the 21st century. *Synlett.* 1225–1240.
133. Seeberger PH, Werz DB (2007) Synthesis and medical applications of oligosaccharides. *Nature* 446:1046–1051.
134. Bongat AFG, Demchenko AV (2007) Recent trends in the synthesis of O-glycosides of 2-amino-2-deoxysugars. *Carbohydr. Res.* 342:374–406.
135. Litjens REJN, van den Bos LJ, Codee JDC, Overkleeft HS, van der Marel GA (2007) The use of cyclic bifunctional protecting groups in oligosaccharide synthesis—an overview. *Carbohydr. Res.* 342:419–429.
136. Carmona AT, Moreno-Vargas AJ, Robina I (2008) Glycosylation methods in oligosaccharide synthesis. Part 2. *Curr. Org. Synth.* 5:81–116.
137. Carmona AT, Moreno-Vargas AJ, Robina I (2008) Glycosylation methods in oligosaccharide synthesis. Part 1. *Curr. Org. Synth.* 5:33–60.

138. Toshima K (2000) Glycosyl fluorides in glycosidations. *Carbohydr. Res.* 327:15–26.
139. Schmidt RR, Kinzy W (1994) Anomeric-oxygen activation for glycoside synthesis—the trichloroacetimidate method. *Adv. Carbohydr. Chem. Biochem.* 50:21–123.
140. Garegg PJ (1997) Thioglycosides as glycosyl donors in oligosaccharide synthesis. *Adv. Carbohydr. Chem. Biochem.* 52:179–205.
141. Danishefsky SJ, Bilodeau MT (1996) Glycals in organic synthesis: The evolution of comprehensive strategies for the assembly of oligosaccharides and glycoconjugates of biological consequence. *Angew. Chem. Int. Ed.* 35:1380–1419.
142. Gildersleeve J, Smith A, Sakurai K, Raghavan S, Kahne D (1999) Scavenging by product in the sulfoxide glycosylation reaction: Application to the synthesis of Cillamycin 0. *J. Am. Chem. Soc.* 121:6176–6182.
143. Boebel TA, Gin DY (2005) Probing the mechanism of sulfoxide-catalyzed hemiacetal activation in dehydrative glycosylation. *J. Org. Chem.* 70:5818–5826.
144. Codee JDC, Hossain LH, Seeberger PH (2005) Efficient installation of beta-mannosides using a dehydrative coupling strategy. *Org. Lett.* 7:3251–3254.
145. Boebel TA, Gin DY (2003) Sulfoxide covalent catalysis: Application to glycosidic bond formation. *Angew. Chem. Int. Ed.* 42:5874–5877.
146. Honda E, Gin DY (2002) C2-hydroxyglycosylation with glycal donors. Probing the mechanism of sulfonium-mediated oxygen transfer to glycal enol ethers. *J. Am. Chem. Soc.* 124:7343–7352.
147. Nguyen HM, Chen YN, Duron SG, Gin DY (2001) Sulfide-mediated dehydrative glycosylation. *J. Am. Chem. Soc.* 123:8766–8772.
148. Garcia BA, Poole JL, Gin DY (1997) Direct glycosylations with 1-hydroxy glycosyl donors using trifluoromethanesulfonic anhydride and diphenyl sulfoxide. *J. Am. Chem. Soc.* 119:7597–7598.
149. Wang CC, Lee JC, Luo SY, Kulkarni SS, Huang YW, Lee CC, Chang KL, Hung SC (2007) Regioselective one-pot protection of carbohydrates. *Nature* 446:896–899.
150. Francais A, Urban D, Beau JM (2007) Tandem catalysis for a one-pot regioselective protection of carbohydrates: The example of glucose. *Angew. Chem. Int. Ed.* 46:8662–8665.
151. Koeller KM, Wong CH (2000) Synthesis of complex carbohydrates and glycoconjugates: Enzyme-based and programmable one-pot strategies. *Chem. Rev.* 100:4465–4493.
152. Douglas NL, Ley SV, Lucking U, Warriner SL (1998) Tuning glycoside reactivity: New tool for efficient oligosaccharide synthesis. *J. Chem. Soc., Perkin Trans. 1* 51–65.
153. Codee JDC, Litjens R, van den Bos LJ, Overkleeft HS, van der Marel GA (2005) Thioglycosides in sequential glycosylation strategies. *Chem. Soc. Rev.* 34:769–782.
154. Wang YH, Zhang LH, Ye XS (2006) Oligosaccharide synthesis and library assembly by one-pot sequential glycosylation strategy. *Comb. Chem. High Throughput Screen* 9:63–75.
155. Wang YH, Ye XS, Zhang LH (2007) Oligosaccharide assembly by one-pot multi-step strategy. *Org. Biomol. Chem.* 5:2189–2200.
156. Tanaka H, Yamada H, Takahashi T (2007) Rapid synthesis of oligosaccharides based on one-pot glycosylation. *Trends Glycosci. Glycotechnol.* 19:183–193.
157. Seeberger PH, Haase WC (2000) Solid-phase oligosaccharide synthesis and combinatorial carbohydrate libraries. *Chem. Rev.* 100:4349–4394.

158. Seeberger PH (2008) Automated oligosaccharide synthesis. *Chem. Soc. Rev.* 37:19–28.
159. Plante OJ, Palmacci ER, Seeberger PH (2003) Development of an automated oligosaccharide synthesizer. *Adv. Carbohydr. Chem. Biochem.* 58:35–54.
160. Buskas T, Ingale S, Boons GJ (2006) Glycopeptides as versatile tools for glycobiology. *Glycobiology* 16:113R–136R.
161. Kunz H, Unverzagt C (1988) Protecting group-dependent stability of intersaccharide bonds—Synthesis of a fucosyl-chitobiose glycopeptide. *Angew. Chem. Int. Ed. Engl.* 27:1697–1699.
162. Unverzagt C, Kunz H (1994) Synthesis of glycopeptides and neoglycoproteins containing the fucosylated linkage region of N-glycoproteins. *Bioorg. Med. Chem.* 2:1189–1201.
163. Peters S, Lowary TL, Hindsgaul O, Meldal M, Bock K (1995) Solid-phase synthesis of a fucosylated glycopeptide of human factor IX with a fucose-a-(1-O)-serine linkage. *J. Chem. Soc., Perkin Trans. 1* 3017–3022.
164. Sjolin P, Elofsson M, Kihlberg J (1996) Removal of acyl protective groups from glycopeptides: Base does not epimerize peptide stereocenters, and beta-elimination is slow. *J. Org. Chem.* 61:560–565.
165. Seitz O (2000) Glycopeptide synthesis and the effects of glycosylation on protein structure and activity. *ChemBioChem* 1:215–246.
166. Grogan MJ, Pratt MR, Marcaurelle LA, Bertozzi CR (2002) Homogeneous glycopeptides and glycoproteins for biological investigation. *Ann. Rev. Biochem.* 71:593–634.
167. Carpino LA, Han GY (1972) 9-Fluorenylmethoxycarbonyl amino-protecting group. *J. Org. Chem.* 37:3404–3409.
168. Albericio F, Carpino LA (1997) Coupling reagents and activation. In Fields GB, *Methods in Enzymology—Solid-Phase Peptide Synthesis*. Academic, New York, pp. 104–126.
169. Konig W, Geiger R (1970) A new method for synthesis of peptides—activation of carboxyl group with dicyclohexylcarbodiimide using 1-hydroxybenzotriazoles as additives. *Chem. Ber.* 103:788–791.
170. Coste J, Lenguyen D, Castro B (1990) Pybop—A new peptide coupling reagent devoid of toxic by-product. *Tetrahedron Lett.* 31:205–208.
171. Knorr R, Trzeciak A, Bannwarth W, Gillessen D (1989) New coupling reagents in peptide chemistry. *Tetrahedron Lett.* 30:1927–1930.
172. Carpino LA, El-Faham A, Minor CA, Albericio F (1994) Advantagous applications of azabenzotriazole (triazolopyridine)-based coupling reagents to solid-pase peptide synthesis. *J. Chem. Soc., Chem. Commun.* 201–203.
173. Carpino LA (1993) 1-Hydroxy-7-azabenzotriazole—an efficient peptide coupling additive. *J. Am. Chem. Soc.* 115:4397–4398.
174. Kihlberg J, Vuljanic T (1993) Piperidine is preferable to morpholine for Fmoc cleavage in solid-phase synthesis of O-linked glycopeptides. *Tetrahedron Lett.* 34:6135–6138.
175. Vuljanic T, Bergquist KE, Clausen H, Roy S, Kihlberg J (1996) Piperidine is preferred to morpholine for Fmoc cleavage in solid phase glycopeptide synthesis as exemplified by preparation of glycopeptides related to HIV gp120 and mucins. *Tetrahedron* 52:7983–8000.
176. Guo ZW, Shao N (2005) Glycopeptide and glycoprotein synthesis involving unprotected carbohydrate building blocks. *Med. Res. Rev.* 25:655–678.

177. Paulsen H, Holck JP (1982) Synthesis of the glycopeptide O-beta-D-galactopyranosyl-(1-]3)-O-(2-acetamido-2-desoxy-alpha-D-galactopyranosyl)-(1-]3)-L-serine and L-threonine. *Carbohydr. Res.* 109:89–107.

178. Kunz H, Birnbach S, Wernig P (1990) Synthesis of glycopeptides with the Tn and T-antigen structures, and their coupling to bovine serum-albumin. *Carbohydr. Res.* 202:207–223.

179. Nakahara N, Iijima H, Sibayama S, Ogawa T (1990) A highly stereoselective synthesis of dimeric and trimeric sialosyl-TN epitope—a partial structure of glycophorin A. *Tetrahedron Lett.* 31:6897–6900.

180. Paulsen H, Bielfeldt T, Peters S, Meldal M, Bock K (1994) A new strategy for the solid-phase synthesis of O-glycopeptides via 2-azido-glycopeptides. *Liebigs Ann. Chem.* 369–379.

181. Liebe B, Kunz H (1997) Solid-phase synthesis of a tumor-associated sialyl-T-N antigen glycopeptide with a partial sequence of the "tandem repeat" of the MUC-1 mucin. *Angew. Chem. Int. Ed.* 36:618–621.

182. Chen XT, Sames D, Danishefsky SJ (1998) Exploration of modalities in building alpha-O-linked systems through glycal assembly: A total synthesis of the mucin-related F1 alpha antigen. *J. Am. Chem. Soc.* 120:7760–7769.

183. Kuduk SD, Schwarz JB, Chen XT, Glunz PW, Sames D, Ragupathi G, Livingston PO, Danishefsky SJ (1998) Synthetic and immunological studies on clustered modes of mucin-related Tn and TF O-linked antigens: The preparation of a glycopeptide-based vaccine for clinical trials against prostate cancer. *J. Am. Chem. Soc.* 120:12474–12485.

184. Paulsen H, Rauwald W, Weichert U (1988) Building units of oligosaccharides. Glycosidation with thioglycosides of oligosaccharides to segments of O-glycoproteins. *Liebigs Ann. Chem.* 75–86.

185. Elofsson M, Salvador LA, Kihlberg J (1997) Preparation of Tn and sialyl Tn building blocks used in Fmoc solid-phase synthesis of glycopeptide fragments from HIV gp120. *Tetrahedron* 53:369–390.

186. Cato D, Buskas T, Boons GJ (2005) Highly efficient stereospecific preparation of Tn and TF building blocks using thioglycosyl donors and the Ph_2SO/Tf_2O promotor system. *J. Carbohydr. Chem.* 24:503–516.

187. Mathieux N, Paulsen H, Meldal M, Bock K (1997) Synthesis of glycopeptide sequences of repeating units of the mucins MUC 2 and MUC 3 containing oligosaccharide side-chains with core 1, core 2, core 3, core 4 and core 6 structure. *J. Chem. Soc., Perkin Trans. 1* 2359–2368.

188. Hanson S, Best M, Bryan MC, Wong C-H (2004) Chemoenzymatic synthesis of oligosaccharides and glycoproteins. *Trends Biochem. Sci.* 29:656–663.

189. Ragupathi G, Koide F, Livingston PO, Cho YS, Endo A, Wan Q, Spassova MK, Keding SJ, Allen J, Ouerfelli O, Wilson RM, Danishefsky SJ (2006) Preparation and evaluation of unimolecular pentavalent and hexavalent antigenic constructs targeting prostate and breast cancer: A synthetic route to anticancer vaccine candidates. *J. Am. Chem. Soc.* 128:2715–2725.

190. Hermanson GT (1996) *Bioconjugate Techniques.* Academic, San Diego, p. 785.

191. Tam JP, Yu QT, Miao ZW (1999) Orthogonal ligation strategies for peptide and protein. *Biopolymers* 51:311–332.

192. Shao J, Tam JP (1995) Unprotected peptides as building-blocks for the synthesis of peptide dendrimers with oxime, hydrazone, and thiazolidine linkages. *J. Am. Chem. Soc.* 117:3893–3899.
193. Zeng W, Ghosh S, Macris M, Pagnon J, Jackson DC (2001) Assembly of synthetic peptide vaccines by chemoselective ligation of epitopes: Influence of different chemical linkages and epitope orientations on biological activity. *Vaccine* 19:3843–3852.
194. Geoghegan KF, Stroh JG (1992) Site-directed conjugation of nonpeptide groups to peptides and proteins via periodate-oxidation of a 2-amino alcohol—application to modification at N-terminal serine. *Bioconjugate Chem.* 3:138–146.
195. Melnyk O, Fehrentz JA, Martinez J, Gras-Masse H (2000) Functionalization of peptides and proteins by aldehyde or keto groups. *Biopolymers* 55:165–186.
196. Cremer GA, Bureaud N, Lelievre D, Piller V, Piller F, Delmas A (2004) Synthesis of branched oxime-linked peptide mimetics of the MUC1 containing a universal T-helper epitope. *Chemistry* 10:6353–6360.
197. Marcaurelle LA, Shin Y, Goon S, Bertozzi CR (2001) Synthesis of oxime-linked mucin mimics containing the tumor-related T(N) and sialyl T(N) antigens. *Org. Lett.* 3:3691–3694.
198. Dawson PE, Muir TW, Clarklewis I, Kent SBH (1994) Synthesis of proteins by native chemical ligation. *Science* 266:776–779.
199. Dawson PE, Kent SB (2000) Synthesis of native proteins by chemical ligation. *Annu. Rev. Biochem.* 69:923–960.
200. Kolb HC, Finn MG, Sharpless KB (2001) Click chemistry: Diverse chemical function from a few good reactions. *Angew. Chem. Int. Ed.* 40:2004–2021.
201. Kolb HC, Sharpless KB (2003) The growing impact of click chemistry on drug discovery. *Drug Dis. Today* 8:1128–1137.
202. Huisgen R (1968) On mechanism of 1,3-dipolar cycloadditions. A reply. *J. Org. Chem.* 33:2291–2297.
203. Brase S, Gil C, Knepper K, Zimmermann V (2005) Organic azides: An exploding diversity of a unique class of compounds. *Angew. Chem. Int. Ed.* 44:5188–5240.
204. Johnson ECB, Kent SBH (2006) Insights into the mechanism and catalysis of the native chemical ligation reaction. *J. Am. Chem. Soc.* 128:6640–6646.
205. Camarero JA, Mitchell AR (2005) Synthesis of proteins by native chemical ligation using Fmoc-based chemistry. *Protein Peptide Lett.* 12:723–728.
206. Warren JD, Miller J, Keding SJ, Danishefsky SD (2004) Toward fully synthetic glycoproteins by ultimately convergent routes: A solution to a long-standing problem. *J Am. Chem. Soc.* 126:6576–6578.
207. Heidler P, Link A (2005) N-Acyl-N-alkyl-sulfonamide anchors derived from Kenner's safety-catch linker: Powerful tools in bioorganic and medicinal chemistry. *Bioorg. Med. Chem.* 13:585–599.
208. Chen JH, Warren JD, Wu B, Chen G, Wan Q, Danishefsky SJ (2006) A route to cyclic peptides and glycopeptides by native chemical ligation using in situ derived thioesters. *Tetrahedron Lett.* 47:1969–1972.
209. Nakamura K, Mori H, Kawakami T, Hojo H, Nakahara Y, Aimoto S (2007) Peptide thioester synthesis via an auxiliary-mediated N-S acyl shift reaction in solution. *Int. J. Pept. Res. Ther.* 13:191–202.

210. Yamamoto N, Tanabe Y, Okamoto R, Dawson PE, Kajihara Y (2008) Chemical synthesis of a glycoprotein having an intact human complex-type sialyloligosaccharide under the Boc and Fmoc synthetic strategies. *J. Am. Chem. Soc.* 130:501–510.
211. Hojo H, Murasawa Y, Katayama H, Ohira T, Nakaharaa Y, Nakahara Y (2008) Application of a novel thioesterification reaction to the synthesis of chemokine CCL27 by the modified thioester method. *Org. Biomol. Chem.* 6:1808–1813.
212. Tan XH, Zhang XH, Yang RL, Liu CF (2008) A simple method for preparing peptide C-terminal thioacids and their application in sequential chemoenzymatic ligation. *ChemBiochem* 9:1052–1056.
213. Ficht S, Fayne RJ, Guy RT, Wong CH (2008) Solid-phase synthesis of peptide and glycopeptide thioesters through side-chain-anchoring strategies. *Chem. Eur. J.* 14: 3620–3629.
214. Kochendoerfer GG, Chen SY, Mao F, Cressman S, Traviglia S, Shao H, Hunter CL, Low DW, Cagle EN, Carnevali M, Gueriguian V, Keogh PJ, Porter H, Stratton SM, Wiedeke MC, Wilken J, Tang J, Levy JJ, Miranda LP, Crnogorac MM, Kalbag S, Botti P, Schindler-Horvat J, Savatski L, Adamson JW, Kung A, Kent SB, Bradburne JA (2003) Design and chemical synthesis of a homogeneous polymer-modified erythropoiesis protein. *Science* 299:884–887.
215. Clayton D, Shapovalov G, Maurer JA, Dougherty DA, Lester HA, Kochendoerfer GG (2004) Total chemical synthesis and electrophysiological characterization of mechanosensitive channels from Escherichia coli and Mycobacterium tuberculosis. *Proc. Natl. Acad. Sci. U.S.A.* 101:4764–4769.
216. Avery OT, Goebel WF (1931) Chemo-immunological studies on conjugated carbohydrate proteins. V. Immunologic specificity of an antigen prepared by combining the capsular polysaccharide of type III pneumococcus with foreign protein. *J. Exp. Med.* 54:437–447.
217. Musselli C, Livingston PO, Ragupathi G (2001) Keyhole limpet hemocyanin conjugate vaccines against cancer: The Memorial Sloan Kettering experience. *J. Cancer Res. Clin. Oncol.* 127(Suppl 2):R20–R26.
218. Nyame AK, Kawar ZS, Cummings RD (2004) Antigenic glycans in parasitic infections: Implications for vaccines and diagnostics. *Arch. Biochem. Biophys.* 426:182–200.
219. Jones C (2005) Vaccines based on the cell surface carbohydrates of pathogenic bacteria. *An. Acad. Bras. Cienc.* 77:293–324.
220. Helling F, Shang A, Calves M, Zhang S, Ren S, Yu RK, Oettgen HF, Livingston PO (1994) GD3 vaccines for melanoma: Superior immunogenicity of keyhole limpet hemocyanin conjugate vaccines. *Cancer Res.* 54:197–203.
221. Bilodeau MT, Park TK, Hu SH, Randolph JT, Danishefsky SJ, Livingston PO, Zhang SL (1995) Total synthesis of a human breast-tumor associated antigen. *J. Am. Chem. Soc.* 117:7840–7841.
222. Allen JR, Danishefsky SJ (1999) New applications of the *n*-pentenyl glycoside method in the synthesis and immunoconjugation of fucosyl GM1: A highly tumor-specific antigen associated with small cell lung carcinoma. *J. Am. Chem. Soc.* 121:10875–10882.
223. Danishefsky SJ, Behar V, Randolph JT, Lloyd KO (1995) Application of the glycal assembly method to the concise synthesis of neoglycoconjugates of Le(Y) and Le(B) blood-group determinants and of H-Type-I and H-Type-II oligosaccharides. *J. Am. Chem. Soc.* 117:5701–5711.

224. Kudryashov V, Kim HM, Ragupathi G, Danishefsky SJ, Livingston PO, Lloyd KO (1998) Immunogenicity of synthetic conjugates of Lewis(y) oligosaccharide with proteins in mice: Towards the design of anticancer vaccines. *Cancer Immunol. Immunother.* 45:281–286.
225. Glunz PW, Hintermann S, Schwarz JB, Kuduk SD, Chen XT, Williams LJ, Sames D, Danishefsky SJ, Kudryashov V, Lloyd KO (1999) Probing cell surface "glycoarchitecture" through total synthesis. Immunological consequences of a human blood group determinant in a clustered mucin-like context. *J. Am. Chem. Soc.* 121:10636–10637.
226. Randolph JT, Danishefsky SJ (1994) An interactive strategy for the assembly of complex, branched oligosaccharide domains on a solid support: a concise synthesis of the Lewisb domain in bioconjugatable form. *Angew. Chem. Int. Ed.* 33:1470–1473.
227. Randolph JT, McClure KF, Danishefsky SJ (1995) Major simplifications in oligosaccharide syntheses arising from a solid-phase based method: an application to the synthesis of the Lewis b antigen. *J. Am. Chem. Soc.* 117:5712–5719.
228. Deshpande PP, Danishefsky SJ (1997) Total synthesis of the potential anticancer vaccine KH-1 adenocarcinoma antigen. *Nature* 387:164–166.
229. Deshpande PP, Kim HM, Zatorski A, Park TK, Ragupathi G, Livingston PO, Live D, Danishefsky SJ (1998) Strategy in oligosaccharide synthesis: An application to a concise total synthesis of the KH-1(adenocarcinoma) antigen. *J. Am. Chem. Soc.* 120:1600–1614.
230. Zhang S, Graeber LA, Helling F, Ragupathi G, Adluri S, Lloyd KO, Livingston PO (1996) Augmenting the immunogenicity of synthetic MUC1 peptide vaccines in mice. *Cancer Res.* 56:3315–3319.
231. Kagan E, Ragupathi G, Yi SS, Reis CA, Gildersleeve J, Kahne D, Clausen H, Danishefsky SJ, Livingston PO (2005) Comparison of antigen constructs and carrier molecules for augmenting the immunogenicity of the monosaccharide epithelial cancer antigen Tn. *Cancer Immunol. Immunother.* 54:424–430.
232. Ragupathi G, Howard L, Cappello S, Koganty RR, Qiu D, Longenecker BM, Reddish MA, Lloyd KO, Livingston PO (1999) Vaccines prepared with sialyl-Tn and sialyl-Tn trimers using the 4-(4-maleimidomethyl)cyclohexane-1-carboxyl hydrazide linker group result in optimal antibody titers against ovine submaxillary mucin and sialyl-Tn-positive tumor cells. *Cancer Immunol. Immunother.* 48:1–8.
233. Sugimoto M, Ogawa T (1985) Synthesis of a hematoside (GM3-ganglioside) and a stereoisomer. *Glycoconjugate J.* 2:5–9.
234. Sato S, Ito Y, Nukada T, Nakahara Y, Ogawa T (1987) Total synthesis of X-hapten, III3 Fuc-alpha-nLc$_4$ Cer. *Carbohydr. Res.* 167:197–210.
235. Sato S, Ito Y, Ogawa T (1988) A total synthesis of dimeric Le(X) antigen, III(3)V(3) Fuc$_2$nLc$_6$ cer—pivaloyl auxiliary for stereocontrolled glycosylation. *Tetrahedron Lett.* 29:5267–5270.
236. Ito Y, Numata M, Sugimoto M, Ogawa T (1989) Highly stereoselective synthesis of ganglioside GD3. *J. Am. Chem. Soc.* 111:8508–8510.
237. Nicolaou KC, Caulfield TJ, Kataoka H, Stylianides NA (1990) Total synthesis of the tumor-associated Le^x family of glycosphingolipids. *J. Am. Chem. Soc.* 112:3693–3695.
238. Kameyama A, Ishida H, Kiso M, Hasegawa A (1991) Synthetic studies on sialoglycoconjugates. 22. Total synthesis of tumor-associated ganglioside, sialyl Lewis X. *J. Carbohydr. Chem.* 10:549–560.

239. Nicolaou KC, Hummel CW, Iwabuchi Y (1992) Total synthesis of sialyl dimeric Le^x. *J. Am. Chem. Soc.* 114:3126–3128.

240. Ishida H, Ohta Y, Tsukada Y, Isogai Y, Ishida H, Kiso M, Hasegawa A (1994) A facile total synthesis of ganglioside GD2. *Carbohydrate Res.* 252:283–290.

241. Iida M, Endo A, Fujita S, Numata M, Suzuki K, Nunomura S, Ogawa T (1996) Total synthesis of glycononaosyl ceramide with a sialyl dimeric Le(x) sequence. *Glycoconjugate J.* 13:203–211.

242. Lassaletta JM, Schmidt RR (1996) Synthesis of the hexasaccharide moiety of globo H (human breast cancer) antigen. *Liebigs Ann.* 1417–1423.

243. Zhu T, Boons GJ (1999) A two-directional and highly convergent approach for the synthesis of the tumor-associated antigen Globo-H. *Angew. Chem. Int. Ed.* 38:1629–1632.

244. Zhu T, Boons GJ (2001) A highly efficient synthetic strategy for polymeric support synthesis of Le(x), Le(y), and H-type 2 oligosaccharides. *Chem. Eur. J.* 7:2382–2389.

245. Burkhart F, Zhang ZY, Wacowich-Sgarbi S, Wong CH (2001) Synthesis of the Globo H hexasaccharide using the programmable reactivity-based one-pot strategy. *Angew. Chem. Int. Ed.* 40:1274–1277.

246. Bosse F, Marcaurelle LA, Seeberger PH (2002) Linear synthesis of the tumor-associated carbohydrate antigens Globo-H, SSEA-3, and Gb3. *J. Org. Chem.* 67:6659–6670.

247. Buskas T, Li YH, Boons GJ (2004) The immunogenicity of the tumor-associated antigen Lewis(y) may be suppressed by a bifunctional cross-linker required for coupling to a carrier protein. *Chem. Eur. J.* 10:3517–3524.

248. Buskas T, Li YH, Boons GJ (2005) Synthesis of a dimeric Lewis antigen and the evaluation of the epitope specificity of antibodies elicited in mice. *Chem. Eur. J.* 11:5457–5467.

249. Werz DB, Castagner B, Seeberger PH (2007) Automated synthesis of the tumor-associated carbohydrate antigens Gb-3 and Globo-H: Incorporation of alpha-galactosidic linkages. *J. Am. Chem. Soc.* 129:2770–2771.

250. Wang Z, Zhou LY, El-Boubbou K, Ye XS, Huang XF (2007) Multi-component one-pot synthesis of the tumor-associated carbohydrate antigen Globo-H based on preactivation of thioglycosyl donors. *J. Org. Chem.* 72:6409–6420.

251. Livingston PO, Wong GY, Adluri S, Tao Y, Padavan M, Parente R, Hanlon C, Calves MJ, Helling F, Ritter G, Oettgen HF, Old LJ (1994) Improved survival in stage III melanoma patients with GM2 antibodies: A randomized trial of adjuvant vaccination with GM2 ganglioside. *J. Clin. Oncol.* 12:1036–1044.

252. Mordoh J, Leis S, Bravo AI, Podhajcer OL, Ballare C, Capurro M, Kairiyama C, Bover L (1994) Description of a new monoclonal antibody, FC-2.15, reactive with human breast cancer and other human neoplasias. *Int. J. Biol. Markers* 9:125–134.

253. Zhang S, Walberg LA, Ogata S, Itzkowitz SH, Koganty RR, Reddish M, Gandhi SS, Longenecker BM, Lloyd KO, Livingston PO (1995) Immune sera and monoclonal antibodies define two configurations for the sialyl Tn tumor antigen. *Cancer Res.* 55:3364–3368.

254. Helling F, Zhang S, Shang A, Adluri S, Calves M, Koganty R, Longenecker BM, Yao TJ, Oettgen HF, Livingston PO (1995) GM2-KLH conjugate vaccine: Increased immunogenicity in melanoma patients after administration with immunological adjuvant QS-21. *Cancer Res.* 55:2783–2788.

255. Goydos JS, Elder E, Whiteside TL, Finn OJ, Lotze MT (1996) A phase I trial of a synthetic mucin peptide vaccine. Induction of specific immune reactivity in patients with adenocarcinoma. *J. Surg. Res.* 63:298–304.

256. MacLean GD, Reddish MA, Koganty RR, Longenecker BM (1996) Antibodies against mucin-associated sialyl-Tn epitopes correlate with survival of metastatic adenocarcinoma patients undergoing active specific immunotherapy with synthetic STn vaccine. *J. Immunother. Emphasis Tumor Immunol.* 19:59–68.

257. Zhang H, Zhang S, Cheung NK, Ragupathi G, Livingston PO (1998) Antibodies against GD2 ganglioside can eradicate syngeneic cancer micrometastases. *Cancer Res.* 58:2844–2849.

258. Alonso DF, Gabri MR, Guthmann MD, Fainboim L, Gomez DE (1999) A novel hydrophobized GM3 ganglioside/*Neisseria meningitidis* outer-membrane-protein complex vaccine induces tumor protection in B16 murine melanoma. *Int. J. Oncol.* 15:59–66.

259. Slovin SF, Ragupathi G, Adluri S, Ungers G, Terry K, Kim S, Spassova M, Bornmann WG, Fazzari M, Dantis L, Olkiewicz K, Lloyd KO, Livingston PO, Danishefsky SJ, Scher HI (1999) Carbohydrate vaccines in cancer: Immunogenicity of a fully synthetic globo H hexasaccharide conjugate in man. *Proc. Natl. Acad. Sci. U.S.A.* 96:5710–5715.

260. Sabbatini PJ, Kudryashov V, Ragupathi G, Danishefsky SJ, Livingston PO, Bornmann W, Spassova M, Zatorski A, Spriggs D, Aghajanian C, Soignet S, Peyton M, O'Flaherty C, Curtin J, Lloyd KO (2000) Immunization of ovarian cancer patients with a synthetic Lewis(Y)-protein conjugate vaccine: A phase 1 trial. *Int. J. Cancer* 87:79–85.

261. Ragupathi G, Meyers M, Adluri S, Howard L, Musselli C, Livingston PO (2000) Induction of antibodies against GD3 ganglioside in melanoma patients by vaccination with GD3-lactone-KLH conjugate plus immunological adjuvant QS-21. *Int. J. Cancer* 85:659–666.

262. Shepherd FA (2001) Alternatives to chemotherapy and radiotherapy in the treatment of small cell lung cancer. *Semin. Oncol.* 28:30–37.

263. Snijdewint FGM, von Mensdorff-Pouilly S, Karuntu-Wanamarta AH, Verstraeten AA, Livingston PO, Hilgers J, Kenemans P (2001) Antibody-dependent cell-mediated cytotoxicity can be induced by MUC-1 peptide vaccination of breast cancer patients. *Int. J. Cancer* 93:97–106.

264. Lay L, Panza L, Poletti L, Prosperi D, Canevari S, Perico ME (2001) Improvement of the synthesis of immunological carbohydrate vaccines containing the tumour associate antigen CaMBr1. *Eur. J. Org. Chem.* 4331–4336.

265. Kirkwood JM, Ibrahim JG, Sosman JA, Sondak VK, Agarwala SS, Ernstoff MS, Rao U (2001) High-dose interferon alfa-2b significantly prolongs relapse-free and overall survival compared with the GM2-KLH/QS-21 vaccine in patients with resected stage IIB-III melanoma: Results of intergroup trial E1694/S9512/C509801. *J. Clin. Oncol.* 19:2370–2380.

266. Gilewski T, Ragupathi G, Bhuta S, Williams LJ, Musselli C, Zhang XF, Bornmann WG, Spassova M, Bencsath KP, Panageas KS, Chin J, Hudis CA, Norton L, Houghton AN, Livingston PO, Danishefsky SJ (2001) Immunization of metastatic breast cancer patients with a fully synthetic Globo H conjugate: A phase I trial. *Proc. Natl. Acad. Sci. U.S.A.* 98:3270–3275.

267. Slovin SF, Ragupathi G, Musselli C, Olkiewicz K, Verbel D, Kuduk SD, Schwarz JB, Sames D, Danishefsky S, Livingston PO, Scher HI (2003) Fully synthetic carbohydrate-based vaccines in biochemically relapsed prostate cancer: Clinical trial results with alpha-N-acetylgalactosamine-O-serine/threonine conjugate vaccine. *J. Clin. Oncol.* 21:4292–4298.
268. Ragupathi G, Livingston PO, Hood C, Gathuru J, Krown SE, Chapman PB, Wolchok JD, Williams LJ, Oldfield RC, Hwu WJ (2003) Consistent antibody response against ganglioside GD2 induced in patients with melanoma by a GD2 lactone-keyhole limpet hemocyanin conjugate vaccine plus immunological adjuvant QS-21. *Clin. Cancer Res.* 9:5214–5220.
269. Slovin SF, Ragupathi G, Fernandez C, Jefferson MP, Diani M, Wilton AS, Powell S, Spassova M, Reis C, Clausen H, Danishefsky S, Livingston P, Scher HI (2005) A bivalent conjugate vaccine in the treatment of biochemically relapsed prostate cancer: A study of glycosylated MUC-2-KLH and Globo H-KLH conjugate vaccines given with the new semi-synthetic saponin immunological adjuvant GPI-0100 OR QS-21. *Vaccine* 23:3114–3122.
270. Gilewski TA, Ragupathi G, Dickler M, Powell S, Bhuta S, Panageas K, Koganty RR, Chin-Eng J, Hudis C, Norton L, Houghton AN, Livingston PO (2007) Immunization of high-risk breast cancer patients with clustered sTn-KLH conjugate plus the immunologic adjuvant QS-21. *Clin. Cancer Res.* 13:2977–2985.
271. Sabbatini PJ, Ragupathi G, Hood C, Aghajanian CA, Juretzka M, Iasonos A, Hensley ML, Spassova MK, Ouerfelli O, Spriggs DR, Tew WP, Konner J, Clausen H, Abu Rustum N, Dansihefsky SJ, Livingston PO (2007) Pilot study of a heptavalent vaccine-keyhole limpet hemocyanin conjugate plus QS21 in patients with epithelial ovarian, fallopian tube, or peritoneal cancer. *Clin. Cancer Res.* 13:4170–4177.
272. Miles DW, Towlson KE, Graham R, Reddish M, Longenecker BM, Taylor-Papadimitriou J, Rubens RD (1996) A randomised phase II study of sialyl-Tn and DETOX-B adjuvant with or without cyclophosphamide pretreatment for the active specific immunotherapy of breast cancer. *Br. J. Cancer* 74:1292–1296.
273. Sandmaier BM, Oparin DV, Holmberg LA, Reddish MA, MacLean GD, Longenecker BM (1999) Evidence of a cellular immune response against sialyl-Tn in breast and ovarian cancer patients after high-dose chemotherapy, stem cell rescue, and immunization with Theratope STn-KLH cancer vaccine. *J. Immunother.* 22:54–66.
274. Falugi F, Petracca R, Mariani M, Luzzi E, Mancianti S, Carinci V, Melli ML, Finco O, Wack A, Di Tommaso A, De Magistris MT, Costantino P, Del Giudice G, Abrignani S, Rappuoli R, Grandi G (2001) Rationally designed strings of promiscuous CD4(+) T cell epitopes provide help to Haemophilus influenzae type b oligosaccharide: A model for new conjugate vaccines. *Eur. J. Immunol.* 31:3816–3824.
275. Ni J, Song H, Wang Y, Stamatos NM, Wang LX (2006) Toward a carbohydrate-based HIV-1 vaccine: synthesis and immunological studies of oligomannose-containing glycoconjugates. *Bioconjug. Chem.* 17:493–500.
276. Tam JP (1996) Recent advances in multiple antigen peptides. *J. Immun. Methods* 196:17–32.
277. Moyle PM, Toth I (2008) Self-adjuvanting lipopeptide vaccines. *Curr. Med. Chem.* 15:506–516.
278. Zauner W, Lingnau K, Mattner F, von Gabain A, Buschle M (2001) Defined synthetic vaccines. *Biol. Chem.* 382:581–595.

279. Frezard F (1999) Liposomes: From biophysics to the design of peptide vaccines. *Braz. J. Med. Biol. Res.* 32:181–189.
280. Apostolopoulos V, Barnes N, Pietersz GA, McKenzie IF (2000) Ex vivo targeting of the macrophage mannose receptor generates anti-tumor CTL responses. *Vaccine* 18:3174–3184.
281. Karanikas V, Thynne G, Mitchell P, Ong CS, Gunawardana D, Blum R, Pearson J, Lodding J, Pietersz G, Broadbent R, Tait B, McKenzie IF (2001) Mannan mucin-1 peptide immunization: Influence of cyclophosphamide and the route of injection. *J. Immunother.* 24:172–183.
282. Bessler WG, Cox M, Lex A, Suhr B, Wiesmuller KH, Jung G (1985) Synthetic lipopeptide analogs of bacterial lipoprotein are potent polyclonal activators for murine lymphocytes B. *J. Immunol.* 135:1900–1905.
283. Hoffmann P, Wiesmuller KH, Metzger J, Jung G, Bessler WG (1989) Induction of tumor cyto-toxicity in murine bone marrow-derived macrophages by 2 synthetic lipopeptide analogs. *Biolog. Chem. Hoppe-Seyler* 370:575–582.
284. Metzger J, Jung G, Bessler WG, Hoffmann P, Strecker M, Lieberknecht A, Schmidt U (1991) Lipopeptides containing 2-(palmitoylamino)-6,7-bis(palmitoyloxy) heptanoic acid: Synthesis, stereospecific stimulation of B-lymphocytes and macrophages, and adjuvanticity *in vivo* and *in vitro*. *J. Med. Chem.* 34:1969–1974.
285. Spohn R, Buwitt-Beckmann U, Brock R, Jung G, Ulmer AJ, Wiesmuller KH (2004) Synthetic lipopeptide adjuvants and Toll-like receptor 2—structure-activity relationships. *Vaccine* 22:2494–2499.
286. Toyokuni T, Dean B, Cai SP, Boivin D, Hakomori S, Singhal AK (1994) Synthetic vaccines—Synthesis of a dimeric Tn antigen-lipopeptide conjugate that elicits immune-responses against Tn-expressing glycoproteins. *J. Am. Chem. Soc.* 116:395–396.
287. Toyokuni T, Hakomori S, Singhal AK (1994) Synthetic carbohydrate vaccines: Synthesis and immunogenicity of Tn antigen conjugates. *Bioorg. Med. Chem.* 2:1119–1132.
288. Glunz PW, Hintermann S, Williams LJ, Schwarz JB, Kuduk SD, Kudryashov V, Lloyd KO, Danishefsky SJ (2000) Design and synthesis of Le(y)-bearing glycopeptides that mimic cell surface Le(y) mucin glycoprotein architecture. *J. Am. Chem. Soc.* 122:7273–7279.
289. Kudryashov V, Glunz PW, Williams LJ, Hintermann S, Danishefsky SJ, Lloyd KO (2001) Toward optimized carbohydrate-based anticancer vaccines: Epitope clustering, carrier structure, and adjuvant all influence antibody responses to Lewis(y) conjugates in mice. *Proc. Natl. Acad. Sci. U.S.A.* 98:3264–3269.
290. Keding SJ, Danishefsky SJ (2004) Prospects for total synthesis: A vision for a totally synthetic vaccine targeting epithelial tumors. *Proc. Natl. Acad. Sci. U.S.A.* 101:11937–11942.
291. Warren JD, Geng XD, Danishefsky SJ (2007) Synthetic glycopeptide-based vaccines. *Top. Curr. Chem.* 267:109–141.
292. Allen JR, Harris CR, Danishefsky SJ (2001) Pursuit of optimal carbohydrate-based anticancer vaccines: Preparation of a multiantigenic unimolecular glycopeptide containing the Tn, MBr1, and Lewis(y) antigens. *J. Am. Chem. Soc.* 123:1890–1897.
293. Ragupathi G, Coltart DM, Williams LJ, Koide F, Kagan E, Allen J, Harris C, Glunz PW, Livingston PO, Danishefsky SJ (2002) On the power of chemical synthesis: Immunological evaluation of models for multiantigenic carbohydrate-based cancer vaccines. *Proc. Natl. Acad. Sci. U.S.A.* 99:13699–13704.

294. Biswas K, Coltart DM, Danishefsky SJ (2002) Construction of carbohydrate-based antitumor vaccines: Synthesis of glycosyl amino acids by olefin cross-metathesis. *Tetrahedron Lett.* 43:6107–6110.
295. Wan Q, Cho YS, Lambert TH, Danishefsky SJ (2005) Olefin cross-metathesis: A powerful tool for constructing vaccines composed of multimeric antigens. *J. Carbohydr. Chem.* 24:425–440.
296. Keding SJ, Endo A, Danishefsky SJ (2003) Synthesis of non-natural glycosylamino acids containing tumor-associated carbohydrate antigens. *Tetrahedron* 59:7023–7031.
297. Geraci C, Consoli GM, Galante E, Bousquet E, Pappalardo M, Spadaro A (2008) Calix[4]arene decorated with four Tn antigen glycomimetic units and P3CS immunoadjuvant: synthesis, characterization, and anticancer immunological evaluation. *Bioconjug. Chem.* 19:751–758.
298. Dziadek S, Hobel A, Schmitt E, Kunz H (2005) A fully synthetic vaccine consisting of a tumor-associated glycopeptide antigen and a T-Cell epitope for the induction of a highly specific humoral immune response. *Angew. Chem. Int. Ed.* 44:7630–7635.
299. Cremer GA, Bureaud N, Piller V, Kunz H, Piller F, Delmas AF (2006) Synthesis and biological evaluation of a multiantigenic Tn/TF-containing glycopeptide mimic of the tumor-related MUC1 glycoprotein. *ChemMedChem* 1:965–968.
300. Bay S, Lo-Man R, Osinaga E, Nakada H, Leclerc C, Cantacuzene D (1997) Preparation of a multiple antigen glycopeptide (MAG) carrying the Tn antigen. *J. Peptide Res.* 49:620–625.
301. Lo-Man R, Bay S, Vichier-Guerre S, Deriaud E, Cantacuzene D, Leclerc C (1999) A fully synthetic immunogen carrying a carcinoma-associated carbohydrate for active specific immunotherapy. *Cancer Res.* 59:1520–1524.
302. Lo-Man R, Vichier-Guerre S, Bay S, Deriaud E, Cantacuzene D, Leclerc C (2001) Antitumor immunity provided by a synthetic multiple antigenic glycopeptide displaying a tri-Tn glycotope. *J. Immunol.* 166:2849–2854.
303. Lo-Man R, Vichier-Guerre S, Perraut R, Deriaud E, Huteau V, BenMohamed L, Diop OM, Livingston PO, Bay S, Leclerc C (2004) A fully synthetic therapeutic vaccine candidate targeting carcinoma-associated Tn carbohydrate antigen induces tumor-specific antibodies in nonhuman primates. *Cancer Res.* 64:4987–4994.
304. Dumy P, Eggleston IM, Cervigni S, Sila U, Sun X, Mutter M (1995) A convenient synthesis of cyclic-peptides as regioselectively addressable functionalized templates (Raft). *Tetrahedron Lett.* 36:1255–1258.
305. Grigalevicius S, Chierici S, Renaudet O, Lo-Man R, Deriaud E, Leclerc C, Dumy P (2005) Chemoselective assembly and immunological evaluation of multiepitopic glycoconjugates bearing clustered Tn antigen as synthetic anticancer vaccines. *Bioconjugate Chem.* 16:1149–1159.
306. Veprek P, Hajduch M, Dzubak P, Kuklik R, Polakova J, Bezouska K (2006) Comblike dendrimers containing Tn antigen modulate natural killing and induce the production of Tn specific antibodies. *J. Med. Chem.* 49:6400–6407.
307. Reichel F, Ashton PR, Boons GJ (1997) Synthetic carbohydrate-based vaccines: Synthesis of an L-*glycero*-D-*manno*-heptose antigen-T-epitope-lipopetide conjugate. *Chem. Commun.* 21:2087–2088.

308. Buskas T, Ingale S, Boons GJ (2005) Towards a fully synthetic carbohydrate-based anticancer vaccine: synthesis and immunological evaluation of a lipidated glycopeptide containing the tumor-associated Tn antigen. *Angew. Chem. Int. Ed.* 44:5985–5988.
309. Krikorian D, Panou-Pomonis E, Voitharou C, Sakarellos C, Sakarellos-Daitsiotis M (2005) A peptide carrier with a built-in vaccine adjuvant: Construction of immunogenic conjugates. *Bioconjug. Chem.* 16:812–819.
310. Ingale S, Wolfert MA, Gaekwad J, Buskas T, Boons GJ (2007) Robust immune responses elicited by a fully synthetic three-component vaccine. *Nat. Chem. Biol.* 3:663–667.
311. Ingale S, Buskas T, Boons GJ (2006) Synthesis of glyco(lipo) peptides by liposome-mediated native chemical ligation. *Org. Lett.* 8:5785–5788.
312. Dziadek S, Jaques S, Bundle DR (2008) A novel linker methodology for the synthesis of tailored conjugate vaccines composed of complex carbohydrate antigens and specific TH-cell peptide epitopes. *Chem. Eur. J.* 14:5908–5917.
313. Renaudet O, BenMohamed L, Dasgupta G, Bettahi I, Dumy P (2008) Towards a self-adjuvanting multivalent B and T cell epitope containing synthetic glycolipopeptide cancer vaccine. *ChemMedChem* 3:737–741.

10

GLYCOENGINEERING OF CELL SURFACE SIALIC ACID AND ITS APPLICATION TO CANCER IMMUNOTHERAPY

Zhongwu Guo

Department of Chemistry, Wayne State University, Detroit, Michigan 48202

10.1 INTRODUCTION

Sialic acids are a family of naturally occurring carbohydrates having a common core structure **1** (Fig. 10.1), known as neuraminic acid. The most prominent member of this family of natural products is *N*-acetylneuraminic acid (NeuNAc **2**), which has an acetyl group attached to the 5-amino group of **1**. NeuNAc is usually located at the glycan nonreducing end of natural glycoproteins and glycolipids. Consequently, NeuNAc is typically exposed on the cell surface to be involved in various biological and pathological processes [1–4].

It has been established that oncogensis is generally accompanied by NeuNAc overexpression [5]. Thus, among the large number of tumor-associated carbohydrate antigens (TACAs) identified so far, many are sialooligosaccharides [4, 6–8]. For example, α-2,8-polysialic acid, sTn antigen, sialyl Lewis antigens, and the GM and GD gangliosides are all sialo-TACAs that are abundantly expressed by a number of tumors. Sialo-TACAs are therefore very important molecular templates and targets in the development of novel diagnostic and therapeutic methods for cancer [9–11].

Carbohydrate-Based Vaccines and Immunotherapies. Edited by Zhongwu Guo and Geert-Jan Boons
Copyright © 2009 John Wiley & Sons, Inc.

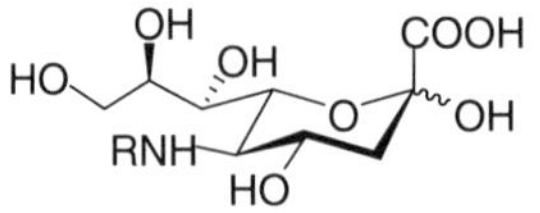

Figure 10.1 Structures of neuraminic acid (**1**) and NeuNAc (**2**).

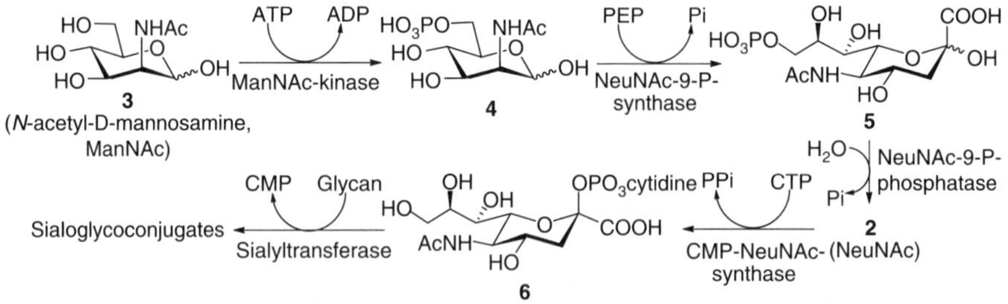

1 R = H(neuraminic acid)
2 R = Ac(N-acetylneuraminic acid, NeuNAc)

Figure 10.2 Biosynthetic pathway of NeuNAc and sialoglycoconjugates.

As briefly described in Chapter 1, the biosynthesis of NeuNAc and sialoglycoconjugates (Fig. 10.2) is achieved through a series of enzymatic reactions with N-acetyl-D-mannosamine (ManNAc **3**) as one of the key intermediates. The enzymes involved can tolerate certain forms of structural modification of the substrates, providing the opportunity to make use of the sialic acid biosynthetic pathway to deliver unnatural sialic acids and unnatural sialoglycoconjugates onto cell surfaces by exposing cells to artificial sialic acid precursors. This technique, known as sialic acid glycoengineering, was pioneered and established by the Reutter [12] and Bertozzi [13] groups. Other groups have also used the technique in various glycobiological and biomedical investigations. This chapter is focused on the application of sialic acid glycoengineering to cancer research, especially its application to molecular labeling of cancer cells and to the development of novel cancer immunotherapies.

10.2 ENGINEERING OF CELL SURFACE SIALIC ACIDS

As outlined in Figure 10.2, ManNAc, which is derived from physiological N-acetyl-D-glucosamine (GlcNAc), is the natural precursor of NeuNAc and sialoglycoconjugates. However, in 1992, Reutter and co-workers [12] found that the pheochromocytoma cell PC12, a neuroendocrine tumor cell line, took up N-propanoylmannosamine (ManNProp **7**), a synthetic analog of ManNAc, from the cultural medium to biosynthesize unnatural N-propanoylneuramic acid (NeuNProp **8**), incorporate NeuNProp in glycosphingolipids, and present the resultant NeuNProp-containing sialoglycoconjugates onto cell surfaces (Fig. 10.3). Later, it was disclosed in rats that ManNProp could also be incorporated by cells in vivo and that the resultant unnatural sialoglycoconjugates had an organ-specific distribution with the highest incorporation observed in liver, lung, and kidney [14]. On the other hand, it was found that cells

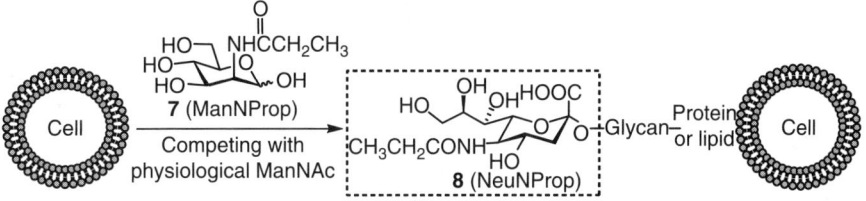

Figure 10.3 Expression of unnatural NeuNProp-containing sialoglycoconjugates on cell surfaces.

could tolerate high concentrations (as high as millimolars) of ManNProp. Moreover, administration of 200 mg/kg body weight of ManNProp to Wistar rats for a period of 3 weeks had no apparent side effects [15]. These results suggest that ManNProp can effectively glycoengineer sialic acid on cells both in vitro and in vivo and that ManNProp has relatively low cytotoxicity to cells and animals.

In addition to ManNProp, other N-acyl derivatives of D-mannosamine (Fig. 10.4), including N-butanoyl-D-mannosamine (ManNBut **9**) [16, 17], N-pentanoyl-D-mannosamine (ManNPent **10**) [17], N-hexanoyl-D-mannosamine (ManNHex **11**) [15], N-levulinoyl-D-mannosamine (ManNLev **12**) [13, 18], N-azidoacetyl-D-mannosamine (ManNAzidAc **13**) [19], N-phenylacetyl-D-mannosamine (ManNPhAc **14**) [20], N-(2-thioacetyl)-D-mannosamine (ManNThioAc **15**) [21], or their O-acylated

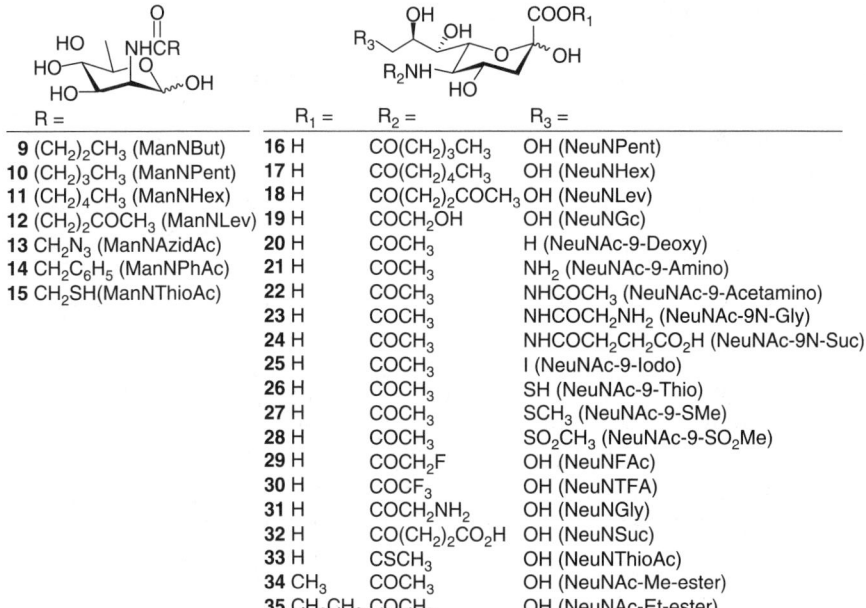

Figure 10.4 Structures of unnatural precursors employed to glycoengineer sialoglycoconjugates on cell surfaces.

derivatives [22–26], were also examined as unnatural sialic acid precursors for cell glycoengineering in various studies. It has been further demonstrated that a broad spectrum of mammalian cells could use unnatural derivatives of ManNAc, metabolize them in the biosynthetic pathway of sialic acid, and incorporate unnatural sialic acids into glycoconjugates in vitro and in vivo [13, 14, 16–36]. The studied mammalian cells include various cell lines of human B and T lymphoma, leukemia, cervix carcinoma, and melanoma, the human T lymphocytes and vein endothelium cells, the rodent neuronal, leukemia, and melanoma cells, and the kidney epithelium and fibroblast cells from different species.

Quantitative analysis of the metabolic incorporation of ManNProp, ManNBut, and ManNPent by human B lymphoma cell BJA-B and African green monkey kidney epithelium cell Vero showed that roughly 50% of physiological NeuNAc was replaced by unnatural N-acyl neuraminic acids in both cell lines after they were incubated with 5 mM of the unnatural ManNAc analogs for 48 h [17]. On the other hand, Horstkorte et al. [37] found that about 7.7, 5.3, and 17.2% of physiological NeuNAc was replaced by unnatural sialic acids in human promyelocytic leukemia cell line HL-60 treated with 5 mM of ManNProp, ManNBut, and ManNPent, respectively.

Despite the fact that high concentrations of unnatural ManNAc analogs can effectively glycoengineer sialic acids on cells in vitro, a significant challenge for practical application of this technique is to achieve efficient glycoengineering in vivo in the presence of physiologically acceptable concentrations of unnatural precursors. To address this issue, O-acetylated derivatives of N-acyl mannosamine have been utilized to improve the metabolic incorporation and the glycoengineering efficiency [22–26]. Yarema and co-workers [25] have demonstrated that cells could utilize O-acetylated ManNAc analogs with up to a 900-fold increase of efficiency compared with the free monosaccharide counterparts. Moreover, increasing the carbon chain length of the acyl groups attached to the hydroxyl groups of ManNAc analogs could further improve the glycoengineering efficiency [24], though a potential problem is that these derivatives decreased cell viability. After studying a series of N-acyl derivatives of D-mannosamine, our group has recently established that ManNPhAc is a particularly effective precursor for glycoengineering of cell surface sialic acids [20]. For example, high levels of NeuNPhAc-containing glycoconjugates were detected on several tumor cell lines incubated with low micromolar concentrations of ManNPhAc. Moreover, ManNPhAc showed no significant influence on cell viability at concentrations as high as 20 mM and caused no obvious side effects in animals treated with high dosages of ManNPhAc (100–150 mg/kg body weight) for more than 2 months (unpublished results).

Unnatural analogs of NeuNAc (Fig. 10.4) were also used to glycoengineer sialoglycoconjugates on cell surfaces [34, 38–40]. An advantage of these analogs over ManNAc analogs for this application is that the former have circumvented several steps of the biosynthetic pathway to avoid certain potential problems [38]. Bertozzi, Gibson, and co-workers [34, 40] discovered that exogenous NeuNAc and its unnatural N-acyl analogs, such as **16–19**, could be efficiently incorporated into cell surface lipooligosaccharides (LOS) by gram-negative bacterium *Haemophilus ducreyi* in a dose-dependent fashion, whereas exogenous ManNAc analogs were not converted to LOS-associated sialosides at a detectable level. Moreover, approximately 1.3 μM total exogenous sialic acid was sufficient to obtain about 50% of the maximum production of sialic acid-containing glycoforms under in vitro growth conditions. These results suggest that to properly exploit the sialic acid biosynthetic pathway for metabolic oligosaccharide engineering in *H. ducreyi* and possibly other prokaryotes that share similar pathways, precursors based on sialic acid and not mannosamine must be used [40]. On the other hand, Pawlita and co-workers [38] prepared 16 sialic acid analogs **20–35** with modifications at *C*-1, *N*-5, and *C*-9, respectively, and investigated their incorporation in the sialoglycoconjugate biosynthetic pathway. Despite the significant structural difference of these compounds, especially *C*-9-modified analogs **20–28**, from the structure of NeuNAc, 10 of them (**20, 25–30, 33–35**) were readily taken up and incorporated by cells. Quantitative analysis of the incorporation of NeuNFAc (**29**) and NeuNAc-9-Iodo (**25**) by endogeneously hyposialylated BJA-B K20 cells revealed that about 95 and 92% of the total membrane-associated sialic acid fractions consisted of the unnatural sialic acid analogs. In contrast, the 6-deoxy-6-iodo analog of ManNAc was not incorporated by BJA-B K20 cells to express NueNAc-9-Iodo-containing sialoglycoconjugates, probably due to the blockade of the essential 6-*O*-phosphorylation step. Varki and co-workers [35] have proved that humans could uptake dietary NeuNGc (**19**), a nonhuman and immunogenic sialic acid, and incorporate it into endogenous glycoproteins, which may be related to certain diseases. Moreover, since human cancer cells displayed relatively high levels of NeuNGc compared to normal cells, the high antigenicity of NeuNGc may be useful for cancer immunotheapy.

Treating cells with unnatural sialic acid precursors can significantly affect the expression levels of sialoglycoconjugates on cell surfaces to influence the biological properties of cells [16, 41–43]. For example, Bertozzi and co-workers [16] reported that ManNBut could inhibit α-2,8-polysialic acid expression in cell types such as teratocarcinoma (NT2) neurons and human neuroblastoma, small-cell lung carcinoma, and cervical carcinoma cells to influence their adhesion behavior. Addition of unnatural ManNAc analogs to the culture medium of cells deficient in *N*-acetylglucosaminyl uridine diphosphate (UDP-GlcNAc) 2-epimerase, which is necessary for the biosynthesis of natural NeuNAc, could dramatically increase the expression of unnatural sialic acid-containing sialoglycoconjugates on cell surfaces [41, 43]. Horstkorte and co-workers [41] showed that ManNProp led to the expression of unnatural sialyl Lewis X on HL-60 cells and increased binding of treated cells to selectins. Bertozzi and co-workers [43] further revealed that ManNAzidAc resulted in a drastic expression of *N*-azidoacetylneuraminic acid (NeuNAzidAc) on K20, HL-60, and Lec3 cells, and the expressed NeuNAzidAc was not recognized by *Limax flavus* agglutinin, a sialic-acid-specific lectin. The sialyl Lewis X-specific antibody HECA-452 and the macrophage expressed sialic-acid-specific lectin Siglec-1 did not recognize or bind NeuNAzidAc on cell surfaces either. However, the more promiscuous sialic-acid-binding lectins *Maackia amurensis* agglutinin and *Sambucus nigra* agglutinin could recognize NeuNAzidAc on cells.

Moreover, since cell surface sialic acid and sialioglycoconjugates play important roles in various biological and pathological processes [1–4], glycoengineering of cell surface sialic acids has shown a significant impact on these events [15, 17, 44–47]. For example, studies using unnatural precursors have revealed that the sialic acid N-acyl group is a critical determinant of sialic acid-dependent polyoma virus–receptor interactions [17, 45]. Cultivation of BJA-B cells with ManNProp, ManNBut, or ManNPent reduced the capacity of these cells to bind B lymphotropic papovavirus (LPV) by 75, 86, and 85%, respectively, and the susceptibility of cells to LPV infection by 97%. More interestingly, Vero cells pretreated with ManNProp and ManNBut were infected by human polyoma virus BK (BKV) more efficiently (up to 7-fold) than controls, while ManNPent treatment rendered cells nearly resistant to BKV infection, suggesting the importance of the length of the sialic acid N-acyl chain [17]. In the murine fibroblast cell line 3T6, unnatural mannosamine derivative treatment decreased the number of polyoma virus receptors on cell surfaces by 28% to reduce virus binding and infection significantly (by 64–78%) [45]. Similarly, sialic acid glycoengineering could inhibit influenza A virus binding to and infection of host cells by up to 80% [28]. ManNProp treatment also had a significant influence on neuron cells [30, 46]. In vitro studies showed that ManNProp could stimulate the proliferation of neonatal rat astrocytes and microglia [30] and induce neurite outgrowth of PC12 and cerebellar neurons [46]. ManNProp did not affect oligodendrocyte progenitor proliferation, but it indeed increased the number of oligodendrocyte progenitor cells expressing the early oligodendroglial surface marker A2B5 epitope [30]. Moreover, ManNProp stimulated human T-cell proliferation in a dose-dependent manner and induced the secretion of interleukin-2, which is the hallmark of T-cell activation, without obvious cytotoxicity even at very high concentrations [15]. Horstkorte and co-workers [48] further observed that ManNProp could lead to an increased release of intracellular calcium, which promoted the proliferation of HL-60 cells.

In summary, taking advantage of the remarkable promiscuity of the sialic acid biosynthetic pathway for modified substrates, a number of unnatural analogs of ManNAc and NeuNAc have been used to successfully glycoengineer cell surface sialic acids and deliver unnatural sialoglycoconjugates onto cell surfaces. Sialic acid glycoengineering has exhibited a significant impact on the properties and functions of cells, thus providing a powerful tool for investigation of various issues in glycobiology using cell-based systems. Similarly, glycoengineering of cell surface N-acetylglucosamine [49, 50], N-acetylgalactosamine [51, 52], and fucose has been achieved for probing protein glycosylation and related issues [53–55].

10.3 SIALIC ACID ENGINEERING FOR MODULATION OF CELL SURFACE REACTIVITY

Sialic acid glycoengineering has been successfully employed to deliver unnatural sialic acids with biologically orthogonal functionalities onto cell surfaces. Through specific reactions with the unique functionalities delivered onto cells, it was feasible to selectively target glycoengineered cells for cell surface molecular landscaping in various biological investigations [54, 56], such as the development of novel diagnostic and therapeutic methods for diseases [57, 58].

10.3 SIALIC ACID ENGINEERING FOR MODULATION OF CELL SURFACE REACTIVITY

In 1997, Bertozzi and co-workers [13] incorporated ManNLev (**12**) in the sialic acid biosynthetic pathway to present levulinoyl (Lev) groups on Jurkat, HL-60, and HeLa cells (Fig. 10.5). The ketone functionality of Lev groups was utilized as a molecular handle to attach various structures onto cell surfaces. For example, ManNLev-treated cells were selectively biotinylated via reaction of the ketone group with biotin hydrazide. Then, cytotoxic ricin toxin A (RTA) conjugated to avidin was employed to selectively target and kill the biotinylated cells, which took advantage of specific recognition and binding between biotin and avidin (Fig. 10.5a). Moreover, they observed that only cancer cells that expressed high levels of NeuNLev (>700,000 copies per cell) were sensitive to the RTA conjugate with lethal doses (LD_{50}) in the range of 1–10 nM, while the RTA conjugate showed no toxicity against cells expressing fewer NeuNLev molecules (<50,000 copies per cell). The biotinylated cells could also selectively interact with other avidin-containing molecules for various biological and biomedical applications, such as fluorescence labeling of cells, targeted drug delivery, and the like. Similarly, specific carbohydrate antigens were attached to ManNLev-glycoengineered cells via reaction of hydrazide or aminooxy derivatives of carbohydrate antigens, such as **36** and **37**, with the ketone groups on cell surfaces (Fig. 10.5b) [18]. These carbohydrate antigens could bind with certain cytotoxic lectins to kill the glycoengineered cells. Through selective reactions with the delivered ketones on cells, magnetic resonance imaging contrast reagents [59] and artificial virus receptors [60] were also selectively attached to cell surfaces for tumor detection and virus-mediated gene transfer, respectively. These studies have established the foundation for modification of cell surface reactivity through sialic acid glycoengineering for a myriad of applications.

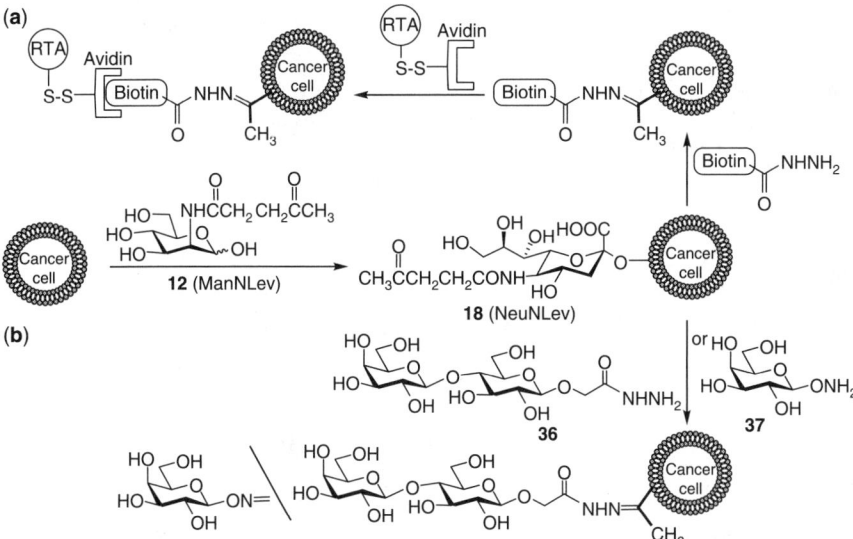

Figure 10.5 Glycoengineered delivery of Lev groups onto cell surfaces for attachment of (a) biotin and (b) carbohydrate antigens for selective targeting and killing of cancer cells.

The biologically orthogonal azido functionality was also delivered onto cell surfaces, both in vitro [19] and in vivo [26], through sialic acid glycoengineering with ManNAzidAc (**13**) as the sialic acid precursor (Fig. 10.6). Staudinger ligation (Figs. 10.6a and 10.6b) and click chemistry (Fig. 10.6c) were employed to attach molecular probes onto cell surfaces based on selective reactions of phosphines or alkynes with the azido group of N-azidoacetylneuraminic acid (NeuNAzidAc **38**) [54]. Because Staudinger ligation using **39** as the reaction substrate afforded products that contain the potentially intervening triarylphosphine oxide group, Bertozzi and co-workers [32] have developed a "traceless" Staudinger ligation method with **40** as the reaction substrate. Thus, the molecular probes could be directly linked to sialic acids through an amide bond. Click chemistry, namely the [3 + 2] cycloaddition reactions between azides and alkynes in the presence of a copper catalyst [61], can be achieved at physiological temperature, and it has been proved to be a very useful method for the selective modification of biological molecules and viral particles [62–64]. Unfortunately, the mandatory copper catalyst is toxic to cells. To overcome this problem, Bertozzi and co-workers [65] developed a strain-promoted and copper-free version of click chemistry, which took advantage of the high reactivity of cyclooctyne derivatives (**41**). The molecular markers attached to cells in vitro and in vivo by these methods include biotin [19, 65], various fluorescent tags [66], and special

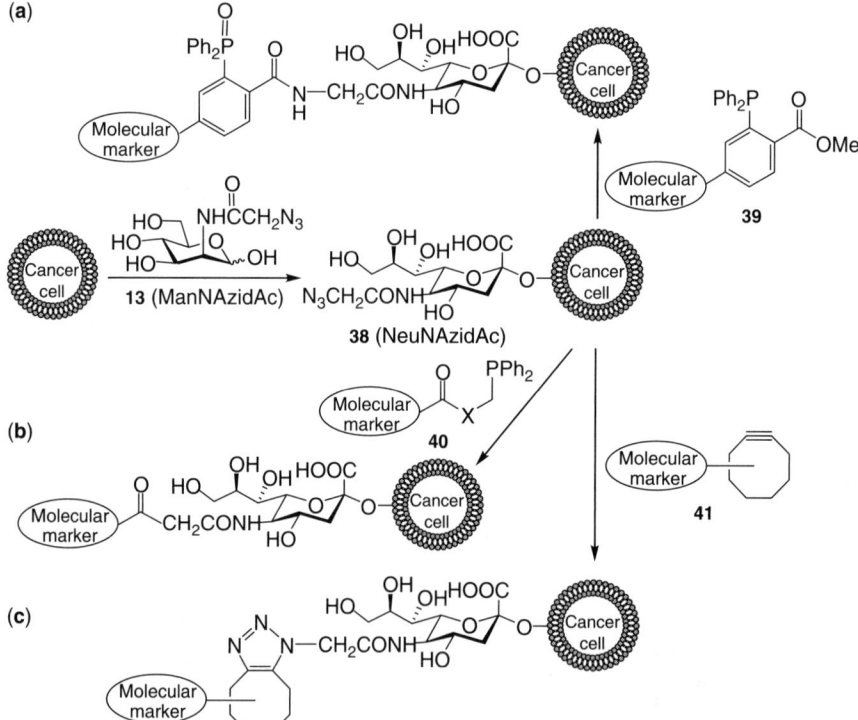

Figure 10.6 Glycoengineered delivery of NeuNAzidAc onto cell surfaces for attachment of (a and b) molecular probes by selective Staudinger ligation and (c) click chemistry.

10.4 SIALIC ACID ENGINEERING FOR CANCER IMMUNOTHERAPY

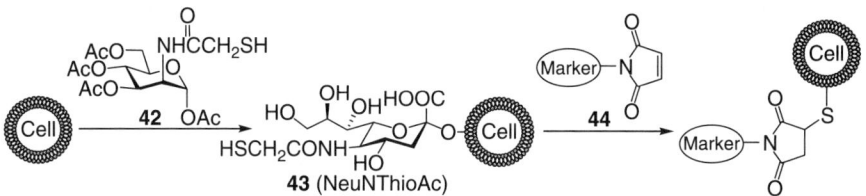

Figure 10.7 Glycoengineered delivery of NeuNThioAc onto cell surfaces for attachment of molecular probes by thiol addition to maleimides.

peptide epitopes [26], enabling a number of interesting biological investigations. For example, this technique has been successfully used for dynamic in vivo imaging [66]. Similarly, the strategy has been used to probe various cell surface glycans [67].

Recently, Yarema and co-workers [21] have engineered cells to express N-thioacetylneuraminic acid (NeuNThioAc **43**) with O-peracetylated ManNThioAc (**42**) as the glycoengineering precursor (Fig. 10.7). The unnatural NeuNThioAc on cell surfaces showed a significant impact on cell adhesion as well as cell proliferation. Meanwhile, the delivered thiol groups were used as molecular handles to attach maleimide-derivatized compounds (**44**) (Fig. 10.7), providing an alternative method for the modulation of cell surface reactivity by cell glycoengineering.

10.4 SIALIC ACID ENGINEERING FOR CANCER IMMUNOTHERAPY

As described in previous chapters, scores of TACAs have been identified and characterized in the past three decades, and these antigens have become important molecular targets for development of new diagnostic and immunotherapeutic strategies for cancer. However, among the identified TACAs, only a few have been successfully employed to construct functional vaccines, and the majority failed to induce antitumor immune response in cancer patients even after being linked to carrier molecules. Therefore, a central issue in cancer vaccine development is that the human immune system is tolerant to many TACAs. This problem is caused by a number of potential mechanisms [68, 69], of which the structural similarity between TACAs and normal carbohydrate antigens play an important role [70]. In fact, TACAs are rarely tumor specific. Most TACAs are either glycans also present on normal cells, though at much lower levels, or intermediates involved in the biosynthesis of normal antigens [70]. Consequently, the immune system can easily ignore these "self" antigens. The flourishing of tumors in cancer patients itself is a clear indication of existing tolerance of cancer patients to tumor cells and tumor-associated antigens. Another issue concerning glycoconjugate cancer vaccines derived from natural TACAs is that even if the glycoconjugates are immunogenic, they often induce IgM antibodies and T-cell-independent immune responses, while a T-cell-dependent antitumor cellular immunity is critical for cancer immunotherapy. These issues have significantly hindered further progress in development of useful cancer vaccines based on TACAs.

To deal with the above-mentioned issues, many approaches have been explored, including use of more immunogenic carrier molecules or use of more potent adjuvants. Another instinctive idea is to modify the structure of TACAs to make them more immunogenic. It has been proved that chemical modification of carbohydrates could indeed improve their immunological properties. For example, α-2,8-polysialic acid (PSA), a bacterial and cancer antigen, is essentially nonimmunogenic in human beings, but Jennings and co-workers [71, 72] have demonstrated that substitution of the N-acetyl groups of PSA by other acyl groups, such as propanoyl and butanoyl groups, resulted in excellent immunogens. In fact, the N-propanoylated form of PSA (PSANProp) has been employed to construct functional conjugate vaccines against group B meningitis [71, 72]. The keyhole limpet hemocyanin (KLH) conjugate of PSANProp was also shown to be an effective vaccine against small-cell lung carcinoma in clinical trials [73]. Chemically modified TACAs are also more immunogenic than their natural counterparts, and they often induce IgG antibodies and T-cell-dependent immune responses. For example, GD3, a tetrasaccharide TACA, is essentially nonimmunogenic, but its derivative with a lactonized sialyl residue elicited a strong and specific immune response in animals [9, 69, 74, 75].

For vaccines made of unnatural TACA analogs to function, the immune response elicited has to be able to recognize and interact with their natural counterparts on the target cells. In the case of PSANProp conjugates, fortunately, the provoked immune system and the PSANProp-specific antibodies could cross-react with natural PSA on bacterial and cancer cells to exhibit cytotoxicity, even though the PSANProp-specific antibody did not bind with pure PSA [71, 72]. However, the chemically modified GD3 was not so lucky. It has been revealed that the unnatural GD3-specific antibodies did not interact with natural GD3 on tumor cells and thus did not exhibit obvious cytotoxicity to cancer cells [9, 69, 74, 75]. Consequently, the conjugates of chemically modified GD3 could not be used as vaccines to treat cancer. On the other hand, these results indicate that the immune response provoked by chemically modified TACAs can be highly specific.

Recently, we have proposed and explored a novel strategy for cancer immunotherapy that can overcome the above problem, namely that the specific immune response induced by an unnatural antigen does not recognize the natural counterpart on cancer cells. The strategy combines cancer vaccines made of unnatural TACA analogs with glycoengineering of TACAs on cancer cells. Its basic principle is described in Figure 10.8. First, a cancer vaccine made of an unnatural analog of a TACA, termed neo-TACA, is prepared and employed to immunize animals or cancer patients. Once an immune response specific for the neo-TACA is established, the animals or patients will be treated with an identically modified biosynthetic precursor of the target TACA to induce the expression of the neo-TACA on tumor cell surfaces. Subsequently, the pre-provoked immune system will recognize and eradicate the glycoengineered tumor cells that express the neo-TACA.

The new cancer immunotherapy can also be achieved by a passive immunization protocol shown in Figure 10.9. First, animals or patients are treated with an unnatural carbohydrate precursor to engineer the expression of a neo-TACA on cancer cells. Meanwhile, a healthy individual is immunized with a synthetic vaccine made of the neo-TACA for the preparation of neo-TACA-specific antibodies. Finally, the

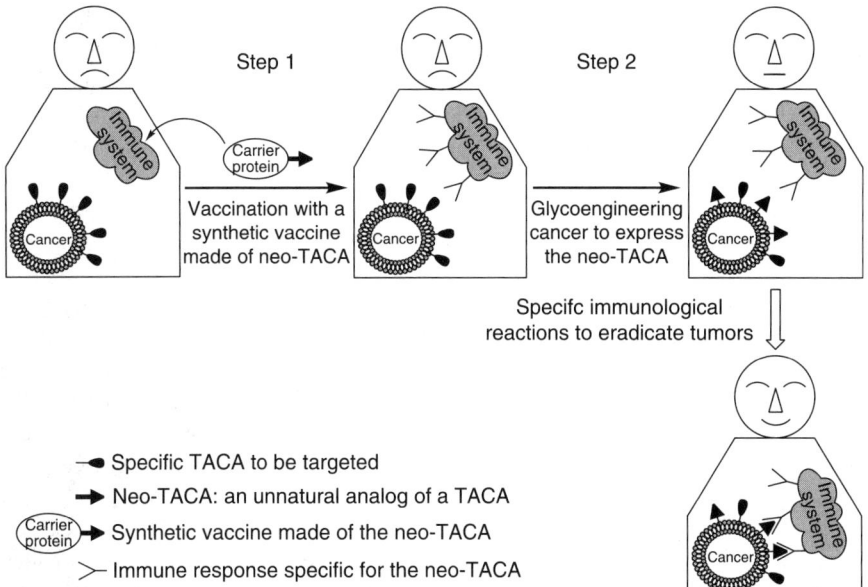

Figure 10.8 Cancer immunotherapy based on cell glycoengineering and an active immunization protocol.

antibodies will be used to treat animals or patients for targeting the glycoengineered tumors, which can be employed for both diagnosis and immunotherapy of cancer.

For the strategy to work, it has to meet two conditions. First, there must be a proper target TACA on cancer cells that is suitable for glycoengineering as well as an appropriate biosynthetic precursor that can effectively glycoengineer the target TACA both

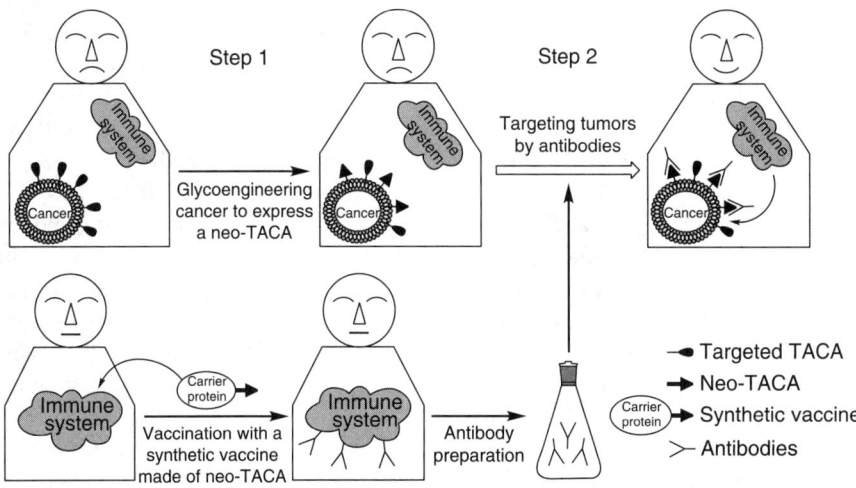

Figure 10.9 Cancer immunotherapy based on cell glycoengineering and a passive immunization protocol.

in vitro and in vivo. Second, there must be a functional vaccine made of the correspondingly modified analog of the target TACA that can induce a specific and effective immune response. For this purpose, sialo-TACAs are ideal targets. As discussed above, cell surface sialic acids can be effectively glycoengineered by a number of unnatural derivatives of ManNAc, and carbohydrate antigens containing unnatural sialic acid derivatives can induce strong and specific immune responses.

There may be some questions concerning the selectivity of cell glycoengineering and/or the new immunotherapy for cancer in the presence of normal tissues. The selectivity would derive from a number of factors. First, faster proliferation of cancer cells than normal cells will guarantee higher precursor uptake proficiencies of cancer cells than normal cells, which will be similar to traditional chemotherapies. Second, higher expression of sialic acids on cancer cells suggests that cancer cells have higher enzymatic activity for sialoglycoconjugate biosynthesis than normal cells, which will guarantee more effective metabolism of the modified precursor in cancer cells. Most importantly, the immune response or the neo-TACA-specific antibodies raised by the synthetic vaccine are specific for glycoengineered tumor cells that express high concentrations of the neo-TACA. Although normal cells may also use the precursor, they express oligosaccharides that have different sequences, or they express a low level of the neo-TACA, which will cause little or no immune reaction. For example, Hakomori and co-workers [76, 77] observed that TACA-raised antibodies would selectively bind to cancer cells in the presence of normal cells, even if normal cells also expressed a low level of the same antigen. They hypothesized that the immune reaction was antigen density dependent [76] and that higher concentrations of TACA on cancer cells might help to create some special microdomains to facilitate the discrimination of cancer cells from normal cells [10].

In our first proof-of-principle study, a lymphatic leukemia cell line RMA was utilized as a tumor model, and the target antigen was PSA [78]. Naturally, ManNProp was employed as the precursor for cell glycoengineering because literature results have revealed that ManNProp was very efficient in glycoengineering cells in vitro and in vivo [12, 17, 78, 79] and that the protein conjugates of PSANProp were effective vaccines. Moreover, some PSANProp-specific cytotoxic monoclonal antibodies (mAb) have already been identified [71, 72, 80]. Our in vitro results have demonstrated that RMA cells were efficiently glycoengineered by ManNProp to express PSANProp and that glycoengineered RMA cells were selectively targeted by PSANProp-specific mAb 13D9. Moreover, mAb 13D9 showed strong and specific cytotoxicity to glycoengineered cancer cells, and the cytotoxicity was dependent on the concentrations of ManNProp used. The new cancer immunotherapy was also evaluated in vivo. In these experiments, solid RMA tumors were first established in mice, and the animals then received treatment with mAb 13D9 and ManNProp. The results have demonstrated that the treatment could inhibit tumor growth but did not eliminate the established solid tumor completely. However, it was discovered that the treatment could completely inhibit tumor metastasis.

Similarly, a melanoma cell line SK-MEL-28 was effectively glycoengineered with ManNBut to express unnatural N-butanoyl GD3 (GD3NBut), and glycoengineered cancer cells were selectively targeted [81]. For example, in the presence of complements, glycoengineered cancer cells was lysed by GD3NBut-specific mAb 2A and GD3Bu antiserum obtained by immunizing mice with GD3NBut-KLH. Though less effective in the control of existing large tumors in mice, mAb 2A in combination with ManNBu could effectively protect mice from SK MEL-28 tumor grafting.

The above two designs were based on well-established glycoengineering precursors, ManNProp and ManNBut, which are structurally very close to ManNAc. However, neither ManNProp combined with a NeuNProp-containing TACA analog nor ManNBut combined with a NeuNBut-containing TACA analog was necessarily the optimal design for the new immunotherapeutic strategy. For example, an obvious problem with these designs is that the required ManNProp and ManNBut concentrations for effective cell glycoengineering were very high (at millimolar levels). Furthermore, the structural similarity of N-propanoylated and N-butanoylated TACA analogs to their natural counterparts may not result in the best vaccines or the most desirable immune reactions.

In order to optimize the therapeutic design, a series of N-acyl derivatives of mannosamine and the KLH conjugates of a series of unnatural N-acyl sialic acids or TACA analogs containing unnatural N-acyl sialic acids were synthesized and examined as cell glycoengineering precursors and as vaccines, respectively [20, 82–87]. It was disclosed that low micromolar concentrations of ManNPhAc could effectively glycoengineer an array of cancer cell lines [20, 87]. This is a significant achievement because low micromolar concentration can be easily achievable in vivo. Furthermore, it was revealed that NeuNPhAc [84] and TACA analogs containing NeuNPhAc [85, 86] were highly immunogenic to provoke specific immune responses. Consequently, combinations of ManNPhAc with vaccines made of NeuNPhAc-containing TACA analogs were investigated in great detail.

NeuNPhAc-containing GM3 analog (GM3NPhAc) [85] and NeuNPhAc-containing sTn analog (sTnNPhAc) [86] were both demonstrated to induce high titers of antigen-specific IgG antibodies in mice, indicating T-cell-dependent immune response that is critical for anticancer activity of cancer immunotherapy. Moreover, in the presence of complements, GM3NPhAc antisera and GM3NPhAc-specific mAb 2H3 exhibited strong and specific cytotoxicity to several melanoma cell lines after the cells were treated with ManNPhAc [20, 87]. Under the same conditions, normal cells were unaffected, suggesting the high selectivity of the new cancer immunotherapy. It was concluded that combined treatment using GM3NPhAc and GM3NPhAc-specific antibodies or using GM3NPhAc and vaccines derived from GM3NPhAc is worthy of further investigation as a potentially useful immunotherapy for melanoma and other tumors that express GM3 [88].

10.5 SUMMARY

Glycoengineering of cell surface sialic acids has been proved to be a powerful technique for cell surface landscaping. For example, it has been employed to successfully modify the surface of many cell types to facilitate detailed studies of the biological functions of cell surface sialic acids. The technique was also used to deliver biologically orthogonal functionalities onto cell surfaces to enable the attachment of various molecular probes, which is useful for many studies in cell glycobiology. Glycoengineering of sialo-TACAs was combined with specific immune reactions to target the resultant TACA analogs on cancer cells for the development of novel cancer immunotherapies. This strategy should be generally applicable to various TACAs, provided that proper target TACAs and efficient methods to engineer the target TACAs are available. The strategy can also help to overcome the immune tolerance problem associated with many TACAs and is thus very useful for studies in cancer immunology.

ACKNOWLEDGMENT

The author thanks the National Institutes of Health (NIH) for financial support (R01 CA095142).

REFERENCES

1. Schauer R (1988) Sialic acids as antigenic determinants of complex carbohydrates. *Adv. Exp. Med. Biol.* 228:47.
2. Reutter W, Stasche R, Stehling P, Baum O (1997) The biology of sialic acids: Insights into their structure, metabolism and function in particular during viral infection. In Gabius H-J, Gabius S, Eds. *Glycosciences*. Chapman & Hall, Weinheim, p. 245.
3. Rosenburg AE (1995) *Biology of the Sialic Acid*. Plenum, New York.
4. Troy FA (1992) Polysialylation: From bacteria to brain. *Glycobiology* 2:5.
5. Takano R, Muchmore E, Dennis JW (1994) Sialylation and malignant potential in tumor cell glycosylation mutants. *Glycobiology* 4:665.
6. Hakomori S (1996) Tumor-associated carbohydrate antigens and modified blood group antigens. In Montreuil J, Vliegenthart JFG, Schachter H (Eds) *Glycoproteins and Disease*. Elsevier, Amsterdam, p. 243.
7. Fukuda M (1996) Possible roles of tumor-associated carbohydrate antigens. *Cancer Res.* 56:2237.
8. Scheidegger EP, Lackie PM, Papay J, Roth J (1994) In vitro and in vivo growth of clonal sublines of human SCLC is modulated by polysialic acid of NCAM. *Lab. Invest.* 70:95.
9. Ragupathi G (1998) Carbohydrate antigens as targets for active specific immunotherapy. *Cancer Immunol. Immunother.* 46:82.
10. Hakomori S, Zhang Y (1997) Glycosphingolipid antigens and cancer therapy. *Chem. Biol.* 3:97.
11. Helling F, Shang A, Calves M, Zhang S, Ren S, Yu RK, Oettgen HF, Linvingston PO (1994) GD3 vaccines for melanoma: Superior immunogenicity of keyhole limpet hemocyanin conjugate vaccines. *Cancer Res.* 54:197.
12. Kayser H, Geile CC, Paul C, Zeitler R, Reutter W (1992) Incorporation of N-acyl-2-amino-2-deoxy-hexoses into glycosphingolipids of pheochromocytoma cell line PC12. *FEBS* 301:137.

13. Mahal KL, Yarema KJ, Bertozzi CR (1997) Engineering chemical reactivity on cell surfaces through oligosaccharide biosynthesis. *Science* 276:1125.
14. Kayser H, Zeitler R, Kannicht C, Grunow D, Nuck R, Reutter W (1992) Biosynthesis of a nonphysiological sialic acid in different rat organs, using N-propanoyl-D-hexosamines as precursors. *J. Biol. Chem.* 267:16934.
15. Keppler OT, Horstkorte R, Pawlita M, Schmidt C, Reutter W (2001) Biochemical engineering of the N-acyl side chain of sialic acid: Biological implications. *Glycobiology* 11:11R.
16. Mahal LK, Charter NW, Angata K, Fukuda M, Koshland DE, Bertozzi CR (2001) A small-molecule modulator of poly-alpha 2,8-sialic acid expression on cultured neurons and tumor cells. *Science* 294:380.
17. Keppler OT, Stehling P, Herrmann M, Kayser H, Grunow D, Reutter W, Pawlita M (1995) Biosynthetic modulation of sialic acid-dependent virus-receptor interactions of two primate polyoma viruses. *J. Biol. Chem.* 270:1308.
18. Yarema KJ, Mahal LK, Bruehl RE, Rodriguez EC, Bertozzi CR (1998) Metabolic delivery of ketone groups to sialic acid residues. Application to cell surface glycoform engineering. *J. Biol. Chem.* 273:31168.
19. Saxon E, Bertozzi CR (2000) Cell surface engineering by a modified Staudinger reaction. *Science* 287:2007.
20. Chefalo P, Pan Y, Nagy N, Harding C, Guo Z (2006) Effective metabolic engineering of GM3 on tumor cells by N-phenylacetyl-D-mannosamine. *Biochemistry* 45:3733.
21. Sampathkumar S-G, Li AV, Jones MB, Sun Z, Yarema KJ (2006) Metabolic installation of thiols into sialic acid modulates adhesion and stem cell biology. *Nature Chem. Biol.* 2:149.
22. Collins BE, Fralich TJ, Itonori S, Ichikawa Y, Schnaar RL (2000) Conversion of cellular sialic acid expression from N-acetyl- to N-glycolylneuraminic acid using a synthetic precursor, N-glycolylmannosamine pentaacetate: Inhibition of myelin-associated glycoprotein binding to neural cells. *Glycobiology* 10:11.
23. Jacobs CL, Goon S, Yarema KJ, Hinderlich S, Hang HC, Chai DH, Bertozzi CR (2001) Substrate specificity of the sialic acid biosynthetic pathway. *Biochemistry* 40:12864.
24. Kim EJ, Sampathkamur S-G, Jones MB, Rhee JK, Baskaran G, Goon S, Yarema KJ (2004) Characterization of the metabolic flux and apoptotic effects of O-hydroxyl and N-acyl-modified N-acetylmannosamine (ManNAc) analogs in Jurkat (human T-lymphoma-derived) cells. *J. Biol. Chem.* 279:18342.
25. Kim EJ, Jones MB, Rhee JK, Sampathkumar SG, Yarema KJ (2004) Establishment of N-acetylmannosamine (ManNAc) analogue-resistant cell lines as improved hosts for sialic acid engineering applications. *Biotech. Prog.* 20:1674.
26. Prescher JA, Dube DH, Bertozzi CR (2004) Chemical remodelling of cell surfaces in living animals. *Nature* 430:873.
27. Kayser H, Ats C, Lehmann J, Reutter W (1993) New amino sugar analogs are incorporated at different rates into glycoproteins of mouse organs. *Experientia* 49:885.
28. Keppler OT, Herrmann M, Von Der Lieth CW, Stehling P, Reutter W, Pawlita M (1998) Elongation of the N-acyl side chain of sialic acids in MDCK II cells inhibits influenza A virus infection. *Biochem. Biophys. Res. Commun.* 253:437.
29. Wieser JR, Heisner A, Stehling P, Oesch F, Reutter W (1996) In vivo modulated N-acyl side chain of N-acetylneuraminic acid modulates the cell contact-dependent inhibition of growth. *FEBS Lett.* 395:170.

30. Schmidt C, Stehling P, Schnitzer J, Reutter W, Horstkorte R (1998) Biochemical engineering of neural cell surfaces by the synthetic N-propanoyl-substituted neuraminic acid precursor. *J. Biol. Chem.* 273:19146.
31. Lemieux GA, Bertozzi CR (2001) Modulating cell surface immunoreactivity by metabolic induction of unnatural carbohydrate antigens. *Chem. Biol.* 8:265.
32. Saxon E, Armstrong JI, Bertozzi CR (2000) A "traceless" Staudinger ligation for the chemoselective synthesis of amide bonds. *Org. Lett.* 2:2141.
33. Luchansky SJ, Goon S, Bertozzi CR (2004) Expanding the diversity of unnatural cell-surface sialic acids. *ChemBioChem* 5:371.
34. Goon S, Schilling B, Tullius MV, Gibson BW, Bertozzi CR (2003) Metabolic incorporation of unnatural sialic acids into Haemophilus ducreyi lipooligosaccharides. *Proc. Natl. Acad. Sci. USA* 100:3089.
35. Tangvoranuntakul P, Gagneux P, Diaz S, Bardor M, Varki N, Varki A, Muchmore E (2003) Human uptake and incorporation of an immunogenic nonhuman dietary sialic acid. *Proc. Natl. Acad. USA.* 100:12045.
36. Daubeuf S, Aucher A, Sampathkumar S, Reville X, Yarema KJ, Hudrisier D (2007) Chemical labels metabolically installed into the glycoconjugates of the target cell surface can be used to track lymphocyte/target cell interplay via trogocytosis: Comparisons with lipophilic dyes and biotin. *Immunol. Invest.* 36:687.
37. Horstkorte R, Muehlenhoff M, Reutter W, Noehring S, Zimmermann-Kordmann M, Gerardy-Schahn R (2004) Selective inhibition of polysialyltransferase ST8SiaII by unnatural sialic acids. *Exp. Cell Res.* 298:268.
38. Oetke C, Brossmer R, Mantey LR, Hinderlich S, Isecke R, Reutter W, Keppler OT, Pawlita M (2002) Versatile biosynthetic engineering of sialic acid in living cells using synthetic sialic acid analogues. *J. Biol. Chem.* 6688.
39. Oetke C, Hinderlich S, Brossmer R, Reutter W, Pawlita M, Keppler OT (2001) Evidence for efficient uptake and incorporation of sialic acid by eukaryotic cells. *Euro. J. Biochem.* 268:4553.
40. Schilling B, Goon S, Samuels NM, Gaucher SP, Leary JA, Bertozzi CR, Gibson BW (2001) Biosynthesis of sialylated lipooligosaccharides in Haemophilus ducreyi is dependent on exogenous sialic acid and not mannosamine. Incorporation studies using N-acylmannosamine analogues, N-glycolylneuraminic acid, and ^{13}C-labeled N-acetylneuraminic acid. *Biochemistry* 40:12666.
41. Horstkorte R, Rau K, Reutter W, Nohring S, Lucka L (2004) Increased expression of the selectin ligand sialyl-Lewisx by biochemical engineering of sialic acids. *Exp. Cell Res.* 295:549.
42. Bork K, Gagiannis D, Orthmann A, Weidemann W, Kontou M, Reutter W, Horstkorte R (2007) Experimental approaches to interfere with the polysialylation of the neural cell adhesion molecule in vitro and in vivo. *J. Neurochem.* 103(Suppl.1):65.
43. Luchansky SJ, Bertozzi CR (2004) Azido sialic acids can modulate cell-surface interactions. *ChemBioChem* 5:1706.
44. Schwarzkopf M, Knobeloch K, Rohde E, Hinderlich S, Wiechens N, Lucka L, Horak I, Reutter W, Horstkorte R (2002) Sialylation is essential for early development in mice. *PNAS* 99:5267.
45. Herrmann M, von der Lieth CW, Stehling P, Reutter W, Pawlita M (1997) Consequences of a subtle sialic acid modification on the murine polyomavirus receptor. *J. Virol.* 71:5922.

46. Buttner B, Kannicht C, Schmidt C, Loster K, Reutter W, Lee H, Nohring S, Horstkorte R (2002) Biochemical engineering of cell surface sialic acids stimulates axonal growth. *J. Neurosci.* 22:8869.
47. Keppler OT, Peter ME, Hinderlich S, Moldenhauer G, Stehling P, Schmitz I, Schwartz-Albiez R, Reutter W, Pawlita M (1999) Differential sialylation of cell surface glycoconjugates in a human B lymphoma cell line regulates susceptibility for CD95 (APO-1/Fas)-mediated apoptosis and for infection by a lymphotropic virus. *Glycobiology* 9:557.
48. Horstkorte R, Rau K, Laabs S, Danker K, Reutter W (2004) Biochemical engineering of the N-acyl side chain of sialic acid leads to increased calcium influx from intracellular compartments and promotes differentiation of HL60 cells. *FEBS Lett.* 57:99.
49. Saxon E, Luchansky SJ, Hang HC, Yu C, Lee SC, Bertozzi CR (2002) Investigating cellular metabolism of synthetic azidosugars with the Staudinger ligation. *J. Am. Chem. Soc.* 124:14893
50. Vocadlo DJ, Hang HC, Kim E, Hanover JA, Bertozzi CR (2003) A chemical approach for identifying O-GlcNAc-modified proteins in cells. *PNAS* 100:9116.
51. Hang HC, Yu C, Kato DL, Bertozzi CR (2003) A metabolic labeling approach toward proteomic analysis of mucin-type O-linked glycosylation. *PNAS* 100:14846.
52. Dube DH, Prescher JA, Quang CN, Bertozzi CR (2006) Probing mucin-type O-linked glycosylation in living animals. *PNAS* 103:4819.
53. Prescher JA, Bertozzi CR (2006) Chemical technologies for probing glycans. *Cell* 126: 851.
54. Laughlin ST, Agard NJ, Baskin JM, Carrico IS, Chang PV, Ganguli AS, Hangauer MJ, Lo A, Prescher JA, Bertozzi CR (2006) Metabolic labeling of glycans with azido sugars for visualization and glycoproteomics. *Methods Enzym.* 415:230.
55. Czlapinski JL, Bertozzi CR (2006) Synthetic glycobiology: Exploits in the Golgi compartment. *Curr. Opin. Chem. Biol.* 10:645.
56. Mahal LK, Bertozzi CR (1997) Engineering cell surfaces: Fertile ground for molecular landscaping. *Chem. Biol.* 4:415.
57. Prescher JA, Bertozzi CR (2005) Chemistry in living systems. *Nature Chem. Biol.* 1:13.
58. Bertozzi CR, Kiessling LL (2001) Chemical glycobiology. *Science* 291:2357.
59. Lemieux GA, Yarema KJ, Jacobs CL, Bertozzi CR (1999) Exploiting differences in sialoside expression for selective targeting of MRI contrast reagents. *J. Am. Chem. Soc.* 121: 4278.
60. Lee JH, Baker TJ, Mahal LK, Zabner J, Bertozzi CR, Wiemer DF, Welsh MJ (1999) Engineering novel cell surface receptors for virus-mediated gene transfer. *J. Biol. Chem.* 274:21878.
61. Huisgen R (1963) 1,3-Dipolar cycloadditions. Past and future. *Angew. Chem. Int. Ed.* 565.
62. Wang Q, Chan TR, Hilgraf R, Fokin VV, Sharpless KB, Finn MG (2003) Bioconjugation by copper(I)-catalyzed azide-alkyne [3 + 2] cycloaddition. *J. Am. Chem. Soc.* 125:3192.
63. Seo TS, Li Z, Ruparel H, Ju J (2003) Click chemistry to construct fluorescent oligonucleotides for DNA sequencing. *J. Org. Chem.* 68:609.
64. Speers AE, Cravatt BF (2004) Profiling enzyme activities in vivo using click chemistry methods. *Chem. Biol.* 11:535.

65. Agard NJ, Prescher JA, Bertozzi CR (2004) A strain-promoted [3 + 2] azide-alkyne cycloaddition for covalent modification of biomolecules in living systems. *J. Am. Chem. Soc.* 126:15046.
66. Baskin JM, Prescher JA, Laughlin ST, Agard NJ, Chang PV, Miller IA, Lo A, Codelli JA, Bertozzi CR (2007) Copper-free click chemistry for dynamic in vivo imaging. *PNAS* 104:16793.
67. Rabuka D, Hubbard SC, Laughlin ST, Argade SP, Bertozzi CR (2006) A chemical reporter strategy to probe glycoprotein fucosylation. *J. Am. Chem. Soc.* 128:12078.
68. Livingston PO (1992) Construction of cancer vaccines with carbohydrate and protein tumor antigens. *Curr. Opin. Immunol.* 4:624.
69. Livingston PO (1995) Approaches to augmenting the immunogenicity of melanoma gangliosides: From the whole melanoma cells to ganglioside-KLH conjugate vaccines. *Immunol. Rev.* 145:147.
70. Hakomori S (1991) Possible functions of tumor-associated carbohydrate antigens. *Curr. Opin. Immunol.* 3:646.
71. Jennings HJ, Sood RK (1994) Synthetic glycoconjugates as human vaccines. In YC Lee, RT Lee (Eds) *Neoglycoconjugates: Preparation and Applications.* Academic, San Diego, p. 325.
72. Jennings H (1997) N-propionylated group B meningococcal polysaccharide glycoconjugate vaccine against group B meningococcal meningitis. *Int. J. Infect. Dis.* 1:158.
73. Krug LM, Ragupathi G, Ng KK, Hood C, Jennings HJ, Guo Z, Kris MG, Miller V, Pizzo B, Tyson L, Baez V, Livingston PO (2004) Vaccination of small cell lung cancer patients with polysialic acid or N-propionylated polysialic acid conjugated to keyhole limpet hemocyanin. *Clin. Cancer Res.* 10:916.
74. Ritter G, Boosfeld E, Asluri S, Calves MJ, Oettgen HF, Old LJ, Livingston PO (1991) Antibody response after immunization with ganglioside GD3, GD3 congeners (lactones, amide and gangliosidol) in patients with malignant melanoma. *Int. J. Cancer* 48:379.
75. Ritter G, Boosfeld E, Calves MJ, Oettgen HF, Old LJ, Livingston PO (1990) Antibody response after immunization with ganglioside GD3, GD3 lactones, GD3 amide and GD3 gangliosidol in the mouse. GD3 lactone I induces antibodies reactive with human melanoma. *Immunobiology* 182:32.
76. Nores GA, Dohi T, Taniguchi M, Hakomori S (1987) Density dependent recognition of cell surface GM3 by a certain anti-melanoma antibody, and GM3 lactone as a possible immunogen: Requirements for tumor-associated antigen and immunogen. *J. Immunol.* 139:3171.
77. Rosenfelder G, Young WWJ, Hakomori S (1977) Association of the glycolipid pattern with antigenic alternations in mouse fibroblasts transformed by murine sarcoma virus. *Cancer Res.* 37:1333.
78. Liu T, Guo Z, Yang Q, Sad S, Jennings HJ (2000) Biochemical engineering of surface a(2-8)polysialic acid for immunotargeting cancer cells. *J. Biol. Chem.* 275:32832.
79. Schmidt C, Stehling P, Schnitzer J, Reutter W, Horskorte R (1998) Biological engineering of neural cell surfaces by the synthetic N-propanoyl-substituted neuraminic acid precursor. *J. Biol. Chem.* 273:19146.
80. Pon R, Lussier M, Yang Q, Jennings HJ (1997) N-Propionylated group B meningococcal polysaccharide mimics a unique bactericidal capsular epitope in group B neisseria meningitides. *J. Exp. Med.* 185:1929.

81. Zou W, Borrelli S, Gilbert M, Liu T, Pon RA, Jennings HJ (2004) Bioengineering of surface GD3 ganglioside for immunotargeting human melanoma cells. *J. Biol. Chem.* 279:25390.
82. Pan Y, Ayani T, Nadas J, Guo Z (2004) Accessibility of N-acyl derivatives of D-mannosamine to N-acetylneuraminic acid aldolase. *Carbohydr. Res.* 339:2091.
83. Xue J, Pan Y, Guo Z (2002) Neoglycoprotein cancer vaccines: Synthesis of an azido derivative of GM3 and its efficient coupling to proteins through a new linker. *Tetrahedron Lett.* 43:1599.
84. Chefalo P, Pan YB, Nagy N, Harding CV, Guo Z (2004) Preparation and immunological studies of protein conjugates of N-acylneuraminic Acids. *Glycoconjugate J.* 20:407.
85. Pan YB, Chefalo P, Nagy N, Harding CV, Guo Z (2005) Synthesis and immunological properties of N-modified GM3 antigens as therapeutic cancer vaccines. *J. Med. Chem.* 48:875.
86. Wu J, Guo Z (2006) Improving the antigenicity of sTn antigen by modification of its sialic acid residue for development of glycoconjugate cancer vaccines. *Bioconjugate Chem.* 17:1537.
87. Wang Q, Zhang J, Guo Z (2007) Efficient glycoengineering of GM3 on melanoma cell and monoclonal antibody-mediated selective killing of the glycoengineered cancer cell. *Bioorg. Med. Chem.* 15:7561.
88. Bitton RJ, Guthmann MD, Babri MR, Carnero AJL, Alonso DF, Fainboim L, Gomez DE (2002) Cancer vaccines: An update with special focus on ganglioside antigens (Review). *Oncol. Rep.* 9:267.

11

THERAPEUTIC CANCER VACCINES: CLINICAL TRIALS AND APPLICATIONS

Hans H. Wandall and Mads A. Tarp

Department of Cellular and Molecular Medicine, Faculty of Health Sciences, University of Copenhagen, Copenhagen N, Denmark

11.1 INTRODUCTION

Cancer remains a major cause of death worldwide. Current treatments primarily rely on a combination of therapies, in many cases reduction of the tumor mass through surgery followed by adjuvant radio- and chemotherapy targeting rapidly dividing cells. Therefore, these treatments do not only affect tumor cells but also normal cells, which cause severe side effects limiting treatment. Recent breakthroughs in our understanding of the immune system and tumor biology have opened the door to the development of new approaches to the treatment of cancer, including the use of passive and active therapeutic cancer vaccines. Therapeutic vaccines aim to treat an already existing disease, in this case cancer, differing from the traditional prophylactic vaccines. Collectively, vaccines are a promising important advance in the care for cancer patients, offering the potential advantage of controlling systemic disease using the body's own innate ability to destroy unwanted cells with minimal toxicity.

Cancer immunotherapy boosts the patient's own immune system to treat the cancer. Pioneer work in immunotherapy started more than 100 years ago by the New York City surgeon and Sloan-Kettering researcher, Dr. William B. Coley. Coley published a series of articles describing the use of a mixture of attenuated bacterial strains in the

---

Carbohydrate-Based Vaccines and Immunotherapies. Edited by Zhongwu Guo and Geert-Jan Boons
Copyright © 2009 John Wiley & Sons, Inc.

treatment of patients with malignant tumors. This mixture was called Coley's toxin, and its active substance was later found to be lipopolysaccharides (LPS) [1–5] known to be a strong stimulus for the immune system. Other examples of the use of attenuated bacteria in today's cancer therapy exist, such as the use of bacille Calmette–Guérin (BCG) in the treatment of bladder cancer [6]. With the development of modern molecular biology and immunology in particular, more specific strategies have been devised, using tumor-associated antigens in generating specific immune responses in the fight against cancer. For this approach well-defined tumor-associated antigens need to be identified. Interestingly, many of the original studies demonstrating cancer-specific immune responses found that some of the most prominent changes during carcinogenesis were the appearance of carbohydrate changes on the surface of cancer cells [7].

Therefore, there has been a great interest in characterizing and synthesizing cancer-specific or cancer-associated carbohydrate structures both for passive and active cancer immunotherapy [8, 9]. The many studies conducted in the field have resulted in a large number of preclinical data suggesting the efficacy and relevance of using carbohydrates to direct the immune response to cancer cells. However, only a few vaccine candidates have been applied in appropriate clinical trials and only a minority has proven efficacious.

In the following we will initially address general aspects of clinical trials of cancer vaccines, including trial design, endpoints, and challenges related to the technical development. Next, we will describe the results of the many early-phase vaccine trials conducted to date and highlight critical steps in the path from laboratory-based experiments into a clinical setting. This includes: (1) identification of vaccine targets that ensure the generation of responses specific for cancer cells without harming normal cells, (2) development of methods that overcome self-tolerance and induce a sufficient and specific immune response, and finally, (3) strategies to prevent the tumor from escaping the induced immune response of the host.

11.2 INNATE AND ADAPTIVE IMMUNITY IN RELATION TO CANCER IMMUNOTHERAPY

Several key steps in the immunological response to cancer cells are important for the development of effective cancer vaccines. The immune system recognizes tumor-specific antigens usually through the help of antigen-presenting cells (APCs) such as dendritic cells [10, 11] (Fig. 11.1). In cancer immunology dendritic cells are pivotal in initiating T-cell anticancer responses by processing and presenting tumor antigens to T lymphocytes. After initial exposure to antigens, dendritic cells can be stimulated via co-stimulatory molecules (CD40L or agonistic anti-CD40) [12], as well as lipopolysaccharides and bacterial DNA (deoxyribonucleic acid), to induce maturation and a potent anticancer immune response [13–16]. Another cell type belonging to the innate immune system is natural killer (NK) cells, which are important in cancer immunotherapy because they are capable of directly killing cancer cells. Several stimulatory and inhibitory molecules exist on NK cells, and direct activation of relevant receptors can stimulate cytotoxicity.

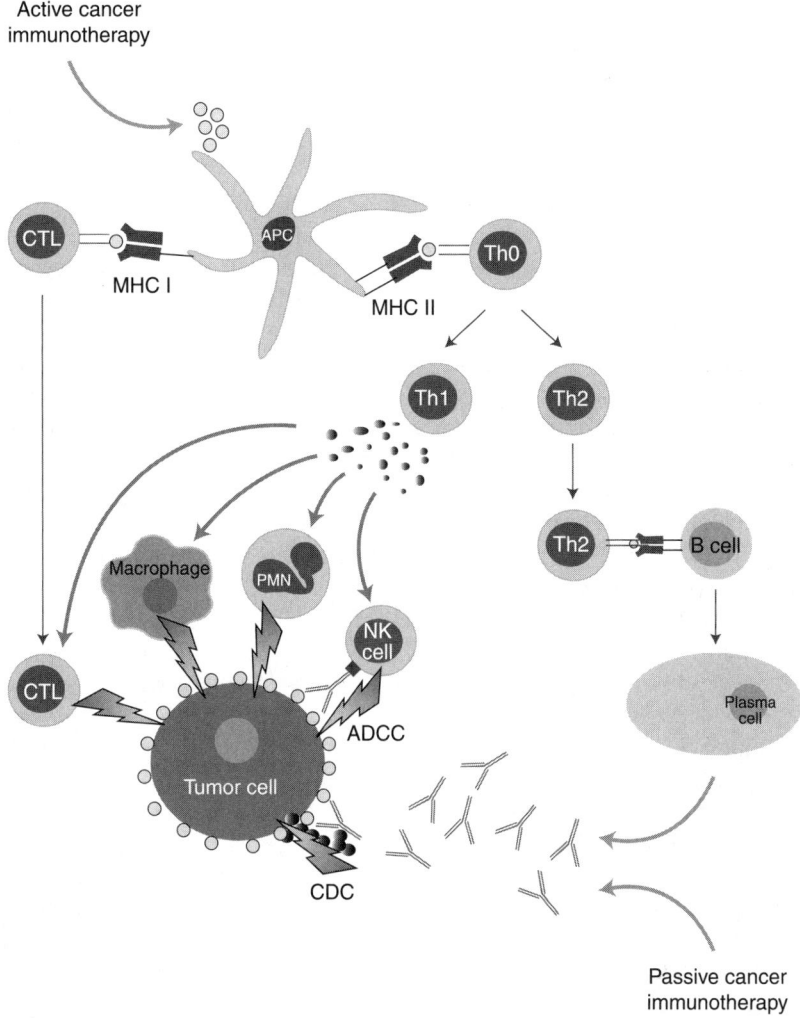

Figure 11.1 Principles of active and passive cancer immunotherapy. (See color insert.)

The cells of the innate immune system bridge to the adaptive immune system by activating and presenting antigens to T and B lymphocytes. In contrast to the innate immune system, T and B lymphocytes are able to develop immunological memory, making them a desirable target for immune therapy in cancer [17]. T-cells are activated either by direct recognition of tumor antigens, or if antigens are presented to them by APCs. Activated T-cells then produce interleukins and chemokines, which mediate killing of tumor cells [18, 19]. In contrast, the primary function of B lymphocytes is to differentiate into plasma cells that produce antibodies, which can facilitate elimination of tumor cells by complement-dependent cytotoxicity (CDC) and/or antibody-dependent cell-mediated cytotoxicity (ADCC) performed by NK cells and macrophages. In the context of carbohydrate-based vaccines, these reactions depend on both the antibody class and the carbohydrate target type. For example, it has been demonstrated that while glycosphingolipids induce CDC, this is not the case for mucin antigens [20]. Another example comes from the demonstration that G_{D3} presented by CD1 MHC (major histocompatibility complex) molecules could represent a direct target for NK cells [21].

Before the adaptive immune system is able to recognize tumor antigens, these antigens are digested into short peptide fragments, which are subsequently complexed with the MHC proteins located on the surface of the tumor cell or APC. $CD8^+$ cytotoxic T lymphocytes (CTLs) recognize MHC class I restricted antigens via the T-cell antigen receptor. Such antigens have been processed by the endogenous processing pathway and presented to the CTLs as peptide–MHC complexes. In order to mount a robust response against tumor-cells carrying tumor-specific antigens, the CTLs must receive stimulatory signals, such as interleukin-2 and interferon-gamma released by a subgroup of $CD4^+$ T-helper (Th) cells, designated Th1 cells. Dendritic cells and macrophages also secrete critical cytokines important for stimulating CTL activation, such as IL-12 and IL-15. Once activated, CTLs can migrate into tumor masses and cause cytolysis of tumor cells. While CTLs recognize MHC class I restricted antigens, Th cells recognize antigens presented by MHC class II molecules, which are expressed on lymphoid cells [22] (Fig. 11.1). Activation of Th cells is also relevant for the recognition of tumor antigens [17, 23], as another subgroup of Th cells, Th2 cells, can activate B cells leading to the production of antibodies targeting cancer antigens. Th1 and Th2 cells both originate from a population of naïve $CD4^+$ T-cells (Th0). The factors determining whether Th0 cells differentiate into Th1 or Th2 cells have not been identified, but it has been suggested that the concentration of antigen and the type of cell presenting the antigen influence the choice. Interestingly, an oxidative or reductive vaccine conjugation method has been shown to select for Th1 or Th2 responses [24].

An important issue is the conditions needed to generate a powerful immune response and to avoid the induction of tolerance. If co-stimulatory molecules are not present when the immune system is being activated, the immune system will likely become tolerant to tumor cells and will not recognize or destroy them. When the immune system no longer recognizes the tumor cells as nonself or dangerous, an immune response will not be elicited and the tumor will escape immune detection. The magnitude of the immune response to a cancer vaccine is only one element needed to ensure the success of therapeutic vaccines. Maintenance of the elicited response is also vital, particularly in cancer where a sustained immune response may well be more clinically relevant than a strong but transient response. Cancer-associated carbohydrates represent ligands for several receptor systems on the surface of antigen-presenting cells, and thus hold promise to increase the likelihood of inducing a long-lasting immune response overcoming tolerance.

In conclusion, some vaccines tend to elicit humoral responses, whereas others generate cellular responses. The general opinion is that the more components of the immune system that are involved in the anticancer immune response the better. Moreover, sustainability of the immune response is an important factor to be considered.

11.3 DESIGN ISSUES FOR CLINICAL CANCER VACCINE TRIALS

A traditional clinical drug development paradigm involves three phases of clinical trials. Phase I trials are the first stage of testing in human subjects. Normally, a small group of healthy volunteers will be selected. Alternatively, the drug can be tested in patients with end-stage disease who lack other treatment options. This last option is usually the case with antineoplastic treatments. Conventional chemotherapeutic phase I trials are designed to assess the safety, tolerability, pharmacokinetics, and pharmacodynamics of a drug and normally include dose ranging, also called dose escalation studies, allowing for finding the appropriate therapeutic dose. The tested range of doses will usually be a fraction of the dose that causes harm in animal testing. Once the initial safety of the study drug has been confirmed in phase I trials, phase II trials are performed on larger groups (hundreds) and are designed to assess efficacy as well as to continue safety assessments in a larger group of volunteers and patients. Phase III studies are randomized controlled multicenter trials on large patient groups (hundreds to thousands) and are aimed at being the definitive assessment of how effective the drug is, in comparison with current state-of-the-art treatment. It is typically expected that there be at least two successful phase III trials, demonstrating a drug's safety and efficacy, in order to obtain approval from the appropriate regulatory agencies, such as the U.S. Food and Drug Administration (FDA) and European Medicines Agency (EMEA) (European Union), to market the drug.

11.4 CLINICAL DEVELOPMENT OF CANCER VACCINES

In traditional phase I chemotherapeutic trials, it is presumed that higher doses have greater antitumor activity and therefore will be more efficacious, but also inherently linked to more serious side effects. Therefore, the requirement to identify the maximum tolerated dose (MTD) motivates the design of conventional phase I trials. In contrast, cancer vaccines are supposed to be much safer than cytotoxic agents, and the dose that yields sufficient immunogenicity is unlikely to confer significant toxicity. As a result, the justification of dose in cancer vaccine trials is not based on safety profiles but rather on biological and clinical activity [25, 26]. Another goal is to characterize the pharmacokinetics of the drug. In conventional chemotherapeutic trials, the absorption, distribution, and metabolism or excretion parameters can be reliably measured and are often related to the toxicity profile. This cannot be done with vaccines and is unlikely to reflect the toxicity profile of the agent.

In contrast to conventional trials in oncology, in which patients with various tumor types are enrolled, it is important in cancer vaccine trials to identify the appropriate target patient population for a given specific tumor antigen. Furthermore, in conventional trials immediate response criteria based on shrinkage of established tumor mass are pivotal. Such measures are not always applicable to cancer vaccines or immunotherapeutic agents because the absence of tumor shrinkage does not necessarily reflect lack of relevant biological or clinical activity. The explanation for this discrepancy is that conventional chemotherapeutic agents cause immediate damage to the target cells, while an active vaccine strategy needs time to develop an immune response and induce the desired effect. In this way active cancer vaccines are more suited to induce long-term effects such as disease stabilization and survival improvements. For the same reasons, it is expected that cancer vaccines are less efficient to affect bulk disease, but more likely to prevent the growth of small quantities of cancer cells, thereby being ideal for minimal residual disease (MRD). These considerations, however, might not apply to passive vaccines.

Therefore, it is strongly advised that early-stage cancer vaccine trials addressing active vaccines primarily should use parameters of biological activity as readout and only secondly use clinical endpoints. If clinical endpoints are used, they should be adjusted and compared with the obtained immune response or other biological parameters.

Another clear difference between traditional chemotherapeutic trials and active cancer vaccine trials is the relation between the dose of therapeutic agent and effect. While there is often a linear relation between chemotherapeutic agents and their effect, this is rarely the case in active cancer vaccines. Here there may not be any linear associations between dose, immunogenicity, and clinical endpoints. Similar levels of immune response may be induced at distinctly different dose levels, whereas the immune response may not be associated with clinical outcome. Therefore, it is often practical concerns that dictate the selection of dose for later phase trials.

In conclusion, the primary goal for any cancer vaccine should be establishment of an active dose regimen providing proof of principle and generation of data for immediate safety, permitting a rational design of randomized trials generating data to determine clinical efficacy of the vaccine in the target population [25].

Due to these differences between traditional pharmacological trials and cancer vaccine trials, the Cancer Vaccine Clinical Trial Working Group (CVCTWG), which represents all key players in the development of cancer vaccines, has recently suggested to use a new paradigm for the clinical development of cancer vaccines with only two different phases: *Proof of principle trials* and *efficacy trials* [25]. While proof of principle trials are exploratory trials combining some aspects of conventional phases I and II trials, efficacy trials are focused solely on clinical efficacy. Such a change in paradigm is believed to secure a more flexible, expeditious, and focused clinical development process.

11.5 PROOF OF PRINCIPLE TRIALS

Proof of principle trials should investigate disease-specific biological parameters in defined patient populations with a minimum of 20 subjects. The primary aim of such trials is to assess safety and presence and frequency of biological effects. The nature of any toxicity issue should be documented and the likelihood of the problem estimated. As discussed above, it is important that the trial population does not have rapidly progressive disease, to allow for sufficient time for biological and potential clinical activity to develop. As we will see in the following, this important message has often been neglected due to various other concerns. Also, it is important to be cautious when selecting the patient population to be studied, in such a way that patient withdrawal can be minimized. Furthermore, the existence of alternative and efficacious treatments for patients in early phases represents a potential problem since it is difficult to predict how the different treatments will influence each other. Of particular concern in relation to immunotherapy is the effect of chemotherapy and other treatments inducing immune suppression.

11.5.1 Toxicity and Pharmacokinetics

It is obviously of great importance to assess the safety of any biological agent used in clinical practice. The following simple steps were suggested by CVCTWG to allow for adequate toxicity testing: A standard safety panel of examinations covering major organ systems and securement of serum and other samples to allow for investigation in case of unexpected toxicities. For most cancer vaccines conventional pharmacokinetic measurements are of limited use. However, vaccines that involve measurable drug products from transfected tumor cells releasing active pharmacological agents, such as interleukins, would be important.

11.5.2 Dose and Administration Schedule

Studies in preclinical models poorly reflect on the dose and schedule to be used in humans, and mostly serve as a help in defining the starting dose. In the first-in-human studies, one can use a cohort design similar to the regime used in many phase I studies with conventional pharmaceuticals in order to estimate the maximum tolerated dose. With cancer vaccines, however, the aim would be to estimate the relationship between the dose and schedule and the biological outcome. Depending on the design and statistics used, six patients per cohort is suggested as a starting point, and the safety and the biological outcome in each cohort are evaluated before initiation of the next cohort [25, 26].

11.5.3 Endpoints: Biological Activity and Clinical Activity

Measurement of biological activity should include the demonstration of a specific immune response to the vaccine target and influence on the disease under investigation. A specific humoral and cellular immune response can be measured in a quantitative manner by several immunoassays. Importantly, however, the immunoassays to be used should ideally be standardized to allow for comparison of results from different clinical trial centers. Several techniques can be used such as ELISA (enzyme-linked immunosorbent assay), ELISPOT (enzyme-linked immunosorbent spot) assay, cytotoxicity assays, intracellular cytokine assays, and tetramer assays [27, 28], but only few of these are technically validated in the laboratory where they are used, and, importantly, there is no established standardization in the immunotherapy community. Therefore, there is still considerable variability of results between laboratories. Because of the lack of validated assays, it is important to secure sufficient amount of material and to follow some specific strategies to minimize the risk of nonreproducible results. Also, samples should be taken sequentially in order to investigate development over time, and a minimum of three assay time points should be used: Baseline and at least two follow-up time points. To prevent incorrect test results, a minimum of two assays should be applied. These assays should be able to prospectively define both the fraction of the immunized patients developing an immune response, and the level of the produced response for the population under study. Importantly, an immune response is identified if observed in at least two assays at two consecutive follow-up time points after the baseline assessment [25].

Alternatively, a conventional tumor response measure such as clinical activity could be used, but such a response, although superior in many ways, cannot always be expected in early proof of principle trials. Clinical activity represents improvement in clinical outcome such as tumor regression, progression-free survival, recurrence, and overall survival. However, many first-in-human studies are still performed in patients with end-stage disease due to safety and ethical considerations. In such patients, cancer vaccines are unlikely to demonstrate clinical efficacy, and importantly, as mentioned above, the failure to detect a clinical response at this stage in the clinical development program could lead to the termination of a potentially effective vaccine. In conclusion, assessment of biological activity should be included and used as a measure of success in every proof of principle trial.

11.6 EFFICACY TRIALS

Decision criteria to advance into clinical efficacy trials are based on the demonstration of proof of principle and that the strategy is safe. Efficacy trials aim to establish clinical benefit or the likelihood thereof and are in general randomized clinical trials intended to be a direct follow-up to the proof of principle trials. They can be designed as comparative randomized phase II trials, conventional phase III trials, or other setups yielding reliable clinical data demonstrating efficacy.

Depending on the clinical path of the product under study and findings from earlier trials, more than one efficacy trial may be needed. The concept of collecting efficacy data from more than one phase II trial allows for an early assessment of vaccine efficacy and a more rapid development of cancer vaccines. Phase III trials are large-size randomized clinical trials with definitive or surrogate clinical endpoints designed to demonstrate superior efficacy or noninferiority of the vaccine under investigation with the present standard care. They are statistically powered to deliver a final conclusion on potential clinical efficacy, and they often include the specification of secondary analysis for the full exploration of the vaccine's effect on the treated population.

11.7 CLINICAL ENDPOINTS IN EFFICACY TRIALS

Survival is the gold standard for products to demonstrate clinical benefit for the treatment of cancer. The problem is that applying this endpoint often requires long follow-up periods when testing patient groups with early-stage cancers, who are most likely to respond to a vaccine. Therefore, it might be useful to apply other endpoint measures reflecting clinical activity, including tumor regression, progression-free survival, and recurrence. Since delay in recurrence or progression of cancer naturally are expected to lead to improvements in survival, time-to-event endpoints such as disease-free survival and progression-free survival in the advanced and metastatic disease settings are accepted as surrogate markers for clinical benefit. When using time-to-event endpoints, it is of particular importance to select the appropriate follow-up time and disease stage when designing the trial. For example, cancer vaccines may fail to induce immediate tumor shrinkage yet still be effective in slowing the rate of progression over time.

Many other confounding factors need to be taken into consideration when designing efficacy trials. Among these it is important to acknowledge the effect of the standard treatment in the selected patient group and its impact on the immune system. Furthermore, it is a problem that progression of disease would normally mandate patients taken off the experimental treatment. This may result in a response rate that does not accurately reflect the efficacy of the vaccine under study.

Because of the long time before firm clinical effects of a given cancer vaccine can be evaluated, it would be advantageous with other measures that reflect disease activity. Biomarkers are such a factor that are objectively measured and evaluated as indicators of disease. It is currently not acceptable to base a product registration efficacy trial on nonvalidated biomarkers as surrogates for efficacy. Nevertheless, it can sometimes be the best second choice, and in those instances it is important to use meaningful biomarkers, which can be measured with established and reproducible tests. Cancer vaccines are currently expected to have the best clinical benefit in populations with minimal residual disease. Therefore, molecular markers reflecting minimal residual disease may function as a measure of biological and/or clinical activity in clinical trials. Notably, biomarker information should be incorporated into clinical trial designs wherever possible, to retrospectively analyze if particular biomarkers were associated with clinical benefit, yielding important information for future clinical trials. Frequent collection of patient sample material should therefore be considered, and informed consent of the patients to collect and store material should be obtained.

11.8 CHALLENGES IN VACCINE DEVELOPMENT

A number of major decision points in the development process should be highlighted: (1) defining the target tumor-associated antigens (product candidate), (2) initiation of nonclinical development, such as production and storage, (3) initiation of proof of principle trials, and (4) initiation of efficacy trials. Each decision to move forward involves a large increase in the required resources. For instance, several points are important to consider in relation to the manufacturing of the vaccine, including commercial implications and the scale at which the product should be produced. A clear plan for validation of the manufacturing process, the product characterization, and the release criteria tests should be developed. What is the product manufacturability at the desirable scale? Such issues need to be taken into consideration at an early time point before continuation of the development process and starting clinical trials. It will, furthermore, be important to ensure that the development program is in line with the regulatory requirements, and it is essential to have a close dialog with the relevant regulatory agencies throughout the development process. Importantly, such a dialog should be initiated early in the process to avoid unnecessary misunderstandings, with the beginning and finalization of trial activities being relevant times for interactions.

11.9 DEFINING THE TARGET TUMOR-ASSOCIATED ANTIGENS

Aberrantly glycosylated cell-surface-exposed carbohydrates and mucin antigens are attractive targets for immunotherapy. Aberrant glycosylation occurs in essentially all types of experimental and human cancers, and many results indicate that aberrant glycosylation is an outcome of initial oncogenic transformation, as well as a key event in induction of invasion and metastasis. In fact, several tumor-associated glycans have been shown to be involved in cancer development, metastasis, and the invasive process. The mechanisms behind such functions are believed to be due to the influence of glycans on specific cell–cell or cell–matrix interactions, as well as their role in cell-signaling processes important for tumor motility and growth as well as tumor-associated angiogenesis [29–40]. Several studies have identified a broad range of carbohydrate antigens residing on both glycolipids and glycoproteins as potential target antigens [29, 41–45]. Such carbohydrate antigens can be categorized in different groups dependent on the nature of their carrier: The mucin-associated epitopes Tn (GalNAcα-Ser/Thr), T (Galβ1-3GalNacα-Ser/Thr), and STn (NeuAcα2-6GalNacα-Ser/Thr) and glycosphingolipids, such as the gangliosides G_{M2}, G_{D2}, G_{D3}, and Fucosyl–G_{M1} (Fig. 11.2). Furthermore, an important category is the blood group antigens, which reside on either O-linked, N-linked, or glycosphingolipids-based glycans [46].

One of the most common types of aberrant glycosylation observed in human cancer is enhanced expression of the β1-6GlcNAc antenna in N-linked structures due to overexpression of β1,6 N-acetyl-glucosaminyltransferase (MGATV) [47]. Immunohistological studies examining the presence of the β1-6GlcNAc branch in N-linked glycans have demonstrated a correlation with metastatic potential of various tumors [48, 49]. However, the presence of the β1-6GlcNAc branch in many normal tissues makes it an unsuitable choice for targeting cancer, serving as an example for the importance of cancer-selective specific expression of the target antigen. In contrast, the expression of Tn and STn antigens caused by simplification of the O-linked structures has been demonstrated specifically on tumor cells [44]. Furthermore, several studies have demonstrated the overexpression of lacto series type 1 and 2 chains structures, often in the form of poly-LacNAc with a variety of fucosylation and sialylation. The expression of H/Le^y resulting from precursor accumulation is also a constant finding in several cancers by many independent studies [50–58].

11.9 DEFINING THE TARGET TUMOR-ASSOCIATED ANTIGENS

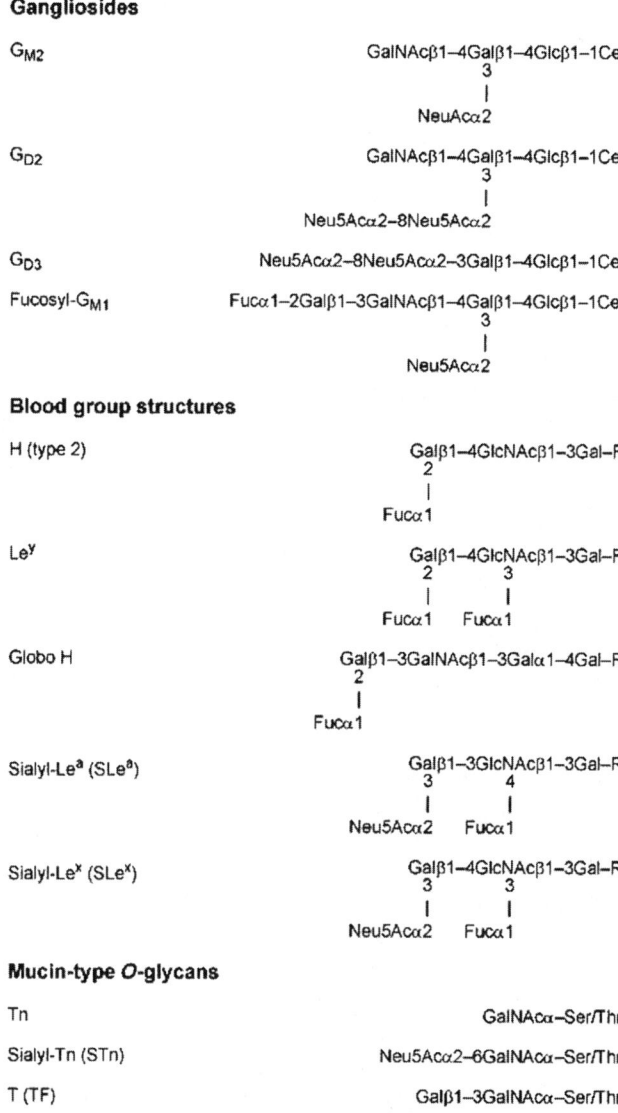

Figure 11.2 Tumor-associated carbohydrate antigens (TACAs).

In addition, several glycosphingolipids are differentially expressed on numerous carcinomas and have thus been considered important markers and targets in human cancers. One of the gangliosides, G_{D2}, is abundantly expressed on cell surfaces in melanomas, sarcomas, and neuroblastomas. An acetylated form of G_{D3} is expressed in melanomas, whereas G_{M2} is strongly expressed in several cancers of neuroectodermal origin. Another example is fucosyl-G_{M1}, which is expressed in an aggressive form of lung carcinoma and not in normal cells [59]. Also, one of the structures to be mentioned is Globo H, which is a complex carbohydrate structure expressed on ovary, prostate, colon, and renal cell carcinomas. Additionally, Lewis structures are expressed on a variety of epithelial tumors such as colon, lung, pancreas, and ovarian cancers [60]. Further refinement in targeted immunotherapy has been suggested by a growing number of studies demonstrating the importance of targeting cancer stem cells in cancer treatment [61]. Interestingly, several findings indicate that carbohydrate signatures might aid in the targeting of cancer stem cells in the future.

An association between the serum concentration of several cancer-associated carbohydrate antigens and tumor progression and metastasis has been seen. As an example the serum levels of SLe^x [62, 63] and SLe^a [64] correlate with tumor burden and prognosis in gastrointestinal, pancreatic, and prostate cancer, whereas the Tn and STn levels may have prognostic values for breast, ovarian, pancreatic, gastric, and biliary cancer patients [65–72]. These findings might have relevance for immunotherapy because they reflect that the target antigen is present on circulating molecules and not only on the target tumor.

Although such glycans are predominantly expressed by tumor cells, several studies have shown the existence of many of these antigens on normal cells, especially during differentiation and proliferation [41, 42]. The glycolipids G_{M2}, G_{D2}, and G_{D3} are expressed in neural tissues, while G_{D2} is found on a subpopulation of B lymphocytes in the spleen and in lymph nodes. G_{M2} is expressed at the secretory borders of most epithelial tissues and G_{D2} and G_{D3} are found at low levels in connective tissues. Globo H, Le^y, SLe^x, Le^x, T, Tn, and STn are expressed of a variety of epithelial tissues. In addition SLe^x is found on polymorphonuclear (PMN) leucocytes. These findings have raised the possibility that targeting a number of such carbohydrate antigens could induce autoimmune responses when used in active and passive immunotherapy. This is exemplified by the sporadic induction of vitiligo in some trials against melanoma cancer [73]. At the same time, the presence of tumor-associated structures on normal cells also raises the opposite concern that it will be difficult to break tolerance and induce a sufficient immune response to eradicate the cancer.

Nevertheless, immunotherapy either generating or using mAbs against G_{M2}, G_{D2}, G_{D3}, Globo H, TF, Tn, and STn have not induced side effects of an autoimmune nature (all trials described below). Also, as we shall see, several examples exist demonstrating that carbohydrate vaccines induce high titers of IgM and IgG antibodies, as well as CDC of tumor cells in vitro, and that carbohydrate-directed antibodies as well as carbohydrate-based antigens induce protection from tumor challenge in mice and prolong the disease-free interval. Therefore, cell surface antigens have shown their value as suitable targets for immune attack against cancer.

11.10 PRODUCTION AND STORAGE ISSUES

Besides the obvious requirement for biological efficacy of cancer vaccine candidates, production and stability issues are also important to consider before pursuing clinical trials. As with all material to be used in clinical practice, it is important to have a clean and homogeneous pharmaceutical in order to consistently generate the desired clinical effect without harming the patients. Therefore, for any production of medical products, standardization and characterization of the pharmaceutical is important. As a first critical step in product standardization, it is essential to have control over the vaccine component materials, including derivation information. It will be advantageous to avoid undesirable, unstable, or poorly characterized raw materials, cell lines, or process intermediates. Also, the characterization of purity and potency is important in the product characterization, which can be particularly challenging in the context of vaccines. Establishment of potency assays is especially difficult for cancer vaccines due to the lack of established biological surrogates of efficacy and/or appropriate in vivo models. Furthermore, cost and ability for the production to meet the required scale are important issues to be addressed.

Such production considerations are of particular importance when addressing carbohydrate-related drugs. One possibility is to purify the cancer-associated carbohydrate structures from natural sources such as tumor cell lines. Due to the microheterogeneity inherent to the in vivo glycosylation process in cancer cell lines this is difficult. Also, it can be troublesome to generate sufficient material by such methods. Therefore, there has been a great interest in producing homogenous and clean cancer vaccine candidates by other methods [8].

One major method to tackle such problems is obviously chemical synthesis, with its many advantages in providing reliable and clean products. Automated chemical synthesis of carbohydrates is, however, particularly difficult because of the complexity of the carbohydrate structures. This mandates multiple synthetic steps to generate the product of interest and often with low production yields. Recently, there has been immense development of alternative methods to generate the desired carbohydrate material for various clinical applications, such as chemoenzymatic synthesis and bioengineering. Several chemoenzymatic strategies have lately been used to generate compounds tested in clinical practice. The advantages of chemical production, in terms of purity and homogeneity, combined with the high production yield and low cost of enzymatic synthesis, have proven efficacious. Chemoenzymatic strategies have been used to generate several glycoderivatives including G_{M3} [74, 75], Tn, STn, and ST [74, 76–79], as well as a series of more complex structures. Bioengineering is another option for the production of protein tumor antigens in cancer cell lines or for the production of tumor-associated glycans in bacteria. Production of cancer-associated O-glycoproteins has been performed in several cases [80–84]. Furthermore, novel strategies seek to overcome the need for large amounts of recombinant enzymes and activated donor monosaccharides in the production of tumor-associated glycans, by genetically manipulating biosynthetic pathways of microorganisms. Large-scale production of G_{M2}, G_{M1}, G_{D3}, and sialyllactose (NeuAcα2-3Galβ1-4Glc) in bacteria has been reported [85–88].

11.11 CLINICAL TRIALS

The relevance of several of the mentioned tumor-associated carbohydrate antigens (TACAs) has been addressed in clinical investigations (Table 11.1), the majority of

Table 11.1 Clinical Trials with Cancer-Associated Carbohydrate Antigens

Target(s)	Vaccine	Reference	Cancer Type
Globo H	Globo H–KLH	92	Prostate
		94	Breast
G_{M2}	G_{M2}	95	Melanoma
		96	
		97	
G_{M2}	G_{M2}–KLH	98	Melanoma
		99	
		100	
		101	
G_{D2}	G_{D2}–KLH	102	Melanoma
G_{D2}	G_{D2} lactone–KLH	103	Melanoma
G_{D3}	G_{D3}–KLH	104	Melanoma
G_{M3}	Ganglioside proteoliposomes	73	Melanoma
G_{M2}, G_{D2},	Monoclonal antibodies	112	
G_{D3}, G_{M3}		113	
		114	
		115	
		116	
Fuc–G_{M1}	Fuc–G_{M1}–KLH	120	Lung
		121	
Tn	Tn/T from O RBC membranes	122	Breast
Tn, STn	Partially desialylated OSM	123	Colorectal
T	T–KLH	124	Ovarian
T	T–KLH	125	Colorectal
STn	STn–KLH	126	Breast
		127	Breast
		128	Breast
		129	Breast
		130	Breast
			Ovarian
			Colorectal
		131	Breast
STn	STn–KLH	132	Ovarian
	High-dose chemotherapy		Breast
	Stem cell rescue		
		133	Ovarian
			Breast
Le^y	Le^y–KLH	134	Ovarian
STn	STn(c)–KLH	135	Breast
		136	Breast
Tn	Tn(c)–KLH	137	Prostate
	Tn(c)–PAM		
T	T(c)–KLH	138	Prostate
PolySA	PolySA–KLH	139	Small cell lung
	NP–polySA–KLH		

(Continued)

Table 11.1 *Continued*

Target(s)	Vaccine	Reference	Cancer Type
Tn-MUC2	Tn–MUC-2–KLH	138	Prostate
Globo H	Globo H–KLH		
G_{M2}, Globo H, Le^y, T, Tn, STn Tn–MUC-1	G_{M2}–KLH Globo H–KLH Le^y–KLH T(c)–KLH Tn(c)–KLH STn(c)–KLH Tn–MUC-1–KLH	140	Ovarian (epithelial) Fallopian tube Peritoneal
MUC1	Mannan-MUC-1	141	Breast Colon Stomach Rectal
		142	Breast
MUC1	Mannan-MUC-1 pulsed DCs	143	Adenocarcinomas

which are phase I trials, addressing safety and dose needed to induce an immunological response. A few targets have been further investigated in efficacy trials, most of these in minor randomized phase II trials, while only a few have been taken into phase III. Another aspect of these trials has been testing of several adjuvants as well as coupling requirements and methodologies. In the following we will go through trials conducted with glycosphingolipid, peptide, or protein carriers.

11.11.1 Glycosphingolipid-Based Vaccines

Some of the first cancer vaccines involved whole-cell vaccines derived from irradiated autologous or allogeneic cancer cell or cell lysate vaccines [89, 90]. Immunized patients developed humoral responses, of which many were directed to irrelevant antigens, but some were directed to surface markers on the cancer cells including G_{M2} and G_{D2}. The response was moderate and consisted of IgM antibodies. Not only was the immune response weak, but it was also difficult to prepare and administer the vaccine. However, from these initial experiments an interest in using purified surface-associated glycosphingolipids in the generation of specific antitumor responses has developed.

Globo H   Globo H is selectively expressed on breast, ovarian, prostate, colon, and renal cell carcinomas. Because there is no available natural source of Globo H, it has been chemically synthesized both as the glycoside and attached to ceramide. In preclinical murine studies Globo H conjugated to keyhole limpet hemocyanin (KLH) induced high IgM and IgG titers [91]. KLH is a potent immunoactivating carrier protein purified from mollusks. The vaccine has been tested in patients with relapsed prostate cancer after primary therapies such as radiotherapy or surgery [92], initially in a safety and dose escalation study in cohorts receiving 3, 10, 30, and 100 μg of Globo H. As with most of the carbohydrate vaccines tested to date, the vaccine was safe and no significant toxicity was detected. A potent immune response was seen, most prominently with the 30-μg dose, with high and specific IgM and IgG titers against Globo H as tested with ELISA. Affinity-purified antibodies were of IgG_1 and IgG_4 subclasses [93]. These trials also included an example of the use of surrogate markers as a potential indication of clinical efficacy in early trials. Using prostate-specific antigen (PSA) as a surrogate marker for clinical efficacy, it was interestingly suggested that the vaccine had a clinical effect, based on the finding that the level of PSA was stabilized or declined in a third of the patients when compared to pretreatment PSA levels [92].

The Globo H–KLH vaccines have also been tested in breast cancer patients [94]. In this trial two different synthetic Globo H–KLH conjugates were used: Conventional Globo H–KLH vaccine directly conjugated to KLH and Globo H–KLH generated by using the bi-functional crosslinker 4-(4-N-maleimidomethyl)cyclohexane-1-carboxyl hydrazide (MMCCH). Both vaccines were well tolerated, and no definite differences were observed among the two formulations. Serologic analyses demonstrated the generation of specific IgM antibody titers in most patients, and only minimal IgG antibody stimulation was seen [94].

On the basis of these data, Globo H has been suggested to constitute one component of a polyvalent vaccine as will be described below.

G_{M2}, G_{D2}, and G_{D3} G_{M2}, G_{D2}, and G_{D3} are often overexpressed on malignant melanomas, astrocytomas, neuroblastomas, and sarcomas. Initial human experiments focused on the immunization of melanoma patients with nonconjugated G_{M2} in combination with BCG, and pretreatment with low doses of cyclophosphamide (CX) were found to induce significant IgM antibody responses in most patients, and IgG antibodies in some [95, 96]. Based on these early proof of principle or phase I trials, a double-blinded randomized phase II trial was therefore conducted on 122 patients with stage III melanoma to test the clinical efficacy of the vaccine [97]. Patients who were free of disease after surgery were randomized to receive treatment with the G_{M2}/BCG vaccine or with BCG alone. All patients were pretreated with low-dose CX in order to prime the immune system. The study verified that the vaccine induced anti-G_{M2} IgM antibodies in the majority of the patients treated with G_{M2}/BCG, and interestingly showed a highly significant increase in the disease-free interval and a 17% increase in overall survival in patients, when only looking at the patients that were antibody positive. However, the size of the trial and the effect of the vaccine were not large enough to demonstrate a statistically significant improvement in disease-free interval or survival for patients treated with G_{M2}/BCG vaccines.

In order to increase immunogenicity, covalent conjugation to KLH was performed and the G_{M2}–KLH conjugate given again to melanoma patients. High titers of both IgG and IgM were noted [98]. Next, different adjuvants (BCG, Detox, or QS-21) were evaluated, with QS-21 being the most effective to induce high and longer lasting titers of an IgM and IgG response [99, 100]. As for previous vaccines no serious toxicity was observed. Based on these findings many of the following studies with other conjugate vaccines were performed with gangliosides conjugated to KLH and used with QS-21. Also, dose escalation studies have been performed with G_{M2}–KLH and QS-21 defining the lowest dose that induced a consistent, high titer IgM and IgG response against G_{M2} [101].

The gangliosides G_{D2}/G_{D3} are also abundantly expressed in not only melanomas but also sarcomas and neuroblastomas, and preclinical models have demonstrated the ability of a G_{D2}–KLH conjugate vaccine to induce antibodies that eliminate micrometastases [102]. One problem has been that the detected antibody levels have not been consistent. In order to improve the immunogenicity of the vaccine, an alternative G_{D2} lactone–KLH (G_{D2}L–KLH) has been synthesized and tested in a clinical trial. First, dose escalation studies were conducted in melanoma cancer patients, with 3 cohorts of 6 patients receiving 3, 10, or 30 μg G_{D2}L–KLH plus immunological adjuvant QS-21 [103]. The vaccine was well tolerated, and the majority of patients in all three dosing groups produced anti-G_{D2} antibodies detectable by ELISA, with no statistical differences between the titers of the different groups, but the patients receiving the 30-μg dose level had a longer lasting immune response. Furthermore, the most potent CDC was induced in patients receiving 30 μg of G_{D2}L–KLH. In conclusion the introduction of a lactone dramatically improved the G_{D2}–KLH strategy, and based on these findings G_{D2}L–KLH was an obvious choice to be incorporated into a more complex multivalent vaccine, as will briefly be discussed below.

Another approach to augment the immune response to a certain target is using anti-idiotypic antibodies. This strategy has been applied by Chapman and co-workers immunizing with a combination of G_{D3}, the anti-idiotypic antibody BEC2, which mimicks G_{D3}, and the adjuvant QS-21 to induce anti-G_{D3} ganglioside antibodies in melanoma patients [104, 105]. The results indicate that mainly a humoral immune response was found, and, although the trial was not designed to address the efficacy of the vaccine, there was no difference in clinical outcome between the group of immunized patients and the nonimmunized group. Thus, there is currently no conclusion on the clinical efficacy of this strategy.

G_{M3} Like many of the other gangliosides, G_{M3} is highly expressed on human malignant melanoma [106], and a clear antibody response has been seen after immunization with the ganglioside in mice [107]. However, a major problem in the design of ganglioside-based cancer vaccines is to overcome natural tolerance. Solutions include alternative coupling methods as well as modifications of the antigen. Other solutions are the use of different presentation methods, such as inserting the antigen in liposomes, in which different danger signals are included. An interesting study has shown that when gangliosides are incorporated into very small proteoliposomes together with outer membrane protein complexes from *N. meningitidis*, a strong immune response was induced in mice, chickens, and monkeys. Production of both IgM and IgG responses toward the gangliosides G_{D3} and G_{M3}, was induced, and included T-cell-dependent isotype switch to IgG_1, IgG_{2a}, and particularly IgG_{2b} [108], and the vaccine was proven to mediate antitumor protection in mouse melanoma models [109]. Given the change in the methodology and the introduction of membrane proteins from *N. meningitidis*, a 12-month toxicity study was performed in primates (monkeys) and proved that the vaccine was safe and only elicited transitory irritation at the injection site without systemic toxicities [110]. Next, a phase I study was performed in 51 patients with metastatic melanoma. Sporadic clinical effect on tumor mass and disease stabilization was seen in two patients, although clinical efficacy was not the primary aim of the study [73]. Importantly, vitiligo was noted in one of the patient with partial clinical effect highlighting the potential risk of inducing autoimmune problems in cancer immunotherapy.

Other methods to surmount the immunotolerance to G_{M3} have also been tested. In order to see if changes in the antigen itself could overcome the problem, a series of G_{M3} derivatives have been tested in murine models. While G_{M3}–KLH elicited low levels of immune response, the KLH conjugates of N-propionyl, N-butanoyl, N-iso-butanoyl, and N-phenylacetyl G_{M3} induced robust immune reactions with antibodies of multiple isotypes, indicating significantly improved and T-cell-dependent immune responses that lead to isotype switching and affinity maturation. These derivatives are currently awaiting further clinical development [111].

As an alternative to using active vaccination protocols, passive vaccination has also been used in the context of antiganglioside antibodies, and mAbs against G_{M2}, G_{D2}, G_{D3}, and G_{M3} have resulted in clinical responses in a proportion of treated patients [112–116].

Fucosyl–G_{M1} Fucosyl–G_{M1} [Fuc–G_{M1}, Fucα1–2Galβ1-3GalNAcβ1-4(NeuAc-α2-3)Galβ1-4Glcβ1-1Cer] is present on most small-cell lung cancer (SCLC) cells [91, 117, 118], and has been used as target for directed therapies against residual SCLC after chemotherapy and irradiation. Fuc–G_{M1} conjugated to KLH together with the immunological adjuvant QS-21 induced a strong immune response in mice [119]. Phase I trials have been conducted in SCLC patients, who have been immunized after initial chemotherapy with either bovine-derived or synthetic Fuc–G_{M1} conjugated to KLH mixed with QS-21 [120, 121]. Like with most of the other ganglioside vaccines, a clear serological IgM response was seen in both trials. In contrast, the IgG response was primarily seen in patients immunized with the bovine-derived Fuc–G_{M1} vaccine and not the synthetic vaccine. The reason for this discrepancy is unclear, but an interesting observation is that while the ceramide part of Fuc–G_{M1} is part of the bovine-derived antigen, this is not the case with the synthetic analog, suggesting that the proximal part of the ceramide chain plays a role in generating antibody maturation [121]. In both trials the induced antibodies reacted with tumor cells and mediated CDC. Dose escalation studies using 1, 10, or 30 µg of antigen furthermore demonstrated that 30 µg of antigen yielded an immune response in 5 out of 6 individuals, while 10 µg induced response in 3/5, emphasizing the relevance of performing dose escalation studies before addressing potential clinical efficacy.

However, the evaluation of immune response is often dependent on in vitro methodologies, which do not necessarily have the sufficient sensitivity to detect the response in all patients. Furthermore, the immune response measured might not be a biologically relevant readout for predicting clinical efficacy, further complicating the evaluation and relevance of dose escalation studies.

11.11.2 *O*-glycan-Based Vaccines

Tumor-associated carbohydrate antigens may also be expressed on membrane-associated mucin-type *O*-glycoproteins in epithelial tumors. Attempts to augment immunogenicity to tumor cells carrying these antigens have been pursued by vaccinating patients with carbohydrate-based vaccines presenting TACAs on their natural backbones.

Vaccines Based on Biological Sources Georg Springer and colleagues pioneered the field by immunizing breast cancer patients with a Tn/T vaccine derived from O red blood cell (RBC) membranes plus calcium phosphate and traces of typhoid vaccine (containing polysaccharide phosphoglycolipid A hyperantigen) as adjuvants [122]. More than 23 years follow-up on 32 breast cancer patients vaccinated against recurrence has been reported, demonstrating a potential effect of the vaccine. However, the study did not include control patients, and therefore the clinical effect reported was assessed by statistical comparison to national survival rates reported by the U.S. National Cancer Institute (NCI).

Another related strategy has been applied in 20 patients with Duke B, C, and D colorectal cancer, where the patients were immunized with 200 µg partially desialylated ovine submaxillary gland mucin (dOSM) containing both Tn and STn epitopes [123]. This was done either without adjuvant or in combination with Detox or BCG. Patients that received the vaccine without adjuvant did not develop any increase in antibody titers, whereas patients receiving the vaccines with Detox or BCG developed high IgM and low IgG titers. The trial demonstrated that such antigens are immunogenic in humans and can be administered safely with these adjuvants.

Synthetic Vaccines One of the challenges encountered when going from preclinical to clinical trials is the requirement for a consistent vaccine product. Since there can be high batch variation of vaccines based on biological sources, attempts to develop chemically defined synthetic vaccines have been pursued. Nevertheless, most of the synthetic *O*-glycan-based vaccines produced until today are still based on KLH.

Monovalent Glycoconjugates A number of vaccines have been based on single TACAs chemically coupled to a carrier protein such as KLH. In a phase I study, 10 ovarian cancer patients were immunized with 100 or 500 µg synthetic T antigen coupled to KLH plus Detox [124]. In 9 of 10 patients a significant increase in IgM titer was noted, but also IgG and IgA responses were recorded in these patients. The study also reported an increase in cytotoxic antibodies against tumor cells expressing the T antigen. Noteworthy, the lower dose vaccine elicited equal or stronger humoral responses compared to the high dose. The study demonstrated that KLH is a safe carrier for carbohydrate haptens in humans.

T–KLH and STn–KLH have been compared as vaccines against recurrence in colorectal cancer patients [125]. Forty-four patients free from disease after surgical resection received 100 µg of the vaccine alone or in combination with Detox or

QS-21. All patients receiving T–KLH or STn–KLH plus adjuvant, in particular QS-21, generated high levels of IgM and IgG antibodies against the synthetic carbohydrate haptens, but these antibodies did not react with the corresponding natural disaccharide epitopes as they are expressed by human cells.

For more than a decade clinical trials up to phase III with the vaccine Theratope (STn–KLH) enrolling more than 1200 patients were carried out in breast, ovarian, and colorectal cancer patients [126–133]. In most of the trials, patients received 100 μg STn–KLH in combination with Detox, and the majority of these patients generated IgM and IgG antibodies specific for the synthetic STn hapten as well as CDC antibodies. Patients pretreated with cyclophosphamide, which is important for modulation of the immune response possibly by inhibiting activity of $CD4^+CD25^+$ regulatory T cells, generated significantly higher levels of IgM antibodies [129]. Some trials have combined the vaccine with high-dose chemotherapy and autologous/syngeneic stem cell rescue [132, 133]. In a fraction of these patients anti-STn Th1 antigen-specific T-cell responses were demonstrated. However, any statistically significant clinical effect has never been demonstrated, and the clinical trials have now been terminated.

The pentasaccharide Le^y is highly expressed on the majority of serous and endometroid carcinomas and has been targeted in a phase I trial in ovarian cancer patients with a Le^y–KLH vaccine [134]. Twenty-five patients were immunized with 3, 10, 30, or 60 μg Le^y–KLH together with QS-21 adjuvant. This trial demonstrated that low titers of predominantly IgM antibodies could safely be induced by this vaccine. Notably, the two higher doses induced the most consistent responses, but the clinical effect of these responses was not evaluated.

Clustered Glycoconjugates Since cancer-associated carbohydrate antigens are likely to be presented in clusters at the surface of cancer cells, vaccines mimicking this pattern have been developed by chemically combining single carbohydrate antigens in clusters. The first TACA to be combined in such a pattern and tested in a clinical setting was the STn antigen [135]. Five out of 9 breast cancer patients vaccinated with synthetic clusters (c) of 3 STn epitopes conjugated to KLH developed high IgM titers against STn(c) and 4 patients developed IgG antibodies against STn(c). In contrast, only low titers were developed against the unclustered STn antigen, suggesting an importance of clustered antigen presentation [135]. The 9 patients described were part of a clinical trial including 27 patients in which antibody reactivity with human cell lines was also studied [136]. Twenty-one and 13 patients developed IgM antibodies reactive with cancer cell lines (LSC) and MCF-7, respectively, whereas no IgG reactivity was observed. Furthermore, the short O-linked glycan Tn has been used in a clinical trial comparing KLH and PAM (palmitic acid) as carrier proteins [137]. Twenty-five prostate cancer patients were immunized with 3–15 μg Tn(c)–KLH or 100 μg Tn(c)–PAM plus QS-21. The KLH-conjugated vaccine was superior to the PAM-conjugated vaccine at safely inducing humoral responses (IgM and IgG), especially at the lowest doses. In a similar way, a clustered glycoconjugate vaccine using T as antigen was tested in a trial. Twenty patients with biochemically relapsed prostate cancer were immunized with 1, 3, 10, or 30 μg T(c)–KLH together with QS-21 [138]. As reported earlier, the highest titers of IgM/IgG antibodies were generated in the patients receiving the lowest doses. However, the generated IgG antibodies did not react with T epitopes as they are expressed on the surface of tumor cells.

Polysialic acid is abundantly expressed in SCLC but is subject to immunological tolerance since it is expressed in the embryo and to limited extent in adult neural tissues. In an attempt to elicit an immune response against this structure, a vaccine based on an N-propionylated polysialic acid vaccine (NP–polySA–KLH) was compared to an unmodified vaccine (polySA–KLH) [139]. The study was carried out in 11 SCLC patients receiving 30 μg vaccine with QS-21. In contrast to the unmodified polySA, NP–polySA–KLH effectively induced IgM antibodies, which cross-reacted with unmodified polySA. However, the IgG antibodies generated to NP–polySA did not cross-react with polySA, indicating that this vaccine will not target SCLC in vivo.

Multivalent Vaccines The rationale behind multivalent vaccines is targeting of more TACAs with a single vaccine. This concept has been tested in 10 prostate cancer patients using a bivalent vaccine containing 5 μg Tn–MUC-2 (mucin-2)–KLH and 10 μg Globo H–KLH [138]. While keeping the vaccine dose fixed, two different adjuvants, GPI-0100 and QS-21, were compared at different doses. At the highest doses GPI-0100 safely induced comparable antibody titers to those induced by QS-21. The bivalent conjugate vaccine induced both IgM and IgG antibodies to Tn–MUC-2, whereas only IgM antibodies against Globo H were generated.

In a similar clinical trial enrolling 11 patients with epithelial ovarian, fallopian tube, or peritoneal cancer, a heptavalent vaccine was tested [140]. The vaccine contained 10 μg G_{M2}, 10 μg Globo H, 10 μg Le^y, 3 μg T(c), 3 μg Tn(c), 3 μg STn(c), and 3 μg Tn–MUC-1 (mucin-1), all individually conjugated to KLH and administered with QS-21. All antigens induced IgM responses with Le^y and G_{M2} being the least immunogenic. In contrast, the only antigen that induced IgG antibodies was Tn–MUC-1.

Recombinant Peptide Glycoconjugates The mannan-MUC-1 vaccine aims to target mannose receptors on APCs with mannan (polymer of mannose) and MUC-1 overexpressed at cell surfaces in many adenocarcinomas. So far, two clinical trials have been conducted. Without use of adjuvant and with doses ranging from 10 to 500 μg, 25 patients with metastatic cancers in breast, colon, stomach, and rectum were immunized with an oxidized mannan-MUC-1 fusion protein (M-FP) [141]. In 12 out of 25 patients anti-MUC-1 IgG_1 antibodies were generated, whereas only weak IgM reactivity and no IgA reactivity were observed. In contrast to what has been reported for many of the vaccines based exclusively on carbohydrate haptens, the level of antibodies positively correlated with immunization dose. Preclinical studies have demonstrated differences in cellular responses generated with oxidized M-FP versus reduced M-FP [24]. Where reduced M-FP tends to generate Th2 responses, oxidized M-FP tends to generate Th1 responses in mice. Nevertheless, the restricted IgG_1 response reported in this clinical study is indicative of a Th2 response. Proliferative cellular responses occurred in 4 out of 15 patients tested, whereas no antitumor activity could be demonstrated.

The clinical trial described above included advanced cancer patients. In a pilot phase II trial 16 early-stage breast cancer patients were treated with oxidized M-FP, and 15 controls were treated with placebo [142]. Nine of 13 patients vaccinated with M-FP generated IgG antibodies against MUC-1 (seroconverted from IgM antibodies), and 4 of 10 M-FP-treated patients developed MUC-1-specific T-cell responses. No humoral or cellular anti-MUC-1 responses were detected in the control patients receiving placebo. Of the 15 controls 4 had recurrence after 5 years, whereas none of the vaccinated patients had recurrence.

Cell-Based Vaccines Dendritic cells (DCs) are professional APCs that under normal conditions capture, process, and present foreign antigens to T and B cells of the immune system. Since TACAs are self-antigens that are also expressed in normal tissues to some degree, one of the challenges in the development of cancer vaccines is how to break tolerance. Vaccination with autologous monocyte-derived DCs pulsed ex vivo with mannan-MUC-1 fusion protein (M-FP) is one way of "teaching" dendritic cells to target MUC-1-expressing tumor cells. Uptake and subsequent processing of M-FP is facilitated by targeting mannose receptors on DCs via mannan. This strategy has been undertaken in a phase I study including 10 patients with MUC-1-positive adenocarcinomas, where the patients underwent leukapheresis and reinjection at monthly intervals [143]. Strong Th1 responses were generated in the 9 patients who completed the treatments, and stabilization of disease was observed in 2 patients. An alternative way of targeting MUC-1 with DC-based vaccines is transduction of DCs with recombinant viruses encoding MUC-1. However, limited control with glycan structures on the processed peptide fragments presented on MHC class I molecules is available with this strategy. Nevertheless, this may be of high importance when the aim is to target cancer cells expressing a self-protein carrying TACAs, but not healthy cells carrying the same protein with a normal glycosylation pattern. Some studies indicate that the choice of vector is crucial for the glycosylation status of the presented antigens. As an example, MUC-1 was expressed with fully processed *O*-glycans when DCs were transduced with a MUC-1/IL-2 encoding modified vaccinia Ankara (MVA) vector [144]. In contrast, by using a fiber-modified adenoviral vector (rAd5F35) as vehicle, MUC-1 was expressed with the TACAs Tn and T [145].

11.12 CONCLUSIONS

Cancer immunotherapy represents a promising advancement in the care of cancer patients and offers the potential advantage of controlling systemic disease using the body's own innate ability to destroy unwanted cells, with minimal toxicity. However, immunological responses to cancer vaccines have rarely been robust enough to achieve measurable clinical responses or disease regression, why strategies to optimize the activity of cancer vaccines are necessary. Carbohydrate-based cancer vaccines have the potential to overcome this problem, and several attempts have been made to produce therapeutic cancer vaccines that specifically target cancer-associated aberrant carbohydrate structures. The majority of clinical cancer vaccine trials have been conducted in patients with widespread disease. These patients are often immunocompromised, and the burden of tumor cells outmatches the magnitude of the vaccine-induced immune response. For these reasons, it is expected that cancer vaccines are less efficient to effect bulk disease, but more likely to prevent the growth of small quantities of cancer cells, thereby being ideal for minimal residual disease.

Therapeutic carbohydrate-based cancer vaccines can be either active or passive vaccines. While passive vaccines function by infusion of typically humanized antibodies targeting cancer-specific structures, active vaccines generate an immune response by immunization with cancer-associated antigens. Some active cancer vaccines tend to elicit humoral responses, whereas others generate cellular responses. The general opinion is that the more components of the immune system that are involved in the anticancer immune response the better. Sustainability of the immune response is another important factor to be considered.

As we point out, there are many ways of evaluating the effect of cancer vaccines. In contrast to conventional chemotherapeutics, cancer vaccines are presumed to be much safer, changing the scope of early trials to monitor both safety and the development of an immunological recognition of the target antigen. Because, active cancer vaccines are more suited to induce long-term effects such as disease stabilization and survival improvements, very long follow-up times are required to conclude on the clinical efficacy. Therefore, the introduction of a new paradigm of clinical development of therapeutic cancer vaccines has been suggested, with two individual phases: *proof of principle trials* and *efficacy trials*. While proof of principle trials are exploratory trials, efficacy trials are focused on clinical efficacy. Such a change in paradigm is believed to secure a more flexible, expeditious, and focused clinical development process.

Several carbohydrate-based active and passive vaccines targeting glycosphingolipids as well as mucin-associated carbohydrate antigens have been tested in the clinic with mixed results. In general, humoral responses have been generated, while cellular immune responses are less frequently observed. When it comes to clinical efficacy data, only very few trials have shown promising results; however, most trials have been conducted in terminal cancer patients. Hopefully, the future will demonstrate clinical efficacy among the new and interesting carbohydrate-based vaccines currently under development.

ACKNOWLEDGMENT

This work was supported by the Danish Medical Research Council, The Lundbeck Foundation, and The Program of Excellence, Copenhagen Center of Glycomics, Faculty of Health Sciences, University of Copenhagen.

REFERENCES

1. Coley WB (1893) A preliminary note on the treatment of inoperable sarcoma by the toxic product of erysipelas. *Post-graduate* 8:278–286.
2. Coley WB (1926) The Cancer Symposium at Lake Mohonk. *Am. J. Surg. (New Series)* 1:222–225.

3. Nauts HC (1977) Bacterial vaccine therapy of cancer. *Dev. Biol. Stand.* 38487–38494.
4. Nauts HC (1982) Bacterial pyrogens: Beneficial effects on cancer patients. *Prog. Clin. Biol. Res.* 107:687–696.
5. Nauts HC (1989) Bacteria and cancer—antagonisms and benefits. *Cancer Surv.* 8:713–723.
6. Perabo FG, Willert PL, Wirger A, Schmidt DH, Wardelmann E, Sitzia M, von RA, Mueller SC (2005) Preclinical evaluation of superantigen (staphylococcal enterotoxin B) in the intravesical immunotherapy of superficial bladder cancer. *Int. J. Cancer* 115:591–598.
7. Hakomori S (1985) Aberrant glycosylation in cancer cell membranes as focused on glycolipids: Overview and perspectives. *Cancer Res.* 45:2405–2414.
8. Freire T, Bay S, Vichier-Guerre S, Lo-Man R, Leclerc C (2006) Carbohydrate antigens: Synthesis aspects and immunological applications in cancer. *Mini. Rev. Med. Chem.* 6:1357–1373.
9. Slovin SF, Keding SJ, Ragupathi G (2005) Carbohydrate vaccines as immunotherapy for cancer. *Immunol. Cell. Biol.* 83:418–428.
10. Banchereau J, Steinman RM (1998) Dendritic cells and the control of immunity. *Nature* 392:245–252.
11. Sallusto F, Lanzavecchia A (2001) Exploring pathways for memory T cell generation. *J. Clin. Invest.* 108:805–806.
12. Ahlers JD, Belyakov IM, Terabe M, Koka R, Donaldson DD, Thomas EK, Berzofsky JA (2002) A push-pull approach to maximize vaccine efficacy: Abrogating suppression with an IL-13 inhibitor while augmenting help with granulocyte/macrophage colony-stimulating factor and CD40L. *Proc. Natl. Acad. Sci. U.S.A.* 99:13020–13025.
13. Zimmermann S, Egeter O, Hausmann S, Lipford GB, Rocken M, Wagner H, Heeg K (1998) CpG oligodeoxynucleotides trigger protective and curative Th1 responses in lethal murine leishmaniasis. *J. Immunol.* 160:3627–3630.
14. Hoebe K, Janssen E, Beutler B (2004) The interface between innate and adaptive immunity. *Nat. Immunol.* 5:971–974.
15. Medzhitov R (2001) CpG DNA: Security code for host defense. *Nat. Immunol.* 2:15–16.
16. Hemmi H, Takeuchi O, Kawai T, Kaisho T, Sato S, Sanjo H, Matsumoto M, Hoshino K, Wagner H, Takeda K, Akira S (2000) A Toll-like receptor recognizes bacterial DNA. *Nature* 408:740–745.
17. Ostrand-Rosenberg S (2005) CD4+ T lymphocytes: A critical component of antitumor immunity. *Cancer Invest.* 23:413–419.
18. van den Broek ME, Kagi D, Ossendorp F, Toes R, Vamvakas S, Lutz WK, Melief CJ, Zinkernagel RM, Hengartner H (1996) Decreased tumor surveillance in perforin-deficient mice. *J. Exp. Med.* 184:1781–1790.
19. Russell JH, Ley TJ (2002) Lymphocyte-mediated cytotoxicity. *Annu. Rev. Immunol.* 20:323–370.
20. Ragupathi G, Liu NX, Musselli C, Powell S, Lloyd K, Livingston PO (2005) Antibodies against tumor cell glycolipids and proteins, but not mucins, mediate complement-dependent cytotoxicity. *J. Immunol.* 174:5706–5712.
21. Wu DY, Segal NH, Sidobre S, Kronenberg M, Chapman PB (2003) Cross-presentation of disialoganglioside GD3 to natural killer T cells. *J. Exp. Med.* 198:173–181.

22. Trombetta ES, Mellman I (2005) Cell biology of antigen processing in vitro and in vivo. *Annu. Rev. Immunol.* 23:975–1028.
23. Knutson KL, Disis ML (2005) Tumor antigen-specific T helper cells in cancer immunity and immunotherapy. *Cancer Immunol. Immunother.* 54:721–728.
24. Apostolopoulos V, Pietersz GA, Loveland BE, Sandrin MS, McKenzie IF (1995) Oxidative/reductive conjugation of mannan to antigen selects for T1 or T2 immune responses. *Proc. Natl. Acad. Sci. U.S.A.* 92:10128–10132.
25. Hoos A, Parmiani G, Hege K, Sznol M, Loibner H, Eggermont A, Urba W, Blumenstein B, Sacks N, Keilholz U, Nichol G (2007) A clinical development paradigm for cancer vaccines and related biologics. *J. Immunother.* 30:1–15.
26. Petroni GR, Disis ML (2008) *Design Issues for Early-Stage Clinical Trials for Cancer Vaccines "Immunotherapy of Cancer."* Humana Press, Inc., Totowa, NJ, pp. 479–485.
27. Hobeika AC, Morse MA, Osada T, Ghanayem M, Niedzwiecki D, Barrier R, Lyerly HK, Clay TM (2005) Enumerating antigen-specific T-cell responses in peripheral blood: A comparison of peptide MHC Tetramer, ELISpot, and intracellular cytokine analysis. *J. Immunother.* 28:63–72.
28. Keilholz U, Weber J, Finke JH, Gabrilovich DI, Kast WM, Disis ML, Kirkwood JM, Scheibenbogen C, Schlom J, Maino VC, Lyerly HK, Lee PP, Storkus W, Marincola F, Worobec A, Atkins MB (2002) Immunologic monitoring of cancer vaccine therapy: Results of a workshop sponsored by the Society for Biological Therapy. *J. Immunother.* 25:97–138.
29. Hakomori S (2002) Glycosylation defining cancer malignancy: New wine in an old bottle. *Proc. Natl. Acad. Sci. U.S.A.* 99:10231–10233.
30. Khaldoyanidi SK, Glinsky VV, Sikora L, Glinskii AB, Mossine VV, Quinn TP, Glinsky GV, Sriramarao P (2003) MDA-MB-435 human breast carcinoma cell homo- and heterotypic adhesion under flow conditions is mediated in part by Thomsen-Friedenreich antigen-galectin-3 interactions. *J. Biol. Chem.* 278:4127–4134.
31. Glinsky VV, Huflejt ME, Glinsky GV, Deutscher SL, Quinn TP (2000) Effects of Thomsen-Friedenreich antigen-specific peptide P-30 on beta-galactoside-mediated homotypic aggregation and adhesion to the endothelium of MDA-MB-435 human breast carcinoma cells. *Cancer Res.* 60:2584–2588.
32. Zou J, Glinsky VV, Landon LA, Matthews L, Deutscher SL (2005) Peptides specific to the galectin-3 carbohydrate recognition domain inhibit metastasis-associated cancer cell adhesion. *Carcinogenesis* 26:309–318.
33. Phillips ML, Nudelman E, Gaeta FC, Perez M, Singhal AK, Hakomori S, Paulson JC (1990) ELAM-1 mediates cell adhesion by recognition of a carbohydrate ligand, sialyl-Lex. *Science* 250:1130–1132.
34. Takada A, Ohmori K, Takahashi N, Tsuyuoka K, Yago A, Zenita K, Hasegawa A, Kannagi R (1991) Adhesion of human cancer cells to vascular endothelium mediated by a carbohydrate antigen, sialyl Lewis A. *Biochem. Biophys. Res. Commun.* 179:713–719.
35. Hiller KM, Mayben JP, Bendt KM, Manousos GA, Senger K, Cameron HS, Weston BW (2000) Transfection of alpha(1,3)fucosyltransferase antisense sequences impairs the proliferative and tumorigenic ability of human colon carcinoma cells. *Mol. Carcinog.* 27:280–288.

36. Weston BW, Hiller KM, Mayben JP, Manousos GA, Bendt KM, Liu R, Cusack JC, Jr (1999) Expression of human alpha(1,3)fucosyltransferase antisense sequences inhibits selectin-mediated adhesion and liver metastasis of colon carcinoma cells. *Cancer Res.* 59:2127–2135.
37. Julien S, Krzewinski-Recchi MA, Harduin-Lepers A, Gouyer V, Huet G, Le BX, Delannoy P (2001) Expression of sialyl-Tn antigen in breast cancer cells transfected with the human CMP-Neu5Ac: GalNAc alpha2,6-sialyltransferase (ST6GalNac I) cDNA. *Glycoconj. J.* 18:883–893.
38. Zeng G, Gao L, Birkle S, Yu RK (2000) Suppression of ganglioside GD3 expression in a rat F-11 tumor cell line reduces tumor growth, angiogenesis, and vascular endothelial growth factor production. *Cancer Res.* 60:6670–6676.
39. Zeng G, Gao L, Yu RK (2000) Reduced cell migration, tumor growth and experimental metastasis of rat F-11 cells whose expression of GD3-synthase is suppressed. *Int. J. Cancer* 88:53–57.
40. Granovsky M, Fata J, Pawling J, Muller WJ, Khokha R, Dennis JW (2000) Suppression of tumor growth and metastasis in Mgat5-deficient mice. *Nat. Med.* 6:306–312.
41. Zhang S, Cordon-Cardo C, Zhang HS, Reuter VE, Adluri S, Hamilton WB, Lloyd KO, Livingston PO (1997) Selection of tumor antigens as targets for immune attack using immunohistochemistry: I. Focus on gangliosides. *Int. J. Cancer* 73:42–49.
42. Zhang S, Zhang HS, Cordon-Cardo C, Reuter VE, Singhal AK, Lloyd KO, Livingston PO (1997) Selection of tumor antigens as targets for immune attack using immunohistochemistry: II. Blood group-related antigens. *Int. J. Cancer* 73:50–56.
43. Dabelsteen E (1996) Cell surface carbohydrates as prognostic markers in human carcinomas. *J. Pathol.* 179:358–369.
44. Hakomori S (1984) Tumor-associated carbohydrate antigens. *Annu. Rev. Immunol.* 2:103–126.
45. Hakomori S (2000) Traveling for the glycosphingolipid path. *Glycoconj. J.* 17:627–647.
46. Clausen H, Hakomori S (1989) ABH and related histo-blood group antigens; immunochemical differences in carrier isotypes and their distribution. *Vox Sang.* 56:1–20.
47. Dennis JW, Granovsky M, Warren CE (1999) Glycoprotein glycosylation and cancer progression. *Biochim. Biophys. Acta* 1473:21–34.
48. Dennis JW, Laferte S (1989) Oncodevelopmental expression of–GlcNAc beta 1—6Man alpha 1-6Man beta 1—branched asparagine-linked oligosaccharides in murine tissues and human breast carcinomas. *Cancer Res.* 49:945–950.
49. Fernandes B, Sagman U, Auger M, Demetrio M, Dennis JW (1991) Beta 1-6 branched oligosaccharides as a marker of tumor progression in human breast and colon neoplasia. *Cancer Res.* 51:718–723.
50. Dabelsteen E, Vedtofte P, Hakomori S, Young WW, Jr (1983) Accumulation of a blood group antigen precursor in oral premalignant lesions. *Cancer Res.* 43:1451–1454.
51. Orntoft TF, Meldgaard P, Pedersen B, Wolf H (1996) The blood group ABO gene transcript is down-regulated in human bladder tumors and growth-stimulated urothelial cell lines. *Cancer Res.* 56:1031–1036.
52. Coon JS, Weinstein RS (1986) Blood group-related antigens as markers of malignant potential and heterogeneity in human carcinomas. *Hum. Pathol.* 17:1089–1106.

53. Juhl BR, Hartzen SH, Hainau B (1986) A, B, H antigen expression in transitional cell carcinomas of the urinary bladder. *Cancer* 57:1768–1775.
54. Lee JS, Ro JY, Sahin AA, Hong WK, Brown BW, Mountain CF, Hittelman WN (1991) Expression of blood-group antigen A—a favorable prognostic factor in non-small-cell lung cancer. *N. Engl. J. Med.* 324:1084–1090.
55. Matsumoto H, Muramatsu H, Shimotakahara T, Yanagi M, Nishijima H, Mitani N, Baba K, Muramatsu T, Shimazu H (1993) Correlation of expression of ABH blood group carbohydrate antigens with metastatic potential in human lung carcinomas. *Cancer* 72:75–81.
56. Orntoft TF (1990) Expression and biosynthesis of ABH-related carbohydrate antigens in normal and pathologic human urothelium. *APMIS Suppl* 17:1–34.
57. Miyake M, Hakomori SI (1991) A specific cell surface glycoconjugate controlling cell motility: Evidence by functional monoclonal antibodies that inhibit cell motility and tumor cell metastasis. *Biochemistry* 30:3328–3334.
58. Miyake M, Taki T, Hitomi S, Hakomori S (1992) Correlation of expression of H/Le(y)/Le(b) antigens with survival in patients with carcinoma of the lung. *N. Engl. J. Med.* 327:14–18.
59. Vangsted A, Drivsholm L, Andersen E, Pallesen T, Zeuthen J, Wallin H (1994) New serum markers for small-cell lung cancer. I. The ganglioside fucosyl-GM1. *Cancer Detect. Prev.* 18:221–229.
60. Hakomori S (1996) Tumor malignancy defined by aberrant glycosylation and sphingo (glyco)lipid metabolism. *Cancer Res.* 56:5309–5318.
61. Romano G (2005) The role of adult stem cells in carcinogenesis. *Drug News Perspect.* 18:555–559.
62. Hoff SD, Matsushita Y, Ota DM, Cleary KR, Yamori T, Hakomori S, Irimura T (1989) Increased expression of sialyl-dimeric LeX antigen in liver metastases of human colorectal carcinoma. *Cancer Res.* 49:6883–6888.
63. Nakamori S, Kameyama M, Imaoka S, Furukawa H, Ishikawa O, Sasaki Y, Kabuto T, Iwanaga T, Matsushita Y, Irimura T (1993) Increased expression of sialyl Lewis x antigen correlates with poor survival in patients with colorectal carcinoma: Clinicopathological and immunohistochemical study. *Cancer Res.* 53:3632–3637.
64. Nakayama T, Watanabe M, Katsumata T, Teramoto T, Kitajima M (1995) Expression of sialyl Lewis(a) as a new prognostic factor for patients with advanced colorectal carcinoma. *Cancer* 75:2051–2056.
65. Itzkowitz SH, Bloom EJ, Kokal WA, Modin G, Hakomori S, Kim YS (1990) Sialosyl-Tn. A novel mucin antigen associated with prognosis in colorectal cancer patients. *Cancer* 66:1960–1966.
66. Werther JL, Rivera-MacMurray S, Bruckner H, Tatematsu M, Itzkowitz SH (1994) Mucin-associated sialosyl-Tn antigen expression in gastric cancer correlates with an adverse outcome. *Br. J. Cancer* 69:613–616.
67. Terasawa K, Furumoto H, Kamada M, Aono T (1996) Expression of Tn and sialyl-Tn antigens in the neoplastic transformation of uterine cervical epithelial cells. *Cancer Res.* 56:2229–2232.
68. Kobayashi H, Terao T, Kawashima Y (1992) Serum sialyl Tn as an independent predictor of poor prognosis in patients with epithelial ovarian cancer. *J. Clin. Oncol.* 10:95–101.
69. Brooks SA, Leathem AJ (1991) Prediction of lymph node involvement in breast cancer by detection of altered glycosylation in the primary tumour. *Lancet* 338:71–74.

70. Galea MH, Ellis I, Bell J, Elston CW, Blarney RW, Baum M (1991) Prediction of lymph node involvement in breast cancer. *Lancet* 338:392–393.
71. Taylor CW, Anbazhagan R, Jayatilake H, Adams A, Gusterson BA, Price K, Gelber RD, Goldhirsch A (1991) Helix pomatia in breast cancer. *Lancet* 338:580–581.
72. Hirao T, Sakamoto Y, Kamada M, Hamada S, Aono T (1993) Tn antigen, a marker of potential for metastasis of uterine cervix cancer cells. *Cancer* 72:154–159.
73. Guthmann MD, Bitton RJ, Carnero AJ, Gabri MR, Cinat G, Koliren L, Lewi D, Fernandez LE, Alonso DF, Gomez DE, Fainboim L (2004) Active specific immunotherapy of melanoma with a GM3 ganglioside-based vaccine: A report on safety and immunogenicity. *J. Immunother.* 27:442–451.
74. Nishimura SI, Yamada K (1997) Transfer of ganglioside GM3 oligosaccharide from a water soluble polymer to ceramide by ceramide glycanase. A novel approach for the chemical-enzymatic synthesis of glycosphingolipids. *J. Am. Chem. Soc.* 119:10555–10556.
75. Lee KJ, Mao S, Sun C, Gao C, Blixt O, Arrues S, Hom LG, Kaufmann GF, Hoffman TZ, Coyle AR, Paulson J, Felding-Habermann B, Janda KD (2002) Phage-display selection of a human single-chain fv antibody highly specific for melanoma and breast cancer cells using a chemoenzymatically synthesized G(M3)-carbohydrate antigen. *J. Am. Chem. Soc.* 124:12439–12446.
76. George SK, Schwientek T, Holm B, Reis CA, Clausen H, Kihlberg J (2001) Chemoenzymatic synthesis of sialylated glycopeptides derived from mucins and T-cell stimulating peptides. *J. Am. Chem. Soc.* 123:11117–11125.
77. Blixt O, Allin K, Pereira L, Datta A, Paulson JC (2002) Efficient chemoenzymatic synthesis of O-linked sialyl oligosaccharides. *J. Am. Chem. Soc.* 124:5739–5746.
78. Sorensen AL, Reis CA, Tarp MA, Mandel U, Ramachandran K, Sankaranarayanan V, Schwientek T, Graham R, Taylor-Papadimitriou J, Hollingsworth MA, Burchell J, Clausen H (2006) Chemoenzymatically synthesized multimeric Tn/STn MUC-1 glycopeptides elicit cancer-specific anti-MUC-1 antibody responses and override tolerance. *Glycobiology* 16:96–107.
79. Freire T, Lo-Man R, Piller F, Piller V, Leclerc C, Bay S (2006) Enzymatic large-scale synthesis of MUC6-Tn glycoconjugates for antitumor vaccination. *Glycobiology* 16:390–401.
80. Muller S, Hanisch FG (2002) Recombinant MUC-1 probe authentically reflects cell-specific O-glycosylation profiles of endogenous breast cancer mucin. High density and prevalent core 2-based glycosylation. *J. Biol. Chem.* 277:26103–26112.
81. Backstrom M, Link T, Olson FJ, Karlsson H, Graham R, Picco G, Burchell J, Taylor-Papadimitriou J, Noll T, Hansson GC (2003) Recombinant MUC-1 mucin with a breast cancer-like O-glycosylation produced in large amounts in Chinese-hamster ovary cells. *Biochem. J.* 376:677–686.
82. Olson FJ, Backstrom M, Karlsson H, Burchell J, Hansson GC (2005) A MUC-1 tandem repeat reporter protein produced in CHO-K1 cells has sialylated core 1 O-glycans and becomes more densely glycosylated if coexpressed with polypeptide-GalNAc-T4 transferase. *Glycobiology* 15:177–191.
83. Muller S, Alving K, Peter-Katalinic J, Zachara N, Gooley AA, Hanisch FG (1999) High density O-glycosylation on tandem repeat peptide from secretory MUC1 of T47D breast cancer cells. *J. Biol. Chem.* 274:18165–18172.

84. Link T, Backstrom M, Graham R, Essers R, Zorner K, Gatgens J, Burchell J, Taylor-Papadimitriou J, Hansson GC, Noll T (2004) Bioprocess development for the production of a recombinant MUC1 fusion protein expressed by CHO-K1 cells in protein-free medium. *J. Biotechnol.* 110:51–62.
85. Antoine T, Heyraud A, Bosso C, Samain E (2005) Highly efficient biosynthesis of the oligosaccharide moiety of the GD3 ganglioside by using metabolically engineered *Escherichia coli*. *Angew. Chem. Int. Ed. Engl.* 44:1350–1352.
86. Antoine T, Priem B, Heyraud A, Greffe L, Gilbert M, Wakarchuk WW, Lam JS, Samain E (2003) Large-scale in vivo synthesis of the carbohydrate moieties of gangliosides GM1 and GM2 by metabolically engineered Escherichia coli. *Chembiochem.* 4:406–412.
87. Priem B, Gilbert M, Wakarchuk WW, Heyraud A, Samain E (2002) A new fermentation process allows large-scale production of human milk oligosaccharides by metabolically engineered bacteria. *Glycobiology* 12:235–240.
88. Endo T, Koizumi S, Tabata K, Kakita S, Ozaki A (1999) Large-scale production of N-acetyllactosamine through bacterial coupling. *Carbohydr. Res.* 316:179–183.
89. Mitchell MS, Harel W, Kempf RA, Hu E, Kan-Mitchell J, Boswell WD, Dean G, Stevenson L (1990) Active-specific immunotherapy for melanoma. *J. Clin. Oncol.* 8:856–869.
90. Livingston PO (1995) Approaches to augmenting the immunogenicity of melanoma gangliosides: From whole melanoma cells to ganglioside-KLH conjugate vaccines. *Immunol. Rev.* 145:147–166.
91. Brezicka T, Bergman B, Olling S, Fredman P (2000) Reactivity of monoclonal antibodies with ganglioside antigens in human small cell lung cancer tissues. *Lung Cancer* 28:29–36.
92. Slovin SF, Ragupathi G, Adluri S, Ungers G, Terry K, Kim S, Spassova M, Bornmann WG, Fazzari M, Dantis L, Olkiewicz K, Lloyd KO, Livingston PO, Danishefsky SJ, Scher HI (1999) Carbohydrate vaccines in cancer: Immunogenicity of a fully synthetic globo H hexasaccharide conjugate in man. *Proc. Natl. Acad. Sci. U.S.A.* 96:5710–5715.
93. Wang ZG, Williams LJ, Zhang XF, Zatorski A, Kudryashov V, Ragupathi G, Spassova M, Bornmann W, Slovin SF, Scher HI, Livingston PO, Lloyd KO, Danishefsky SJ (2000) Polyclonal antibodies from patients immunized with a globo H-keyhole limpet hemocyanin vaccine: Isolation, quantification, and characterization of immune responses by using totally synthetic immobilized tumor antigens. *Proc. Natl. Acad. Sci. U.S.A.* 97:2719–2724.
94. Gilewski T, Ragupathi G, Bhuta S, Williams LJ, Musselli C, Zhang XF, Bornmann WG, Spassova M, Bencsath KP, Panageas KS, Chin J, Hudis CA, Norton L, Houghton AN, Livingston PO, Danishefsky SJ (2001) Immunization of metastatic breast cancer patients with a fully synthetic globo H conjugate: A phase I trial. *Proc. Natl. Acad. Sci. U.S.A.* 98:3270–3275.
95. Livingston PO, Natoli EJ, Calves MJ, Stockert E, Oettgen HF, Old LJ (1987) Vaccines containing purified GM2 ganglioside elicit GM2 antibodies in melanoma patients. *Proc. Natl. Acad. Sci. U.S.A.* 84:2911–2915.
96. Livingston PO, Ritter G, Srivastava P, Padavan M, Calves MJ, Oettgen HF, Old LJ (1989) Characterization of IgG and IgM antibodies induced in melanoma patients by immunization with purified GM2 ganglioside. *Cancer Res.* 49:7045–7050.

97. Livingston PO, Wong GY, Adluri S, Tao Y, Padavan M, Parente R, Hanlon C, Calves MJ, Helling F, Ritter G (1994) Improved survival in stage III melanoma patients with GM2 antibodies: A randomized trial of adjuvant vaccination with GM2 ganglioside. *J. Clin. Oncol.* 12:1036–1044.

98. Kitamura K, Livingston PO, Fortunato SR, Stockert E, Helling F, Ritter G, Oettgen HF, Old LJ (1995) Serological response patterns of melanoma patients immunized with a GM2 ganglioside conjugate vaccine. *Proc. Natl. Acad. Sci. U.S.A.* 92:2805–2809.

99. Helling F, Zhang S, Shang A, Adluri S, Calves M, Koganty R, Longenecker BM, Yao TJ, Oettgen HF, Livingston PO (1995) GM2-KLH conjugate vaccine: Increased immunogenicity in melanoma patients after administration with immunological adjuvant QS-21. *Cancer Res.* 55:2783–2788.

100. Livingston PO, Adluri S, Helling F, Yao TJ, Kensil CR, Newman MJ, Marciani D (1994) Phase 1 trial of immunological adjuvant QS-21 with a GM2 ganglioside-keyhole limpet haemocyanin conjugate vaccine in patients with malignant melanoma. *Vaccine* 12:1275–1280.

101. Chapman PB, Morrissey DM, Panageas KS, Hamilton WB, Zhan C, Destro AN, Williams L, Israel RJ, Livingston PO (2000) Induction of antibodies against GM2 ganglioside by immunizing melanoma patients using GM2-keyhole limpet hemocyanin + QS21 vaccine: A dose-response study. *Clin. Cancer Res.* 6:874–879.

102. Zhang H, Zhang S, Cheung NK, Ragupathi G, Livingston PO (1998) Antibodies against GD2 ganglioside can eradicate syngeneic cancer micrometastases. *Cancer Res.* 58:2844–2849.

103. Ragupathi G, Livingston PO, Hood C, Gathuru J, Krown SE, Chapman PB, Wolchok JD, Williams LJ, Oldfield RC, Hwu WJ (2003) Consistent antibody response against ganglioside GD2 induced in patients with melanoma by a GD2 lactone-keyhole limpet hemocyanin conjugate vaccine plus immunological adjuvant QS-21. *Clin. Cancer Res.* 9:5214–5220.

104. Chapman PB, Wu D, Ragupathi G, Lu S, Williams L, Hwu WJ, Johnson D, Livingston PO (2004) Sequential immunization of melanoma patients with GD3 ganglioside vaccine and anti-idiotypic monoclonal antibody that mimics GD3 ganglioside. *Clin. Cancer Res.* 10:4717–4723.

105. Yao TJ, Meyers M, Livingston PO, Houghton AN, Chapman PB (1999) Immunization of melanoma patients with BEC2-keyhole limpet hemocyanin plus BCG intradermally followed by intravenous booster immunizations with BEC2 to induce anti-GD3 ganglioside antibodies. *Clin. Cancer Res.* 5:77–81.

106. Hamilton WB, Helling F, Lloyd KO, Livingston PO (1993) Ganglioside expression on human malignant melanoma assessed by quantitative immune thin-layer chromatography. *Int. J. Cancer* 53:566–573.

107. Livingston PO, Ritter G, Calves MJ (1989) Antibody response after immunization with the gangliosides GM1, GM2, GM3, GD2 and GD3 in the mouse. *Cancer Immunol. Immunother.* 29:179–184.

108. Estevez F, Carr A, Solorzano L, Valiente O, Mesa C, Barroso O, Sierra GV, Fernandez LE (1999) Enhancement of the immune response to poorly immunogenic gangliosides after incorporation into very small size proteoliposomes (VSSP). *Vaccine* 18:190–197.

109. Neninger E, Diaz RM, de la TA, Rives R, Diaz A, Saurez G, Gabri MR, Alonso DF, Wilkinson B, Alfonso AM, Combet T, Perez R, Vazquez AM (2007) Active

immunotherapy with 1E10 anti-idiotype vaccine in patients with small cell lung cancer: Report of a phase I trial. *Cancer Biol. Ther.* 6:145–150.

110. Bada A, Casaco PA, Arteaga M, Martinez J, Leon A, Santana E, Hernandez O, Orphee R, Gonzalez A, Mesa C, Gonzalez C, Montero E, Fernandez LE (2002) Toxicity of a GM3 cancer vaccine in Macaca fascicularis monkey: A 12-month study. *Hum. Exp. Toxicol.* 21:263–267.

111. Pan Y, Chefalo P, Nagy N, Harding C, Guo Z (2005) Synthesis and immunological properties of N-modified GM3 antigens as therapeutic cancer vaccines. *J. Med. Chem.* 48:875–883.

112. Houghton AN, Mintzer D, Cordon-Cardo C, Welt S, Fliegel B, Vadhan S, Carswell E, Melamed MR, Oettgen HF, Old LJ (1985) Mouse monoclonal IgG3 antibody detecting GD3 ganglioside: A phase I trial in patients with malignant melanoma. *Proc. Natl. Acad. Sci. U.S.A.* 82:1242–1246.

113. Cheung NK, Lazarus H, Miraldi FD, Abramowsky CR, Kallick S, Saarinen UM, Spitzer T, Strandjord SE, Coccia PF, Berger NA (1987) Ganglioside GD2 specific monoclonal antibody 3F8: A phase I study in patients with neuroblastoma and malignant melanoma. *J. Clin. Oncol.* 5:1430–1440.

114. Saleh MN, Khazaeli MB, Wheeler RH, Dropcho E, Liu T, Urist M, Miller DM, Lawson S, Dixon P, Russell CH (1992) Phase I trial of the murine monoclonal anti-GD2 antibody 14G2a in metastatic melanoma. *Cancer Res.* 52:4342–4347.

115. Irie RF, Matsuki T, Morton DL (1989) Human monoclonal antibody to ganglioside GM2 for melanoma treatment. *Lancet* 1:786–787.

116. Irie RF, Ollila DW, O'Day S, Morton DL (2004) Phase I pilot clinical trial of human IgM monoclonal antibody to ganglioside GM3 in patients with metastatic melanoma. *Cancer Immunol. Immunother.* 53:110–117.

117. Brezicka FT, Olling S, Bergman B, Berggren H, Engstrom CP, Hammarstrom S, Holmgren J, Larsson S, Lindholm L (1992) Coexpression of ganglioside antigen Fuc-GM1, neural-cell adhesion molecule, carcinoembryonic antigen, and carbohydrate tumor-associated antigen CA 50 in lung cancer. *Tumour. Biol.* 13:308–315.

118. Brezicka FT, Olling S, Bergman B, Berggren H, Engstrom CP, Holmgren J, Larsson S, Lindholm L (1991) Immunohistochemical detection of two small cell lung carcinoma-associated antigens defined by MAbs F12 and 123C3 in bronchoscopy biopsy tissues. *APMIS* 99:797–802.

119. Cappello S, Liu NX, Musselli C, Brezicka FT, Livingston PO, Ragupathi G (1999) Immunization of mice with fucosyl-GM1 conjugated with keyhole limpet hemocyanin results in antibodies against human small-cell lung cancer cells. *Cancer Immunol. Immunother.* 48:483–492.

120. Dickler MN, Ragupathi G, Liu NX, Musselli C, Martino DJ, Miller VA, Kris MG, Brezicka FT, Livingston PO, Grant SC (1999) Immunogenicity of a fucosyl-GM1-keyhole limpet hemocyanin conjugate vaccine in patients with small cell lung cancer. *Clin. Cancer Res.* 5:2773–2779.

121. Krug LM, Ragupathi G, Hood C, Kris MG, Miller VA, Allen JR, Keding SJ, Danishefsky SJ, Gomez J, Tyson L, Pizzo B, Baez V, Livingston PO (2004) Vaccination of patients with small-cell lung cancer with synthetic fucosyl GM-1 conjugated to keyhole limpet hemocyanin. *Clin. Cancer Res.* 10:6094–6100.

122. Springer GF (1997) Immunoreactive T and Tn epitopes in cancer diagnosis, prognosis, and immunotherapy. *J. Mol. Med.* 75:594–602.
123. O'Boyle KP, Zamore R, Adluri S, Cohen A, Kemeny N, Welt S, Lloyd KO, Oettgen HF, Old LJ, Livingston PO (1992) Immunization of colorectal cancer patients with modified ovine submaxillary gland mucin and adjuvants induces IgM and IgG antibodies to sialylated Tn. *Cancer Res.* 52:5663–5667.
124. MacLean GD, Bowen-Yacyshyn MB, Samuel J, Meikle A, Stuart G, Nation J, Poppema S, Jerry M, Koganty R, Wong T (1992) Active immunization of human ovarian cancer patients against a common carcinoma (Thomsen-Friedenreich) determinant using a synthetic carbohydrate antigen. *J. Immunother. (1991)* 11:292–305.
125. Adluri S, Helling F, Ogata S, Zhang S, Itzkowitz SH, Lloyd KO, Livingston PO (1995) Immunogenicity of synthetic TF-KLH (keyhole limpet hemocyanin) and sTn-KLH conjugates in colorectal carcinoma patients. *Cancer Immunol. Immunother.* 41:185–192.
126. MacLean GD, Reddish M, Koganty RR, Wong T, Gandhi S, Smolenski M, Samuel J, Nabholtz JM, Longenecker BM (1993) Immunization of breast cancer patients using a synthetic sialyl-Tn glycoconjugate plus Detox adjuvant. *Cancer Immunol. Immunother.* 36:215–222.
127. Longenecker BM, Reddish M, Koganty R, MacLean GD (1993) Immune responses of mice and human breast cancer patients following immunization with synthetic sialyl-Tn conjugated to KLH plus detox adjuvant. *Ann. N. Y. Acad. Sci.* 690:276–291.
128. Longenecker BM, Reddish M, Koganty R, MacLean GD (1994) Specificity of the IgG response in mice and human breast cancer patients following immunization against synthetic sialyl-Tn, an epitope with possible functional significance in metastasis. *Adv. Exp. Med. Biol.* 353:105–124.
129. Miles DW, Towlson KE, Graham R, Reddish M, Longenecker BM, Taylor-Papadimitriou J, Rubens RD (1996) A randomised phase II study of sialyl-Tn and DETOX-B adjuvant with or without cyclophosphamide pretreatment for the active specific immunotherapy of breast cancer. *Br. J. Cancer* 74:1292–1296.
130. MacLean GD, Miles DW, Rubens RD, Reddish MA, Longenecker BM (1996) Enhancing the effect of THERATOPE STn-KLH cancer vaccine in patients with metastatic breast cancer by pretreatment with low-dose intravenous cyclophosphamide. *J. Immunother. Emphasis. Tumor Immunol.* 19:309–316.
131. Miles D, Cameron D, Dodwell D, Glaspy J, Guillem V, Ibrahim N, Martin M, Perren T, Roche H, Tres A (2003) Pre-treatment chemotherapy (CX) regimens and responses in 1028 metastatic breast cancer (MBC) patients from an international, randomized phase III clinical trial of STn-KLH therapeutic vaccine. *Proc. Am. Soc. Clin. Oncol.* 22:abstract 221.
132. Sandmaier BM, Oparin DV, Holmberg LA, Reddish MA, MacLean GD, Longenecker BM (1999) Evidence of a cellular immune response against sialyl-Tn in breast and ovarian cancer patients after high-dose chemotherapy, stem cell rescue, and immunization with Theratope STn-KLH cancer vaccine. *J. Immunother.* 22:54–66.
133. Holmberg LA, Oparin DV, Gooley T, Lilleby K, Bensinger W, Reddish MA, MacLean GD, Longenecker BM, Sandmaier BM (2000) Clinical outcome of breast and ovarian cancer patients treated with high-dose chemotherapy, autologous stem cell rescue and THERATOPE STn-KLH cancer vaccine. *Bone Marrow Transplant.* 25:1233–1241.

134. Sabbatini PJ, Kudryashov V, Ragupathi G, Danishefsky SJ, Livingston PO, Bornmann W, Spassova M, Zatorski A, Spriggs D, Aghajanian C, Soignet S, Peyton M, O'Flaherty C, Curtin J, Lloyd KO (2000) Immunization of ovarian cancer patients with a synthetic Lewis(y)-protein conjugate vaccine: A phase 1 trial. *Int. J. Cancer* 87:79–85.

135. Dickler M, Gilewski T, Ragupathi G, Adluri S, Koganty R, Houghton A, Norton L, Livingston P (1997) Vaccination of breast cancer patients (pts) with no evidence of disease (NED) with sialyl Tn cluster (sTa(c))-keyhole limpet hemocyanin (KLH) conjugate plus adjuvant QS-21: Preliminary results (Meeting abstract). ASCO Annual Meeting.

136. Gilewski TA, Ragupathi G, Dickler M, Powell S, Bhuta S, Panageas K, Koganty RR, Chin-Eng J, Hudis C, Norton L, Houghton AN, Livingston PO (2007) Immunization of high-risk breast cancer patients with clustered sTn-KLH conjugate plus the immunologic adjuvant QS-21. *Clin. Cancer Res.* 13:2977–2985.

137. Slovin SF, Ragupathi G, Musselli C, Olkiewicz K, Verbel D, Kuduk SD, Schwarz JB, Sames D, Danishefsky S, Livingston PO, Scher HI (2003) Fully synthetic carbohydrate-based vaccines in biochemically relapsed prostate cancer: Clinical trial results with alpha-N-acetylgalactosamine-O-serine/threonine conjugate vaccine. *J. Clin. Oncol.* 21:4292–4298.

138. Slovin SF, Ragupathi G, Fernandez C, Jefferson MP, Diani M, Wilton AS, Powell S, Spassova M, Reis C, Clausen H, Danishefsky S, Livingston P, Scher HI (2005) A bivalent conjugate vaccine in the treatment of biochemically relapsed prostate cancer: A study of glycosylated MUC-2-KLH and Globo H-KLH conjugate vaccines given with the new semi-synthetic saponin immunological adjuvant GPI-0100 OR QS-21. *Vaccine* 23:3114–3122.

139. Krug LM, Ragupathi G, Ng KK, Hood C, Jennings HJ, Guo Z, Kris MG, Miller V, Pizzo B, Tyson L, Baez V, Livingston PO (2004) Vaccination of small cell lung cancer patients with polysialic acid or N-propionylated polysialic acid conjugated to keyhole limpet hemocyanin. *Clin. Cancer Res.* 10:916–923.

140. Sabbatini PJ, Ragupathi G, Hood C, Aghajanian CA, Juretzka M, Iasonos A, Hensley ML, Spassova MK, Ouerfelli O, Spriggs DR, Tew WP, Konner J, Clausen H, Abu RN, Dansihefsky SJ, Livingston PO (2007) Pilot study of a heptavalent vaccine-keyhole limpet hemocyanin conjugate plus QS21 in patients with epithelial ovarian, fallopian tube, or peritoneal cancer. *Clin. Cancer Res.* 13:4170–4177.

141. Karanikas V, Hwang LA, Pearson J, Ong CS, Apostolopoulos V, Vaughan H, Xing PX, Jamieson G, Pietersz G, Tait B, Broadbent R, Thynne G, McKenzie IF (1997) Antibody and T cell responses of patients with adenocarcinoma immunized with mannan-MUC-1 fusion protein. *J. Clin. Invest* 100:2783–2792.

142. Apostolopoulos V, Pietersz GA, Tsibanis A, Tsikkinis A, Drakaki H, Loveland BE, Piddlesden SJ, Plebanski M, Pouniotis DS, Alexis MN, McKenzie IF, Vassilaros S (2006) Pilot phase III immunotherapy study in early-stage breast cancer patients using oxidized mannan-MUC 1 [ISRCTN71711835]. *Breast Cancer Res.* 8:R27.

143. Loveland BE, Zhao A, White S, Gan H, Hamilton K, Xing PX, Pietersz GA, Apostolopoulos V, Vaughan H, Karanikas V, Kyriakou P, McKenzie IF, Mitchell PL (2006) Mannan-MUC-1-pulsed dendritic cell immunotherapy: A phase I trial in patients with adenocarcinoma. *Clin. Cancer Res.* 12:869–877.

144. Trevor KT, Hersh EM, Brailey J, Balloul JM, Acres B (2001) Transduction of human dendritic cells with a recombinant modified vaccinia Ankara virus encoding MUC-1 and IL-2. *Cancer Immunol. Immunother.* 50:397–407.

145. Van Leeuwen EB, Cloosen S, Senden-Gijsbers BL, Agervig TM, Mandel U, Clausen H, Havenga MJ, Duffour MT, Garcia-Vallejo JJ, Germeraad WT, Bos GM (2006) Expression of aberrantly glycosylated tumor mucin-1 on human DC after transduction with a fiber-modified adenoviral vector. *Cytotherapy* 8:24–35.

12

CARBOHYDRATES AS UNIQUE STRUCTURES FOR DISEASE DIAGNOSIS

Kate Rittenhouse-Olson

Department of Biotechnical and Clinical Laboratory Sciences and Department of Microbiology, University at Buffalo, Buffalo, New York 14214

12.1 INTRODUCTION

Surface carbohydrates are involved in adhesion and recognition in the interaction between cells and viruses, bacteria, fungi, parasites, and tumor cells. This interaction can involve carbohydrate to lectin binding or carbohydrate to carbohydrate interaction and is necessary to the infection, colonization, or metastasis in the various diseases. Since the presence of these surface carbohydrates is integral to the disease state, the surface molecule is usually a stable part of the pathogenic process and as such can be used as a diagnostic tool.

Indeed, many cell surface carbohydrates of viruses, bacteria, fungi, parasites, and tumor cells have been used diagnostically. The reason that not every disease-associated surface carbohydrate is used for diagnosis is that in order for a marker to be developed as a diagnostic tool it must meet several criteria. The disease that is being diagnosed must represent a significant health problem; the marker must be elevated in an assayable fluid in most individuals with the disease (high sensitivity, low false negatives); it must be at background in most individuals without the disease (high specificity, low false positives), and utilization of the diagnostic marker

---

Carbohydrate-Based Vaccines and Immunotherapies. Edited by Zhongwu Guo and Geert-Jan Boons
Copyright © 2009 John Wiley & Sons, Inc.

must change outcome in some way (treatment decisions, prognostic decisions, epidemiologic indications) so that diagnosis is important.

In the development of a diagnostic assay for any disease, an analysis must be performed to determine the specificity, sensitivity, and positive predictive value for the groups that will be tested. In addition, the accuracy and precision of the test must be determined. The specificity is the number of subjects with the disease that test positive divided by the total number of subjects that test positive multiplied by 100%. This gives us the likelihood of someone without the disease being diagnosed with the disease, so 85% specificity indicates that 15% of the time we would tell a patient that he has the disease when he does do not. The sensitivity is the number of subjects with the disease that test positive divided by the total number of subjects with the disease multiplied by 100%. In this case, 90% sensitivity would indicate that 10% of the time we would tell a patient that she does not have the disease when she do. Sensitivity and specificity can be predicted by the answers to three questions: (1) Is the antigen unique, that is, is it present in any other diseases or in any individuals without disease? (2) Is the antigen present in all individuals with the disease, and (3) are the limits set on the laboratory values so that they clearly indicate a true positive from a true negative?

Figure 12.1 shows that in the creation of a laboratory test, the placement of the cut-off limit of what is normal has drastic and opposite effects on sensitivity and specificity. Raising a cut-off value will decrease sensitivity but increase specificity. Which is more important in diagnosis, sensitivity or specificity? The answer is dependent on the disease; a false negative in a treatable but life-threatening neonatal disease

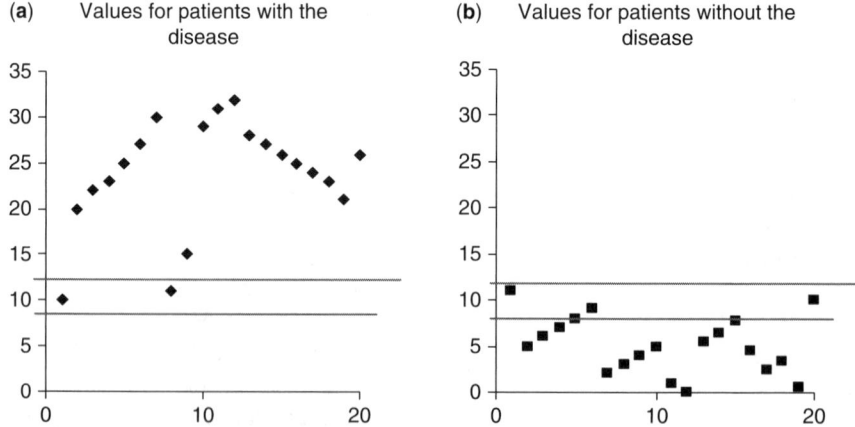

Figure 12.1 Trade-off in diagnostic testing of sensitivity and specificity. Using a cutoff as indicated by the upper line would give 18 out of 20 true positive values for patients with disease, or a 90% sensitivity. This same cutoff would, as indicated in (b), give 100% specificity, 20 of the 20 individuals measured have a true negative value. If the lower line is used as the cutoff, the sensitivity becomes 100%, meaning that no person with the disease is missed. However, in the specificity graph, 3 individuals without the disease in question will have false-positive results, giving a specificity of 17/20 or 85%.

or in a cancer diagnosis may cause the delay of treatment until significant morbidity occurs. A false positive in a disease diagnosis may lead to additional testing that causes morbidity in itself. An example of this would be a false positive in pancreatic cancer screening leading to laproscopic surgery to confirm the diagnosis.

Another important value in the development and use of a diagnostic test, especially if it is used to screen for the disease, is the positive predictive value. The positive predictive value is a combination of the specificity and the prevalence of the disease in the population. It gives the percent of time that a positive result in an assay, when done on a population, gives a true positive result. The positive predictive value can be a daunting figure to individuals trying to develop a diagnostic assay. An especially pointed example of this would occur in trying to develop a routine screening test for a relatively rare disease, such as renal cancer (incidence of about 12/100,000), using an assay that had a sensitivity of 100%, a specificity of 96%. A 96% specificity indicates that 4 false positives occur for every 100 people screened, a prevalence of 12 in 100,000 indicates that about 12 people will be found as true positives in screening of 100,000. In this 100,000 population, 4% or 4000 people would have a false-positive diagnosis. The positive predictive value is percent of the time a positive test will be a true positive. This value is determined as 100 times the true positives divided by the total number of positives, so it would be (100)12/4012 or 0.3% of the time a positive test result would indicate disease. Not all is lost with this type of assay; if one does not use it for generalized screening, and uses it for diagnosis of individuals that are showing clinical indicators of the disease or are at higher risk, the positive predictive value will improve as the incidence of the disease in these individuals will be higher.

12.2 VIRUSES

12.2.1 Infectious Mononucleosis

A classical example for diagnostic uses of carbohydrate antigens for viral infections is the detection of heterophile antibodies in the diagnosis of infectious mononucleosis. The clinical symptoms of patients with infectious mononucleosis are fatigue, lymphadenopathy, sore throat, and hematologic irregularities. As some of these features are similar to those of lymphocytic leukemia, the heterophile antibody test is important for the correct diagnosis.

Heterophile antibodies were defined as antibodies that react with a substance from more than one species, which seemed rather paradoxical at the time of their discovery. Most antibodies developed at that time were to protein antigens, and they reacted and cross-reacted in a manner that could be predicted by the evolutionary relatedness of the animals from which the protein was taken. Heterophile antibody reactivity was, however, not related to evolutionary relatedness. We now know that this was due to the fact that heterophile antibodies reacted with carbohydrate antigens rather than protein antigens. Interestingly, the heterophile antibodies do not react with antigens on or in the Epstein-Barr virus that causes infectious mononucleosis; instead, they react with the heterophile antigens through other mechanisms of the disease process, possibly due to polyclonal activation of B cells during the disease [1].

The diagnosis of infectious mononucleosis involves a presumptive test using sheep red blood cells (SRBCs), followed by the Davidsohn differential assay using bovine red blood cells and guinea pig kidney. The original presumptive test was developed by Paul and Bunnell in 1932, and 90% of patients with infectious mononucleosis reacted positively in this test. The Davidsohn differential assay differentiates three possible reasons for a positive result in the presumptive test, namely, individuals with infectious mononucleosis, individuals with serum sickness, and those healthy individuals who are positive for Forsmann antibodies. The heterophile antibody diagnostic for infectious mononucleosis reacts with the Paul–Bunnell antigen, having a proposed structure of Neu5Gcα2 → 3Galβ1 → 3(Neu5Gcα2 → 6)GalNAcα1-Thr-Pro-Gly-Pro-ProAsx [2]. Antibodies in serum sickness individuals react with N-glycolylneuramic acid [3]. Forssman antibodies react with a glycosphingolipid, having the structure of GalNAcα1 → 3GalNAcβ1 → 3Galα1 → 4Galβ1 → 4Glc-ceramide [4]. Sheep red blood cells are agglutinated upon incubation with sera containing any of the above three antibodies; however, guinea pig kidney reacts with the Forsmann antibodies and serum sickness antibodies, while bovine erythrocytes react with serum sickness antibodies and infectious mononucleosis antibodies.

In a Davidsohn diferential assay, the patient's serum is mixed with SRBCs alone on one spot on the Monospot differential card (the presumptive test); on the next spot the serum is premixed with guinea pig kidney and then mixed with the SRBCs, and on the last spot the serum is premixed with bovine red blood cell extract and then mixed with the SRBCs. After treatment with guinea pig kidney only infectious mononucleosis serum (anti-Paul-Bunnell antigen) will still agglutinate SRBCs. After treatment with bovine erythrocytes only Forsmann antibodies will still agglutinate SRBCs. Therefore, after a positive presumptive test, the Davidsohn diferential assay can be used to differentiate among the three potential positive results.

It was once also important to distinguish infectious mononucleosis from serum sickness, an immune complex disease observed in patients receiving therapies with foreign antibodies. For example, when diphtheria patients were treated with passive transfer of horse antitoxin immunoglobulin, they frequently produced antihorse immunoglobulin antibody to cause serum sickness. Serum sickness is no longer a common event, so a new presumptive test is adopted, which involves the use of bovine erythrocyte antigen only. A positive result in this test indicates infectious mononucleosis. Although serum sickness rarely occurs nowadays, serum sickness antibodies may be formed in patients receiving monoclonal antibody therapy. This will depend, however, on how humanized the mouse monoclonal was, and efforts are being made to minimize the possibility of serum sickness due to monoclonal antibody therapy.

12.2.2 Influenza A and B

The next example for the diagnosis of viruses is the ZstatFlu, which received Food and Drug Administration (FDA) approval for test of influenza in the late 1990s [5]. Although this is not an antigenic test, it is included here because a neuraminidase and its carbohydrate substrate are integral to this diagnostic test. This assay detects the viral neuraminidase produced by influenza type A and B, using a substrate that is specific for this neuraminidase. Upon reaction with the viral neuraminidase obtained in a throat swab, this carbohydrate substrate changes color to blue and precipitates if the viral neuraminidase is present. The specificity of the reaction is due to the ability of the influenza type A and B viral neuraminidase to hydrolyze carbohydrate substrates that contain α-ketosidically linked N-acetylneuraminic acid. Parainfluenza virus and mumps virus also contain a viral neuraminidase, but the tailored substrate employed in the influenza type A and B diagnostic test does not fit in the binding site of the enzymes from parainfluenza and mumps viuses [6].

12.3 BACTERIA

12.3.1 *Streptococcus pyogenes*

Several bacterial diseases are diagnosed using either measurement of bacterial carbohydrates from patient throat swabs or cultures or measurement of patient antibodies to these bacterial carbohydrates. This section will begin with the most common tests, those on streptococcal pathogens. Probably the most common carbohydrate antigen test performed in terms of bacterial disease diagnosis is the rapid assay of *S. pyogenes*, also referred to as group A strep. This causes 15–30% of all sore throats in children, and diagnosis is important to prevent the unnecessary use of antibiotics, to allow timely treatment, to reduce transmission [7], and to reduce the sequelae that can result from untreated strep throat.

While a throat culture is considered the gold standard diagnostic assay, the rapid direct antigen test was found to be very useful with 80% sensitivity and 100% specificity. The 100% specificity indicates that the positive predictive value would be 100%. Routinely, a negative rapid test is followed by a throat culture to eliminate the potential problem of false-negative results. Several different rapid kits are available for the detection of the carbohydrate from group A strep, but the protocol, a sandwich immunoassay, is very similar for all of them. The first step involves getting a throat specimen on a swab. The swabs are processed as soon as possible using an extraction buffer or two extraction buffers that extract the carbohydrate antigens. A capture immunochromatography assay is often used in which the extracted carbohydrate antigen is mixed with an antibody that is labeled with a colloid. The bound antigen travels across a membrane until it encounters a second antibody, the capture antibody. The capture antibody has been immobilized as a dot or stripe on a membrane, and a colored line resulting from the capture of the antigen-labeled antibody complex indicates a positive test. A control line is present that indicates that sufficient fluid was used [8]. The range of sensitivity for the various assays is 80–90% with specificity generally above 95% [9]. N-acetylglucosamine and rhamnose are present in the polymer that forms the group A polysaccharide [10].

12.3.2 Groups A, B, C, D, F, and G *Streptococcus*

A diagnostic latex agglutination test is develpoed to differentiate several types of β-hemolytic streptococci, including Lancefield streptococcal groups A, B, C, D, F, and G, which are classified based on their surface carbohydrates. This test is also used in combination with other bacteriological assays to differentiate group C strep and *Streptococcus pneumoniae* [11]. Group A streptococci are well-known human pathogens, while group B streptococci were defined in the original Lancefield listing [12] as strains from cows, which are now known to contain a severe human neonatal pathogen. Group C streptococci contain strains from other animals; together with group G they are known to cause pharyngitis, endocarditis, and bacteremia. Group D streptococci are in cheeses and can occasionally cause bacteremia and endocarditis. In terms of carbohydrate antigens, group B is known to contain rhamnose-glucosamine, and group C is known to contain rhamnose-*N*-acetylgalactosamine. The latex agglutination test for streptococci involves the reaction of carbohydrate antigens with specific antibody-coated beads.

The latex particle agglutination test can be used to detect either antigens or antibodies. When this method is employed to detect a specific antigen, such as in the case of streptococci detection, an antibody for the antigen is attached to the latex bead. In the test, usually an IgG antibody is used. As shown in Figure 12.2a, in the presence of the antigen, the antibody-coated beads are drawn together through the binding of an antigen with an antibody on one bead and an antibody on another bead. There are two potential problems with this assay. One is a prozone effect (Fig. 12.2b)—that is when excess antigen is present, all the antibody binding sites

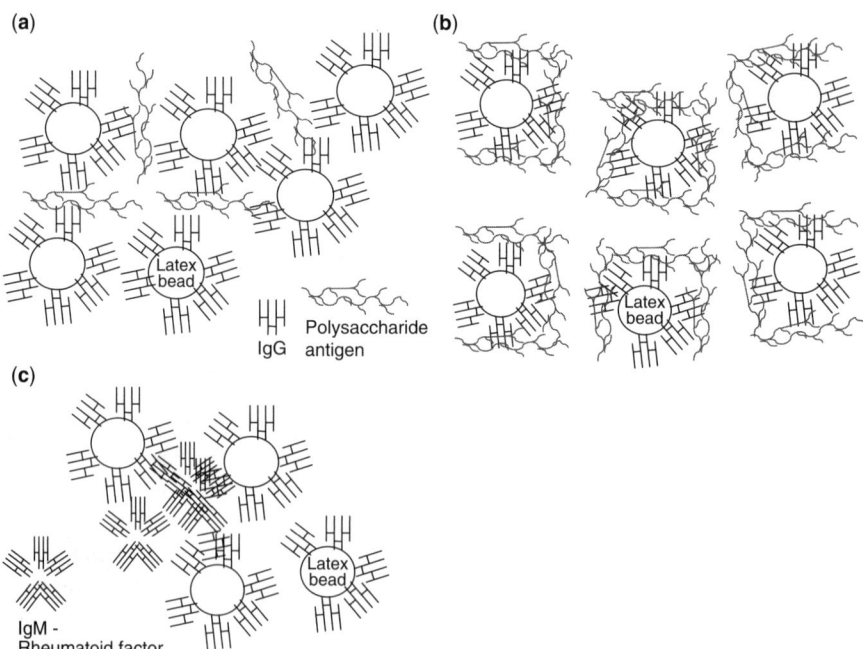

Figure 12.2 Schematic presentation of the latex agglutination assays. (a) True positive result, (b) a prozone effect—a false-negative result—due to excess antigen presence that inhibits agglutination; (c) a false-positive result due to rheumatoid factor (IgM) that binds to the Fc portion of IgG antibody.

are saturated, and the beads do not share an antigen, so they are not bridged together, and a false-negative result is observed. This problem can be overcome by diluting the sample and rerunning the assay. The dilution test is performed when the clinical picture indicates that the disease is present but the latex agglutination test is negative. The other problem with this assay, as well as with all capture assays that use antibodies, is that if the patient has rheumatoid factor the assay will show a false-positive result. Rheumatoid factor is an IgM antibody, which is one of the antibodies present in rheumatoid arthritis, and it binds to the Fc portion of IgG antibody. The rheumatoid factor can also bring the beads together by binding to the Fc portion of the coating antibodies (Fig. 12.2c). To overcome this problem, the latex particle agglutination test-based IgM antibody was developed.

12.3.3 Streptococcus pneumoniae

Streptococcus pneumoniae is one of the major causes of pneumonia that must be treated because if untreated it can result in more severe illness and complications, including death [13]. A diagnostic kit is available for *S. pneumoniae*, which is based on the detection of cell wall C-polysaccharide antigen present in all strains of *S. pneumoniae* [14]. It is similar to the diagnostic kit for *S. pyogenes* in that both are two-site immunochromagraphic methods that involve a capture antibody that captures the antigen on the membrane and a second detecting antibody. The *S. pneumoniae* assay is an indirect assay in that the second detecting antibody is not labeled, but a third species-specific antibody, which is labeled, is used for the detection. This assay has been developed for the testing of *S. pneumoniae* in urine and cerebral spinal fluid. It is intended to be used with culture and together with other bacteriologic methods. Culture is only positive in 40–70% of the cases, so an immunologic assay for this disease is of particular importance [14]. In various populations the sensitivity of the test ranged from 86 to 90%, and the specificity from 71 to 94% [15]. The test has also been proposed for use in severe otitis media using middle ear fluid. For this purpose it was found to have an 85% sensitivity and 100% specificity [16].

A slide agglutination diagnostic kit is also available for the diagnosis of *S. pneumoniae*, which uses detection of the serotype-specific capsular antigens by agglutination of latex beads coated with anticapsular antigen antibody of 83 different serotype specificities. It is meant to be performed either on isolated colonies from plate culture or broth cultures of the organism. It was found to be 98.4% sensitive and 93% specific on isolated colonies and 95.7% sensitive and 88% specific on broth cultures.

12.3.4 Meningitis

Latex agglutination tests have also been developed to help differentiate between the different bacteria that can cause meningitis. This combination kit includes latex agglutination kits for *S. pneumoniae* and group B *Streptococcus* as mentioned above and also a latex agglutination kit for the polysaccharide antigens of *Hemophilus influenzae* type b and *Niesseria meningitidis* groups A, B, C, Y, and W135. *H. influenzae* type b, *N. meningitidis*, and *S. pneumoniae* infections account for about 84% of cases of bacterial meningitis. Diagnosis is important to choose the correct antibiotic. These agglutination assays all rely on the use of specific antibody-coated beads that agglutinate in the presence of a multiple epitope antigen that binds to the antibody. The *N. meningitidis* test sensitivity was 83–100% and specificity was 92–100% [17]. Group B *N. meningitidis* polysaccharide cross reacts with *Escherichia coli* K1 meningitidis polysaccharide, but treatment that is effective for both can be used.

12.3.5 *Chlamydia trachomatis*

Chlamydia are obligate intercellular bacteria. A diagnostic test kit is available to detect the genus-specific lipopolysaccharide (LPS) produced by the three different species of *Chlamydia*, *C. pneumoniae*, *C. trachomatis*, and *C. psittaci*, which are implicated in human diseases. The only samples that these kits have been approved for are endocervical samples, so the test is designed for the diagnosis of *C. trachomatis*. *C. trachomatis* is a sexually transmitted bacterium that can lead to pelvic inflammatory disease and infertility in women [18]. The assay is an Optical ImmunAssay in which the LPS extracted from a cervical swab binds to a specific antibody that has been coated onto a thin molecular film. An antibody enzyme conjugate is then added to catalyze a reaction that changes the optical properties of the film for detection. The sensitivity of this assay is 83.6% and the specificity is 100%.

12.3.6 Future

A recent study by Thirumalapura et al. [19] describes the development of a printed microarray format for detection of gram-negative and gram-positive bacteria on nitrocellulose glass slides. This allows the detection of antibodies to multiple bacteria with a small sample in a one through-put slide. The conclusion of this study is that such a microarray could be applicable to serodiagnosis. A microarray may be a rapid method of diagnosis, but whole bacterial cells are not optimum for serodiagnosis for two reasons: (1) The Fc receptors present on some bacteria would result in a false-positive reaction, and (2) it is important to use carefully selected antigens instead of whole bacteria so that multiple cross-reactions are not measured. Nevertherless, with the improved synthetic technologies, microarrays spotted with purified or synthetic carbohydrate and protein bacterial antigens may be the diagnostic media of the future.

12.4 FUNGI

The incidence of fungal disease has increased with the effected populations including cancer patients, transplant recipients, individuals with acquired immune deficiency syndrome (AIDS), and other immunocompromised individuals. The use of rapid diagnostic techniques allows for timely initiation of the antifungal therapy [20]. Diagnosis includes growth and microscopic analysis of fungi grown from body fluids, tissue biopsy, and histopathologic analysis. When the fungal infection is invasive, it may be difficult to get a sample for culture, so serodiagnosis for antibodies or antigens of the fungi are important. The problem in looking for antibodies is that many of the effected populations are immunocompromised and so may not produce enough antibodies for detection. Detection of antigens may require a large body burden of the disease, but antigenic assays have been developed that are diagnostically useful [21].

12.4.1 Aspergillus fumigatus

Invasive aspergillosis is a pulmonary infection, causing pneumonia that can have an allergic component. The causative organism of this genus is most often *A. fumigatus*. The symptoms are shortness of breath, fever, and coughing, which can progress to weight loss, bone pain, and decreased urinary output. Diagnosis of invasive aspergillosis by culture is often difficult because biopsy is often precluded due to the debilitated state of the patients, and antibody formation may not occur in the immunocompromised patients that most often are infected with this pathogen. These two factors increase the importance of diagnostic tests that involve the use of capture enzyme immunoassays, radioimmunoassay, or latex particle agglutination tests for the galactomannan fungal antigen. The most sensitive assay developed is a sandwich enzyme immunoassay (EIA), which can detect as little as 0.5–1.0 ng/ml of the antigen, with a sensitivity of at least 95% and a specificity of 82–99%. Because of the range of false-positive rates, confirmation should be made by radiological (chest X ray) or microbiological methods [21].

12.4.2 Invasive Candidiasis

Invasive candidiasis is a fungal infection in which *Candida* species have systemically infected the individual, entering and growing in the individual's bloodstream. The symptoms are nonspecific and include fever and chills that do not respond to antibiotics and, as a result, may progress to deep-organ infections of kidney, liver, or bone, which can result in additional symptoms. Hospitalized patients are most often infected, with high-risk individuals including low-birth-weight infants, surgical patients, and individuals that are immunocompromised [20].

Antibody detection is not useful in this disease because antibody can form due to nonpathogenic gastrointestinal colonization of healthy individuals, and because the individuals that get systemic colonization are too immunocompromised. Direct detection of *Candida* antigens does have clinical utility but is still investigational. An assay has been developed that measures the β-(1-3)-D-glucan, the cell wall component of *Candida*. This is a genus-specific but not species-specific assay. A mannan-specific and species-specific test has also been developed but suffers from false negatives due to more rapid clearance of this antigen [22]. A combination of several test antigens may in the future create the best diagnostic test.

12.4.3 Cryptococcus neoformans

Cryptococcus neoformans usually manifests as a systemic infection that begins in the lungs and can progress to meningoencephalitis. It is associated with bird droppings and is found in the soil, spreading in dusty conditions through inhalation. It has an incidence of about 0.4–1.3 in 100,000 healthy individuals and 200–700 in 100,000 individuals with AIDS [23]. Both an IgG- and an IgM-based latex immunoassay have been developed to detect the cryptococcal capsular polysaccharide antigen. There are also methods involving pretreatment of the sera to eliminate the influence of rheumatoid factor. An enzyme immunoassay for cryptococcal antigens has also been developed, the advantage of which is that it does not have prozone, it does not react with rheumatoid factor and it yields fewer false-positive results.

12.4.4 *Histoplasma capsulatum*

Histoplasma capsulatum causes histoplasmosis. Alhough most people infected by *H. capsulatum* are asymptomatic or have self-limiting infections, it can cause severe illnesses especially in the immunocompromised [24]. *H. capsulatum* primarily causes pulmonary infection that leaves scars, which on X ray can look like tuberculosis or lung cancer, so a differential diagnosis is important.

The organism is in soil contaminated with bat or bird droppings, and people are infected when they inhale the spores [25]. Histoplasmosis is diagnosed through identification of the organism in tissues or secretions, looking for antibodies, or testing for *H. capsulatum* carbohydrate antigen in urine, serum, or other body fluids. Sometimes a transbronchial biopsy is needed for diagnosis. Culturing is not commonly used, and a polymerase chain reaction (PCR) method has been developed to identify the pathogen but is not currently the first choice for diagnosis [26]. A sandwich enzyme immunoassay amplified with biotin and strepavidin (Fig. 12.3) has been developed and been proven valuable in rapid diagnosis. It uses polyclonal antibody to the *H.-capsulatum*-specific carbohydrate. This immunoassay has a 92% sensitivity using urine from patients with disseminated disease as the antigen source and a 75% sensitivity in patients with acute pulmonary disease. Specificity is reported to be 98%. False positives can occur in patients with rheumatoid factor. This test has also been useful in monitoring the efficacy of therapy [24].

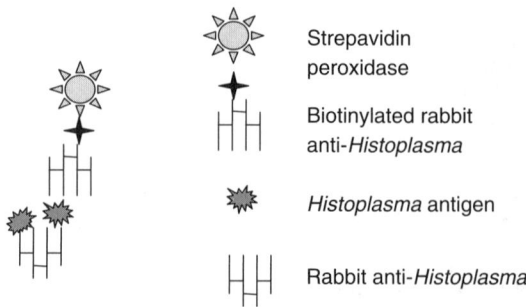

Figure 12.3 Sandwich assay for *Histoplasma* antigen using biotin strepavidin amplification.

12.5 PARASITES

Diagnosis of most parasites is made by the microscopic examination of host stools or tissue for the parasite. Antibody diagnosis is used for detection of infections in deep tissues including toxoplasmosis, toxicariasis, cystercosis, and echinococcosis, but the antigen used in this antibody diagnosis is frequently crude parasite extract or parasite ova extract [27]. An excellent review on the unique carbohydrates expressed by parasites and the importance of these structures to the pathogenesis was reported by Mendonça-Previato et al. [28]. Characteristic carbohydrate motifs for different parasites listed in this review are as follows: *Leishmania* (visceral, cutaneous, and mucocutaneous leishmaniases)—glycan core/phosphorylated disaccharide unit backbone and Manα1-PO$_3$-Ser; *T. brucei* blood form—glycosylphosphatidylinositol (GPI) containing α-Gal$p$; *T. brucei* procyclic form—sialylated and branched poly-*N*-acetyllactosamine; *T. cruzi*—GPI with 2-aminoethylphosphonate and β-Gal$f$ substituents, O-glycan with Gal$f$, Manα1 $\rightarrow$ 2Manα1 $\rightarrow$ 2Manα1-PO$_3$-Ser; *Plasmodium falciparum* (malaria)—GPI with tetramannose; and *Entamoeba histolytica*—GPI with α1-2Gal$p$ and Galα1-PO$_3$-Ser. Partial carbohydrate structures of parasite-specific glycoconjugates were also presented. This review article suggests that these novel carbohydrates may be useful for diagnosis or vaccine development, but widespread use of carbohydrates or any other well-defined structures remains limited [28]. Another recent review article [29] described the cancer-associated mucin antigen expression (Tn, TF, sialyl-Tn and Tk ag) in parasites and the potentials of these antigens, but they have not yet been utilized. The following diagnostic tests for parasites use extracts that have been characterized as containing carbohydrates.

12.5.1 *Echinococcus multilocularis*

Echinococcus multilocularis causes human alveolar echinococcosis. The primary hosts for this parasite are mainly foxes, coyotes, dogs, and cats and the intermediate hosts are vole and other rodents [30]. The primary hosts get the infection from eating intermediate hosts. Infection in humans is rare. This disease can be lethal in up to 100% of untreated patients, while therapy can cost hundreds of thousands of dollars. The parasite has been found from Alaska, Canada, and eastern Montana to central Ohio, and Europe, China, and Japan. This is a very small tapeworm, and contamination of humans occurs by ingesting stools from infected primary hosts. This may occur because domestic dogs or cats may have stool of their own or of other animals on their fur due to recent rolling in it. Human infections can be asymptomatic for years, and then liver involvement can cause abdominal pain, or lung involvement can cause chest pain. Rupture of enlarging cysts can cause anaphylaxis and death. To test humans for this parasite, a blood test is done looking for antibodies to the parasitic antigens. The best test utilizes the carbohydrate antigen EM-2 and a recombinant protein II/3-10 as coating antigen in an indirect enzyme immunoassay. The carbohydrate antigen EM2 is a major antigen of the cellular laminated layer that is the physical barrier between the parasite and the host tissue. The test with these two antigens is available in Europe but not in the United States [30–32]. Yamano et al. [32] have synthesized Galβ1 → 6(Fucα1 → 3)Galβ1 → 6Galβ1-ceramide and found it to be useful diagnostically for *E. multilocularis*, but it is not yet used clinically.

12.5.2 Clonorchis sinensis

Clonorchis sinensis is a human liver fluke, endemic in Asia, that feeds on bile and can cause liver inflammation and carcinoma of the liver. A carbohydrate antigen was identified as diagnostically useful in 1996 [33], but it has not been characterized. Currently, however, fecal examination for eggs is the most common method of diagnosis [34].

12.5.3 Trichinella

Trichinellosis is acquired by eating undercooked meat of omnivorous/carniverous animals that contains encysted larvae of the *Trichinella* species. The most common transmission is from infected pork. Symptoms can include gastrointestinal effects and muscle weakness where the larvae have formed cysts. Antibody to β tyvelose has been studied as a diagnostic marker for *Trichinella*. Although a diagnostic assay has been developed utilizing the excretory-secretory protein antigen, there is diagnostic potential for an assay utilizing the β-tyvelose antigen. β tyvelose may also have vaccine potential [35–37].

12.5.4 Schistosoma mansoni

Schistosoma mansoni is a chronic disease infecting 200 million people worldwide. It can be asymptomatic but can cause abdominal pain, liver and spleen enlargement, diarrhea, fever, and cough. It is usually acquired by wading or swimming in water that contains the eggs of this parasite. Aquatic snails are involved as intermediate hosts in the life cycle of this parasite and are one of the targets in attempts to eliminate the disease. Normally, diagnosis is made by fecal examination for the eggs. In patients with a negative fecal screen but a clinical picture that indicates schistosomiasis, screening for antibody to *S. mansoni* adult microsomal antigens is performed. The reaction of this antigen depends on both the carbohydrate and the protein components. The carbohydrate component has not been characterized [38, 39].

Although unique carbohydrate antigens have been found on parasites (see review articles cited above), they have not made a diagnostic or therapeutic impact. It is clear that diagnosis by fecal examination for eggs is time consuming and requires skill, thus increased use of specific parasite-associated carbohydrate antigens should be explored for more rapid and cost-effective diagnoses.

12.6 AUTOIMMUNITY

12.6.1 Diabetes

When blood glucose levels are high, proteins will become glycosylated in a nonenzymatic way. This occurs by the attachment of free aldehyde groups of glucose to the free amino groups of proteins. Since red blood cells are fully permeable to glucose, in this process some of the hemoglobin will become glycosylated, and this is called hemoglobin A1c (HbA1c). The possible glycation sites on the hemoglobin A1 molecule include the four valines that are at N-terminal on the polypeptide chains and all the free ε-amino groups of the lysines. The level of HbA1c is measured as a percentage of total hemoglobin, and this gives an indication of the average blood glucose levels. Hemoglobin A is the adult-type hemoglobin, and hemoglobin A1 is more negatively charge and includes HbA1a, HbA1b, and HBA1c. The glycosylated HbA1c is measured by high-pressure liquid chromatography [40, 41]. A table has been developed that correlates the average blood glucose levels over the last 4 months with the percent of HbA1c in total hemoglobin. The reason that the HbA1c levels correlate with the 4-month average blood glucose is that once hemoglobin becomes glycosylated it stays that way, and hemoglobin lasts the life of the red blood cell, which is about 4 months. Normal levels of HbA1c are 7% or lower. Elevations or decreases that do not correlate with the amount of blood glucose can be due to uremia, chronic excessive alcohol use, elevated triglycerides, and sickle cell disease. Vitamins A and E can also affect test results. Diabetics who keep their HbA1c levels under 7% are less likely to experience side effects including vascular, neurologic, and fertility issues [40, 41].

12.6.2 Cold Agglutinin Disease

Cold agglutinin disease is an autoimmune disease in which the patient has developed antibodies to the red blood cell I or i antigen, which results in an autoimmune hemolytic anemia. This occurs in about 1 in every 100,000 individuals with an increase in disease incidence with age to a maximum in patients in their 70 s. The reason that this is called cold agglutinin is that these antibodies are usually IgM, which bind best at lower temperatures. The autoimmune anemia is worse when the optimal temperature of binding for the antibody is closer to 37 °C. These anti-I or i antibodies can be polyclonal postinfection or monoclonal. The polyclonal antibodies are produced primarily after a respiratory infection with *Mycoplasma pneumoniae* or after an infectious mononucleosis infection. The polyclonal antibody disappears after the infectious disease resolves. The monoclonal antibody IgM to these antigens is not associated with an infectious disease and is more serious because it is a chronic anemia, which does not resolve. IgM antibody is the best antibody class at binding complement, and this trait means that red blood cell lysis is the result of the presence of this antibody to the red cell surface antigens I and i. The i chain antigen is Galβ1 $\rightarrow$ 4GlcNAcβ1 $\rightarrow$ 3Galβ1 $\rightarrow$ 4GlcNAcβ1 $\rightarrow$ 3Gal, and the I chain antigen is a branched addition to the above Galβ1 $\rightarrow$ 4GlcNAcβ1 $\rightarrow$ 3(Galβ1 $\rightarrow$ 4GlcNAcβ1 $\rightarrow$ 6)Galβ1 $\rightarrow$ 4GlcNAcβ1 $\rightarrow$ 3Gal. The assay for this antibody involves adding the serum to red blood cells, incubating at 4 °C, and subsequently showing that the red blood cell agglutination disappears upon warming the cells. Rituximab, which reacts with the CD20 molecule on human B cell, has recently been applied to treat cold agglutinin disease [42, 43].

12.6.3 Inflammatory Bowel Disease

Inflammatory bowel disease (IBD) includes two different clinical diagnoses, Crohn's disease and ulcerative colitis. Both of these diseases cause abdominal pain and diarrhea, with bloody diarrhea more often associated with ulcerative colitis. It is important to differentiate because surgery is effective therapy in ulcerative colitis, but the disease returns after surgery in Crohn's disease. Antibodies to tumor necrosis factor are used as an effective therapy in Crohn's disease, but the efficacy of this therapy has not been proven in ulcerative colitis. Endoscopy can be used to distinguish between the two diseases but there is sometimes an overlap. The autoantibody to the nuclear lamina protein present in neutrophils (pANCA) has been well described as associated with inflammatory bowel disease but will not be discussed here as it is an antibody to a protein antigen. Antibodies against the mannan epitopes of the yeast *Saccharomyces cerevisiae* (gASCA) are also well known to be associated with IBD. The combination of the presence of gASCA and pANCA antibodies has a 95–98% specificity when used for diagnosis of Crohn's disease, but the sensitivity is only 30–60%. To improve the ability to diagnose and differentiate between Crohn's and ulcerative colitis, a recent study assessed the comparative levels of antibodies to other saccharide structures in patients with these two syndromes. Antibodies are quantitated using a GlycoChip (Glycominds Ltd., Lod, Isreal) and enzyme immunoassays [44, 45].

gASCA and pANCA remained the best combination to distinguish between the diseases; for differentiation between people with IBD and those without, gASCA was the best marker. Antibody levels to laminarabioside (ALCA) improved accuracy in the later differentiation. Levels of gASCA, ALCA, antichitobioside, antimannobioside, and antiouter membrane protein are all related to disease severity in Crohn's disease. Although these carbohydrate markers are not yet routinely used for diagnosis and differentiation, the combination of these carbohydrate antigens may prove to be clinically useful [44, 45].

12.7 TUMORS

Since the origin of tumors is the transformation of one of the host's own cells, it is only in virally induced human tumors that we find tumor-specific antigens. What is normally found with cancers is more properly referred to as tumor associated antigens. Tumor-associated antigens are antigens that are found in increased amounts in tumor tissue but can be found in normal tissue at some level or some gestational time. Altered expression of glycans is a feature that has been known to be associated with cancer tissue since 1969 [46]. Tumors have been shown to have changes in the total sialic acid content (both increases and decreases are seen) and changes in the amino sugar and neutral sugar contents of cell membranes. There is evidence of incomplete synthesis of some glycolipids, enhanced fucose labeling patterns of some glycoconjugates from transformed cells, and changes in the amount or activities of enzymes related to carbohydrates (sialyltransferase, glycosyltransferases, fucosidases), and there is altered lectin binding activity. Tumor-associated carbohydrate antigens may be related to an earlier developmental stage of the tissue in which the surface carbohydrate was a differentiation antigen involved in cellular interactions [47–51]. Many of these tumor-associated carbohydrate antigens can be found in fetal tissue. The alteration in cell surface glycans has also been linked to changes in cell–cell adhesion and cancer metastasis [47–49].

The monoclonal antibodies that reacted with tumor-associated antigens were analyzed using different compounds to determine the chemical nature of the antigen, and many were found to be carbohydrates. Examples of carbohydrate tumor-associated antigens include CA15.3 (a marker for breast cancer), CA19.9 which is also called sialyl Le^a (a marker for colonic and pancreatic cancer), sialyl Lewis x (sLe^x), sialyl Tn (sTn), Globo H, Lewis y (Le^y), polysialic acid, sialyl Lewis a (sLe^a), Tn, Fucosyl GM1, Thomsen–Friedenreich antigen (TF-Ag), and the gangliosides GD2, GD3, and fucosyl GM1 [47]. However, the presence of these markers on a tumor cell does not mean that this particular antigen will be diagnostically useful. Since Chapter 8 of this volume has presented a detailed discussion about tumor-associated carbohydrate antigens, this section will be limited to those tumor-associated carbohydrate antigens that are used diagnostically or have real potential to be used diagnostically.

12.7.1 Bladder

According the American Cancer Society (ACS) bladder cancer occurs in 1 of every 30 men and 1 of every 90 women. There are 4 types of bladder cancer based on their cells of original. Diagnosis of bladder cancer usually follows the patient finding blood in the urine and reporting this to the physician. Blood in the urine does not mean the patient has cancer, as it can mean the patient has an infection. Thus, the physician must perform additional tests to proceed with the diagnosis. The patient may also complain of the feeling that there is a need to urinate, even when there is not. A cystoscopy will be performed to diagnose, and cancer cells in the urine will be looked for. The tumor-associated proteins, bladder tumor-associated antigen (BTA) and nuclear matrix protein 22 (NMP-22) are used in conjunction with cystoscopy for bladder cancer diagnosis. Tn, TF, and sTn have been found in bladder cancer but are not used in its diagnosis [47]. Two carbohydrate antigen tests are under investigation for diagnostic use. The commercial test called Immunocyte measures expression of sulfated mucin glycoproteins and glycosylated forms of CEA in the urine. The HA-HAase test measures the level of hyaluronic acid (HA) and hyaluronidase (HAase) in an enzyme immunoassay. HA is a glycosaminoglycan ($\rightarrow$ 4GlcUAβ1 $\rightarrow$ 3GlcNAcβ1 $\rightarrow$ polymer) that is elevated in bladder and prostate cancer. HAase is an enzyme that breaks down HA to angiogenic fragments. The sensitivity, specificity, and negative and positive predictive values of the HA-HAase test are promising [52].

12.7.2 Breast

According to the Centers for Disease Control and Prevention (CDC), breast cancer is the most commonly diagnosed cancer in women, with nearly 200,000 new cases a year diagnosed in the United States, and it is the second most common cause of cancer deaths in women [53]. Although much attention has been given to hereditary forms of breast cancer, they account for only 5–10% of the cases (ACS). The two most common types of invasive breast cancer include infiltrating ductal carcinoma (IDC) (80% of the cases) and infiltrating lobular carcinoma (ILC).

The currently accepted methods of diagnosis of breast cancer are annual mammograms, clinical breast exams, and breast self-exams. The strongest risk factor for developing breast cancer is being a woman over the age of 50, so an annual mammogram is necessary after this age [53]. Early diagnosis is important because patients with tumors less than 2 cm in diameter have a greater than 90% chance of surviving 5 years, while patients with tumors over 5 cm in diameter have only a 60% chance of surviving 5 years. Tn, TF, sTn, sLe^a, sLe^x, and Le^y have been found in breast cancer but are not used in its diagnosis [47]. The protein progesterone, estrogen, and Her2/neu receptor are used to determine potential responsiveness to certain therapeutic regimes. There are carbohydrate antigens, such as CA15-3 (CA stands for carbohydrate antigen) and CEA, which are not currently used for diagnosis because their use does not represent an improvement on the above diagnostic methods, but may be useful in detecting recurrences or monitoring therapy. CA15-3 is a mucin containg 28% of carbohydrate. CEA is a glycoprotein containing 50% of carbohydrate by weight. CEA is different from its normal counterparts in terms of their glycosylation patterns; CEA has increased expression of Le^x and Le^y [54, 55]. CA15-3 and CEA are used mainly during active therapy in monitoring response with data from physical exam, imaging techniques, and history. The clinical assays involve a sandwich immunoassay to detect serum concentrations of the antigen.

12.7.3 Colon

According to the ACS, colon cancer in most cases takes years to form, and in most cases it begins as a small polyp. These polyps can be removed and this significantly decreases cancer incidence. Thus, early diagnosis and screening is especially important because the cancer or precursor polyp can be removed before it is a problem. The screening and diagnostic tests for colorectal cancer include the fecal occult blood test, sigmoidoscopy, colonoscopy, double-contrast barium enema, and the digital rectal examination [56]. Tn, TF, STn, H blood group, sLe^a, sLe^x, Le^y, Globo H, and Sd^a have been found in colon cancer but are not used in its diagnosis [47]. Ninety-five percent of colon cancers are adenocarcinomas, that is, they are formed from the glandular cells that line the colon and rectum. CEA (see above) is the molecule most often used for the monitoring of treatment efficacy for colon cancer. If CEA is not elevated, CA19-9, an antibody to sialyl Le^a, is utilized for monitoring therapy [56].

12.7.4 Liver

Liver cancer can either be a hepatocellular carcinoma (HCC) (75% of all liver cancers), a cholangiocarcinoma, an angiocarcinoma, a hemangiocarcinoma, or a heptoblastoma. Many tumors from other sites metastasize to the liver, but their diagnosis and treatment is as of their tumor of origin. Risk factors for developing hepatocellular carcinoma include chronic infection with either hepatitis B or hepatitis C, liver cirrhosis, diabetes, obesity, aflatoxin ingestion, steroid use, and arsenic contamination [57]. For suspected liver cancer, a tumor marker, α-fetoprotein (AFP), is used for diagnosis. Liver cancer is relatively uncommon in the United States, so AFP is not used as a general screening test but a tool to help in the diagnosis of individuals with symptoms or risk factors. Elevated AFP values are in two-thirds of the individuals with liver cancer. Ultrasound is also used diagnostically in this disease. AFP, like CEA, is a tumor antigen that is also present in fetal tissue [57]. It is a glycoprotein with one linked N-glycan, which differs in the fetal AFP and the AFP produced by a tumor. The AFP in liver carcinoma has differential binding with Lens culinaris agglutinin (LCA), a lectin that binds mannose or N-acetylglucosamine with an attached fucose. Total AFP from HCC can be separated into three fractions based on binding to LCA. AFP-L1 does not bind LCA; AFP-L2 is expressed on yolk sac tumors and is in maternal serum during pregnancy, and AFP-L3 binds to LCA and is related to HCC. This different lectin binding ability is due to the addition of a fucose $\alpha 1-6$ linked to the N-acetylglucosamine at the reducing terminal end of the sugar chain. The AFP-L3% is the ratio of AFP-L3 to total AFP. AFP-L3% is particularly useful when the patient has an AFP value between 10 and 200 ng/ml, where an additional marker is useful for diagnosis. In this range, using an AFP-L3% cut-off of 10%, AFP-L3% had a sensitivity of 71% and a specificity of 63% for diagnosis of HCC; or using a cut-off of 35%, the specificity is 100%. It has been found that earlier treatment of HCC has prognostic significance, so the ability of AFP-L3 to diagnose patients in lower AFP range has clinical significance [58–61].

12.7.5 Lung

Lung cancer has the highest mortality rate of all cancers, resulting in more deaths each year than breast, prostate, and colorectal cancers combined. According to the ACS, an estimated 160,440 Americans die each year from lung cancer, accounting for 28% of all cancer deaths [62]. Over 173,000 new cases of lung cancer are diagnosed each year, accounting for 13% of new cancer cases. Lung tumors are divided into two major categories: non-small-cell lung carcinoma (NSCLC) and small-cell lung carcinoma (SCLC). Eighty percent of lung cancers are of the NSCLC type, and this category includes 3 tumor subtypes: adenocarcinoma (ADC), squamous cell carcinoma (SqCC), and large cell carcinoma. SqCC is the second most common form of lung cancer, accounting for one-third of all bronchogenic carcinomas and is strongly linked to a history of tobacco use. This tumor arises in the bronchial epithelium of the central lung [63].

In the last 30 years, all available clinical data show that for lung cancer patients, outcome is dramatically better when the disease is detected at an early stage and surgically treated. The potential of a nearly 100% cure exists in patients diagnosed in the earliest stages where the tumor diameter is 3 mm or less, and when there is no invasion of the bronchial wall [64]. Since it is a fact that early-stage lung cancer can be cured and that death is virtually certain without curative treatment, implementation of early diagnosis in groups at highest risk is imperative.

Screening protocols for lung cancer are not currently used in the general population, and even in the high-risk smoking population screening is controversial. Mass screening analyses in the United States in the 1970s and 1980s with chest X ray or chest X ray plus sputum cytology in randomly controlled trials led to the conclusion by the Mayo Clinic that this type of screening was not sensitive enough to save lives [65]. Other studies agreed with this conclusion [66–70], however, a more recent case-controlled study in Japan in the last decade again using mass screening with an annual chest X ray or chest X ray plus sputum cytology indicated that this type of screening could reduce lung cancer deaths by 32–60% [70]. These conflicting results could be due to improvements in chest X rays or to the differences between a randomly controlled trial and a case-controlled trial [71–73]. Whatever the reason for the difference in the studies, the results are not ideal, and a better way to screen and diagnose lung cancer patients is needed to prevent the many lung cancer fatalities. It is also necessary to define a collection of reliable biomarkers that can be used for diagnosis of early-stage disease to assist this effort.

The Early Detection Research Network (EDRN) [74] was developed to create a large collaborative effort to find biomarkers for lung, colon, ovarian, breast, pancreatic, cervical, bladder, and prostate cancers. In the area of lung cancer, several avenues are being studied by the EDRN. Due to the heterogeneity of lung cancer, the EDRN's focus is on the development of a panel of markers that will be used together to create the optimal specificity and sensitivity for lung cancer early detection [74–76]. Markers that the EDRN is currently looking at to aid in early diagnosis of lung cancer are: (1) at the chromosomal level: aneuploidy in sputa as detected by fluorescent in-situ hybridization (FISH) [77]; (2) at the DNA (deoxyribonucleic acid) level: aberrant methylation patterns [78]; and (3) at the protein level [79]: computed tomography (CT) with sputum molecular markers [80] and autoantibodies to tumor antigens [81].

Although difucosyl Le (A) X (Galβ1 $\rightarrow$ 3[Fucα1 $\rightarrow$ 4]GlcNAcβ1 $\rightarrow$ 3Galβ1 $\rightarrow$ 4[Fucα1 $\rightarrow$ 3] GlcNAcβ1 $\rightarrow$ 3Galβ1 $\rightarrow$ 4Glc) was found to have a lung squamous cell carcinoma association [49] and another paper [82] indicated that a mucin containing this carbohydrate antigen is secreted by SqCC cells, a suitable diagnostic assay was not developed. Lewis Y [83], sialyl Lewis X-I [84], Fuc-GM1 [85], sialylated and fucosylated type 1 antigens [86], gangliosides with NeuAcα(2,6) linkages [87], GD3 [87], GM2, GD2, and 9-O-acetylGD3 [88, 89] have all been shown to have a squamous cell lung cancer association; however, clinical utility of any of these as a diagnostic marker has not been shown. There is potential that renewed research concerning these markers singly or together will yield a useful diagnostic test.

12.7.6 Melanoma

Melanoma occurs due to the malignant transformation of the skin melanocytes. It comprises only 3% of all skin cancers but is the most serious. Approximately 60,000 cases occur annually in the United States, with over 8000 deaths due to this disease a year. Too much unprotected exposure to Ultraviolet (UV) light is the major risk factor along with fair skin and the presence of certain kinds of moles on your skin. The tumor cells are often brown or black because of their production of melanin, but they can also be colorless. Surgery is usually curative if the disease is found early; however, there is no general diagnostic marker that will help find melanoma early. The diagnosis is based on self-exams and examinations followed by biopsies by health care professionals [90]. Moles that change in size, shape, or color should be seen by a health care professional right away. The ABCD rule is used in looking at moles; any of the following characteristics should prompt a call to the doctor: asymmetry, border irregularity, color variation, or diameter over 0.25 inches. Two tumor antigens, TA-90 and S-100, both of which are protein antigens, can be used in immunohistochemical analysis to diagnose suspicious biopsies. The gangliosides GD2, GD3, GM2, and GM3 [47] have been found to be expressed in melanoma but thus far have not been found diagnostically useful. Vaccination trials for therapy are being developed, and these may include the attempts to develop immune responses to the associated gangliosides.

12.7.7 Ovarian

A women's lifetime risk of getting ovarian cancer is 1 out of 67, with the risk of death due to this disease of 1 out of 95. Most of the women ($\sim$66%) that get ovarian cancer are over 55. There are three types of malignant ovarian cancers, epithelial, germ cell, and stromal cell, named for the cell type of origin. Cancers that develop from the epithelial cells are called carcinomas, and about 85% of ovarian tumors are of this type. Epithelial ovarian cancers are further divided into serous, mucinous, endometrioid, and clear cell. If the cells cannot be classified into one of these groups, the tumor is called undifferentiated [91].

Primary peritoneal carcinoma (PPC) is not an epithelial carcinoma but is closely related to it. About 5% of ovarian cancers are germ cell tumors, either teratomas, dysgerminomas, endodermal sinus tumors, or choriocarcinomas. Five to 7% of ovarian tumors are derived from the ovarian stromal cells and are granulosa cell tumors, granulose-theca tumors, and Sertoli–Leydig cell tumors [91]. The 5-year survival rates for ovarian cancer is about 45%, but if diagnosis occurs before the cancer has left the ovary, the 5-year survival rate is 93%. Only 20% of ovarian cancers are found at this early stage, thus it is clear that earlier diagnostic methods would have clinical impact [91].

Risk factors for ovarian cancer include age (above 40), obesity, nulliparity, fertility drug use, family history of breast or ovarian cancer, personal history of breast cancer, and talcum powder use in the genital area. The symptoms of ovarian cancer are vague and often ignored. The most common symptoms are bloating, feeling full quickly, abdominal pain, urinary frequency, fatigue, back pain, and constipation. These symptoms can happen in many disorders, so again, a marker for ovarian cancer would be important [91].

Diagnosis begins with a pelvic exam, and then imaging is performed to look for a pelvic mass, and the tumor marker CA-125 is used to aid in diagnosis. CA-125 was discovered in 1981 by immunization with a human ovarian carcinoma cell line. It is another tumor antigen found also on fetal tissue and on some normal tissues, including endometrium, peritoneum, and pleura, so it is not a tumor-specific antigen. Used as an aid in diagnosis, CA-125 cannot be used in generalized screening the general population because: (1) the specificity is too low since in premenopausal women there are other conditions that can cause its elevation, and (2) the sensitivity is too low since CA-125 levels are elevated in only 58% of stage 1 disease and 80% of total epithelial ovarian cancers. Studies are being done utilizing CA-125 followed by sonography to improve the specificity, and screening only postmenopausal women to improve the sensitivity [92]. CA-125 is mainly used to monitor the course of the disease after initial treatment. Like CEA and AFP, there appears to be glycosylation changes between the CA-125 from tumors and the CA-125 from normal tissues [93]. The glycan content of CA-125 is unique in that it contains branched core 1 structures, with sialic acid or fucose as the branches, as well as N-linked glycans. In addition, although the antibody recognizes the peptide sequence, removal of the N-glycans reduces binding of the OC125 antibody. CA-125 from tumors has high mannose and biantennary complex N-linked glycans [94]. In addition to CA-125, Tn, TF, STn, sLe^a, sLe^x, Le^y, and Globo H have been found in ovarian cancer, however, these markers are not used in the diagnosis [47].

12.7.8 Pancreatic

Cancer of the pancreas occurs in about 1 out of 79 people over their lifetime, but it has a high mortality rate, so even with a relatively low occurrence rate, it is the fourth leading cause of cancer-related death. The pancreas contains two types of glands: 95% of the glandular cells are exocrine glands, making enzymes released into the intestine that help digest food, and 5% of the cells are endocrine glands that make insulin and glucagons and release it into the blood [95].

Each of these cell types can give rise to a tumor, and these tumors are not alike. Exocrine tumors are the most common. The endocrine tumors are called islet cell tumors and are named according to the hormone production of the cell of origin (i.e., insulinomas). Exocrine and endocrine tumors have different risk factors, signs, symptoms, prognosis, and are diagnosed differently. CA19-9 is used to help in the diagnosis of exocrine pancreatic tumors. Again, this is not a screening molecule for the general population; it is used as a confirmatory aid when other signs, symptoms, and data support the conclusion that the patient has pancreatic cancer. One of the symptoms of pancreatic cancer is jaundice, and another is darkening of the urine and back pain. However, cancer is not the only thing that causes these symptoms. Imaging tests, blood tests for bilirubin, and CA19-9 and CEA analysis are performed to aid in diagnosis [95]. CA19-9 is sialylated Lewis a blood group antigen. The sensitivity of CA19-9 is 69–92%, and the specificity is 82%, with the false-positive results due to jaundice of other causes. The higher the level of CA19-9, the more likely it is that the cancer has metastasized. Some of the problems in sensitivity are due to the fact that Le^{a-b-} patients cannot manufacture CA19-9 and testing of these individuals can result in false negatives, but screening the population for Le^{a-b-} could reduce the false-negative reports [96, 97]. Finally, a biopsy performed by laparoscopy is done to confirm diagnosis. Only 10% of people with pancreatic cancer are diagnosed when they only have local disease, but even for these, the 5-year survival rate is 20%. The rest of the patients have a 5-year survival rate of 2%. Tn, TF, sTn, sLe^a, sLe^x, and Le^y have been found in pancreatic cancer but are not used in its diagnosis [47].

12.7.9 Prostate

The prostate is a walnut-sized gland between the rectum and the bladder responsible for producing some of the fluid in ejaculate. The seminal vesicles make most of the fluid and these are located behind the prostate. The prostate increases in size with age in older men, and this increase is called benign prostatic hypertrophy (BPH), a noncancerous condition that can increase the need to urinate as the gland presses on the bladder. In prostatic cancer the tumor causes pressure and an increased need to urinate can also be a result. Most prostate cancers grow slowly. About 1 out of 6 men will develop prostate cancer but only 1 out of 7 of these will die from it; however, it is the second leading cause of cancer death in men, behind lung cancer. Ninety percent of prostate cancers are found while they are still localized to the region, for these men their 5-year survival rates are unaffected. For all men with prostate cancer the 10-year survival rate is 91% and 15-year is 76%. These numbers are improving each year due to better diagnostic and treatment methods. Risk factors are age, race (African-American men have an increased incidence), nationality, and family history [98].

Digital rectal examinations with a blood test for prostate specific antigen (PSA) are used to screen for prostate cancer, and this screening should begin at age 50 for men with no risk factors, age 45 for men with a risk factor, and age 40 for men with multiple risk factors. Noncancer reasons for elevations in PSA include prostatitis, benign prostatic hypertrophy, recent ejaculation, and reasons for falsely low PSA levels include medications to treat BPH and herbal therapies. PSA can occur in a free or a complexed form. The percentage of free PSA is lower in men with prostate cancer, so the free to total PSA levels are used to help determine if someone who has a borderline level of PSA should have a biopsy. If free PSA is below 25%, then a biopsy is suggested. PSA velocity is the rate of change of the PSA levels from 1 year to the next, and this is helpful in the early diagnosis of cancer in individuals with a low baseline PSA. In order to correct for the normal increase in size of the prostate as a man ages, there are age-related cut-off before elevation of PSA levels are determined. PSA is also useful for monitoring whether tumor remains after surgery or radiation, and for monitoring therapy [98]. This very important antigen has been found to be glycosylated differently in tumors from that purified from normal tissues and fluids. These glycosylation differences may be exploitable to help differentiate between the PSA produced in benign prostatic hypertrophy, prostatitis, and prostate cancer [99–101]. This will decrease the apprehension and biopsies that are the result of false-positive PSA levels. Tn, TF, STn, and Globo H have been found in prostate cancer but are not used in its diagnosis [47].

12.8 INHERITED OR ACQUIRED DISORDERS OF GLYCOSYLATION

Alterations in glycosylation can cause diseases that are diagnosable by analysis of their signature glycans [102]. Diseases in this category are rare, including Wiskotts–Aldrich syndrome, hereditary erythroblastic multinuclearity with a positive acidified serum test, leukocyte adhesion deficiency (LAD) type II, and congenital disorders of glycosylation (CDGs).

Wiskotts–Aldrich syndrome is an immunodeficiency disease that is accompanied by eczema and thrombocytopenia. IgG levels in these patients are normal, but their IgM levels are decreased, and IgA and IgE levels elevated. This disease can be diagnosed by observing a decrease in the O-glycosylation of CD43 molecules on the blood mononuclear cell membranes [102]. Hereditary erythroblastic multinuclearity with a positive acidified serum test patients have anemia, enlarged spleens and livers, and liver cirrhosis. Its biochemical diagnosis is based on the detection of an abnormally glycosylated band in the electrophoresis of erythrocyte membranes [102]. The clinical symptoms associated with LAD type II are mental retardation, short stature, and unusual features. It is diagnosed by the absence of sialyl Lewis x on neutrophils [102]. CDGs are a group of related diseases usually associated with neurological disorder, except type Ib, which is assocviated with liver and intestinal problems. Type I CDG patients have defects in glycoprotein assembly, while type II CDG patients have defects in glycoprotein processing. Diagnosis of CDGs involves electrophoresis to show abnormal protein glycosylation [102].

REFERENCES

1. Gilbert GA (1991) Miscellaneous virus infections. In *Monographs in Clinical Pediatrics: Infectious Disease in Pregnancy and the Newborn Infants.* Harwood Academic, Camberwell, Victoria, Australia, pp. 393–413.
2. Patarca R, Fletcher MA (1995) Structure and pathophysiology of the erythrocyte membrane-associated Paul-Bunnell heterophile antibody determinant in Epstein-Barr virus-associated disease. *Crit. Rev. Oncogen.* 6:305–326.
3. Milgrom F, Abeyounis CJ, Eaton R (1984) Immunochemistry of tissue specific and tumor antigens, in *Molecular Immunology*, Zouhair Atassi M, Van Oss CJ, Absolom DR, Eds. Marcel Dekker, New York, pp. 75–90.
4. Hakomori S, Wang S-M, Young WW (1977) Isoantigenic expression of forssman glycolipid in human gastric and colonic mucosa: Its possible identity with "A-like antigen" in human cancer. *PNAS* 74:3023–3027.
5. Slavkin HC (1998) Toward molecularly based diagnostics for the oral cavity. *J. Am. Dent. Assoc.* 129:1138–1143.
6. Zymetx website product page: http://www.zymetx.com/index.cfm?pageid=124&subpageid=116. Accessed 2/7/08.
7. Fox JW, Marcon MJ, Bonsu BK (2006) Diagnosis of streptococcal pharyngitis by detection of *Streptococcus pyogenes* in posterior pharygeal versus oral cavity specimens. *J. Clin. Micro.* 44:2593–2594.

8. BD Chek™ StrpA Test (Dipstick). For detection of group A streptococcal antigen directly from throat swabs. Available at: http://www.bd.com/ds/technicalCenter/inserts/pkgInserts.asp#PF4.
9. Biso AL, Gerber MA, Gwaltney, JM, Kaplan, EL, Scwartz, RH (2002) Practice guidelines for the diagnosis and management of Group A streptococcal pharyngitis. *Clin. Infect. Dis.* 35:113–125.
10. Todar K. *Streptococcus pyogenes* and streptococcal disease. Available at: http://www.bact.wisc.edu/themicrobialworld/strep.html. Accessed 2/7/08.
11. Lancefield R (1933) A serological differentiation of human and other groups of hemolytic streptococci. *J. Exp. Med.* 57(4):571.
12. BBL™ Streptocard™ Enzyme Latex Test. A latex agglutination test for the identification of streptococcal groups A, B, C, D, F and G. Available at: http://www.bd.com/ds/technicalCenter/inserts/X5424(D)(0707).pdf. Accessed 2/7/08.
13. Infectious Disease Epidemiology Section Office of Public Health, Louisiana Dept of Health & Hospitals Streptococcal infections. Available at: http://dhh.louisiana.gov/offices/miscdocs/docs-249/Manual/StreptococcalInfectionsManual.pdf. Accessed 3/4/08.
14. Kobashi Y, Yoshida K, Miyashuta N, Niki Y, Matsushima T (2007) Evaluating the use of a *Streptococcus pneumoniae* urinary antigen detection kit for the management of community-acquired pneumonia in Japan. *Respiration* 74:387–393.
15. Inverness Medical (2006) *Streptococcus pneumoniae* Test. Available at: http://www.invernessmedicalpd.com/poc/downloads/OIA_CHL_PI_05059.08.pdf. Accessed 2/9/08.
16. Gisselsson-Solen M, Bylander A, Wilhelmsson C, Hermansson A, Melhus A (2007) The Binax NOW test as a tool for diagnosis of severe otitis media and associated complications. *J. Clin. Microbiol.* 45:3003–3007.
17. BD Directigen™ Meningitis Combo Test. Available at: http://www.bd.com/ds/technicalCenter/inserts/0214011JAA(0606).pdf. Accessed 2/7/08.
18. BioStar® OIA® Chlamydia product insert. Available at: http://www.invernessmedicalpd.com/poc/downloads/OIA_CHL_PI_05059.08.pdf. Accessed 3/2/08.
19. Thirumalapura NR, Ramachandran A, Morton RJ, Malayer JR (2006) Bacterial cell microarrays for the detection and characterization of antibodies against surface antigens. *J. Immunol. Methods* 309:48–54.
20. Centers for Disease Control and Prevention. WHO Collaborating Center for the Mycoses. Available at: http://www.cdc.gov/ncidod/dbmd/mdb/index.htm. Accessed 2/26/08.
21. Yeo SF, Wong B (2002) Current status of nonculture methods for diagnosis of Invasive Fungal Infections. *Clin. Microbiol. Rev.* 15:465–484.
22. Centers for Disease Control and Prevention. Division of Bacterial and Mycotic Diseases Invasive Cadidiasis. Available at: http://www.cdc.gov/ncidod/dbmd/diseaseinfo/candidiasis_inv_g.htm. Accessed 2/9/08.
23. Centers for Disease Control and Prevention. Division of Bacterial and Mycotic Diseases Cryptococcus. Available at: http://www.cdc.gov/ncidod/dbmd/diseaseinfo/cryptoccosis_t.htm. Accessed 2/9/08.
24. Wheat LJ, Garringer T, Brizendinea E, Connolly P (2002) Diagnosis of histoplasmosis by antigen detection based upon experience at the histoplasmosis reference laboratory. *Diag. Micr. Infec. Dis.* 43:29–37.

25. Centers for Disease Control and Prevention. Division of Bacterial and Mycotic Diseases Histoplasmosis. Available at: http://www.cdc.gov/ncidod/dbmd/diseaseinfo/histoplasmosis_g.htm. Accessed 2/9/08.
26. Centers for Disease Control and Prevention. National Institute for Occupational Health and Safety. Histoplasmosis—Protecting workers at risk. Available at: http://www.cdc.gov/niosh/docs/2005-109/#c. Accessed 2/9/08.
27. Center for Disease Control. Laboratory Identification of Parasites of Public Health Concern. Diagnostic Procedures. Available at: http://www.dpd.cdc.gov/dpdx/HTML/DiagnosticProcedures.htm. Accessed 2/9/08.
28. Mendonça-Previato L, Todeschini AR, Heise N, Previato J (2005) Protozoan parasite-specific carbohydrate structures. *Curr. Opin. Struct. Biol.* 15:499–505.
29. Osinaga E (2007) Expression of cancer-associated simple mucin-type O-glycosylated antigens in parasites. *IUBMB Life* 59(4):269–273.
30. Center for Disease Control. Division of Parasitic Diseases. Parasitic Disease Information. Available at: http://www.cdc.gov/ncidod/dpd/parasites/alveolarechinococcosis/facts ht_alveolarechinococcosis.htm. Accessed 2/9/08.
31. Dai WJ, Hemphill A, Waldvogel A, Ingold K, Deplazes P, Mossman H, Gottstein B (2001) Major carbohydrate antigen of *Echinococcus multilocularis* induces an Immunoglobulin G response independent of CD4T Cells. *Infect. Immun.* 69:6074–6083.
32. Yamano K, Hada N, Yamamura T, Takeda T, Honma H, Sawada Y (2006) Serodiagnostic potential of chemically synthesized glycosphingolipid antigens in an enzyme-linked immunosorbent assay for alveolar echinococcosis. *J. Helminthol.* 80(4):387–391.
33. Yong T-S, Lee J-S, Cho S-N, Seo J-H, Park HA (1996) A carbohydrate antigen of *Clonorchis sinensis* recognized by a species-specific monoclonal antibody. *Korean J. Parasitol.* 34(4):279–281.
34. Lim MK, Ju Y-H, Franceschi S, Oh J-K, Kong H-J, Hwang SS, Park S-K, Cho S-I, Sohn W-M, Kim D-I, Yoo K-Y, Hong S-T, Shin H-R (2006) *Clonorchis sinensis* infection and increasing risk of cholangiocarcinoma in the Republic of Korea. *Am. J. Trop. Med. Hyg.* 75:93–96.
35. Forbes LB, Appleyard GD, Gajadhar AA (2004) Comparison of synthetic tyvelose antigen with excretory-secretory antigen for the detection of trichinellosis in swine using enzyme-linked immunosorbent assay. *J. Parisitol.* 90(4):835–840.
36. Møller LN, Petersen E, Gamble HR, Kapel CMO (2005) Comparison of two antigens for demonstration of *Trichinella* spp. antibodies in blood and muscle fluid of foxes, pigs and wild boars. *Vet. Parasitol.* 132:81–84.
37. Zhang GP, Guo J, Wang X, Yang J, Yang Y, Li Q, Li X, Deng R, Xiao Z, Yang J, Xing G, Zhao D (2006) Development and evaluation of an immunochromatographic strip for trichinellosis detection. *Vet. Parasitol.* 137:286–293.
38. Center for Disease Control. Division of Parasitic Diseases. Parasitic disease information. Available at: http://www.cdc.gov/NCIDOD/DPD/parasites/schistosomiasis/default.htm.
39. Tsang VC, Hancock K, Kelly MA, Wilson BC, Maddison SE (1983) *Schistosoma mansoni* adult microsomal antigens, A serologic reagent II. Specificity of antibody responses to the S. *mansoni* microsomal antigen (MAMA). *J. Immunol.* 130:1303–1136.
40. Krishnamurti U, Steffes MW (2001) Glycohemoglobin: A primary predictor of the development or reversal of complications of diabetes mellitus. *Clin. Chem.* 47:1157–1165.

41. Rohlfing CL, Weidenmeyer H-M, Little R, England JD, Tennill A, Goldstein DE (2002) Defining the relationship between plasma glucose and HbA1c analysis of glucose profiles and HbA1c in the diabetes control and complications. *Trial Diabetes Care* 25:275–278.
42. Feizi T, Kapadia A, Yount W (1980) I and i antigens of human peripheral blood lymphocytes cocap with receptors for concanavalin A. *PNAS* 77(1):376–380.
43. Gertz MA. (2005) Cold agglutinin disease. *Haematologica* 91:439–441.
44. Ferrante M, Henckaerts L, Joossens M, Pierik M, Joossens S, Dotan N, Norman GL, Altstock RT, Van Steen K, Rutgeerts P, Van Assche G, Vermeire S (2007) New serological markers in inflammatory bowel disease are associated with complicated disease behaviour. *Gut* 56:1394–1403.
45. Tresca AJ. The Difference between ulcerative colitis and Crohn's disease. Available at: http://ibdcrohns.about.com/od/ulcerativecolitis/a/diffuccd.htm. Accessed 2/20/08.
46. Meezan E, Wu HC, Black PH, Robbins PW (1969) Comparative studies on the carbohydrate-containing membrane components of normal and virus-transformed mouse fibroblasts. II. Separation of glycoproteins and glycopeptides by sephadex chromatography. *Biochemistry* 8(6):2518–2524.
47. Frieer T, Bay S, Vichier-Guere S, Lo-Man R, Leclerc C (2006) Carbohydrate antigens: Synthesis aspects and immunological applications in cancer. *Mini-Rev. Med. Chem.* 6:1357–1373.
48. Hakomori S (2000) Traveling for the glycosphingolipid path. *Glycoconjugate J.* 17(7–9):627–647.
49. Martensson S, Due C, Pahlsson P, Nilsson B, Eriksson H, Zopf D, Olsson L, Lunblad A (1988) A carbohydrate epitope associated with human squamous lung cancer. *Cancer Res.* 48:2125–2131.
50. Dube DH, Bertozzi CR (2005) Glycans in cancer and inflammation-potential for therapeutics and diagnostics. *Nat. Rev. Cancer* 4:477–487.
51. Center for Disease Control. National Program of Cancer Registries. United States cancer statistics (USCS) (NPCR). Available at: http://www.cdc.gov/cancer/npcr/uscs/2004/facts_major_findings.htm. Accessed 2/28/08.
52. Hautmann S, Toma M, Gomez MFL, Friedrich MG, Jaekel T, Michl U, Schroeder GL, Huland H, Juenemann KP, Lokeshwar VB (2004) Immunocyte and the HA-HAase urine tests for the detection of bladder cancer: A side by side comparison. *Eur. Urol.* 46:466–471.
53. Harris L, Fritsche H, Mennel R, Norton L, Ravdin P, Taube S, Somerfield MR, Hayes, DF, Bast RC (2007) American Society of Clinical Oncology update of recommendations for the use of tumor markers in breast cancer. *J. Clin. Oncol.* 25:1–26.
54. van Gisbergen KPJM, Aarnoudse CA, Meijer GA, Geijtenbeek TBH, van Kooyk Y (2005) Dendritic cells recognize tumor-specific glycosylation of carcinoembryonic antigen on colorectal cancer cells through dendritic cell-specific intercellular adhesion molecule-3-grabbing nonintegrin. *Cancer Res.* 65:5935–5944.
55. Yamashita K, Totani K, Kuroki M, Matsuoka Y, Ueda I, Kobata A (1987) Structural studies of the carbohydrate moieties of carcinoembryonic antigens. *Cancer Res.* 47:3451–3459.
56. American Cancer Society Overview Colon Cancer. Available at: http://www.cancer.org/docroot/lrn/lrn_0.asp. Accessed 3/4/08.
57. American Cancer Society Overview Liver Cancer. Available at: http://www.cancer.org/docroot/CRI/CRI_2_1x.asp?dt=25. Accessed 3/4/08.

58. Mizejewski GJ (2001) Alpha-fetoprotein structure and function: Relevance to isoforms, epitopes, and conformational variants. *Exp. Biol. Med.* 226:377–408.
59. Carr BI, Kanke F, Wise M, Satumura S (2007) Clinical evaluation of Lens culinaris agglutinin-reactive α-fetoprotein and des-γ-carboxy prothrombin in histologically proven hepatocellular carcinoma in the United States. *Digest Dis. Sci.* 52:776–782.
60. Leerapun A, Suravarapu SV, Bida JP, Clark RJ, Sanders EL, Mettler T, Stadheim LM, Aderca I, Moser CD, Nagorney DM, LaRusso NF, de Groen PC, Menon KVN, Lazaridis KN, Gores GJ, Charlton MR, Roberts RO, Therneau TM, Katzmann JA, Roberts LR (2007) The utility of Lens culinaris agglutinin-reactive α-fetoprotein in the diagnosis of hepatocellular carcinoma: Evaluation in a United States referral population. *Clin. Gastroenterol. H* 5:394–402.
61. Li D, Mallory T, Satomura S (2001) AFP-L3: A new generation of tumor marker for hepatocellular carcinoma. *Clin. Chim. Acta* 313:15–19.
62. American Cancer Society Overview Lung Cancer. Available at: http://www.cancer.org/docroot/CRI/content/CRI_2_4_1x_What_Is_Non-Small_Cell_Lung_Cancer.asp?sitearea=. Accessed 3/2/08.
63. Fujii T, Dracheva T, Player A, Chacko S, Clifford R, Strausberg RL, Buetow K, Azumi N, Travis WD, Jen J (2002) A preliminary transcriptome map of non-small cell lung cancer. *Cancer Res.* 62:3340–3346.
64. Kutikova L, Bowman L, Chang S, Long SR, Obasaju C, Crown WH (2005) The economic burden of lung cancer and the associated costs of treatment failure in the United States. *Lung Cancer-J. IASLC* 50:143–154.
65. Fontana RS, Sanderson DR, Woolner LB, Taylor WF, Miller WE, Muhm JR (1986) Lung cancer screening: The Mayo program. *J. Occup. Med.* 28(8):746–750.
66. Melamed MR, Flehinger BJ, Zaman MB, Heelan RT, Perchick WA, Martini N (1984) Screening for early lung cancer. Results of the Memorial Sloan-Kettering study in New York. *Chest* 86(1):44–53.
67. Frost JK, Ball WC, Levin ML, Tockman MS, Baker RR, Carter D, Eggleston JC, Erozan YS, Gupta PK, Khouri NF, Marsh BR, Stitik FP (1984) Early lung cancer detection: Results of the initial (prevalence) radiologic screening in the John Hopkins study. *Am. Rev. Respir Dis.* 130:549–554.
68. Fontana RS, Sanderson DR, Woolner LB, Taylor WF, Miller WE, Muhm JR, Bernatz PE, Payne WS, Pairolero PC, Bergstralh EJ (1991) Screening for lung cancer. A critique of the Mayo Lung Project. *Cancer* 67(4 Suppl):1155–1164.
69. Kubik A, Polak J (1986) Lung cancer detection. Results of a randomized prospective study in Czechoslovakia. *Cancer* 57(12):2427–2437.
70. Sagawa M, Tsubono Y, Saito Y, Sato M, Tsuji I, Takahashi S, Usuda K, Tanita T, Kondo T, Fujimura S (2003) A case control study for evaluating the efficacy of mass screening programs for lung cancer in Miyagi Prefecture, Japan. *Lung Cancer* 41:29–36.
71. Marcus P (2003) Conflicting evidence in lung cancer screening: Randomized controlled trial versus case-control studies. *Lung Cancer-J. IASLC* 41:37–39.
72. Bepler G, Goodridge CD, Djulbegovic B, Clark RA, Tockman MA (2003) Systematic review and lessons learned from early lung cancer detection trials using low dose computed tomography of the chest. *Cancer Control* 10:306–314.
73. Patz EF Jr (2006) Lung cancer screening, overdiagnosis bias, and reevaluation of the Mayo Lung Project. *J. Natl. Cancer I.* 98(11):724–725.

74. Srivastava S, Kruegar K, Johnsey D (2005) The Early Detection Research Network: Translational research to identify early cancer and cancer risk. Third Report. U.S. Department of Health and Human Services, National Institutes of Health, National Cancer Institute, Rockville, MD.
75. Rom WN, Tchou-Wong KM (2003) Functional genomics in lung cancer and biomarker detection. *Am. J. Respir. Cell Mol. Biol.* 29:153–156.
76. Rom WN, Tchou-Wong KM (2003) Molecular and genetic aspects of lung cancer. *Met. Mol. Med.* 75:3–26.
77. Romeo MS, Sokolova IA, Morrison LE, Zeng C, Baron AE, Hirsch FR, et al (2003) Chromosomal abnormalities in non-small cell lung carcinomas and in bronchial epithelia of high-risk smoker detected by multi-target interphase fluorescence in situ hybridization. *J. Mol. Diagn.* 5:103–112.
78. Topaloglu O, Hoque MO, Tokumaru Y, Lee J, Ratovitski E, Sidransky D, Moon CS (2004) Detection of promoter hypermethylation of multiple genes in the tumor and bronchoalveolar lavage of patients with lung cancer. *Clin. Cancer Res.* 10:2284–2288.
79. Zhukov TA, Johanson RA, Cantor AB, Clark RA, Tockman MS (2003) Discovery of distinct protein profiles specific for lung tumors and pre-malignant lung lesions by SELDI mass spectrometry. *Lung Cancer-J. IASLC* 40:267–279.
80. Rossi A, Maione P, Colantuoni G, Gaizo FD, Guerriero C, Nicolella D, Ferrara C, Gridelli C (2005) Screening for lung cancer: New horizons? *Crit. Rev. Oncol-Hemat.* 56(3): 311–320.
81. Zhang L, Liu J, Zhang H, Wu S, Huang L, He D, Xiao X (2005) Discovery and identification of anti-U1-A snRNP antibody in lung cancer. *Science in China. Series C, Life Sci.* 48(6):641–647.
82. Stranahan PL, Howard RB, Pfenninger O, Cowen ME, Johnston MR, Pettibone DE (1992) Mucin gel formed by tumorigenic squamous lung carcinoma cells has Le(A)-X oligosaccharides and excludes antibodies from underlying cells. *Cancer Res.* 52:2923–2930.
83. Tanaka F, Miyahara R, Ohtake Y, Yanagihara K, Fukuse T, Hitomi S, Wada H (1998) Lewis Y antigen expression and postoperative survival in non-small cell lung cancer. *Ann. Thorac. Surg.* 66(5):1745–1750.
84. Satoh H, Ishikawa H, Yamashita YT, Takahashi H, Ishikawa S, Kamma H, Ohtsuka M, Hasegawa S (1998) Predictive value of preoperative serum sialyl Lewis X-i antigen levels in non-small cell lung cancer. *Anticancer Res.* 18(4B):2865–2868.
85. Vangsted AJ, Zeuthen J (1993) Monoclonal antibodies for diagnosis and potential therapy of small cell lung cancer—the ganglioside antigen fucosyl-GM1. *Acta Oncol.* 32(7–8):45–51.
86. Nilsson O (1992) Carbohydrate antigens in human lung carcinoma. *APMIS* 100(S27):149–161.
87. Giaccone G, Debruyne C, Felip E, Chapman PB, Grant SC, Millward M, Thiberville L, D'addario G, Coens C, Rome LS, Zatloukal P, Masso O, Legrand C (2005) Phase III study of adjuvant vaccination with Bec2/bacille Calmette-Guerin in responding patients with limited-disease small-cell lung cancer. *J. Clin. Oncol.* 23(28):6854–6864.
88. Fuentes R, Allman R, Mason MD (1997) Ganglioside expression in lung cancer cell lines. *Lung Cancer-J. IASLC* 18(1):21–33.

89. Rossi A, Maione P, Colantuoni G, Gaizo FD, Guerriero C, Nicolella D, Ferrara C, Gridelli C (2005) Screening for lung cancer: New horizons? *Crit. Rev. Oncol-Hemat.* 56(3): 311–320.
90. American Cancer Society Detailed Guide: Melanoma. Available at: http://www.cancer.org/docroot/CRI/content/CRI_2_2_1X_What_is_melanoma_skin_cancer_50.asp?sitearea=. Accessed 3/4/08.
91. American Cancer Society Detailed Guide: Ovarian Cancer. Available at: http://www.cancer.org/docroot/CRI/content/CRI_2_4_1X_What_is_ovarian_cancer_33.asp?sitearea=. Accessed 3/4/08.
92. Zhang Z, Yu Y, Xu F, Berchuck A, van Haaften-Day C, Havrilesky LJ, de Bruijn HWA, van der Zee AGJ, Woolas RP, Jacobs IJ, Skates S, Chan DW, Bast RC (2007) Combining multiple serum tumor markers improves detection of stage I epithelial ovarian cancer. *Gynecol. Oncol.* 107:526–531.
93. Jankovic MM, Tapuskovic BS (2005) Molecular forms and microheterogenity of the oligosaccharide chains of pregnancy-associated CA125 antigen. *Human Reprod.* 29:2632–2638.
94. Wong NK, Easton RL, Panico M, Sutton-Smith M, Morrison JC, Lattanzio FA, Morris HA, Clark GF, Dell A, Patankar MS (2003) Characterization of the oligosaccharides associated with the human ovarian tumor marker CA-125. *J. Biol. Chem.* 278:28619–28634.
95. American Cancer Society Detailed Guide: Pancreatic Cancer. Available at: http://www.cancer.org/docroot/CRI/content/CRI_2_4_1X_What_is_pancreatic_cancer_34.asp?sitearea=. Accessed 3/2/08.
96. Goonetilleke KS, Siriwardena AK (2007) Systematic review of carbohydrate antigen (CA19-9) as a biochemical marker in the diagnosis of pancreatic cancer. *Eur. J. Surg. Oncol.* 33:266–270.
97. Bouvet M (2004) Tumor markers for pancreatic cancer: What happens when preoperative CA19-9 is undetectable? *Annals Surg. Oncol.* 11:637–638.
98. American Cancer Society Detailed Guide: Prostate Cancer. Available at: http://www.cancer.org/docroot/CRI/CRI_2_3x.asp?dt=36. Accessed 3/2/08.
99. Peracaula R, Tabarés G, Royle L, Harvey DJ, Dwek RA, Rudd PM, de Llorens R (2003) Altered glycosylation pattern allows the distinction between prostate-specific antigen (PSA) from normal and tumor origins. *Glycobiology* 13:457–470.
100. Janković M, Kosanović MM (2005) Glycosylation of urinary prostate-specific antigen in benign hyperplasia and cancer: Assessment by lectin-binding patterns. *Clin. Biochem.* 38:58–65.
101. Drake RR, Schwegler EE, Malik G, Diaz J, Block T, Mehta A, Semmes OJ (2006) Lectin capture strategies combined with mass spectrometry for the discovery of serum glycoprotein biomarkers. *Mol. Cell Proteomics* 5:1957–1967.
102. Durand G, Shea N (2000) Protein glycosylation and diseases: Blood and urinary oligosaccharides as markers for diagnosis and therapeutic monitoring. *Clin. Chem.* 46:795–805.

INDEX

Acquired glycosylation disorders, diagnosis, 388
Acute otitis media, *Streptococcus pneumoniae*, antibacterial vaccines, 140
Adaptive immunity, 22–23. *See also* Immune system; Innate immunity
　carbohydrate-based cancer vaccines, 270–271
　innate immunity and, therapeutic cancer vaccines, 334–336
Adaptive response, initiation and stimulation of, adjuvants, 90–92
Adenocarcinoma (ADC), 383
Adjuvants:
　adaptive response initiation and stimulation, 90–92
　animal and in vitro models, 104–106
　applications, 97–98
　　PRR agonists, 98
　　TLR agonists, 97–98
　carbohydrate-based cancer vaccines, 278
　carbohydrate-based vaccines, 98–100
　　free polysaccharides (Ti antigens), 99–100
　　generally, 98–99
　　glycoconjugate vaccines, 100
　　Td and Ti B-cell responses, 99
　classical adjuvants, 92–95
　　aluminum, 92–93
　　antigen/formulation targeting, 94
　　CD8+ CTL cells, 95
　　emulsions, 93–94
　　liposomes and microparticles, 94
　　saponins, 94
　combinations, 101
　defined, 89
　immunomodulation, 101–102
　innate immunity, 95–97
　　antigen capture and recognition, 97
　　non-toll-like receptor (TLR) agonists, 97
　　toll-like receptor (TLR) agonists, 95–96
　need for, 90
　regulatory factors, 103
　routes of administration, 102–103
　　epidermal or intradermal, 102–103
　　mucosal, 102
　safety versus efficacy factors, 103–104
Administration:
　adjuvants, 102–103
　　epidermal or intradermal, 102–103
　　mucosal, 102
　therapeutic cancer vaccines, 339
Aluminum, classical adjuvant, 92–93
American Cancer Society (ACS), 383
Animal models, adjuvants, 104
Antibacterial vaccines. *See* Carbohydrate-based antibacterial vaccines
Antibody-dependent cellular cytotoxicity (ADCC):
　carbohydrate-based cancer vaccines, 268
　therapeutic cancer vaccines, 335–336
Antibody glycosylation, glycoimmunobiology, 23–24

Antifungal vaccines. *See* Carbohydrate-based antifungal vaccines
Antigen(s), structural classification, 228, 229–231
Antigen capture, innate immunity, 97
Antigen/formulation targeting, classical adjuvants, 94
Antigen-presenting cells (APCs):
  adaptive response initiation and stimulation, adjuvants, 90–92
  carbohydrate-based cancer vaccines, 269
  classical adjuvants, 93
  therapeutic cancer vaccines, clinical trials, 354
Antiparasitic vaccines. *See* Carbohydrate-based antiparasitic vaccines
Antiviral vaccines. *See* Carbohydrate-based antiviral vaccines
Aspergillus spp., 215, 216
Aspergillus fumigatus, 216
  diagnosis, 375
  vaccine for, 221
Autoimmunity disease diagnosis:
  cold agglutinin disease, 379
  diabetes, 378–379
  inflammatory bowel disease, 380

Bacterial carbohydrate-based antigens, 30–32
Bacterial disease diagnosis, 371–374
  Chlamydia trachomatis, 374
  future prospects, 374
  Streptococcus groups A, B, C, D, F, and G, 371–373
  Streptococcus pneumoniae, 373
  Streptococcus pyogenes, 371
Bacterial disease vaccines. *See* Carbohydrate-based antibacterial vaccines
B-cells:
  adaptive response initiation and stimulation, adjuvants, 90–92
  carbohydrate-based cancer vaccines, 268
  infectious mononucleosis, 369–370
  T-cell di-epitope constructs and, carbohydrate-based cancer vaccines, 284–288

Td and Ti B-cell responses, carbohydrate-based vaccines, adjuvants, 99
B-epitope, receptor ligand di-epitope constructs and, carbohydrate-based cancer vaccines, 279–283
Beta-glucan-protein conjugates, glycoconjugate vaccines, carbohydrate-based antifungal vaccines, 221–222
Bilharzia (schistosomiasis), carbohydrate-based antiparasitic vaccines, 205–207
Biological activity, therapeutic cancer vaccines, 339–340
Bladder cancer, diagnosis, 381
Blood type, carbohydrate antigens, 28
Bordetella pertussis, 95
Breast cancer:
  diagnosis, 381–382
  vaccines, 263
Bunyamwera virus, glycosylation, carbohydrate-based antiviral vaccines, 174

Calixarenes, carbohydrate-based cancer vaccines, 283
Cancer, carbohydrate antigens, 29–30
Cancer-associated and related glycosphingolipids, 227–261. *See also* Carbohydrate-based cancer vaccines; Sialic acid glycoengineering; Therapeutic cancer vaccines; Tumor-associated carbohydrate antigens (TACAs); Tumor diagnosis
  antigen structural classification, 228, 229–231
  ganglio-series antigens, 236–240
  globo-series and related antigens, 233–236
  glycosphingolipid (GSL) glycan structures, 228, 232–233
  lacto/neolacto-hybrid and repeating lacto-series antigens, 247–248
  lacto-series (type 1 chain; Lc_n) antigens, 240–241
  neolacto-series (type 2 chain; nLc_x) antigens, 241–247
  overview, 227–228

therapeutic cancer vaccines, clinical
trials, 345–350
Cancer immunotherapy, sialic acid
glycoengineering, 321–325
Cancer Vaccine Clinical Trial Working
Group (VCVTWG), 338, 339
Cancer vaccines. *See* Cancer-associated and
related glycosphingolipids;
Carbohydrate-based cancer
vaccines; Sialic acid
glycoengineering; Therapeutic
cancer vaccines
Candida albicans, antifungal vaccines, 216,
220–221
Candida albicans, glycoconjugate vaccines,
carbohydrate-based antifungal
vaccines, 215, 216, 220–221
Candida neoformans, glycoconjugate
vaccines, carbohydrate-based
antifungal vaccines, 217–220,
221–222
Candidiasis, invasive, diagnosis, 375
Capsular polysaccharide-protein
conjugates, 56–68
Group B *Streptococcus* (GBS), 65–67
Haemophilus influenzae, 56–59
Neisseria meningitidis, 60–64
Salmonella typhi, 64–65
Streptococcus pneumoniae, 59–60
Carbohydrate antigens, 28–38
cancer, 29–30
carbohydrate-based vaccines, 34–37
challenges, 35–37
historical perspective, 34–35
metabolic glylcoengineering, 37
requirements, 35
historical perspective, 28–29
pathogens, 30–34
bacterial carbohydrate-based antigens,
30–32
glycosylation, 30
parasites, 32–33
vaccines, 34
viral glycosulation, 33–34
Carbohydrate-based antibacterial vaccines,
117–166
future trends, 146–147
glycoconjugate vaccines, 121–122
Group B *Streptococcus* (GBS), 140–145

Haemophilus influenzae, 122–125
Hib conjugate vaccines, 123–125
Hib polysaccharides, 122–123
Neisseria meningitidis, 125–133
generally, 125–126
meningococcal conjugate vaccines,
126–133
serogroup A, 126–128
serogroup B, 130–131
serogroup C, 128–130
serogroups A, C, W-135, and Y,
131–133
meningococcal polysaccharide
vaccines, 126
overview, 117–118
polysaccharide and glycoconjugate
immunobiology, 118–120
polysaccharides, human immune
response to, 120–121
Salmonella typhi, 145–146
Streptococcus pneumoniae, 133–140
acute otitis media, 140
generally, 133–139
invasive pneumococcal disease, 139
Carbohydrate-based antifungal vaccines,
215–226
glycoconjugate vaccines, 217–222
beta-glucan-protein conjugates,
221–222
Candida albicans, 220–221
Candida neoformans, 217–220
immune system function, 222–223
overview, 215–216
terminology, 216
Carbohydrate-based antiparasitic vaccines,
195–214
future prospects, 208
leishmaniasis, 201–205
malaria, 197–201
overview, 195–196
schistosomiasis (bilharzia), 205–207
toxoplasmosis and Chagas' disease,
207–208
Carbohydrate-based antiviral vaccines,
167–193
glycosylation, 168–174
ebola virus, 174
hepatitis C virus, 173
HIV, 170–172

Carbohydrate-based antiviral
  vaccines (*Continued*)
    influenza A virus, 172–173
    SARS coronavirus (SARS-CoV), 174
    viral *N*-glycosylation, 168–169
    West Nile virus, 174
  overview, 167–168
  vaccine and drug development, 174–182
    hepatitis C virus, 182
    HIV, 174–181
      biosynthetic vaccines, 180
      carbohydrate-binding agents,
        180–181
      synthetic vaccines, 175–180
      viral envelope, 175
    influenza A virus, 181–182
Carbohydrate-based cancer vaccines, 263–311. See also Cancer-associated and related glycosphingolipids; Sialic acid glycoengineering; Therapeutic cancer vaccines; Tumor-associated carbohydrate antigens (TACAs); Tumor diagnosis
  B- and T-cell di-epitope constructs,
    284–288
  B-epitope and receptor ligand di-epitope constructs, 279–283
  chemical synthesis of tumor-associated carbohydrates and glycopeptides,
    271–276
  development of, 267
  fully synthetic carbohydrate-based
    vaccines, 279
  humoral immune response to
    carbohydrates, 267–269
  major histocompatibility-peptide
    complex (MHC)-mediated
    immune response, to
    glycopeptides, 269–270
  overview, 263–264
  semisynthetic carbohydrate-based
    vaccines, 276–278
  toll-like receptors and immunity,
    270–271
  tricomponent vaccines, 288–292
  tumor-associated carbohydrate antigens
    (TACAs), 264–267
Carbohydrate-based diagnostics, 367–394
  autoimmunity, 378–380
    cold agglutinin disease, 379
    diabetes, 378–379
    inflammatory bowel disease, 380
  bacteria, 371–374
    Chlamydia trachomatis, 374
    future prospects, 374
    meningitis, 373–374
    Streptococcus groups A, B, C, D, F,
      and G, 371–373
    Streptococcus pneumoniae, 373
    Streptococcus pyogenes, 371
  fungi:
    Aspergillus fumigatus, 375
    Cryptococcus neoformans, 375–376
    generally, 374–375
    Histoplasma capsulatum, 376
    invasive candidiasis, 375
  generally, 367–369
  glycosylation disorders, 388
  parasites, 377–378
    Clonorchis sinensis, 378
    Echinococcus multilocularis,
      377–378
    generally, 377
    Schistomsoma mansoni, 378
    Trichinella spp., 378
  tumors, 380–387
    bladder cancer, 381
    breast cancer, 381–382
    colon cancer, 382
    generally, 380–381
    liver cancer, 382–383
    lung cancer, 383–384
    melanoma, 384–385
    ovarian cancer, 385–386
    pancreatic cancer, 386–387
    prostate cancer, 387
  viruses, 369–371
    infectious mononucleosis,
      369–370
    influenza A and B, 370–371
Carbohydrate-based vaccines, 34–37
  adjuvants, 98–100
    free polysaccharides (Ti antigens),
      99–100
    generally, 98–99
    glycoconjugate vaccines, 100
    Td and Ti B-cell responses, 99
  challenges, 35–37

historical perspective, 34–35
metabolic glylcoengineering, 37
requirements, 35
Carbohydrates:
humoral immune response, carbohydrate-based cancer vaccines, 267–269
mammalian, 1–2
tumor-associated glycopeptides and, chemical synthesis of, carbohydrate-based cancer vaccines, 271–276
CD8+ cytotoxic T lymphocytes (CTLs):
classical adjuvants, 95
therapeutic cancer vaccines, 336
Cell surface carbohydrates. See Carbohydrates
Cell surface sialic acid, glycoengineering of, 313–331. See also Sialic acid glycoengineering
Cell- to systems-level control (siglecs), glycoimmunobiology, 26–27
Centers for Disease Control and Prevention (CDC), 381
Cervical carcinoma, 264
Chagas' disease, carbohydrate-based antiparasitic vaccines, 207–208
Chlamydia pneumoniae, diagnosis, 374
Chlamydia psittaci, diagnosis, 374
Chlamydia trachomatis, 75, 374
Classical adjuvants, 92–95
aluminum, 92–93
antigen/formulation targeting, 94
CD8+ CTL cells, 95
emulsions, 93–94
liposomes and microparticles, 94
saponins, 94
Clinical activity, therapeutic cancer vaccines, 339–340
Clinical drug trials:
generally, 337–347
glycosphingolipid-based vaccines, 347–350
therapeutic cancer vaccines, 337–338 (*See also* Therapeutic cancer vaccines)
Clonorchis sinensis, diagnosis, 378
Cluster glycoside effect, 27–28
Cold agglutinin disease, diagnosis, 379
Colon cancer, diagnosis, 382

Complement-dependent cytotoxicity (CDC):
carbohydrate-based cancer vaccines, 268
therapeutic cancer vaccines, 335–336
Crohn's disease, diagnosis, 380
Cryptococcus gattii, 216
Cryptococcus neoformans, 215–216, 375–376
Cyclophosphamide (CY), carbohydrate-based cancer vaccines, 278
Cytomegalovirus, glycosylation, carbohydrate-based antiviral vaccines, 174
Cytosolic proteins, O-GlcNAc modification, 15–16

Dendritic cells:
adaptive response initiation and stimulation, adjuvants, 90–92
therapeutic cancer vaccines, clinical trials, 354
Diabetes, diagnosis, 378–379
Diagnosis, 367–394
autoimmunity, 378–380
cold agglutinin disease, 379
diabetes, 378–379
inflammatory bowel disease, 380
bacteria, 371–374
Chlamydia trachomatis, 374
future prospects, 374
meningitis, 373–374
Streptococcus groups A, B, C, D, F, and G, 371–373
Streptococcus pneumoniae, 373
Streptococcus pyogenes, 371
fungi, 374–376
Aspergillus fumigatus, 375
Cryptococcus neoformansj, 375–376
generally, 374–375
Histoplasma capsulatum, 376
invasive candidiasis, 375
glycosylation disorders, 388
overview, 367–369
parasites:
Clonorchis sinensis, 378
Echinococcus multilocularis, 377–378
generally, 377
Schistomsoma mansoni, 378
Trichinella spp., 378

Diagnosis (*Continued*)
  tumors, 380–387
    bladder cancer, 381
    breast cancer, 381–382
    colon cancer, 382
    generally, 380–381
    liver cancer, 382–383
    lung cancer, 383–384
    melanoma, 384–385
    ovarian cancer, 385–386
    pancreatic cancer, 386–387
    prostate cancer, 387
  viruses, 369–371
    infectious mononucleosis, 369–370
    influenza A and B, 370–371
Di-epitope, receptor ligand di-epitope constructs and b-epitope, carbohydrate-based cancer vaccines, 279–283
Disease. *See also* Diagnosis
  carbohydrates, 367–394
  glycosylation, 4–5
Disialosyl-LC$_4$Cer antigen, 240
Disialosyl-Le^a, 241
Dosage, therapeutic cancer vaccines, 339

Ebola virus, glycosylation, carbohydrate-based antiviral vaccines, 174
Echinococcus multilocularis, diagnosis, 377–378
Emulsions, classical adjuvants, 93–94
Endothelial glycocalyx layer (EGL), leukocyte extravasation, 24–25
Enzyme-linked immunosorbent assay (ELISA), therapeutic cancer vaccines, 339–340
Enzyme-linked immunosorbent spot assay (ELISPOT), therapeutic cancer vaccines, 339–340
Epidermal adjuvants, routes of administration, 102–103
Epithelial cancers, carbohydrate-based cancer vaccines, 266
Epitope constructs:
  B- and T-cell di-epitope constructs, carbohydrate-based cancer vaccines, 284–288

B-epitope and receptor ligand di-epitope constructs, carbohydrate-based cancer vaccines, 279–283
tricomponent vaccines, 288–292
Escherichia coli, 75
  diagnosis, 374
  Lipopolysaccharide (LPS) conjugates, 69–70
  mucosal adjuvants, 102
European Medicines Agency (EMEA), 337

Filoviridae family (virus), glycosylation, carbohydrate-based antiviral vaccines, 174
Food and Drug Administration (FDA), 103–104, 337, 370
Free polysaccharides (Ti antigens), carbohydrate-based vaccines, adjuvants, 99–100
Freund's adjuvant, 93–94
Fucosylated *N*-glycan, schistosomiasis (bilharzia), 205–207
Fucosyl-G$_{M1}$ ANTIGEN:
  cancer-associated and related glycosphingolipids, 239–240
  clinical trials, 350
Fungal disease diagnosis. *See also* Carbohydrate-based antifungal vaccines
  Aspergillus fumigatus, 375
  Cryptococcus neoformansj, 375–376
  generally, 374–375
  Histoplasma capsulatum, 376
  invasive candidiasis, 375
Fungal disease vaccines. *See* Carbohydrate-based antifungal vaccines

Ganglio-series antigens, cancer-associated and related glycosphingolipids, 236–240
Globo-series and related antigens:
  cancer-associated and related glycosphingolipids, 233–236
  clinical trials, 347–350
Glucuronoxylomannan (GXM), *Candida neoformans*, glycoconjugate vaccines, carbohydrate-based antifungal vaccines, 217–220
Glycan biosynthesis, glycosylation, 6–7

Glycobiology, 3–19
  glycan biosynthesis, 6–7
  glycoproteins, 7–16
    N-linked glycans, 12–13
    O-GlcNAc modification, 15–16
    O-linked glycans, 13–15
    protein modification, 7–11
  glycosylation, 3–6
    disease, 4–5
    generally, 3–4
    toxins and infectious agents, 5–6
  immunology and, 2
  lipid-based glycans, 16–18
    glycospingolipids, 16–18
    glycosylphosphatidylinositol membrane anchors, 18
  polysaccharides, 18–19
  term of, 1
  toxins and infectious agents, 5–6
Glycoconjugate immunobiology, carbohydrate-based antibacterial vaccines, 118–120
Glycoconjugate vaccines, 55–88
  adjuvants, carbohydrate-based vaccines, 100
  capsular polysaccharide-protein conjugates, 56–68
    Group B *Streptococcus* (GBS), 65–67
    Haemophilus influenzae, 56–59
    Neisseria meningitidis, 60–64
    Salmonella typhi, 64–65
    Staphylococcus aureus (Types 5 and 6), 67–68
    Streptococcus pneumoniae, 59–60
  carbohydrate-based antibacterial vaccines, 121–122
  carbohydrate-based antifungal vaccines, 216, 217–222
    beta-glucan-protein conjugates, 221–222
    Candida albicans, 220–221
    Candida neoformans, 217–220
  clinical trials, 352–354
  lipopolysaccharide (LPS) and lipooligosaccharide (LOS) conjugates, 69–76
    Escherichia coli, 69–70
    Haemophilus influenzae, 72–76
    Neisseria meningitidis, 71–72
    Shigella dysenteriae, 71–72
    Vibrio cholerae, 70–71
  overview, 55–56
  pure oligosaccharides, 76–79
Glycoengineering, metabolic, carbohydrate-based vaccines, 37
Glycoimmunobiology, 23–27
  antibody glycosylation, 23–24
  cell- to systems-level control (siglecs), 26–27
  field of, 2
  leukocyte extravasation, 24–26
Glycopeptides:
  major histocompatibility-peptide complex (MHC)-mediated immune response to, carbohydrate-based cancer vaccines, 269–270
  tumor-associated carbohydrates and, chemical synthesis of, carbohydrate-based cancer vaccines, 271–276
Glycoproteins, 7–16
  N-linked glycans, 12–13
  O-GlcNAc modification, 15–16
  O-linked glycans, 13–15
  protein modification, 7–11
  therapeutic cancer vaccines, O-glycan-based vaccine, clinical trials, 351–353
Glycosaminoglycans, polysaccharides, 18–19
Glycosphingolipid(s) (GSLs):
  lipid-based glycans, 16–18
  tumor-associated, carbohydrate-based cancer vaccines, 265
Glycosphingolipid-based vaccines, clinical drug trials, 347–350
Glycosphingolipid (GSL) glycan structures, abnormal expression in, 228, 232–233. *See also* Cancer-associated and related glycosphingolipids
Glycosylation, 3–6
  antibodies, glycoimmunobiology, 23–24
  carbohydrate-based antiviral vaccines, 168–174
    ebola virus, 174

Glycosylation (*Continued*)
  hepatitis C virus, 173
  HIV, 171–172
  influenza A virus, 172–173
  SARS coronavirus (SARS-CoV), 174
Glycosylation
  viral *N*-glycosylation, 168–169
  West Nile virus, 174
  disease, 4–5
  generally, 3–4
  pathogens, 30
  protein modification, 7–11
  sugar interactions with, immune system, 27–28
  therapeutic cancer vaccines, target tumor-associated antigens, 342–344
  toxins and infectious agents, 5–6
Glycosylation disorders, diagnosis, 388
Glycosylphosphatidylinositol membrane anchors, lipid-based glycans, 18
Glycosylphosphatidylinositols (GPIs):
  malaria, carbohydrate-based antiparasitic vaccines, 197–201
  tosoplasmosis and Chagas disease, carbohydrate-based antiparasitic vaccines, 197–201
Group B *Streptococcus* (GBS). *See also Streptococcus pneumoniae*
  capsular polysaccharide-protein conjugates, 65–67
  carbohydrate-based antibacterial vaccines, 140–145

Haemophilus spp., 69
Haemophilus influenzae, 32, 38, 55, 56, 118, 121, 217, 278
  carbohydrate-based antibacterial vaccines, 122–125
    Hib conjugate vaccines, 123–125
    Hib polysaccharides, 122–123
  diagnosis, 373
  lipooligosaccharide (LOS) conjugates, 72–76
  vaccines for, 196
Haemophylis ducyreii, 37
Hantaan virus, glycosylation, carbohydrate-based antiviral vaccines, 174

H antigen, carbohydrate antigens, 28
Helicobacter pylori, epidermal or intradermal adjuvants, 103
Hemagglutinin (HA), influenza A virus, glycosylation, 172–173
Hemorragic fever, glycosylation, carbohydrate-based antiviral vaccines, 174
Hendra virus, glycosylation, carbohydrate-based antiviral vaccines, 174
Hepatitis B virus (HBV), carbohydrate-based cancer vaccines, 263
Hepatitis C virus:
  carbohydrate-based antiviral vaccines, vaccine and drug development, 182
  glycosylation, carbohydrate-based antiviral vaccines, 173
Hereditary erythroblastic multinuclearity, diagnosis, 388
Hib conjugate vaccines, *Haemophilus influenzae*, carbohydrate-based antibacterial vaccines, 123–125
Hib polysaccharides, *Haemophilus influenzae*, carbohydrate-based antibacterial vaccines, 122–123
Histoplasma capsulatum, diagnosis, 376
HIV. *See* Human immune deficiency virus (HIV)
Human immune deficiency virus (HIV):
  carbohydrate-based antiviral vaccines, 170–172
  vaccine and drug development, 174–181
    biosynthetic vaccines, 180
    carbohydrate-binding agents, 180–181
    synthetic vaccines, 175–180
    viral envelope, 175
  carbohydrate-based cancer vaccines, 270
  fungi diagnosis, 374–375
  immunosuppression, 215
Human papilloma virus (HPV), carbohydrate-based cancer vaccines, 264
Humoral immune response, to carbohydrates, carbohydrate-based cancer vaccines, 267–269

IgG antibodies, carbohydrate-based cancer vaccines, 268, 276, 278, 281, 289, 291
IgM antibodies, carbohydrate-based cancer vaccines, 268, 276, 281
Immune system, 20–28. *See also* Adaptive immunity; Innate immunity
   adaptive immunity, 22–23
   carbohydrate-based antifungal vaccines, 222–223
   function of, 20–21
   glycoimmunobiology, 23–27
     antibody glycosylation, 23–24
     cell- to systems-level control (siglecs), 26–27
     leukocyte extravasation, 24–26
   glycosylation and sugar interactions, 27–28
   innate immunity, 21–22
Immunization, vaccination contrasted, antifungal vaccines, 216
Immunology, glycobiology and, 2
Immunomodulation, adjuvants, 101–102
Immunostimulating complex (ISCOM), adjuvants, 94
Infectious mononucleosis, diagnosis, 369–370
Inflammatory bowel disease, diagnosis, 380
Influenza A virus:
   carbohydrate-based antiviral vaccines, vaccine and drug development, 181–182
   diagnosis, 370–371
   glycosylation, carbohydrate-based antiviral vaccines, 172–173
Influenza B virus, diagnosis, 370–371
Inherited glycosylation disorders, diagnosis, 388
Innate immunity, 21–22. *See also* Adaptive immunity; Immune system
   adaptive immunity and, therapeutic cancer vaccines, 334–336
   adjuvants, 95–97
     adaptive response initiation and stimulation, 90–92
     non-toll-like receptor (TLR) agonists, 97
     toll-like receptor (TLR) agonists, 95–96
     antigen capture and recognition, 97
   carbohydrate-based cancer vaccines, 270–271
Intradermal adjuvants, routes of administration, 102–103
Invasive candidiasis, diagnosis, 375
Invasive pneumococcal disease, *Streptococcus pneumoniae*, carbohydrate-based antibacterial vaccines, 139
In vitro models, adjuvants, 104–106

KH- antigen, carbohydrate-based cancer vaccines, 266
Klebsiella pneumoniae, 32

Lacto-series (type 1 chain; Lc_n) antigens, cancer-associated and related glycosphingolipids, 240–241
Laminaria digitata, 221
Lancefield streptococcal groups A, B, C, D, F, and G, diagnosis, 371–373
Langerhans cells, epidermal or intradermal adjuvants, 102–103
Large cell carcinoma (ADC), 383
Lassa virus, glycosylation, carbohydrate-based antiviral vaccines, 174
Leishmania spp.:
   glycosylation, 30
   lipophosphonoglycans (LPGs), 201–205
Leishmaniasis:
   carbohydrate antigens, 32–33
   carbohydrate-based antiparasitic vaccines, 201–205
Leukocyte extravasation, glycoimmunobiology, 24–26
Lipid-based glycans, 16–18
   glycospingolipids, 16–18
   glycosylphosphatidylinositol membrane anchors, 18
Lipophosphonoglycans (LPGs), leishmaniasis, carbohydrate-based antiparasitic vaccines, 201–205
Lipopolysaccharide (LPS) and lipooligosaccharide (LOS) conjugates, 69–76
   Escherichia coli, 69–70
   glycoconjugate vaccines, *Shigella dysenteriae*, 71–72

Lipopolysaccharide (LPS) and lipooligosaccharide (LOS) conjugates (*Continued*)
   Haemophilus influenzae, 72–76
   Neisseria meningitidis, 71–72
   Vibrio cholerae, 70–71
Lipopolysaccharides (LPS), carbohydrate antigens, 30–32
Liposomes, microparticles and, classical adjuvants, 94
Liver cancer, diagnosis, 382–383
Lung cancer, diagnosis, 383–384
Lymphatic leukemia cell line RMA, sialic acid glycoengineering, cancer immunotherapy, 324

Major histocompatibility-peptide complexes (MHCs):
   carbohydrate-based cancer vaccines, 269
   therapeutic cancer vaccines, 336
Major histocompatibility-peptide complex (MHC)-mediated immune response, carbohydrate-based cancer vaccines, to glycopeptides, 269–270
Malaria:
   carbohydrate antigens, 32–33
   carbohydrate-based antiparasitic vaccines, 197–201
   glycosylation, 30
Mammalian carbohydrates, 1–2
Melanoma, diagnosis, 384–385
Meningitis, diagnosis, 373–374
Meningococcal conjugate vaccines, *Neisseria meningitidis*, antibacterial vaccines, 126–133
   serogroup A, 126–128
   serogroup B, 130–131
   serogroup C, 128–130
   serogroups A, C, W-135, and Y, 131–133
Meningococcal polysaccharide vaccines, *Neisseria meningitidis*, carbohydrate-based antibacterial vaccines, 126
Metabolic glycoengineering, carbohydrate-based vaccines, 37

Metapneumoniavirus, glycosylation, carbohydrate-based antiviral vaccines, 174
MF59, emulsion, 94
Microarray technology, GPI, malaria vaccines, 200–201
Microparticles, liposomes and, classical adjuvants, 94
Mineral oil, emulsions, 93–94
Monoclonal antibodies (mAbs), carbohydrate-based cancer vaccines, 263
Monosialosyl-Le[a], 240–241
Monosodium urate (MSU), aluminum salts, 93
Montanide ISA51, 93–94
Moraxella catarrhalis, 74
Mucins:
   carbohydrate-based cancer vaccines, 266
   clinical trials, 353–354
Mucosal adjuvants, routes of administration, 102
Multivalent vaccines, clinical trials, 353
Mycobacterium tuberculosis, 93–94
Mycoplasma pneumoniae, cold agglutinin disease, 379

Neisseria meningitidis, 32, 55, 56, 58, 69, 70, 118, 120, 122, 288
   capsular polysaccharide-protein conjugates, 60–64
   carbohydrate-based antibacterial vaccines, 125–133
   generally, 125–126
   meningococcal conjugate vaccines, 126–133
   serogroup A, 126–128
   serogroup B, 130–131
   serogroup C, 128–130
   serogroups A, C, W-135, and Y, 131–133
   meningococcal polysaccharide vaccines, 126
   diagnosis, 373–374
   lipooligosaccharide (LOS) conjugates, 72–76
   vaccines for, 196
Neisseria meningococci, 74

INDEX **405**

Neolacto-series (type 2 chain; nLc$_x$) antigens, cancer-associated and related glycosphingolipids, 241–247
Neuraminidase (HA), influenza A virus, glycosylation, 172–173
Newcastle virus, glycosylation, carbohydrate-based antiviral vaccines, 174
N-glycans, HIV, carbohydrate-based antiviral vaccines, 171–172
N-glycosylation (viral), carbohydrate-based antiviral vaccines, 168–169
Nipah virus, glycosylation, carbohydrate-based antiviral vaccines, 174
N-linked glycans, glycoproteins, 12–13
non-Hodgkin's lymphoma, 263
Non-small-cell lung cancer, diagnosis, 383–384
Non-toll-like receptor (TLR) agonists, innate immunity, 97

O-GlcNAc modification, glycoproteins, 15–16
O-glycan-based vaccines, therapeutic cancer vaccines, clinical trials, 351–353
Oil-in-water (O/W) emulsions, 93–94
Oligosaccharide, leishmaniasis, carbohydrate-based antiparasitic vaccines, 204–205
Oligosaccharides:
  mammalian, 1–2
  pure, glycoconjugate vaccines, 76–79
O-linked glycans, glycoproteins, 13–15
Otitis media, acute, *Streptococcus pneumoniae*, carbohydrate-based antibacterial vaccines, 140
Ovarian cancer, diagnosis, 385–386

Pancreatic cancer, diagnosis, 386–387
Parasites:
  carbohydrate antigens, 32–33 (*See also* Carbohydrate-based antiparasitic vaccines; specific parasites)
  diagnosis, 377–378
  Clonorchis sinensis, 378

Echinococcus multilocularis, 377–378
generally, 377
Schistomsoma mansoni, 378
Trichinella spp., 378
Parasitic disease vaccines. *See* Carbohydrate-based antiparasitic vaccines
Pathogen-associated molecular patterns (PAMP), carbohydrate-based cancer vaccines, innate immunity, 271
Pathogens. *See also* specific pathogens carbohydrate antigens:
  bacterial carbohydrate-based antigens, 30–32
  glycosylation, 30
  parasites, 32–33
  vaccines, 34
  viral glycosulation, 33–34
  glycosylation, 30–34
Phamacokinetics, therapeutic cancer vaccines, 339
Phosphoglycan, leishmaniasis, carbohydrate-based antiparasitic vaccines, 201–205
Plasmodium spp., carbohydrate antigens, 32–33
Plasmodium falciparum, 197–201, 207
Pneumococcal disease, invasive, *Streptococcus pneumoniae*, antibacterial vaccines, 139
Polymorphic epithelial mucin (PEM), carbohydrate-based cancer vaccines, 266–267
Polysaccharide capsule, *Candida neoformans*, glycoconjugate vaccines, carbohydrate-based antifungal vaccines, 217–219
Polysaccharide immunobiology, carbohydrate-based antibacterial vaccines, 118–120
Polysaccharides:
  glycobiology, 18–19
  human immune response to, carbohydrate-based antibacterial vaccines, 120–121
Polysaccharides (Ti antigens), carbohydrate-based vaccines, adjuvants, 99–100

Polysialic acid (PSA), sialic acid glycoengineering, cancer immunotherapy, 322
Preclinical models, adjuvants, 104–106
Production issues, therapeutic cancer vaccines, 344–345
Proof of principle trials, therapeutic cancer vaccines, 338–340
Prostate cancer, diagnosis, 387
Protein modification, glycoproteins, 7–11
Protozoa, carbohydrate-based antiparasitic vaccines, 197–201
PRR agonists, applications, 98
Pseudomonas aeruginosa, 32
Pure oligosaccharides, glycoconjugate vaccines, 76–79

Quillaja saponaria Molina, 94

Receptor ligand di-epitope constructs, b-epitope and, carbohydrate-based cancer vaccines, 279–283
Recombinant peptide glycoconjugate vaccines, clinical trials, 353–354
Regulatory factors, adjuvants, 103
Risk-benefit ratio, adjuvants, 103–104
Routes of administration:
  adjuvants, 102–103
    epidermal or intradermal, 102–103
    mucosal, 102
  therapeutic cancer vaccines, 339

Safety versus efficacy factors, adjuvants, 103–104
Salmonella spp., 69
Salmonella typhi, 32, 56, 118, 120
  capsular polysaccharide-protein conjugates, 64–65
  carbohydrate-based antibacterial vaccines, 145–146
  vaccines for, 196
Saponins:
  carbohydrate-based cancer vaccines, 278
  classical adjuvants, 94
SARS coronavirus (SARS-CoV), glycosylation, carbohydrate-based antiviral vaccines, 174

Scheduling, therapeutic cancer vaccines, 339
Schistosoma spp., 205–207
Schistosoma mansoni, diagnosis, 378
Schistosomiasis (bilharzia), carbohydrate-based antiparasitic vaccines, 205–207
Shigella spp., 69
Shigella dysenteriae, 32, 71–72, 78–79
Shigella flexneri, 71
Shigella sonnei, 71
Sialic acid glycoengineering, 313–331. *See also* Cancer-associated and related glycosphingolipids; Glycoengineering; Therapeutic cancer vaccines; Tumor-associated carbohydrate antigens (TACAs); Tumor diagnosis
  cancer immunotherapy, 321–325
  cell surface reactivity modulation, 318–321
  overview, 313–314
  process of, 314–318
Siglecs (sialic acid-binding Ig-like lectins, cell- to systems-level control), 26–27
Skin cancer, melanoma diagnosis, 384–385
Small-cell lung cancer:
  diagnosis, 383–384
  therapeutic cancer vaccines, clinical trials, 350
Squamous cell carcinoma (ADC), 383
Staphylococcus aureus (Types 5 and 6), capsular polysaccharide-protein conjugates, 67–68
Stem cell research, carbohydrate antigens, 28–29
Storage issues, therapeutic cancer vaccines, 344–345
Streptococcus spp., groups A, B, C, D, F, and G, diagnosis, 371–373, 373. *See also* Group B *Streptococcus* (GBS)
Streptococcus agalactiae, carbohydrate-based antibacterial vaccines, 140–145
Streptococcus pneumoniae, 32, 55, 56, 77–78, 117, 118, 120, 122, 217. *See also* Group B *Streptococcus* (GBS)

capsular polysaccharide-protein
    conjugates, 59–60
carbohydrate-based antibacterial
    vaccines, 133–140
  acute otitis media, 140
  generally, 133–139
  invasive pneumococcal disease, 139
  diagnosis, 372, 373
  vaccines for, 196
Streptococcus pyogenes, diagnosis, 371
Sugar, glycosylation interactions with,
    immune system, 27–28

Target tumor-associated antigens,
    therapeutic cancer vaccines,
    342–344
T-cells:
  adaptive response initiation and
    stimulation, adjuvants, 90–92
  carbohydrate-based cancer vaccines,
    268–269, 270
  T-cell di-epitope constructs and,
    carbohydrate-based cancer
    vaccines, 284–288
Td and Ti B-cell responses, carbohydrate-
    based vaccines, adjuvants, 99
Therapeutic cancer vaccines, 333–366. *See
    also* Cancer-associated and related
    glycosphingolipids; Carbohydrate-
    based cancer vaccines;
    Glycoengineering; Sialic acid
    glycoengineering; Tumor-
    associated carbohydrate antigens
    (TACAs); Tumor diagnosis
  challenges in, 341–342
  clinical development, 337–338
  clinical trials, 345–354
    generally, 345–347
    glycosphingolipid-based vaccines,
      347–350
    O-glycan-based vaccines, 351–353
  design issues, 337
  efficacy trials, 340–341
  innate and adaptive immunity, 334–336
  overview, 333–334
  production and storage issues, 344–345
  proof of principle trials, 338–340
  target tumor-associated antigens,
    342–344

Ti antigens (free polysaccharides),
    carbohydrate-based vaccines,
    adjuvants, 99–100
Toll-like receptor (TLR) agonists:
  applications, 97–98
  carbohydrate-based cancer vaccines,
    270–271, 279, 288–289
  classical adjuvants, 92–95
  innate immunity, 95–96
Toxicity, therapeutic cancer
    vaccines, 339
Toxins and infectious agents, glycosylation,
    5–6
Toxoids, antifungal vaccines, 216
Toxoplasma gondii, 207–208
Toxoplasmosis, carbohydrate-based
    antiparasitic vaccines, 207–208
Trichinella spp., diagnosis, 378
Tricomponent vaccines, carbohydrate-based
    cancer vaccines, 288–292
Triggering receptors expressed on
    myeloid cells (TREMs), innate
    immunity, 97
Trypanosoma brucei VSG, 197
Trypanosoma cruzi, 207–208
Tumor-associated carbohydrate antigens
    (TACAs), 29–30. *See also*
    Sialic acid glycoengineering;
    Therapeutic cancer vaccines;
    Tumor-associated carbohydrate
    antigens (TACAs)
  carbohydrate-based cancer vaccines,
    264–267, 277
  clinical trials, 345–347, 354
  glycoengineering, cell surface sialic acid,
    313
  glycopeptides and, chemical synthesis of,
    271–276
  sialic acid glycoengineering, cancer
    immunotherapy, 321–325
Tumor diagnosis, 380–387. *See also*
    Sialic acid glycoengineering;
    Therapeutic cancer vaccines;
    Tumor-associated carbohydrate
    antigens (TACAs)
  bladder cancer, 381
  breast cancer, 381–382
  colon cancer, 382
  generally, 380–381

Tumor diagnosis (*Continued*)
   liver cancer, 382–383
   lung cancer, 383–384
   melanoma, 384–385
   ovarian cancer, 385–386
   pancreatic cancer, 386–387
   prostate cancer, 387

Ulcerative colitis, diagnosis, 380

Vaccination, immunization contrasted, antifungal vaccines, 216
Vaccines, carbohydrate-based, 34–37
Vibrio cholerae, 69, 78
   lipopolysaccharide (LPS) (LOS) conjugates, 70–71
   mucosal adjuvants, 102

Viral disease diagnosis, 369–371
   infectious mononucleosis, 369–370
   influenza A and B, 370–371
Viral disease vaccines. *See* Carbohydrate-based antiviral vaccines
Viral glycosulation, pathogens, 33–34
Viral N-glycosylation, carbohydrate-based antiviral vaccines, 168–169

Water-in-oil (W/O) emulsions, 93–94
West Nile virus, glycosylation, carbohydrate-based antiviral vaccines, 174
Wiskotts-Aldrich syndrome, diagnosis, 388

Yesernia, 31